Mitochondria

A subject collection from *Cold Spring Harbor Perspectives in Biology*

OTHER SUBJECT COLLECTIONS FROM *COLD SPRING HARBOR PERSPECTIVES IN BIOLOGY*

Signaling by Receptor Tyrosine Kinases
DNA Repair, Mutagenesis, and Other Responses to DNA Damage
Cell Survival and Cell Death
Immune Tolerance
DNA Replication
The Endoplasmic Reticulum
Wnt Signaling
Protein Synthesis and Translational Control
The Synapse
Extracellular Matrix Biology
Protein Homeostasis
Calcium Signaling
The Golgi
Germ Cells
The Mammary Gland as an Experimental Model
The Biology of Lipids: Trafficking, Regulation, and Function
Auxin Signaling: From Synthesis to Systems Biology
The Nucleus
Neuronal Guidance: The Biology of Brain Wiring
Cell Biology of Bacteria
Cell–Cell Junctions
Generation and Interpretation of Morphogen Gradients
Immunoreceptor Signaling
NF-κB: A Network Hub Controlling Immunity, Inflammation, and Cancer
Symmetry Breaking in Biology
The Origins of Life
The p53 Family

SUBJECT COLLECTIONS FROM *COLD SPRING HARBOR PERSPECTIVES IN MEDICINE*

Cystic Fibrosis: A Trilogy of Biochemistry, Physiology, and Therapy
Hemoglobin and its Diseases
Addiction
Parkinson's Disease
Type 1 Diabetes
Angiogenesis: Biology and Pathology
HIV: From Biology to Prevention and Treatment
The Biology of Alzheimer Disease

Mitochondria

A subject collection from *Cold Spring Harbor Perspectives in Biology*

EDITED BY

Douglas C. Wallace

Children's Hospital of Philadelphia
and
University of Pennsylvania

Richard J. Youle

National Institutes of Health

COLD SPRING HARBOR LABORATORY PRESS
Cold Spring Harbor, New York • www.cshlpress.org

Mitochondria

A Subject Collection from *Cold Spring Harbor Perspectives in Biology*
Articles online at www.cshperspectives.org

Executive Editor	Richard Sever
Managing Editor	Maria Smit
Project Manager	Barbara Acosta
Permissions Administrator	Carol Brown
Production Editor	Diane Schubach
Production Manager/Cover Designer	Denise Weiss
Publisher	John Inglis

Front cover artwork: Mitochondria in one cell are distributed throughout the cytosol and excluded from the central nucleus. When microtubule polymerization and mitochondrial divisions are inhibited, mitochondria form a variety of shapes revealing their plasticity. Image from Stephan Frank (Basel University Hospital, Switzerland).

Library of Congress Cataloging-in-Publication Data

Mitochondria / edited by Douglas C. Wallace, Children's Hospital of Philadelphia and University of Pennsylvania and Richard J. Youle, National Institutes of Health.
pages cm
"A subject collection from Cold Spring Harbor perspectives in biology."
Includes bibliographical references and index.
ISBN 978-1-936113-35-4 (hardcover : alk. paper)
1. Mitochondria. 2. Mitochondrial pathology. 3. Neuromuscular diseases. 4. Cell metabolism.
I. Wallace, Douglas C., editor of compilation. II. Youle, Richard, editor of compilation.

QH603.M5M52 2013
571.6'57--dc23

2013011220

All World Wide Web addresses are accurate to the best of our knowledge at the time of printing.

For a complete catalog of all Cold Spring Harbor Laboratory Press publications, visit our website at www.cshlpress.org.

Contents

Preface

LIFE IS THE INTERPLAY BETWEEN STRUCTURE (anatomy), energy (vital force), and the information needed to construct anatomical and energetic systems. To understand biology and the human condition we must understand all three components of this life formula. However, for the past half millennium, biomedical science has been preoccupied with the anatomical differences between organisms and the chromosomal genetics that defines those anatomical differences. Although anatomical differences are essential for understanding differences between species, the important differences between individuals within a species are not anatomical. Hence, the most important factors for human variability are likely to be energetic. Yet only a tiny fraction of biomedical resources have been invested in understanding the energetic differences between people and the importance of energetic variation in health and disease. Perhaps, this is why the common diseases have remained so enigmatic and are routinely referred to as "complex."

This book is about the biology of energy. The prime protagonist is the mitochondrion. Mitochondria are the cellular organelles that produce most of our energy by oxidative phosphoryation (OXPHOS). The eukaryotic cell can be divided into two major functional domains: the nucleus–cytosol, which specializes in anatomy, and the mitochondrion, which specializes in energy. This division of labor is the result of a defining event in the biological history of higher organisms, the symbiosis that brought an oxidative bacterium, the α-proteobacterium, together with an anaerobic archaebacterium about two to three billion years ago, which ultimately yielded the eukaryotic cell. The increased energy provided by the oxidative bacterium permitted the archaebacterium to develop an increasingly large genome, which ultimately coalesced into the nucleus. The presence of multiple cytosolic bacteria provides large amounts of energy for the nucleus–cytosol such that the modern human cell cytoplasm harbors 100s to 1000s of oxidative bacteria and the nucleus encompasses more than 20,000 genes (see the chapter by Gray).

Over time the α-proteobacterium surrendered most of its structural genes to the nucleus. This decreased the energy requirements of the bacterium, permitting more of its energy to be contributed to the nucleus to elaborate an ever-increasing array of nuclear genes. By the time the fungal–animal lineage of multicellular animals evolved, the α-proteobacterium DNA had been reduced to about 14 polypeptide genes plus the rRNAs and tRNAs to sustain the residual bacterial protein-synthesis apparatus, thus creating the modern mitochondrion.

Today's mitochondrial DNA (mtDNA) codes for the core elements of mitochondrial OXPHOS, and thus represents the wiring diagram of the mitochondrial power plant. All of the remaining genes needed for mitochondrial bioenergetics, biogenesis, dynamics, and metabolism now reside in the eukaryotic nuclear DNA (nDNA). The dispersion of the mitochondrial genome between the 100s to 1000s of copies of the mtDNA plus the 1000 to 2000 nDNA genes creates the uniquely complex genetics of mitochondrial energy metabolism.

The cytosolic mitochondria and mtDNAs are confined to the cytosol and thus must be inherited through the cytoplasm. Hence, they are transmitted through the female oocyte and are maternally inherited. Because each cell has 1000s of mitochondria and mtDNAs, a mutant mtDNA can be mixed with normal mtDNAs within a cell at different ratios, a state known as heteroplasmy. Variation in both the nature of mitochondrial gene mutants and the percentage of mutant mtDNAs can affect energy production and human health (see the chapter by Wallace and Chalkia).

The gene products of the more than 1000 nDNA-coded mitochondrial genes must be returned to the mitochondrion. This is accomplished by a sophisticated system that tags proteins translated in the cytosol that are destined for the mitochondrion and selectively imports them into the mitochondrion (see the chapter by Stojanovski et al.).

Within the mitochondrion the mtDNA is packaged into nucleoids by the mitochondrial packaging protein and transcription factor, TFAM (see the chapter by Gilkerson et al.). The mtDNA is replicated within the mitochondrion by the mitochondrion-specific DNA polymerase γ, which has been found to be mutated in a wide spectrum of complex diseases (see the chapter by Stumpf et al.).

The biosynthetic processes of the mitochondrion and the nucleus–cytosol must be coordinated to coincide with the availability of sufficient energy resources to permit mitochondrial and cellular biogenesis and reproduction (see the chapter by Dominy and Puigserver). Hence, a sophisticated system has evolved for communication between the mitochondria and nucleus, which encompasses the epigenome and the cellular signal transduction pathways. The epigenome modulates nDNA gene expression and replication as well as controlling mtDNA replication and mitochondrial proliferation. The epigenome and the signal transduction systems are modulated by high-energy intermediates generated by the mitochondrion in response to environmental energy supplies. Hence, mitochondrial intermediates provide the barometer for the epigenome and the signal transduction systems to determine when there is sufficient energy and biosynthetic intermediates to permit the cell and the mitochondria to grow and multiply.

As mitochondria evolved within the nucleus–cytosol, this intracellular population of bacteria developed a complex system for coordinating their biochemistries. This is achieved by the exchange of gene products through repeated cycles of mitochondrial fusion and fission (see the chapter by van der Bliek et al.).

Because mitochondria continually replicate in cells, even within postmitotic ones, they also must be removed to avoid their becoming too numerous. Damaged mitochondria and mtDNAs must also be preferentially removed before they can accumulate in the cytoplasm and erode cellular function. This intracellular turnover of mitochondria is known as mitochondrial autophagy or mitophagy (see the chapter by Narendra et al.).

To have the maximum energy benefit, the mitochondria must be moved to the sites of greatest energy demand. This challenge is greatest in the neuron, where it is thought that the mitochondria are produced within the cell body but must migrate to the synaptic boutons to energize neurotransmission. Neuronal antigrade and retrograde transport is now understood to be mediated by a complex array of molecular motors and regulatory elements (see the chapter by Schwarz). Thus, mitochondrial fission, fusion, mitophagy, and transport all work together to maintain mitochondrial integrity and, thus, cellular and human health.

There are many instances in which cells must be removed from a tissue, either as a part of development, viral attack, cancer surveillance, or elimination of dysfunctional cells. This process is frequently mediated by an intracellular autodigestion process known as apoptosis (see the chapter by Bender and Martinou). The intrinsic pathway of apoptosis is initiated by mitochondria through their interaction with the Bcl-2 family of proteins. Apoptosis is an energy-requiring process, thus requiring functional mitochondria, and has the benefit of degrading cellular proteins before they can be released into the extracellular space.

The autodigestion of intracellular proteins and DNA is particularly pertinent for the mitochondrion, because mitochondria are the most prevalent bacteria within our bodies. If the cell simply ruptured it would release the mitochondrial antigens, known as mitochondrial damage-associated molecular patterns (DAMPs), into the circulation. This would activate the innate and adaptive immune systems, resulting in inflammation and disease. In cases of trauma where cells are damaged, mitochondrial energy production capacity is impaired, apoptosis fails, and the cell undergoes necrosis, releasing the mitochondrial DAMPs, which leads to inflammation (see the chapter by Stehling and Lill).

Mitochondria lie at the nexus of a broad spectrum of critical cellular functions. Besides generating much of a cell's energy, mitochondria modulate Ca^{2+} levels, pH levels, multiple catabolic and anabolic pathways, redox homeostasis, reactive oxygen species (ROS) production and signaling, and initiation of the intrinsic pathway of apoptosis through activation of the mitochondrial permeability transition pore (mtPTP). Hence, mitochondrial dysfunction and mutations in mitochondrial genes can perturb a plethora of systems, resulting in a wide range of clinical symptoms (see the chapter by Stehling and Lill). For example, one function of the mitochondrion is to generate iron–sulfur centers for redox enzyme biochemistry. Perturbation of this pathway has been associated with a variety of metabolic diseases (see the chapter by Stehling and Lill). Similarly, mutations in a mitochondrial matrix Fe-containing sulfide dioxygenase block the pathway for detoxifying H_2S. The excess H_2S then inhibits cytochrome *c* oxidase resulting in mitochondrial dysfunction and disease (see the chapter by Tiranti and Zeviani).

Mitochondria also monitor the extracellular physical environment. This was demonstrated by the discovery that mutations in the extracellular matrix protein collagen VI sensitize the mtPTP. The resulting predisposition to activation of the intrinsic mitochondrial apoptosis pathway causes progressive myopathy (see the chapter by Bernardi and Bonaldo).

The central role of the mitochondrion in energy metabolism, intermediary metabolism, cellular homeostasis, and cell death and the unique features of mitochondrial dynamics and genetics have profound implications for both cellular and organismal biology. It is not surprising, then, that our rapidly expanding understanding of mitochondrial biology and genetics is providing new insights into cancer (see the chapter by Gasparre et al.) and aging (see the chapter by Sack and Finkel).

In conclusion, the study of mitochondrial biology and genetics has produced a new bioenergetic systems biology. This is providing powerful new insights into a wide variety of metabolic and degenerative diseases, cancer, and aging. Thus, the biology and genetics of bioenergetics may provide the missing elements necessary for us to elucidate the causes and generate cures for the common "complex" diseases.

Douglas C. Wallace
Richard J. Youle

Mitochondrial Evolution

Michael W. Gray

Centre for Comparative Genomics and Evolutionary Bioinformatics, Department of Biochemistry and Molecular Biology, Dalhousie University, Halifax, Nova Scotia B3M 4R2, Canada

Correspondence: m.w.gray@dal.ca

Viewed through the lens of the genome it contains, the mitochondrion is of unquestioned bacterial ancestry, originating from within the bacterial phylum α-Proteobacteria (*Alphaproteobacteria*). Accordingly, the endosymbiont hypothesis—the idea that the mitochondrion evolved from a bacterial progenitor via symbiosis within an essentially eukaryotic host cell—has assumed the status of a theory. Yet mitochondrial genome evolution has taken radically different pathways in diverse eukaryotic lineages, and the organelle itself is increasingly viewed as a genetic and functional mosaic, with the bulk of the mitochondrial proteome having an evolutionary origin outside *Alphaproteobacteria*. New data continue to reshape our views regarding mitochondrial evolution, particularly raising the question of whether the mitochondrion originated after the eukaryotic cell arose, as assumed in the classical endosymbiont hypothesis, or whether this organelle had its beginning at the same time as the cell containing it.

In 1970, Lynn Margulis published *Origin of Eukaryotic Cells*, an influential book that effectively revived the long-standing but mostly moribund idea that mitochondria and plastids (chloroplasts) evolved from free-living bacteria via symbiosis within a eukaryotic host cell (Margulis 1970). The discovery in the 1960s of DNA within these organelles together with the recognition that they contain a translation system distinct from that of the cytosol were two of the observations that Margulis marshaled in support of the endosymbiont hypothesis of organelle origins. Indeed, throughout her career, Margulis forcefully argued that symbiosis is a potent but largely unrecognized and unappreciated force in evolution (Margulis 1981). Technological developments in DNA cloning and sequencing in the 1970s and 1980s opened the way to the detailed characterization of mitochondrial genomes and genes, and the generation of key molecular data that were instrumental in affirming a bacterial origin of the mitochondrial and plastid genomes, allowing researchers to pinpoint the extant bacterial phyla to which these two organelles are most closely related. Over the past several decades, numerous reviews have documented in detail the biochemical and molecular and cell biological data bearing on the endosymbiont hypothesis of organelle origins (Gray 1982, 1983, 1989a,b, 1992, 1993, 1999; Gray and Doolittle 1982; Wallace 1982; Cavalier-Smith 1987b, 1992; Gray and Spencer 1996; Andersson and Kurland 1999; Gray et al. 1999, 2001, 2004; Lang et al. 1999; Andersson et al. 2003; Burger et al. 2003a; Bullerwell and Gray 2004). Various endosymbiotic

Cite this article as *Cold Spring Harb Perspect Biol* doi: 10.1101/cshperspect.a011403

models proposed over the years have been comprehensively critiqued (Martin et al. 2001), while the debates surrounding the endosymbiont hypothesis have been recounted in an engaging perspective that traces the development of ideas regarding organelle origins (Sapp 1994). Within a historical context, the present article emphasizes more recent data and insights that are relevant to continuing questions regarding how mitochondria originated and have since evolved.

WHAT DO GENETIC, GENOMIC, AND PHYLOGENOMIC DATA TELL US ABOUT THE ORIGIN OF MITOCHONDRIA?

The genetic function of mitochondrial DNA (mtDNA) was first fully revealed by complete sequencing of the ~16-kb mitochondrial genome from several mammalian species (Anderson et al. 1981, 1982; Bibb et al. 1981). This work established the paradigm that mtDNA encodes a small number (13 in mammals) of protein subunits of the mitochondrial electron transport chain and ATP synthase, as well as the ribosomal RNA (rRNA) and transfer RNA (tRNA) components of a mitochondrial translation system. The paradigm was revised when investigations of mtDNA from non-animal species showed quite extraordinary variation in size, physical form, coding capacity, organizational patterns, and modes of expression across the eukaryotic domain (Gray et al. 1998). In some non-animal taxa, additional proteins are encoded in mtDNA, notably extra respiratory proteins and ribosomal proteins (Table 1). On the other hand, some mitochondrial genomes have been reduced drastically in size, losing many of the protein genes encoded in animal mtDNA as well as some or all mtDNA-encoded tRNA genes. At ~6 kb in size, the mitochondrial genome of *Plasmodium falciparum* (human malaria parasite) and related apicomplexans is the smallest known, harboring only three protein genes, highly fragmented and rearranged small subunit (SSU) and large subunit (LSU) rRNA genes, and no tRNA genes (Feagin 2000). In marked contrast, within land plants, mtDNA has expanded substantially in size (>200 kb) if not in coding capacity, with the largest known mitochondrial genome in this lineage (~11,000 kb) exceeding the size of some bacterial and even some nuclear genomes (Sloan et al. 2012). What was evident even early on is that none of the initial mtDNAs investigated by detailed sequencing, including animal mtDNAs, looks anything like a typical bacterial genome in the way in which genes are organized and expressed.

Ribosomal RNA genes are among the few genes universally encoded by mtDNA across eukaryotes, and the corresponding rRNA sequences were exploited early on in phylogenetic reconstructions aimed at elucidating their evolutionary origin. Although mitochondrial large subunit (LSU) and small subunit (SSU) rRNAs are in general very different in size and secondary structure compared with their cytosolic and prokaryotic counterparts, they retain a sufficient degree of primary sequence and secondary structure correspondence that they can be incorporated into the aligned sequence databases on which these phylogenetic reconstructions are based (Gray et al. 1984, 1989; Cedergren et al. 1988). These initial phylogenetic trees showed that mitochondrial rRNA sequences, and presumably the genomes encoding them, emanate from within the domain Bacteria (eubacteria) and not from within Archaea (archaebacteria) or Eucarya (eukaryotes). Plant mitochondrial rRNAs are especially slowly evolving and bacteria-like (Schnare and Gray 1982; Spencer et al. 1984) and were instrumental in pinpointing the α-class of Proteobacteria (*Alphaproteobacteria*) as the specific bacterial lineage from within which they originated (Yang et al. 1985). Subsequent phylogenetic reconstructions using a much larger number of both mitochondrial and bacterial rRNA sequences have consistently confirmed this affiliation and, moreover, pointed to the Rickettsiales, one of six or more orders within *Alphaproteobacteria* (Williams et al. 2007)—and comprising obligate parasites such as *Rickettsia*, *Anaplasma*, and *Ehrlichia*—as especially close relatives of mitochondria (Gray and Spencer 1996; Gray 1998).

These phylogenetic conclusions were considerably bolstered by the sequencing in 1997 of a minimally derived (ancestral) mitochon-

Table 1. Mitochondrial DNA-encoded genes and their functions

Function	Genes
(1) Coupled electron transport–oxidative phosphorylation (ATP synthesis)	
Complex I (NADH:ubiquinone oxidoreductase)	***nad1, 2, 3, 4, 4L, 5, 6***, *7, 8, 9, 10, 11*
Complex II (succinate:ubiquinone oxidoreductase)	*sdh 2, 3, 4*
Complex III (ubiquinol:cytochrome *c* oxidoreductase)	***cob***
Complex IV (cytochrome c:O_2 oxidoreductase)	***cox1, 2, 3***
Complex V (F_1F_0 ATP synthase)	*atp1, 3,* ***4,***[a] ***6, 8,***[b] ***9***
(2) Translation	
Ribosomal RNAs	***rnl*** **(LSU)**, ***rns*** **(SSU)**, *rrr5* (5S)
Ribosomal proteins	
Small subunit (SSU)	*rps1, 2, 3, 4, 7, 8, 10, 11, 12, 13, 14, 19*
Large subunit (LSU)	*rpl1, 2, 5, 6, 10, 11, 14, 16, 18, 19, 20, 27, 31, 32, 34*
Transfer RNAs	***trnA, C, . . . W, Y***
Elongation factor	*tufA*
tm RNA (unstalling of translation)	*ssrA*
(3) Transcription	
Core RNA polymerase	*rpoA, B, C*
Sigma factor	*rpoD*
(4) RNA processing	
RNase P RNA (5′ tRNA processing)	*rnpB*
(5) Protein import	
ABC transporter	*ccmA* (*yejV*), *ccmB* (*yejW*)
Heme delivery	*ccmC* (*yejU*)
SecY-type transporter	*secY*
Sec-independent transporter	*tatA* (*mttA*)[c], *tatC* (*mttB*)
(6) Protein maturation	
Cytochrome oxidase assembly	*cox11*
Heme *c* maturation	*ccmF* (*yejR*)

A subset of the genes is encoded in the mtDNA of various eukaryotes; for example, those in bold are found in human and other mammalian mitochondrial genomes. Only the most ancestral (gene-rich) mtDNAs, for example, those of jakobid flagellates such as *Reclinomonas americana*, encode all or almost all of this gene set. Rare genes of uncertain origin and function include *dpo* (plasmid-derived DNA polymerase), *rpo* (plasmid-derived RNA polymerase), *rtl* (reverse transcriptase), *mutS* (DNA repair), and *dam* (methyltransferase). See Gray et al. (2004) for details.

[a]Also referred to as *orf25* and *ymf39* (Burger et al. 2003b).

[b]Also referred to as *orfB* and *ymf19* (Gray et al. 1998).

[c]Jacob et al. (2004).

drial genome from *Reclinomonas americana*, a member of an obscure group of protists (eukaryotic microbes) termed jakobid flagellates. The ~69-kb, circular-mapping mtDNA of *R. americana* not only contains more genes than any other characterized mitochondrial genome (including genes specifying 5S rRNA and RNase P RNA) (Table 1), but it also displays evident bacterial characteristics not seen in combination in any other mtDNA, such as operon-like gene clusters, highly bacteria-like rRNA and tRNA secondary structures, and putative Shine–Dalgarno motifs upstream of protein-coding genes. Indeed, *R. americana* mtDNA—so different from that of animal, plant, fungal, and many other protist mtDNAs—has been dubbed "a eubacterial genome in miniature," so striking is its resemblance to a typical bacterial genome (Lang et al. 1997).

At the same time, sequencing of the first Rickettsiales genome, that of *Rickettsia prowazekii*, showed it to be a markedly reduced bacterial genome, superficially resembling mitochondria in its dependence on a host cell (Andersson et al. 1998). However, the genomes of the mitochondrion and members of Rickettsiales are clearly the products of independent evolutionary reduction, implying that mitochondria were not derived directly from a Rickettsiales taxon (Andersson et al. 1998; Gray 1998); rather, these two groups share a more remote common ancestor. Although the mitochondria–Rickettsiales connection has been a consistent phylogenetic finding (Viale and Arakaki 1994; Gupta 1995; Sicheritz-Pontén et al. 1998; Lang et al. 1999), it is still not certain whether the two are sister groups, or whether mitochondria actually branch within Rickettsiales, which is composed of two distinct families, Rickettsiaceae and Anaplasmataceae (Williams et al. 2007). In several studies, mitochondria appear to be more closely related to the former family, containing various *Rickettsia* species, than to the latter, comprising the genera *Anaplasma*, *Ehrlichia*, and *Wolbachia* (Karlin and Brocchieri 2000; Emelyanov 2001a,b, 2003a,b).

The specific affiliation of mitochondria and Rickettsiales in phylogenetic trees, although robust, has been questioned (e.g., Esser et al. 2004) on the grounds that this inferred relationship might be a phylogenetic artifact due to the high rate of sequence divergence and elevated A+T content of the genomes of Rickettsiales taxa and mitochondria—in other words, a long-branch-attraction (LBA) artifact. For that reason, there has been considerable interest in expanding the database of both mitochondrial and bacterial sequences and in applying a more comprehensive phylogenomics approach (based on many genes) to the reconstruction of phylogenetic trees, in combination with multiple and more sophisticated algorithms, rigorous statistical evaluation methods, and more realistic evolutionary models.

Of particular interest in broadening taxon sampling has been the identification of free-living members of *Alphaproteobacteria* that may be specifically affiliated with the parasitic group Rickettsiales (Williams et al. 2007). These previously unknown α-proteobacterial lineages came to light through the Global Oceanic Survey, metagenomic data from which revealed that oceanic α-Proteobacteria are abundant, with one particular clade (termed SAR11) comprising 30%–40% of total oceanic cell counts. The 1.3-Mb genome of one SAR11 member, *Candidatus Pelagibacter ubique*, is the smallest known genome, encoding the fewest genes, of any free-living bacterium (Giovannoni et al. 2005). Several recent reports have suggested that the SAR11 clade including *Ca. P. ubique* shares a common ancestor with mitochondria, the two together forming a sister group to Rickettsiales (Georgiades et al. 2011; Thrash et al. 2011). Other investigators reject this affiliation, concluding instead that *Ca. P. ubique* is most closely related to soil and aquatic α-Proteobacteria with large genomes (Viklund et al. 2012) and arguing that a rare oceanic group of α-Proteobacteria, termed Oceanic Mitochondria Affiliated Clade (OMAC), represents the closest free-living relatives to mitochondria (Brindefalk et al. 2011). Still other groups have suggested that free-living members of α-proteobacterial orders other than Rickettsiales should be considered as possible sources of the mitochondrial progenitor (Esser et al. 2004; Atteia et al. 2009).

These differing conclusions emphasize the challenges inherent in these sorts of analyses, which have to contend with various types of systematic error (for review, see Thrash et al. 2011), biased taxon sampling, and the highly restricted gene content of mitochondrial genomes. At present, although we know a great deal regarding the mitochondrial family tree, we have to admit that the identity of the immediate next of kin remains elusive.

Phylogenetic and phylogenomic reconstructions strongly and consistently support a monophyletic mitochondrial assemblage, and therefore a single origin of mitochondria. Two other pieces of genomic evidence support the view that extant mitochondrial genomes share a single common ancestor. First, the genes encoded by mitochondrial genomes are, with few exceptions, a subset of the genes encoded by the most gene-rich mtDNA, that of *R. americana*. Because the mitochondrial genome is considered to be a highly reduced version of a much larger α-proteobacterial progenitor genome, it is extremely unlikely that genome reduction in independently acquired bacterial symbionts would separately converge on the same small repertoire of respiratory-chain and ribosomal-protein genes. Second, in plant and many ancestral protist mtDNAs, ribosomal protein genes are clustered in the same transcriptional order in which they appear in the corresponding bacterial operons. However, some genes present in the bacterial operons are missing from the corresponding mitochondrial gene clusters, either having been moved elsewhere in the mitochondrial genome or to the nuclear genome, as a result of endosymbiotic gene transfer (EGT). The same gene deletions are seen in the homologous mitochondrial ribosomal gene clusters where these occur throughout eukaryotes, arguing that these deletions were already present in the common ancestor of extant mitochondrial genomes (Gray et al. 1999). This conclusion is reinforced by consideration of various nucleus-encoded mitochondrial proteins that are components of the mitochondrial proteome (see below).

HOW DID THE MITOCHONDRIAL SYMBIOSIS HAPPEN?

Endosymbiotic models for the origin of mitochondria (for review, see Martin et al. 2001) are basically variations on two fundamentally different themes that have been referred to, respectively, as the "archezoan scenario" and the "symbiogenesis scenario" (Koonin 2010). In the archezoan scenario, "The host of the proto-mitochondrial endosymbiont was a hypothetical primitive amitochondrial eukaryote, termed archezoan" (Koonin 2010). In contrast, in the symbiogenesis scenario, "A single endosymbiotic event involving the uptake of an α-Proteobacterium by an archaeal cell led to the generation of the mitochondria," followed subsequently "by the evolution of the nucleus and compartmentalization of the eukaryotic cell" (Koonin 2010). The archezoan scenario most closely approximates the classical endosymbiont hypothesis of mitochondrial origin (Margulis 1970; Doolittle 1980), whereas the hydrogen hypothesis (Martin and Müller 1998) is an example of the symbiogenesis scenario. A fundamental difference between these two scenarios is whether the α-Proteobacterial endosymbiosis that gave rise to the proto-mitochondrion happened at the same time as and was integral to the formation of the eukaryotic cell (symbiogenesis scenario) or occurred subsequent to the formation of a primitive, amitochondriate cell that served as host and that was already essentially eukaryotic (archezoan scenario).

Archezoan Scenario

A major boost to the classical endosymbiont hypothesis came from early phylogenetic reconstructions, based initially on SSU rRNA sequences, that showed several eukaryotic lineages branching deeply within the eukaryotic domain of the resulting trees. These early-branching lineages, collectively termed "archezoa," consisted of protists such as microsporidians, diplomonads, and parabasalids, living as parasites in anaerobic environments and characterized by the absence of recognizable mitochondria (Cavalier-Smith 1987a, 1989). Members of archezoa, assumed to be primitively amitochondriate (i.e., never having had mitochondria in their evolutionary history: the so-called "archezoa hypothesis"), could be seen as modern representatives of the sort of ancestral eukaryote that might have played host to an α-Proteobacterial symbiont in a classical endosymbiosis scheme.

Two findings led to the ultimate abandonment of the archezoa hypothesis. First, more rigorous phylogenetic reconstructions combined with other sorts of data convincingly showed that the early-branching position of archezoan taxa in the eukaryotic tree is a methodological artifact, due to a relatively high rate of sequence divergence of the archezoan sequences used in the analysis. This property gives rise to an LBA effect that incorrectly positions these taxa at the base of the eukaryotic clade, closest to the outgroup (prokaryotic) sequences used to root the tree. Microsporidia, for example, were eventually recognized as secondarily degenerate fungi rather than as primitive, early-branching eukaryotes (Hirt et al. 1999; Keeling et al. 2000). Indeed, the current eukaryotic tree is more accurately characterized as a bush, with no one lineage clearly the earliest diverging (Keeling et al. 2005; Koonin 2010). Second, in every apparently amitochondrial lineage that has been carefully investigated, evolutionary remnants of mitochondria ("mitochondrion-related organelles," or MROs) have been identified (see below). Thus, we currently know of no extant eukaryotic lineages that are convincingly amitochondrial and that therefore might have been primitively amitochondriate (Embley and Hirt 1998). This conclusion does not mean that such lineages do not exist. We may simply not have discovered them yet, or they may have existed at some point in evolutionary history but have now all become extinct. Nevertheless, rejection of the archezoa hypothesis on phylogenetic grounds coupled with the apparent absence of any known amitochondriate eukaryotic lineages considerably weakens the case for an acquisition of mitochondria via the classical endosymbiont route.

Symbiogenesis Scenario

An alternative view, that the host cell for the mitochondrial endosymbiosis was a prokaryote—and specifically an archaeon, not a eukaryote—has recently gained in prominence (Koonin 2010). The "hydrogen hypothesis" (Martin and Müller 1998) is perhaps the best-known example of a symbiogenesis scenario. This scheme proposes that eukaryotes arose

> ...through symbiotic association of an anaerobic, strictly hydrogen-dependent, strictly autotrophic archaebacterium (the host) with a eubacterium (the symbiont) that was able to respire, but generated molecular hydrogen as a waste product of anaerobic heterotrophic metabolism. The host's dependence upon molecular hydrogen produced by the symbiont is put forward as the selective principle that forged the common ancestor of eukaryotic cells. (Martin and Müller 1998)

Thus, the hydrogen hypothesis "posits that the origins of the heterotrophic organelle (the symbiont) and the origins of the eukaryotic lineage are identical" (Martin and Müller 1998). A corollary of the hydrogen hypothesis and other symbiogenesis scenarios is that the complexity of the eukaryotic cell and its defining features emerged *after* the mitochondrial symbiosis, rather than *before*.

Several arguments can be advanced against a symbiogenesis scenario for the origin of mitochondria (Koonin 2010). Endocytosis (a eukaryotic hallmark) has long been considered an essential function for incorporating a bacterial symbiont, although it is the case that bacterial endosymbioses (e.g., γ-Proteobacteria inside β-Proteobacteria) have been documented (von Dohlen et al. 2001; Thao et al. 2002). In addition, assuming an α-Proteobacterial symbiont as the mitochondrial progenitor in a partnership that simultaneously gave rise to this organelle and the rest of the eukaryotic cell, one might expect that any eubacterial genetic signal in the nuclear genome would be overwhelmingly α-Proteobacterial. However, although an α-Proteobacterial signal does, in fact, predominate (Pisani et al. 2007), in any given eukaryotic lineage collectively more bacterial-type genes appear to derive from diverse non-α-Proteobacterial lineages or fail to affiliate robustly with any specific bacterial phylum. Nevertheless, it is possible that ancestral lineages contributing to a bacterial–archaeal symbiogenesis might have possessed genomes already "scrambled" to a certain extent by horizontal gene transfer (HGT).

A prominent feature of the hydrogen hypothesis is its claim to account simultaneously for the origins of both aerobic and anaerobic energy metabolism in eukaryotes, the assumption being that both pathways were contained in and contributed to the hybrid cell by the α-Proteobacterial partner. It is supposed that the two pathways would have been differentially expressed in the free-living bacterial symbiont when it encountered the appropriate environmental conditions. The hypothesis further posits that genes for aerobic respiration were lost in those eukaryotic lineages in which the mitochondrion was converted to an MRO, some types of which (e.g., hydrogenosome) function in anaerobic energy metabolism (see below). The hydrogen hypothesis predicts that genes of anaerobic energy metabolism in MROs should have been inherited vertically throughout eukaryotes from a common ancestor, clustering as a monophyletic lineage together with α-Proteobacteria in phylogenetic reconstructions. However, a rigorous study of the

 Cite this article as *Cold Spring Harb Perspect Biol* doi: 10.1101/cshperspect.a011403

phylogenetic distributions and histories of proteins involved in anaerobic pyruvate metabolism in eukaryotes has not provided support for this prediction (Hug et al. 2010). Rather, MROs and the enzymatic machinery they contain for anaerobic energy metabolism appear to reflect a high degree of independent and convergent evolution (see below).

Very recently, Lane and Martin (2010) have argued, from a consideration of the energetics of genome complexity, that because eukaryotes encode and express a substantially larger number of proteins than do prokaryotes, this increased expression demands a level of cellular energy that only the mitochondrion is able to satisfy. Accordingly, these investigators view the mitochondrion as the *sine qua non* to eukaryotic genomic and cellular complexity, concluding, rather definitively, that "the host for mitochondria was a prokaryote."

On balance, it would seem that a symbiogenesis scenario (bacterial endosymbiont in an archaeal host) better accommodates the accumulated data relevant to the question of mitochondrion origin than does an archezoan scenario (bacterial endosymbiont in an amitochondriate but essentially eukaryotic host). However, as emphasized above, the latter scenario cannot be entirely discounted at this point. Each scenario raises its own set of issues that are difficult to rationalize without resorting to ad hoc explanations. In the end, each faces the conundrum that there is no straightforward and compelling way to discern how similar the genomes of the proposed prokaryotic ancestors of the eukaryotic cell were to their modern-day descendants.

MITOCHONDRION-RELATED ORGANELLES (MROs)

An extreme form of mitochondrial genome reduction is found in two types of MROs, hydrogenosomes and mitosomes, which entirely lack mtDNA. These two MRO types are distinguished by the fact that hydrogenosomes retain ATP-generating capacity, whereas mitosomes do not. The hydrogenosome, a double-membrane-bound organelle originally discovered in the parabasalid *Trichomonas vaginalis* (Lindmark and Müller 1973), was subsequently identified in various other anaerobic eukaryotes. The *T. vaginalis* hydrogenosome not only lacks a genome but has an incomplete tricarboxylic acid cycle and electron transport chain and no cytochromes. ATP is generated from pyruvate via a substrate-level pathway comprising a characteristic suite of enzymes, including pyruvate: ferredoxin oxidoreductase and an iron–iron hydrogenase. The organelle takes its name from the fact that molecular hydrogen (H_2) is one of the end products of this pathway. The anaerobic metabolism performed by the hydrogenosome initially suggested that the organelle might have originated through an endosymbiosis with an anaerobic bacterium such as *Clostridium* (Whatley et al. 1979). However, subsequent studies have revealed that the *T. vaginalis* hydrogenosome contains several proteins typical of mitochondria, including chaperonins (Bui et al. 1996), the NADH dehydrogenase module of electron transport complex I (Hrdy et al. 2004), and components of the ISC biosynthesis pathway, the mitochondrial machinery for synthesis of iron–sulfur (Fe–S) clusters (Sutak et al. 2004). These results strongly support the view that the *T. vaginalis* hydrogenosome is a relict mitochondrion.

A second group of double-membrane-bound MROs, in this case unable to generate ATP, has been discovered in a number of anaerobic, parasitic protists that were initially considered to be amitochondriate, including the amoebozoons *Entamoeba histolytica* (Clark and Roger 1995; Mai et al. 1999; Tovar et al. 1999) and *Mastigamoeba balamuthi* (Gill et al. 2007), the microsporidians *Trachipleistophora hominis* (Williams et al. 2002) and *Encephalitozoon cuniculi* (Goldberg et al. 2008; Tsaousis et al. 2008), and the diplomonad *Giardia lamblia* (Tovar et al. 2003). The MROs in these protists are collectively termed "mitosomes" (Embley et al. 2003; Embley 2006; Hjort et al. 2010). Here again, identification of typical mitochondrial proteins in this MRO argues that mitosomes, like hydrogenosomes, are evolutionary derivatives of conventional mitochondria, but are even more highly reduced than hydrogenosomes (see Hjort et al. 2010 for a detailed listing and discussion of relevant data). Very limited metabolic capacity is retained by the mitosome, of which Fe–S cluster formation stands out. This pathway is ubiquitous in both conventional mitochondria and MROs, suggesting that Fe–S cluster formation, rather than oxidative phosphorylation, may be the *raison d'être* of the mitochondrion and its evolutionary derivatives (Tovar et al. 2003).

The discovery of what appear to be transitional forms of MRO in certain anaerobic eukaryotes has recently blurred the distinction between mitochondria, on the one hand, and hydrogenosomes and mitosomes, on the other (Gray 2011). These novel MROs, like conventional hydrogenosomes, are able to generate H_2 via an ATP-producing, hydrogenase-mediated pathway; however, they retain a genome, albeit lacking several typical mtDNA-encoded genes, notably ones specifying components of respiratory complexes III, IV, and V (see Table 1). Thus, these MROs lack the ability to generate ATP via coupled electron transport and oxidative phosphorylation; however, they support a considerably more complex biochemistry than either hydrogenosomes or mitosomes. MROs of this type have been identified in the anaerobic ciliate *Nyctotherus ovalis* (Boxma et al. 2005; de Graaf et al. 2011) and the anaerobic stramenopiles *Blastocystis* sp. (Pérez-Brocal and Clark 2008; Stechmann et al. 2008; Wawrzyniak et al. 2008) and *Proteromonas lacertae* (Pérez-Brocal et al. 2010), relatives of brown algae, diatoms, and oomycetes. In these cases, the MRO genome specifies components of an organellar translation system (rRNAs, tRNAs, ribosomal proteins) as well as subunits of electron transport complexes I and II, suggesting the presence of a partial electron transport chain.

The sporadic phylogenetic distribution of anaerobic MRO-containing eukaryotes and their interspersion with aerobic taxa within particular lineages strongly indicate that MROs have originated independently, multiple times, throughout the eukaryotic domain (Roger and Silberman 2002; Embley et al. 2003; Embley 2006; Hjort et al. 2010). Consequently, many of the characteristics shared between MROs (e.g., between those of *Nyctotherus* and *Blastocystis*) are almost certainly due to convergent evolution rather than vertical inheritance. The continued study of variously evolved MROs will be important for understanding both the pathways and mechanisms involved in the evolutionary conversion of a conventional mitochondrion to an MRO. At the same time, we will gain a better appreciation of the evolutionary flexibility of mitochondria, a theme considered below in the discussion of mitochondrial proteome evolution.

ORIGIN AND EVOLUTION OF THE MITOCHONDRIAL PROTEOME

Only 27 mitochondrial proteins are encoded by the gene-rich *R. americana* mitochondrial genome, and only three by the most gene-poor mtDNA, that of apicomplexan parasites such as *P. falciparum*. The vast majority of mitochondrial proteins are encoded in the nucleus, synthesized in the cytosol, and imported back into the organelle. Overall, the number of mitochondrial proteins is estimated to range from several hundred in *P. falciparum* to more than 3000 in vertebrate animals (Richly et al. 2003). Relocation of functional genes from the mitochondrion to the nucleus via EGT has contributed significantly to the nucleus-encoded cohort (Gray 1999; Adams and Palmer 2003; Timmis et al. 2004). Occasionally, such genes are transferred in pieces (Adams et al. 2001; Pérez-Martínez et al. 2001; Waller and Keeling 2006; Gawryluk and Gray 2009; Morales et al. 2009), sometimes with a portion remaining in the mitochondrial genome (Oudot et al. 1999; Kück et al. 2000; Nedelcu et al. 2000; Gawryluk and Gray 2010a). Notably, only a minority of the protein products of these relocated genes are imported back into the organelle; most now function in other subcellular compartments (Gabaldón and Huynen 2007). Extensive retailoring of the mitochondrial proteome in the course of evolution has also involved addition of novel proteins and new functions, as discussed below.

Identification of the proteins contributed to the eukaryotic cell by the proto-mitochondrial endosymbiont has been approached through comparisons of the gene contents of sequenced α-Proteobacterial and eukaryotic genomes. This strategy has identified at least 840 orthologous groups that bear a clear α-Proteobacterial signature—that is, a close and specific evolutionary relationship to α-Proteobacterial homologs, without evidence of recent HGT (Gabaldón and Huynen 2003, 2007; Szklarczyk and Huynen 2010). Comparisons among α-Proteobacterial genomes suggest that the free-living bacterial ancestor of mitochondria contained about 3000–5000 genes (Boussau et al. 2004), with an upper bound of approximately 1700 ancestral clusters of orthologous genes in the proto-mitochondrial genome (Szklarczyk and Huynen 2010). These estimates indicate that about 1000–3000 genes were lost in the transition from bacterial symbiont to proto-organelle. Significantly, of the more than 800 human genes that display an α-Proteobacterial signature, only about 200 are found in the human mitochondrial proteome, indicating that the proto-mitochondrial contribution to eukaryotic cell evolution and function extends beyond the mitochondrion itself, to encompass other cellular compartments, as well.

In the proto-mitochondrion, the complete electron transport chain and complete pathway for β-oxidation of fatty acids (providing NADH and $FADH_2$ to the former) were likely present, indicating that the mitochondrial endosymbiont had an aerobic metabolism. Also prominently represented were pathways for the synthesis of lipids, biotin, heme, and Fe–S clusters, as well as an abundance of cation transporters. In all, the reconstructed metabolism suggests that the proto-mitochondrion was capable of at least facultative aerobic respiration (Szklarczyk and Huynen 2010). More than half of what remains of this proto-mitochondrial metabolism in modern mitochondria comprises proteins that function in oxidative phosphorylation and translation, including posttranslational modifications, a clear trend toward specialization in energy metabolism. In short, mitochondrial proteome evolution has been characterized by loss of many original functions, retargeting of others to different cellular locations, and wholesale addition of host-derived proteins—a veritable "hijacking of mitochondrial protein synthesis and metabolism" by the eukaryotic cell (Gabaldón and Huynen 2007).

Initial delineation of the mitochondrial proteome relied on various bioinformatics approaches, including homology searches as well as techniques developed to identify amino-terminal mitochondrial targeting signals (MTSs) (e.g., Claros and Vincens 1996; Emanuelsson et al. 2000). Not all imported mitochondrial proteins feature MTSs identifiable in this way, and the accuracy and sensitivity of the search algorithms may be compromised where protein sequences are highly divergent (Richly et al. 2003). An alternative technique that has proven useful is high-throughput immunolocalization of tagged gene products to determine their subcellular localization. Based on these approaches, the yeast mitochondrial proteome has been estimated to comprise about 400–800 proteins, or between ~7% and ~13% of the total yeast proteome of approximately 6100 proteins (Karlberg et al. 2000; Marcotte et al. 2000; Kumar et al. 2002).

Considering the clear α-Proteobacterial signature of the mitochondrial genome, we might have anticipated that the mitochondrial proteome would consist largely of α-Proteobacteria-like proteins plus proteins exclusive to the eukaryotic domain, "invented" during the evolutionary transition from bacterial endosymbiont to organelle. Somewhat surprisingly, this expectation has not been met. In surveys of the yeast mitochondrial proteome (Karlberg et al. 2000; Marcotte et al. 2000), only ~10%–15%—a much smaller proportion than might have been anticipated—proved to originate clearly from the α-Proteobacterial lineage. A larger, generically "prokaryotic" proportion (~40%–50%) consists of proteins that appear to originate outside α-Proteobacteria but are not necessarily affiliated with any specific bacterial or archaeal clade. Another, "eukaryotic" fraction (~20%–30%) contains proteins having no obvious homologs in either Archaea or Bacteria. A final, "unique" subset (~20%) comprises seemingly yeast-specific proteins having no identifiable homologs in other eukaryotes or in prokaryotes. These results indicate that the yeast mitochondrial proteome has multiple evolutionary origins and a complex evolutionary history (Kurland and Andersson 2000; Gray et al. 2001), a conclusion now firmly established for the mitochondria of other eukaryotes (Gabaldón and Huynen 2004; Szklarczyk and Huynen 2010). Bacteriophage-like proteins (notably the mitochondrial RNA polymerase in most eukaryotes) make an additional small contribution to the evolutionary mix (Shutt and Gray 2006).

Comparative mitochondrial genomics, based on complete sequencing of mtDNAs, has told us much regarding the nature of the ancestral mitochondrial genome and about patterns and mechanisms of mitochondrial genome evolution (Gray et al. 1998; Gray 1999). In the same way, comparative mitochondrial proteomics, based on a combination of bioinformatics data mining and direct analysis of whole mitochondria or submitochondrial fractions and complexes (Dreger 2003; Yan et al. 2009), is proving to be an equally incisive approach for unraveling the evolution of the mitochondrial proteome. Investigations using mass spectrometry (MS) have confirmed and extended the initial, bioinformatics-based findings that indicated a mosaic evolutionary origin of the mitochondrial proteome.

MS provides a powerful adjunct to bioinformatics-based approaches in its capacity to reveal novel mitochondrial proteins that have no identifiable sequence homologs and that lack MTSs. At least 13 novel, "ciliate-specific" proteins identified during MS analysis of mitochondria from *Tetrahymena pyriformis* (Smith et al. 2007) were subsequently found as components of the purified mitochondrial F_1F_0-ATP synthase (complex V) of this protist (Nina et al. 2010). These observations highlight an emerging theme of taxon-specific retooling of mitochondrial complexes such as electron transport chain assemblies and ribosomes, only the core α-Proteobacteria-like components of which appear to have been contributed by the proto-mitochondrion. A feature of this retailoring is the addition of novel proteins of generally unknown function, sometimes accompanied by loss of otherwise conserved components. An example is the ATP synthase (complex V) of the chlorophycean green alga, *Chlamydomonas reinhardtii*, in which nine novel subunits of unknown evolutionary origin replace eight subunits that are otherwise conserved in the ATP synthase of other non-chlorophycean green algae, as well as in plants, animals, and fungi (Lapaille et al. 2010).

Other mitochondrial respiratory complexes, such as complex I (CI; NADH:ubiquinone oxidoreductase), clearly exemplify this retailoring theme. Bacterial CI comprises 14 subunits, all of which are found in the corresponding mammalian complex, with seven of these subunits encoded in mammalian mtDNA (Table 1). A further 18 subunits in mammalian CI are assumed to be eukaryote-specific additions already present in the last eukaryotic common ancestor, because they are ubiquitous throughout eukaryotes but are not found in bacteria (but see below). Thirteen other mammalian CI subunits appeared to display a narrow phylogenetic distribution, being identified initially only in metazoan animals (Brandt 2006).

In attempting to define the ancestral state of selected mitochondrial components and pathways, the importance of taking a comprehensive phylogenetic approach (examining mitochondrial proteomes from as wide a range of eukaryotes as possible) needs to be recognized. As noted above, mitochondrial CI has an additional 18 "eukaryote-specific" subunits that are considered to have been incorporated at the earliest stages of mitochondrial CI evolution (Gabaldón et al. 2005; Brandt 2006). In plants (Heazlewood et al. 2003; Parisi et al. 2004) and green algae (Cardol et al. 2004), mitochondrial CI additionally contains proteins with high similarity to γ-type carbonic anhydrases (γCAs). Comparative studies over a relatively narrow phylogenetic range initially suggested that these proteins represented specific additions in the plant lineage (Parisi et al. 2004). However, a more recent study focusing on protists—eukaryotic microbes, wherein lies most of the evolutionary diversity within the domain Eucarya—has revealed a much broader distribution of mitochondrial γCAs, either shown or presumed to be associated with mitochondrial

CI (Gawryluk and Gray 2010b), than previously suspected. It appears likely that γCAs were ancestral components of mitochondrial CI and that they were subsequently lost from CI specifically in the evolutionary line leading to animals and fungi (opisthokonts), rather than added to CI specifically in the line leading to plants and algae. These results emphasize the importance of comprehensive taxon coverage in drawing conclusions regarding mitochondrial proteome evolution. In fact, wider taxon sampling has recently demonstrated that almost all of the 18 supposedly metazoan-specific CI subunits were likely already present in CI of the last eukaryotic common ancestor (Cardol 2011; Gawryluk et al. 2012). In the same way, a more phylogenetically broad analysis of homologous bacterial complexes appears warranted, judging by the recent demonstration that CI from the α-Proteobacterium *Paracoccus dentrificans* contains three of the "eukaryote-specific" CI proteins that had long been considered to be absent from bacterial CI (Yip et al. 2011). (In this regard, it is notable that 35 years ago [before the advent of sequence-based phylogenetic reconstructions], John and Whatley [1975] pointed out that *P. denitrificans* "resembles a mitochondrion more closely than do other bacteria, in that it effectively assembles in a single organism those features of the mitochondrial respiratory chain and oxidative phosphorylation which are otherwise distributed at random among most other aerobic bacteria.")

Another mitochondrial complex that shows a profound degree of evolutionary retailoring is the mitochondrial ribosome. In many eukaryotes (e.g., most animals), the mitochondrial LSU and SSU rRNAs have become much smaller than their bacterial counterparts, while at the same time many new ribosomal proteins have been added, so that what was originally an RNA-rich complex is now a protein-rich one (O'Brien 2002). The most extreme example of this ribosome retailoring is seen in the kinetoplastid protozoa, such as *Trypanosoma brucei* (Ziková et al. 2008) and *Leishmania tarentolae* (Sharma et al. 2009). The *T. brucei* mitochondrial ribosome, for example, contains 56 SSU and 77 LSU proteins, compared with the 21 SSU and 34 LSU proteins present in the *Escherichia coli* ribosome. The novel mitoribosomal proteins identified in these analyses do not have detectable homologs outside of the kinetoplastid protozoa and display only a low degree of sequence conservation within this lineage. These observations reinforce the importance of direct MS analyses of isolated mitochondrial complexes in order to accurately determine their composition, given that so many of these components appear to be new, lineage-specific inventions.

Other studies have shown that the ancestral mitochondrial ribosome in the last eukaryotic common ancestor was already much larger than its bacterial ancestor, containing some 19 additional eukaryote-specific proteins (Smits et al. 2007; Desmond et al. 2011). The fact that these novel mitoribosomal proteins are found in all of the currently recognized eukaryotic supergroups is yet another strong argument in favor of a monophyletic origin of contemporary mitochondria, a conclusion in this case based on eukaryote-specific rather than prokaryote-specific features.

CONCLUDING REMARKS

Comparative mitochondrial genomics has offered us a glimpse of what the ancestral mitochondrial genome was like, and what genes it contained. A monophyletic origin of the mitochondrial genome from within α-Proteobacteria is strongly supported, with Rickettsiales most often identified as the α-Proteobacterial order most closely related to mitochondria. This phylogenetic-based approach has revealed that mitochondrial genome evolution has been characterized by massive expansion in some lineages and extreme reduction and compaction in others, with endosymbiotic gene transfer relocating much of the initial genetic information in the proto-mitochondrial genome to the nucleus. Comparative mitochondrial proteomics has provided evidence of a mosaic evolutionary origin of the protein complement of this organelle, with a much smaller proportion of the proteome than might have been anticipated clearly deriving from an α-Proteobacterial progenitor.

Mitochondri-specific proteins, pathways, and functions emerging within the eukaryotic lineage subsequent to the α-Proteobacterial endosymbiosis are being identified. Retailoring of key mitochondrial complexes relative to their α-Proteobacterial antecedents is seen to have occurred through addition of novel protein components, often in a narrow, lineage-specific manner. We may anticipate that the unabated deluge of genomic and proteomic data will confront us with currently unappreciated aspects of mitochondrial structure and function across the broad range of eukaryotes, with new data and insights continually reshaping and refining our ideas regarding mitochondrial evolution.

ACKNOWLEDGMENTS

This essay is dedicated to the memory of Professor Lynn Margulis (1938–2011), a passionate advocate for the idea that endosymbiosis is a driving force in cell evolution and was the process that led to the origin of mitochondria and plastids. I sincerely thank the agencies that have enabled my research in this area over the past four decades through operating and equipment grants and salary support, most particularly the Canadian Institutes of Health Research (formerly the Medical Research Council of Canada), Natural Sciences and Engineering Research Council of Canada, Canadian Institute for Advanced Research (Program in Evolutionary Biology), Canada Research Chairs Program, Canada Foundation for Innovation, and Genome Canada.

REFERENCES

Adams KL, Palmer JD. 2003. Evolution of mitochondrial gene content: Gene loss and transfer to the nucleus. *Mol Phylogenet Evol* **29:** 380–395.

Adams KL, Ong HC, Palmer JD. 2001. Mitochondrial gene transfer in pieces: Fission of the ribosomal protein gene *rpl2* and partial or complete gene transfer to the nucleus. *Mol Biol Evol* **18:** 2289–2297.

Anderson S, Bankier AT, Barrell BG, de Bruijn MHL, Coulson AR, Drouin J, Eperon IC, Nierlich DP, Roe BA, Sanger F, et al. 1981. Sequence and organization of the human mitochondrial genome. *Nature* **290:** 457–465.

Anderson S, de Bruijn MHL, Coulson AR, Eperon IC, Sanger F, Young IG. 1982. Complete sequence of bovine mitochondrial DNA conserved features of the mammalian mitochondrial genome. *J Mol Biol* **156:** 683–717.

Andersson SGE, Kurland CG. 1999. Origins of mitochondria and hydrogenosomes. *Curr Opin Microbiol* **2:** 535–541.

Andersson SGE, Zomorodipour A, Andersson JO, Sicheritz-Pontén T, Alsmark UCM, Podowski RM, Näslund AK, Eriksson A-S, Winkler HH, Kurland CG. 1998. The genome sequence of *Rickettsia prowazekii* and the origin of mitochondria. *Nature* **396:** 133–140.

Andersson SG, Karlberg O, Canbäck B, Kurland CG. 2003. On the origin of mitochondria: A genomics perspective. *Philos Trans R Soc Lond B Biol Sci* **358:** 177–179.

Atteia A, Adrait A, Brugière S, Tardif M, van Lis R, Deusch O, Dagan T, Kuhn L, Gontero B, Martin W, et al. 2009. A proteomic survey of *Chlamydomonas reinhardtii* mitochondria sheds new light on the metabolic plasticity of the organelle and on the nature of the α-proteobacterial mitochondrial ancestor. *Mol Biol Evol* **26:** 1533–1548.

Bibb MJ, Van Etten RA, Wright CT, Walberg MW, Clayton DA. 1981. Sequence and gene organization of mouse mitochondrial DNA. *Cell* **26:** 167–180.

Boussau B, Karlberg EO, Frank AC, Legault B-A, Andersson SGE. 2004. Computational inference of scenarios for α-proteobacterial genome evolution. *Proc Natl Acad Sci* **101:** 9722–9727.

Boxma B, de Graaf RM, van der Staay GW, van Alen TA, Ricard G, Gabaldon T, van Hoek AH, Moon-van der Staay SY, Koopman WJ, van Hellemond JJ, et al. 2005. An anaerobic mitochondrion that produces hydrogen. *Nature* **434:** 74–79.

Brandt U. 2006. Energy converting NADH:quinone oxidoreductase (complex I). *Annu Rev Biochem* **75:** 69–92.

Brindefalk B, Ettema TJG, Viklund J, Thollesson M, Andersson SGE. 2011. A phylometagenomic exploration of oceanic Alphaproteobacteria reveals mitochondrial relatives unrelated to the SAR11 clade. *PLoS ONE* **6:** e24457.

Bui ET, Bradley PJ, Johnson PJ. 1996. A common evolutionary origin for mitochondria and hydrogenosomes. *Proc Natl Acad Sci* **93:** 9651–9656.

Bullerwell CE, Gray MW. 2004. Evolution of the mitochondrial genome: Protist connections to animals, fungi and plants. *Curr Opin Microbiol* **7:** 528–534.

Burger G, Gray MW, Lang BF. 2003a. Mitochondrial genomes: Anything goes. *Trends Genet* **19:** 709–716.

Burger G, Lang BF, Braun H-P, Marx S. 2003b. The enigmatic mitochondrial ORF *ymf39* codes for ATP synthase chain b. *Nucleic Acids Res* **31:** 2353–2360.

Cardol P. 2011. Mitochondrial NADH:ubiquinone oxidoreductase (complex I) in eukaryotes: A highly conserved subunit composition highlighted by mining of protein databases. *Biochim Biophys Acta* **1807:** 1390–1397.

Cardol P, Vanrobaeys F, Devreese B, Van Beeumen J, Matagne RF, Remacle C. 2004. Higher plant-like subunit composition of mitochondrial complex I from *Chlamydomonas reinhardtii*: 31 conserved components among eukaryotes. *Biochim Biophys Acta* **1658:** 212–224.

Cavalier-Smith T. 1987a. Eukaryotes with no mitochondria. *Nature* **326:** 332–333.

Cavalier-Smith T. 1987b. The simultaneous symbiotic origin of mitochondria, chloroplasts, and microbodies. *Ann NY Acad Sci* **503:** 55–71.

Cavalier-Smith T. 1989. Archaebacteria and Archezoa. *Nature* **339:** 100–101.

Cavalier-Smith T. 1992. The number of symbiotic origins of organelles. *Biosystems* **28:** 107–108.

Cedergren R, Gray MW, Abel Y, Sankoff D. 1988. The evolutionary relationships among known life forms. *J Mol Evol* **28:** 98–112.

Clark AG, Roger AJ. 1995. Direct evidence for secondary loss of mitochondria in *Entamoeba histolytica*. *Proc Natl Acad Sci* **92:** 6518–6521.

Claros MG, Vincens P. 1996. Computational method to predict mitochondrially imported proteins and their targeting sequences. *Eur J Biochem* **241:** 779–786.

de Graaf RM, Ricard G, van Alen TA, Duarte I, Dutilh BE, Burgtorf C, Kuiper JWP, van der Staay GWM, Tielens AGM, Huynen MA, et al. 2011. The organellar genome and metabolic potential of the hydrogen-producing mitochondrion of *Nyctotherus ovalis*. *Mol Biol Evol* **28:** 2379–2391.

Desmond E, Brochier-Armanet C, Forterre P, Gribaldo S. 2011. On the last common ancestor and early evolution of eukaryotes: Reconstructing the history of mitochondrial ribosomes. *Res Microbiol* **162:** 53–70.

Doolittle WF. 1980. Revolutionary concepts in evolutionary biology. *Trends Biochem Sci* **5:** 146–149.

Dreger M. 2003. Proteome analysis at the level of subcellular structures. *Eur J Biochem* **270:** 589–599.

Emanuelsson O, Nielsen H, Brunak S, von Heijne G. 2000. Predicting subcellular localization of proteins based on their N-terminal amino acid sequence. *J Mol Biol* **300:** 1005–1016.

Embley TM. 2006. Multiple secondary origins of the anaerobic lifestyle in eukaryotes. *Philos Trans R Soc Lond B Biol Sci* **361:** 1055–1067.

Embley TM, Hirt RP. 1998. Early branching eukaryotes? *Curr Opin Genet Dev* **8:** 624–629.

Embley TM, van der Giezen M, Horner DS, Dyal PL, Bell S, Foster PG. 2003. Hydrogenosomes, mitochondria and early eukaryotic evolution. *IUBMB Life* **55:** 387–395.

Emelyanov VV. 2001a. Evolutionary relationship of Rickettsiae and mitochondria. *FEBS Lett* **501:** 11–18.

Emelyanov VV. 2001b. Rickettsiaceae, Rickettsia-like endosymbionts, and the origin of mitochondria. *Biosci Rep* **21:** 1–17.

Emelyanov VV. 2003a. Common evolutionary origin of mitochondrial and rickettsial respiratory chains. *Arch Biochem Biophys* **420:** 130–141.

Emelyanov VV. 2003b. Mitochondrial connection to the origin of the eukaryotic cell. *Eur J Biochem* **270:** 1599–1618.

Esser C, Ahmadinejad N, Wiegand C, Rotte C, Sebastiani F, Gelius-Dietrich G, Henze K, Kretschmann E, Richly E, Leister D, et al. 2004. A genome phylogeny for mitochondria among α-Proteobacteria and a predominantly eubacterial ancestry of yeast nuclear genes. *Mol Biol Evol* **21:** 1643–1660.

Feagin JE. 2000. Mitochondrial genome diversity in parasites. *Int J Parasitol* **30:** 371–390.

Gabaldón T, Huynen MA. 2003. Reconstruction of the proto-mitochondrial metabolism. *Science* **302:** 609.

Gabaldón T, Huynen MA. 2004. Shaping the mitochondrial proteome. *FEBS Lett* **1659:** 212–220.

Gabaldón T, Huynen MA. 2007. From endosymbiont to host-controlled organelle: The hijacking of mitochondrial protein synthesis and metabolism. *PLoS Comput Biol* **3:** e219.

Gabaldón T, Rainey D, Huynen MA. 2005. Tracing the evolution of a large protein complex in the eukaryotes, NADH:ubiquinone oxidoreductase (Complex I). *J Mol Biol* **348:** 857–870.

Gawryluk RMR, Gray MW. 2009. A split and rearranged nuclear gene encoding the iron-sulfur subunit of mitochondrial succinate dehydrogenase in Euglenozoa. *BMC Res Notes* **2:** 16.

Gawryluk RMR, Gray MW. 2010a. An ancient fission of mitochondrial *cox1*. *Mol Biol Evol* **27:** 7–10.

Gawryluk RMR, Gray MW. 2010b. Evidence for an early evolutionary emergence of γ-type carbonic anhydrases as components of mitochondrial respiratory complex I. *BMC Evol Biol* **348:** 857–870.

Gawryluk RMR, Chisholm KA, Pinto DM, Gray MW. 2012. Composition of the mitochondrial electron transport chain in *Acanthamoeba castellanii*: Structural and evolutionary insights. *Biochim Biophys Acta* doi: org/10.1016/j.bbabio.2012.06.005.

Georgiades K, Madoui M-A, Le P, Robert C, Raoult D. 2011. Phylogenomic analysis of *Odyssella thessalonicensis* fortifies the common origin of Rickettsiales, *Pelagibacter ubique* and *Reclimonas americana* mitochondrion. *PLoS ONE* **6:** e24857.

Gill EE, Diaz-Triviño S, Barberà MJ, Silberman JD, Stechmann A, Gaston D, Tamas I, Roger AJ. 2007. Novel mitochondrion-related organelles in the anaerobic amoeba *Mastigamoeba balamuthi*. *Mol Microbiol* **66:** 1306–1320.

Giovannoni SJ, Tripp HJ, Givan S, Podar M, Vergin KL, Baptista D, Bibbs L, Eads J, Richardson TH, Noordewier M, et al. 2005. Genome streamlining in a cosmopolitan oceanic bacterium. *Science* **309:** 1242–1245.

Goldberg AV, Molik S, Tsaousis AD, Neumann K, Kuhnke G, Delbac F, Vivares CP, Hirt RP, Lill R, Embley TM. 2008. Localization and functionality of microsporidian iron–sulphur cluster assembly proteins. *Nature* **452:** 624–629.

Gray MW. 1982. Mitochondrial genome diversity and the evolution of mitochondrial DNA. *Can J Biochem* **60:** 157–171.

Gray MW. 1983. The bacterial ancestry of plastids and mitochondria. *BioScience* **33:** 693–699.

Gray MW. 1989a. The evolutionary origins of organelles. *Trends Genet* **5:** 294–299.

Gray MW. 1989b. Origin and evolution of mitochondrial DNA. *Annu Rev Cell Biol* **5:** 25–50.

Gray MW. 1992. The endosymbiont hypothesis revisited. *Int Rev Cytol* **141:** 233–357.

Gray MW. 1993. Origin and evolution of organelle genomes. *Curr Opin Genet Dev* **3:** 884–890.

Gray MW. 1998. Rickettsia, typhus and the mitochondrial connection. *Nature* **396:** 109–110.

Gray MW. 1999. Evolution of organellar genomes. *Curr Opin Genet Dev* **9:** 678–687.

Gray MW. 2011. The incredible shrinking organelle. *EMBO Rep* **12:** 873.

Gray MW, Doolittle WF. 1982. Has the endosymbiont hypothesis been proven? *Microbiol Rev* **46:** 1–42.

Gray MW, Spencer DF. 1996. Organellar evolution. In *Evolution of microbial life* (ed. Roberts D, et al.), pp. 109–126. Cambridge University Press, Cambridge.

Gray MW, Sankoff D, Cedergren RJ. 1984. On the evolutionary descent of organisms and organelles: A global phylogeny based on a highly conserved structural core in small subunit ribosomal RNA. *Nucleic Acids Res* **12:** 5837–5852.

Gray MW, Cedergren R, Abel Y, Sankoff D. 1989. On the evolutionary origin of the plant mitochondrion and its genome. *Proc Natl Acad Sci* **86:** 2267–2271.

Gray MW, Lang BF, Cedergren R, Golding GB, Lemieux C, Sankoff D, Turmel M, Brossard N, Delage E, Littlejohn TG, et al. 1998. Genome structure and gene content in protist mitochondrial DNAs. *Nucleic Acids Res* **26:** 865–878.

Gray MW, Burger G, Lang BF. 1999. Mitochondrial evolution. *Science* **283:** 1476–1481.

Gray MW, Burger G, Lang BF. 2001. The origin and early evolution of mitochondria. *Genome Biol* **2:** 1018.1–1018.5.

Gray MW, Lang BF, Burger G. 2004. Mitochondria of protists. *Annu Rev Genet* **38:** 477–524.

Gupta RS. 1995. Evolution of the chaperonin families (Hsp60, Hsp10 and Tcp-1) of proteins and the origin of eukaryotic cells. *Mol Microbiol* **15:** 1–11.

Heazlewood JL, Howell KA, Millar AH. 2003. Mitochondrial complex I from *Arabidopsis* and rice: Orthologs of mammalian and fungal components coupled with plant-specific subunits. *Biochim Biophys Acta* **1604:** 159–169.

Hirt RP, Logsdon JM Jr, Healy B, Dorey MW, Doolittle WF, Embley TM. 1999. Microsporidia are related to Fungi: Evidence from the largest subunit of RNA polymerase II and other proteins. *Proc Natl Acad Sci* **96:** 580–585.

Hjort K, Goldberg AV, Tsaousis AD, Hirt RP, Embley TM. 2010. Diversity and reductive evolution of mitochondria among microbial eukaryotes. *Philos Trans R Soc Lond B Biol Sci* **365:** 713–727.

Hrdy I, Hirt RP, Dolezal P, Bardonová L, Foster PG, Tachezy J, Embley TM. 2004. *Trichomonas* hydrogenosomes contain the NADH dehydrogenase module of mitochondrial complex I. *Nature* **432:** 618–622.

Hug LA, Stechmann A, Roger AJ. 2010. Phylogenetic distributions and histories of proteins involved in anaerobic pyruvate metabolism in eukaryotes. *Mol Biol Evol* **27:** 311–324.

Jacob Y, Seif E, Paquet P-O, Lang BF. 2004. Loss of the mRNA-like region in mitochondrial tmRNAs of jakobids. *RNA* **10:** 605–614.

John P, Whatley FR. 1975. *Paracoccus denitrificans* and the evolutionary origin of the mitochondrion. *Nature* **254:** 495–498.

Karlberg O, Canbäck B, Kurland CG, Andersson SGE. 2000. The dual origin of the yeast mitochondrial proteome. *Yeast* **17:** 170–187.

Karlin S, Brocchieri L. 2000. Heat shock protein 60 sequence comparisons: Duplications, lateral transfer, and mitochondrial evolution. *Proc Natl Acad Sci* **97:** 11348–11353.

Keeling PJ, Luker MA, Palmer JD. 2000. Evidence from β-tubulin phylogeny that microsporidia evolved from within the fungi. *Mol Biol Evol* **17:** 23–31.

Keeling PJ, Burger G, Durnford DG, Lang BF, Lee RW, Pearlman RE, Roger AJ, Gray MW. 2005. The tree of eukaryotes. *Trends Ecol Evol* **20:** 670–676.

Koonin EV. 2010. The origin and early evolution of eukaryotes in the light of phylogenomics. *Genome Biol* **11:** 209.

Kück U, Jekosch K, Holzamer P. 2000. DNA sequence analysis of the complete mitochondrial genome of the green alga *Scenedesmus obliquus*: Evidence for UAG being a leucine and UCA being a non-sense codon. *Gene* **253:** 13–18.

Kumar A, Agarwal S, Heyman JA, Matson S, Heidtman M, Piccirillo S, Umansky L, Drawid A, Jansen R, Liu Y, et al. 2002. Subcellular localization of the yeast proteome. *Genes Dev* **16:** 707–719.

Kurland CG, Andersson SGE. 2000. Origin and evolution of the mitochondrial proteome. *Microbiol Mol Biol Rev* **64:** 786–820.

Lane N, Martin W. 2010. The energetics of genome complexity. *Nature* **467:** 929–934.

Lang BF, Burger G, O'Kelly CJ, Cedergren R, Golding GB, Lemieux C, Sankoff D, Turmel M, Gray MW. 1997. An ancestral mitochondrial DNA resembling a eubacterial genome in miniature. *Nature* **387:** 493–497.

Lang BF, Gray MW, Burger G. 1999. Mitochondrial genome evolution and the origin of eukaryotes. *Annu Rev Genet* **33:** 351–397.

Lapaille M, Escobar-Ramírez A, Degand H, Baurain D, Rodríguez-Salinas E, Coosemans N, Boutry M, Gonzalez-Halphen D, Remacle C, Cardol P. 2010. Atypical subunit composition of the chlorophycean mitochondrial F_1F_O-ATP synthase and role of Asa7 protein in stability and oligomycin resistance of the enzyme. *Mol Biol Evol* **27:** 1630–1644.

Lindmark DG, Müller M. 1973. Hydrogenosome, a cytoplasmic organelle of the anaerobic flagellate *Tritrichomonas foetus*, and its role in pyruvate metabolism. *J Biol Chem* **248:** 7724–7728.

Mai Z, Ghosh S, Frisardi M, Rosenthal B, Rogers R, Samuelson J. 1999. Hsp60 is targeted to a cryptic mitochondrion-derived organelle ("crypton") in the microaerophilic protozoan parasite *Entamoeba histolytica*. *Mol Cell Biol* **19:** 2198–2205.

Marcotte EM, Xenarios I, van Der Bliek AM, Eisenberg D. 2000. Localizing proteins in the cell from their phylogenetic profiles. *Proc Natl Acad Sci* **97:** 12115–12120.

Margulis L. 1970. *Origin of eukaryotic cells.* Yale University Press, New Haven, CT.

Margulis L. 1981. *Symbiosis in cell evolution.* Freeman, San Francisco.

Martin W, Müller M. 1998. The hydrogen hypothesis for the first eukaryote. *Nature* **392:** 37–41.

Martin W, Hoffmeister M, Rotte C, Henze K. 2001. An overview of endosymbiotic models for the origins of eukaryotes, their ATP-producing organelles (mitochondria and hydrogenosomes), and their heterotrophic lifestyle. *Biol Chem* **382:** 1521–1539.

Morales J, Mogi T, Mineki S, Takashima E, Mineki R, Hirawake H, Sakamoto K, Ōmura S, Kita K. 2009. Novel mitochondrial complex II isolated from *Trypanosoma cruzi* is composed of 12 peptides including a heterodimeric Ip subunit. *J Biol Chem* **284:** 7255–7263.

Nedelcu AM, Lee RW, Lemieux C, Gray MW, Burger G. 2000. The complete mitochondrial DNA sequence of *Scenedesmus obliquus* reflects an intermediate stage in the evolution of the green algal mitochondrial genome. *Genome Res* **10:** 819–831.

Nina PB, Dudkina NV, Kane LA, van Eyk JE, Boekema EJ, Mather MW, Vaidya AB. 2010. Highly divergent mitochondrial ATP synthase complexes in *Tetrahymena thermophila*. *PLoS Biol* **8:** e1000418.

O'Brien TW. 2002. Evolution of a protein-rich mitochondrial ribosome: Implications for human genetic disease. *Gene* **286:** 73–79.

Oudot M-P, Kloareg B, Loiseaux-de Goër S. 1999. The mitochondrial *Pylaiella littoralis nad11* gene contains only the N-terminal FeS-binding domain. *Gene* **235:** 131–137.

Parisi G, Perales M, Fornasari MS, Colaneri A, Gonzalez-Schain N, Gomez-Casati D, Zimmermann S, Brennicke A, Araya A, Ferry JG, et al. 2004. Gamma carbonic anhydrases in plant mitochondria. *Plant Mol Biol* **55:** 193–207.

Pérez-Brocal V, Clark CG. 2008. Analysis of two genomes from the mitochondrion-like organelle of the intestinal parasite *Blastocystis*: Complete sequences, gene content and genome organization. *Mol Biol Evol* **25:** 2475–2482.

Pérez-Brocal V, Shahar-Golan R, Clark CG. 2010. A linear molecule with two large inverted repeats: The mitochondrial genome of the stramenopile *Proteromonas lacertae*. *Genome Biol Evol* **2:** 257–266.

Pérez-Martínez X, Antaramian A, Vázquez-Acevedo M, Funes S, Tolkunova E, d'Alayer J, Claros MG, Davidson E, King MP, González-Halphen D. 2001. Subunit II of cytochrome *c* oxidase in chlamydomonad algae is a heterodimer encoded by two independent nuclear genes. *J Biol Chem* **276:** 11302–11309.

Pisani D, Cotton JA, McInerney JO. 2007. Supertrees disentangle the chimerical origin of eukaryotic genomes. *Mol Biol Evol* **24:** 1752–1760.

Richly E, Chinnery PF, Leister D. 2003. Evolutionary diversification of mitochondrial proteomes: Implications for human disease. *Trends Genet* **19:** 356–362.

Roger AJ, Silberman JD. 2002. Cell evolution: Mitochondria in hiding. *Nature* **418:** 827–829.

Sapp J. 1994. *Evolution by association. A history of symbiosis.* Oxford University Press, New York.

Schnare MN, Gray MW. 1982. 3′-Terminal sequence of wheat mitochondrial 18S ribosomal RNA: Further evidence of a eubacterial evolutionary origin. *Nucleic Acids Res* **10:** 3921–3932.

Sharma MR, Booth TM, Simpson L, Maslov DA, Agrawal RK. 2009. Structure of a mitochondrial ribosome with minimal RNA. *Proc Natl Acad Sci* **106:** 9637–9642.

Shutt TE, Gray MW. 2006. Bacteriophage origins of mitochondrial replication and transcription proteins. *Trends Genet* **22:** 90–95.

Sicheritz-Pontén T, Kurland CG, Andersson SGE. 1998. A phylogenetic analysis of the cytochrome *b* and cytochrome *c* oxidase I genes supports an origin of mitochondria from within the Rickettsiaceae. *Biochim Biophys Acta* **1365:** 545–551.

Sloan DB, Alverson AJ, Chuckalovcak JP, Wu M, McCauley DE, Palmer JD, Taylor DR. 2012. Rapid evolution of enormous, multichromosomal genomes in flowering plant mitochondria with exceptionally high mutation rates. *PLoS Biol* **10:** e1001241.

Smith DG, Gawryluk RM, Spencer DF, Pearlman RE, Siu KW, Gray MW. 2007. Exploring the mitochondrial proteome of the ciliate protozoon *Tetrahymena thermophila*: Direct analysis by tandem mass spectrometry. *J Mol Biol* **374:** 837–863.

Smits P, Smeitink JAM, van den Heuvel LP, Huynen MA, Ettema TJG. 2007. Reconstructing the evolution of the mitochondrial ribosomal proteome. *Nucleic Acids Res* **35:** 4686–4703.

Spencer DF, Schnare MN, Gray MW. 1984. Pronounced structural similarities between the small subunit ribosomal RNA genes of wheat mitochondria and *Escherichia coli*. *Proc Natl Acad Sci* **81:** 493–497.

Stechmann A, Hamblin K, Perez-Brocal V, Gaston D, Richmond GS, van der Giezen M, Clark CG, Roger AJ. 2008. Organelles in *Blastocystis* that blur the distinction between mitochondria and hydrogenosomes. *Curr Biol* **18:** 580–585.

Sutak R, Dolezal P, Fiumera HL, Hrdy I, Dancis A, Delgadillo-Correa M, Johnson PJ, Müller M, Tachezy J. 2004. Mitochondrial-type assembly of FeS centers in the hydrogenosomes of the amitochondriate eukaryote *Trichomonas vaginalis*. *Proc Natl Acad Sci* **101:** 10368–10373.

Szklarczyk R, Huynen M. 2010. Mosaic origin of the mitochondrial proteome. *Proteomics* **10:** 4012–4024.

Thao ML, Gullan PJ, Baumann P. 2002. Secondary (γ-Proteobacteria) endosymbionts infect the primary (β-Proteobacteria) endosymbionts of mealybugs multiple times and coevolve with their hosts. *Appl Environ Microbiol* **68:** 3190–3197.

Thrash JC, Boyd A, Huggett MJ, Grote J, Carini P, Yoder RJ, Robbertse B, Spatafora JW, Rappe MS, Giovannoni SJ. 2011. Phylogenomic evidence for a common ancestor of mitochondria and the SAR11 clade. *Sci Rep* **1:** 13.

Timmis JN, Ayliffe MA, Huang CY, Martin W. 2004. Endosymbiotic gene transfer: Organelle genomes forge eukaryotic chromosomes. *Nat Rev Genet* **5:** 123–135.

Tovar J, Fischer A, Clark CG. 1999. The mitosome, a novel organelle related to mitochondria in the amitochondrial parasite *Entamoeba histolytica*. *Mol Microbiol* **32:** 1013–1021.

Tovar J, León-Avila G, Sánchez LB, Sutak R, Tachezy J, van Der Giezen M, Hernández M, Müller M, Lucocq JM. 2003. Mitochondrial remnant organelles of *Giardia* function in iron-sulphur protein maturation. *Nature* **426:** 172–176.

Tsaousis AD, Kunji ERS, Goldberg AV, Lucocq JM, Hirt RP, Embley Tm. 2008. A novel route for ATP acquisition by the remnant mitochondria of *Encephalitozoon cuniculi*. *Nature* **453:** 553–557.

Viale AM, Arakaki AK. 1994. The chaperone connection to the origins of the eukaryotic organelles. *FEBS Lett* **341:** 146–151.

Viklund J, Ettema TJG, Andersson SGE. 2012. Independent genome reduction and phylogenetic reclassification of the oceanic SAR11 clade. *Mol Biol Evol* **29:** 599–615.

von Dohlen CD, Kohler S, Alsop ST, McManus WR. 2001. Mealybug β-proteobacterial endosymbionts contain γ-proteobacterial symbionts. *Nature* **412:** 433–436.

Wallace DC. 1982. Structure and evolution of organelle genomes. *Microbiol Rev* **46:** 208–240.

Waller RF, Keeling PJ. 2006. Alveolate and chlorophycean mitochondrial *cox2* genes split twice independently. *Gene* **383:** 33–37.

Wawrzyniak I, Roussel M, Diogon M, Couloux A, Texier C, Tan KS, Vivarès CP, Delbac F, Wincker P, El Alaoui H. 2008. Complete circular DNA in the mitochondria-like organelles of *Blastocystis hominis*. *Int J Parasitol* **38:** 1377–1382.

Whatley JM, John P, Whatley FR. 1979. From extracellular to intracellular: The establishment of mitochondria and chloroplasts. *Proc R Soc Lond B Biol Sci* **204:** 165–187.

Williams BA, Hirt RP, Lucocq JM, Embley TM. 2002. A mitochondrial remnant in the microsporidian *Trachipleistophora hominis*. *Nature* **418:** 865–869.

Williams KP, Sobral BW, Dickerman AW. 2007. A robust species tree for the *Alphaproteobacteria*. *J Bacteriol* **189:** 4578–4586.

Yan W, Aebersold R, Raines EW. 2009. Evolution of organelle-associated protein profiling. *J Proteomics* **72:** 4–11.

Yang D, Oyaizu Y, Oyaizu H, Olsen GJ, Woese CR. 1985. Mitochondrial origins. *Proc Natl Acad Sci* **82:** 4443–4447.

Yip C-y, Harbour ME, Jayawardena K, Fearnley IM, Sazanov LA. 2011. Evolution of respiratory complex I. "Supernumerary" subunits are present in the α-proteobacterial enzyme. *J Biol Chem* **286:** 5023–5033.

Ziková A, Panigrahi AK, Dalley RA, Acestor N, Anupama A, Ogata Y, Myler PJ, Stuart K. 2008. *Trypanosoma brucei* mitochondrial ribosomes, affinity purification and component identification by mass spectrometry. *Mol Cell Proteomics* **7:** 1286–1296.

Mitochondrial DNA Genetics and the Heteroplasmy Conundrum in Evolution and Disease

Douglas C. Wallace and Dimitra Chalkia

Center for Mitochondrial and Epigenomic Medicine, The Children's Hospital of Philadelphia, Department of Pathology and Laboratory Medicine, University of Pennsylvania, Philadelphia, Pennsylvania 19104

Correspondence: wallaced1@email.chop.edu

The unorthodox genetics of the mtDNA is providing new perspectives on the etiology of the common "complex" diseases. The maternally inherited mtDNA codes for essential energy genes, is present in thousands of copies per cell, and has a very high mutation rate. New mtDNA mutations arise among thousands of other mtDNAs. The mechanisms by which these "heteroplasmic" mtDNA mutations come to predominate in the female germline and somatic tissues is poorly understood, but essential for understanding the clinical variability of a range of diseases. Maternal inheritance and heteroplasmy also pose major challengers for the diagnosis and prevention of mtDNA disease.

THE GENETIC CHALLENGES OF mtDNA DISEASES

It is has become increasingly clear that mitochondrial dysfunction lies at the nexus of a wide range of metabolic and degenerative diseases, cancer, and aging. Two major reasons for why mitochondrial dysfunction has been overlooked in "complex" diseases is that subtle bioenergetic alterations can have major clinical consequences and mitochondrial defects can be generated by the unique quantitative genetics of the maternally inherited mitochondrial DNA (mtDNA).

The mitochondrial genome encompasses between 1000 to 2000 nuclear DNA (nDNA) genes plus thousands of copies of the maternally inherited mtDNA. The mtDNA codes for the most important bioenergetic genes. So mtDNA defects impinge on a wide spectrum of cellular functions.

A large number of pathogenic mtDNA mutations have been identified and the more severe mutations are frequently mixed with normal mtDNAs within the cell, a state known as heteroplasmy. Heteroplasmic alleles can shift in percentage during both mitotic and meiotic cell division, leading to a potentially continuous array of bioenergetic defects, a process known as replicative segregation. As the percentage of mutant mtDNAs increases, the resulting bioenergetic defect becomes increasingly severe. Because different tissues have different bioenergetic thresholds, as a patient's bioenergetic capacity declines it eventually falls below the minimum threshold for that tissue and symptoms

Cite this article as *Cold Spring Harb Perspect Biol* doi: 10.1101/cshperspect.a021220

ensue. Because the tissues and organs with the highest bioenergetic requirements are also those that are primarily affected in the common metabolic and degenerative diseases, it follows that mitochondrial dysfunction may be a major contributor to complex diseases.

Women that harbor deleterious heteroplasmic mutations have a high probability of having affected children, the nature and severity of the phenotype depending on the mtDNA mutation and the percentage of heteroplasmy. Cells and individuals can accumulate an array of different mtDNA mutations over time, the aggregate of which degrade the energetic capacity of the cell. Such mutations are important in aging and cancer. Given the enormous potential explanatory power of heteroplasmic mtDNA mutations, it is striking that very little is known about the origin, genetics, and phenotypic effects of heteroplasmic mtDNA mutations.

HUMAN mtDNA GENETICS

That mtDNA mutations could cause disease was first reported at the molecular level in 1988 with the demonstration that isolated patients with mitochondrial myopathy could harbor heteroplasmic mtDNA deletions (Holt et al. 1988); that the maternally inherited sudden onset blindness disease, Leber hereditary optic neuropathy (LHON), was caused by a homoplasmic missense mutation in the *ND4* gene at nt 11778G>A (arginine codon 340 to histidine, R340H) (Wallace et al. 1988a); and that myoclonic epilepsy and ragged red fiber disease (MERRF) was caused by a heteroplasmic mutation in the $tRNA^{Lys}$ gene at nt 8344A>G (Wallace et al. 1988b; Shoffner et al. 1990). These discoveries set the stage for investigating and understanding a broad range of enigmatic familial and age-related diseases.

Incidence of mtDNA Mutations and Disease

Mutations in mtDNA are surprisingly common. Genetic epidemiological studies quantifying only the most common pathogenic mtDNA mutations have estimated that the incidence of clinical mitochondrial diseases is about one in 5000 (Schaefer et al. 2004, 2008). More surprising, a survey of newborn cord bloods revealed that one in 200 infants harbored one of 10 common pathogenic mtDNA mutations (Elliott et al. 2008; Chinnery et al. 2012). Hence, pathogenic mtDNA mutations are very common and constantly arising.

Human OXPHOS and the Range of Phenotypes: Conception to Old Age

To understand the clinical implications of mtDNA mutations, it is essential to understand the central role that mitochondrial oxidative phosphorylation (OXPHOS) plays in cellular biology. The mitochondria oxidize the calories in our diet with the oxygen that we breathe to generate ~90% of cellular energy. In OXPHOS, electrons (reducing equivalents) derived from our food flow down the mitochondrial inner membrane electron transport chain (ETC) from reduced to oxidized states, ultimately terminating with reduction of oxygen to water. The ETC is initiated with oxidation of NADH by complex I (NADH:CoQ oxidoreductase or NADH dehydrogenase) or succinate by complex II (succinate:CoQ oxidoreductase or succinate dehydrogenase). The electrons are then transferred to coenzyme Q (CoQ), complex III, cytochrome *c*, complex IV (cytochrome *c* oxidase or COX), and finally to oxygen. As the electrons traverse complexes I, III, and IV, the energy released is used to pump protons from the mitochondrial matrix across the mitochondrial inner membrane to the intermembrane space (Wallace 2005, 2007, 2011). This creates a transmembrane electrochemical gradient of ~0.2 volts. This potential energy can then be used to drive OXPHOS complex V (H^+-translocating ATP synthase) to condense ADP and phosphate (Pi) to generate ATP (Mitchell 1961), thus coupling oxidation by the ETC with phosphorylation by the ATP synthase. The mitochondrial ATP is then exported to the cytosol via the adenine nucleotide translocators (ANTs), where the ATP energizes cellular reactions and drives work.

In addition to generating ATP energy, the mitochondria regulate cytosolic Ca^{2+} levels, which in turn modulate cellular and mitochondrial metabolic pathways, control the cellular REDOX state that regulates a wide array of cellular enzymatic reactions and transcription factors via thiol-disulfide interconversion, regulate mitochondrial ROS production that is both a signal transduction agent that impinges on molecules such as HIF and RAS, and is the major source of oxidative stress that can activate the innate immunity response through NF-κB signaling. When mitochondria experience extreme stress (elevated Ca^{2+} and ROS, depleted adenine nucleotides, and reduced membrane potential), this can activate the mitochondrial permeability transition pore (mtPTP) thus initiating apoptosis and necrosis (Wallace 2005, 2011, 2012, 2013a,b; Wallace et al. 2013).

Mitochondrial Genetics

The mtDNA

The mtDNA (Fig. 1) codes for the 13 most important OXPHOS polypeptides. These include seven of the ~45 polypeptides of OXPHOS complex I (ND1-3, ND4L, ND4-6): one of the 11 polypeptides of complex III (cytochrome *b*, cytb), three of the 13 polypeptides of complex IV (COI-III), and two of the ~15 polypeptides of complex V (ATP6 and 8). In addition, the mtDNA encodes the mitochondrial 16S and 18S rRNAs and 22 tRNAs for mitochondrial protein synthesis. The mtDNA also encompasses an ~1000 nt control region that contains an origin for replication of the G-rich heavy (H) stand and the promoters for transcription of both the H stand and the C-rich light (L) stand. Both mtDNA stands are transcribed into large polycistronic transcripts in which the larger rRNA and mRNA transcripts are punctuated by tRNAs. The tRNAs are processed out and the larger RNA products are polyadenylated. The mtDNA mRNAs are translated on mitochondrial-specific 55S ribosomes, which are sensitive to bacterial ribosomal initiators chloramphenicol (CAP) and aminoglycosides and are initiated with an *N*-formylmethionine just like bacterial protein synthesis (Wallace 2007).

mtDNA Mutations

The mtDNA genes have a very high sequence evolution rate, on the order of 10–20 times that of comparable nDNA genes (Brown et al. 1982; Neckelmann et al. 1987; Wallace et al. 1987). This is the product of both an exceptionally high mutation rate, perhaps 100- to 1000-fold higher than nDNA genes, times an mtDNA mutant fixation rate: $E = \mu F$ where E is the sequence evolution rate, μ the mutation rate, and F the fixation rate. When a new mutation arises in the mtDNA, it creates an intracellular heteroplasmic mixture of mutant and normal mtDNAs, but the mutant mtDNA is but one among thousands of nonmutant mtDNAs. In some manner, the initial mutant mtDNA becomes enriched within certain cells, ultimately coming to predominate and influence the cellular and patient phenotype. The mechanism by which this enrichment occurs in either germline or somatic cells remains a mystery.

Once an mtDNA mutation reaches an appreciable level within cells, the percentage of mutant mtDNAs can drift by replicative segregation. For an embryo generated by the fertilization of a heteroplasmic oocyte, the percentage of mutant and normal mtDNAs in different descendant tissues and organs can have quite different values. This genetic mosaicism results in bioenergetic mosaicism and phenotypic complexity. Added to the stochastic segregation of heteroplasmic mtDNA is the differential sensitivity of different organs to different mitochondrial physiological alterations. The brain is the most sensitive to partial bioenergetic defects followed by heart, muscle, kidney, and endocrine systems (Wallace 2005). Hence, subtle systemic mitochondrial deficiencies can result in organ-specific symptoms.

There are three classes of clinically relevant mtDNA variants: recent deleterious mutations, ancient adaptive mtDNA mutations, and somatic mtDNA mutations.

Maternally Inherited Diseases. The most clinically overt class of mtDNA variants is newly arising maternally inherited disease mutations. Because of the high mtDNA mutation rate, new pathogenic mtDNA mutations are contin-

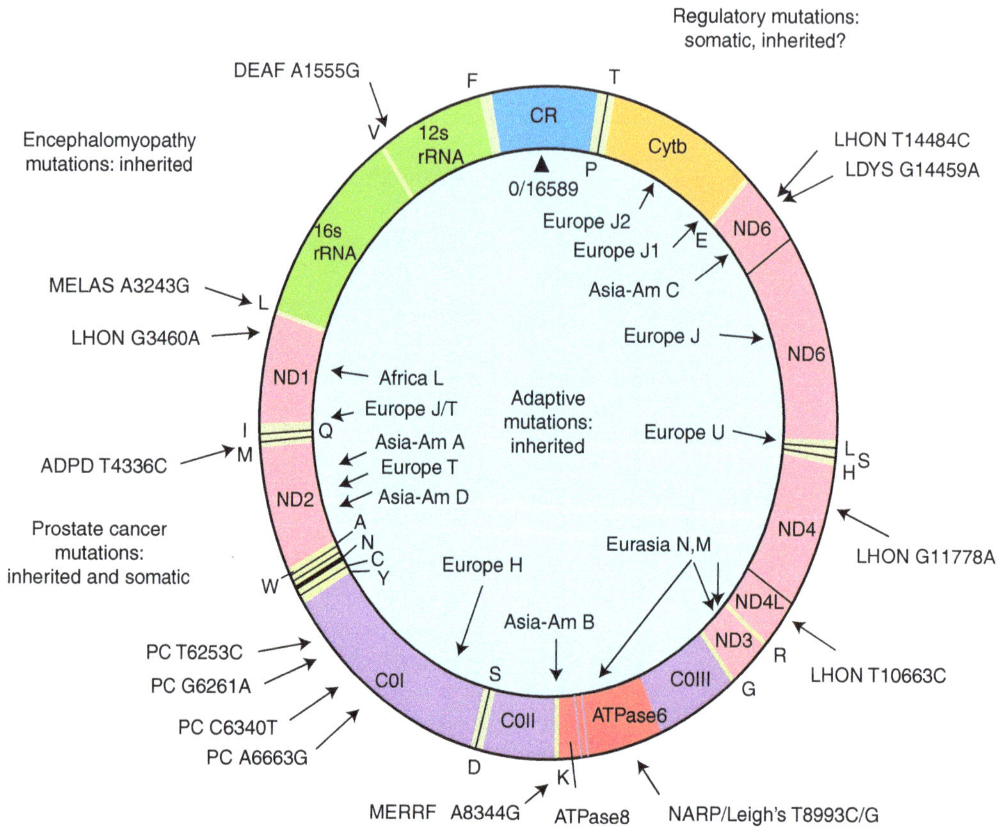

Figure 1. Human mitochondrial DNA map showing representative pathogenic and adaptive base substitution mutations. CR (control region) = D-loop. The letters around the outside perimeter or on the inside circle indicate cognate amino acids of the tRNA genes. Other gene symbols are defined in the text. Arrows followed by continental names and associated letters on the inside of the circle indicate the position of defining polymorphisms of selected region-specific mtDNA lineages. Arrows associated with abbreviations followed by numbers around the outside of the circle indicate representative pathogenic mutations, the number being the nucleotide position of the mutation. The full array of pathogenic mtDNA mutations and polymorphisms are available through Mitomap.org (MITOMAP 2012). DEAF, deafness; MELAS, mitochondrial encephalomyopathy, lactic acidosis, and stroke-like episodes; LHON, Leber hereditary optic neuropathy; ADPD, Alzheimer's disease and Parkinson's disease; MERRF, myoclonic epilepsy and ragged red fiber disease; NARP, neurogenic muscle weakness, ataxia, retinitis pigmentosum; LDYS, LHON + dystonia; PC, prostate cancer. (From Wallace 2007; reproduced, with permission, from the author.)

uously being introduced into the human population. Hundreds of pathogenic mtDNA mutations have now been documented (MITOMAP 2012; Wallace et al. 2013) and these can affect virtually every tissue in the body, depending on the mutation's severity, nature, and heteroplasmy level. Hence, pathogenic mtDNA mutations and mitochondrial dysfunction result in a wide range of multisystem degenerative diseases.

Ancient Adaptive Polymorphisms. There is also substantial mtDNA sequence diversity between individuals and human populations. These ancient mtDNA polymorphisms accumulated along radiating maternal lineages as women migrated out of Africa to colonize the globe. As new mtDNA mutations arose, new branches of the mtDNA tree were generated. If a founder mutation changed mitochondrial physiology in a manner beneficial to individuals within that regional environment, then that mtDNA lineage became enriched in that geographic locality. Subsequent additional base substitutions in descendant mtDNAs generated a group of related regional haplotypes designated a haplogroup. Hence, each continent and geographical region is associated with a distinctive array of mtDNA sequence types.

All African mtDNAs are related and encompassed within one large continent-specific lineage designated "macrohaplogroup L." Macrohaplogroup L arose between 130,000 and 200,000 years before present (YBP) (Fig. 2), founded by mtDNAs similar to haplogroup L0, which is common among the Khoi-San Bushmen of South Africa. In Ethiopia, ~65,000 YBP African haplogroup L3 gave rise to two mtDNAs designated M and N. Only the mtDNA descendants from M and N mtDNAs left Africa to colonize the rest of the world, generating macrohaplogroups M and N. From Africa, macrohaplogroup M moved along tropical Southeast Asia, ultimately reaching Australia. Later, M descendants moved north out of Southeast Asia to form a plethora of central and eastern Asian mtDNA haplogroups including C, D, G, and M1–M20. Out of Africa, macrohaplogroup N went in two directions. In one, N moved through Southeast Asia to Australia and from southern Asia, north into central Asia to generate haplogroups A and Z. In the second, macrohaplogroup N moved north out of Africa to form the European haplogroups I, X, and W. In western Eurasia, N also gave rise to submacrohaplogroup R. R then gave rise to the remaining European haplogroups H, J, Uk, T, U, and V. R also moved east to produce the Asian mtDNA haplogroups B and F (Fig. 3).

Of all the Asian mtDNA variants, only A, C, and D became enriched in northeastern Siberia and were in a position to cross the Bering Land Bridge ~20,000 YBP to establish the Paleo-Indian populations. Later additional migrations brought haplogroups B and X to join haplogroups A, C, and D (Fig. 2) (Wallace et al. 1999, 2013).

The regionality of the mtDNA haplogroups is extraordinary in several respects. First, of all of the African diversity, only two mtDNA lineages (M and N) colonized the rest of the world. Second, of all of the Asian mtDNAs, only three mtDNA lineages (A, C, and D) moved to extreme northeast Siberia to found the Paleo-Indians. Third, and most surprising, the mtDNA sequence evolution rate is such that it produced important mtDNA evolutionary changes that coincide with the major human geographic migrations. Such associations could not have occurred by chance. Rather, it is most likely that mtDNA variation permitted adaptation of our human ancestors to different regional environments, thus being the adaptive system that permitted human colonization of the diverse environments that they encountered around the globe (Ruiz-Pesini et al. 2004; Mishmar et al. 2006; Ruiz-Pesini and Wallace 2006; Wallace 2013a).

The founding mtDNA for macrohaplogroup N, which moved directly from subtropical Africa north into the European temperate zone, harbored two polypeptide variants, *ND3* nt 10398G>A (A114T) and *ATP6* nt 8701G>A (A59T) (Wallace et al. 1999, 2013). These variants have been associated with alterations in the mitochondrial membrane potential and Ca^{2+} metabolism (Kazuno et al. 2006). Presumably, these variants reduced the coupling efficiency of mitochondrial OXPHOS (loose coupling), resulting in an increase in the number of calories burned by the mitochondria to generate the ATP required to perform work. Because a calorie is a unit of heat, burning more calories would increase core body-heat production, rendering these individuals more resistant to the cold stress encountered in more northern environments. By contrast, macrohaplogroup M mtDNAs, which initially remained in the tropics, did not acquire comparable functional mtDNA mutations. Presumably, this mtDNA lineage retained the tight coupling of OXPHOS found in Africa in which ATP production is maximized and heat production is minimized (Ruiz-Pesini et al. 2004; Mishmar et al. 2006; Ruiz-Pesini and Wallace 2006; Wallace 2013a). Consistent with the concept that mtDNA variation has permitted climatic adaptation, mtDNA variation but not nDNA variation has been found to correlate with climatic differences (Balloux et al. 2009). Also, the basal metabolic

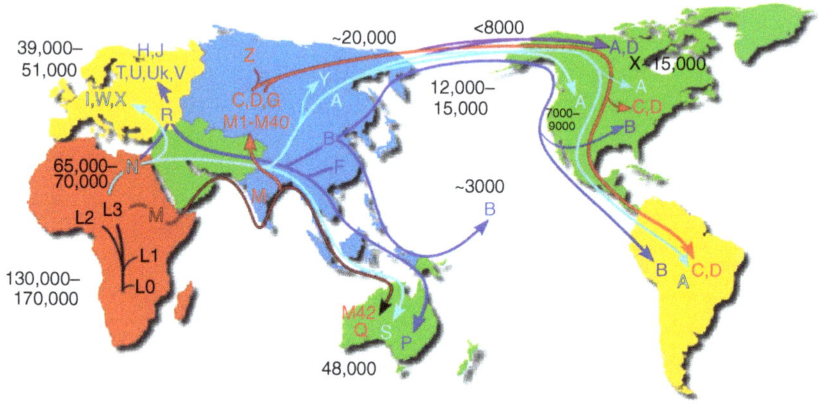

Figure 2. Diagram of the migratory history of the human mtDNA haplogroups. *Homo sapiens* mtDNAs arose in Africa ~130,000–200,000 years before present (YBP), with the first African-specific haplogroup branch being L0, followed by the appearance in Africa of lineages L1, L2, and L3. In northeastern Africa, L3 gave rise to two new lineages, M and N. Only M and N mtDNAs successfully left Africa ~65,000 YBP and colonized all of Eurasia and the Americas. In Eurasia and the Americas, M and N gave rise to a diverse array of mtDNA lineages designated macrohaplogroups M and N. The founders of macrohaplogroup M moved out of Africa through India and along the Southeast Asian coast down along the Malaysian peninsula and into Australia, generating haplogroups Q and M42 ~48,000 YBP. Subsequently, M moved north out of Southeast Asia to produce a diverse array of Central Asian mtDNA lineages including haplogroups C, D, G, and many other M haplogroup lineages. In northeast Asia, haplogroup C gave rise to haplogroup Z. The founders of macrohaplogroup N also moved through Southeast Asia and into Australia, generating haplogroup S. In Asia, macrohaplogroup N mtDNAs also moved north to generate central Asian haplogroup A and Siberian haplogroup Y. In western Eurasia, macrohaplogroup N founders also moved north to spawn European haplogroups I, W, and X, and in western Eurasia, gave rise to submacrohaplogroup R. R moved west to produce the European haplogroups H, J, Uk, T, U, and V and also moved east to generate Australian haplogroup P and eastern Asian haplogroups F and B. By 20,000 YBP, mtDNA haplogroups C and D from M, and A from N, were enriched in northeastern Siberia and thus were positioned to migrate across the Bering land bridge (Beringia) to give rise to the first Native American populations, the Paleo-Indians. Haplogroups A, C, and D migrated throughout North America and on through Central America to radiate into South America. Haplogroup X, which is most prevalent in Europe but is also found in Mongolia though not in Siberia, arrived in North America ~15,000 YBP but remained in northern North America. Haplogroup B, which is not found in Siberia but is prevalent along the coast of Asia, arrived in North America ~12,000–15,000 YBP and moved through North and Central America and into South America, combining with A, C, D, and X to generate the five dominant Paleo-Indian haplogroups (A, B, C, D, X). A subsequent migration of haplogroup A out of the Chukotka peninsula ~7000–9000 YBP gave rise to the Na-Déné (Athabaskins, Navajo, Apache, etc.). Subsequent movement across the Bering Strait, primarily carrying haplogroups A and D after 6000 YBP, produced the Eskimo and Aleut populations. Most recently, eastern Asian haplogroup B migrated south along the Asian coast through Micronesia and out into the Pacific to colonize all of the Pacific islands. Ages of migrations are approximated using mtDNA sequence evolution rates determined by comparing regional archeological or physical anthropological data with corresponding mtDNA sequence diversity. Because selection may have limited the accumulation of diversity in certain contexts, ages for regional migrations were estimated from the diversity encompassed within an individual regional or continental haplogroup lineages. This is because selection would have acted on the haplogroup mtDNA but most subsequent mutations would accumulate by random genetic drift and thus be "clock-like." (From Wallace 2013a,b; reproduced, with permission, from the author.)

rate of Siberian populations is higher than that of more southern populations (Leonard et al. 2002; Snodgrass et al. 2005, 2008).

The portion of the European mtDNA tree encompassing haplogroups J and T provides one example of the importance of functional variants in founding and defining the branches of the mtDNA haplogroup trees (Fig. 3). The J-T lineages were founded by two polypeptide gene amino acid substitution variants, ND1 nt

A

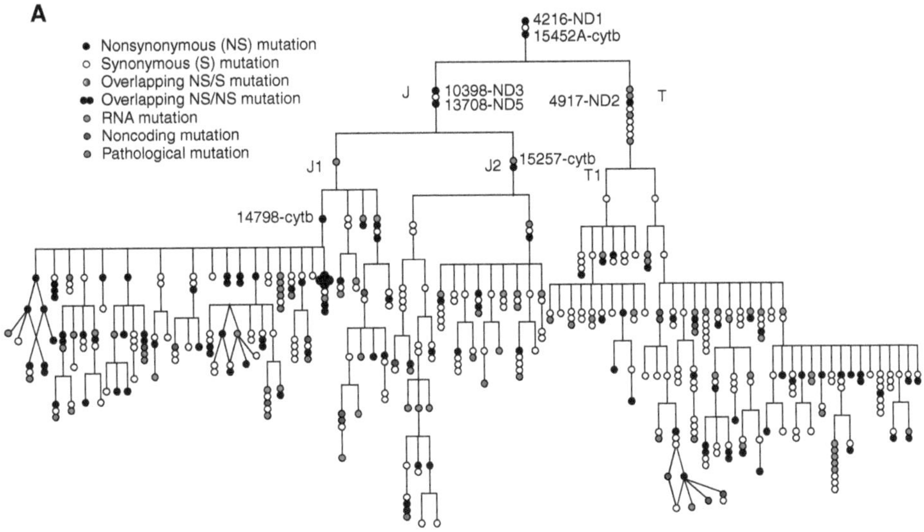

B

Hplgr	Gene	npΔ	CI%	Function
T	ND2	4917A	90	?
J1	Cytb	14798C	79	Qi
J2	Cytb	15257A	95	Qo

Figure 3. (*A*) Phylogeny of the haplogroups J and T demonstrating that each new branch of the mtDNA phylogeny is founded by a functionally significant polypeptide variant that is subsequently transmitted to all downstream descendants. Key internal replacement mutations are designated by the gene name and the nucleotide substitution. (*B*) The table provides function information and interspecific sequence conservation (conservation index = CI) for selected polymorphic amino acid sites. (From Ruiz-Pesini et al. 2004; reproduced, with permission, from the author and the American Association for the Advancement of Science © 2004.)

4216T>C (Y304H) and cytb nt 15452C>A (L236I). The lineages then subdivided into the T and J haplogroups, the T haplogroup being founded by an ND2 amino acid substitution at nt 4917A>G (N150D) and haplogroup J founded by the reversion of the out-of-Africa ND3 10398A>G (T114A) variant and the acquisition of a new ND5 variant at nt 13708G>A (A458T). Haplogroup J then subdivided into two lineages: J1 founded by a 16S rRNA 3010G>A variant followed by cytb variant at nt 14798T>C (F18 L) and J2 founded by a cytb variant at nt 15257G>A (D171N). Each of the founding polypeptide substitutions changes an evolutionarily highly conserved amino acid. The haplogroup T nt 4917 variant changes an amino acid that has been conserved from the first metazoans to man, having an interspecific conservation index (CI) = 90%. The subhaplogroup J1 cytb 14798 variant changes an amino acid conserved to *Caenorhabditis elegans* (CI = 79%), and the J2 cytb 15257 variant alters an amino acid that is conserved down to bacteria (CI = 95%). Thus, mtDNA polymorphic variants have accumulated within our species that change amino acids highly conserved throughout evolution. This phenomenon, in which conserved amino acids among species are polymorphic within a species, is inconsistent with classical neo-Darwinian theory. However, it can be explained through mitochondrial physiology and the high mtDNA sequence evolution rate. Once a species arises and begins to expand its range, it encounters environmental variation that favors bioenergetic alterations in central OXPHOS functions. These demands are met by the high mutation rate of the energetically important genes of the mtDNA. However, when a new species arises, it likely may require a more efficient mitochondrial bioenergetic system. The high mtDNA mutation rate accommodates this by reverting the regional variants back to the more efficient and universal energy production system (Wallace 2013a).

That mtDNA variants alter mitochondrial function has been confirmed by OXPHOS analysis of European haplogroups H and Uk (Rollins et al. 2009; Gomez-Duran et al. 2010) and Asian haplogroups B and F (Ji et al. 2012).

The strict maternal inheritance of the mtDNA means that mtDNAs can never mix and thus do not recombine. Hence, mtDNA single-nucleotide variants that have accumulated throughout human history remain in total linkage disequilibrium. The significance of an mtDNA variant is strongly influenced by the preexisting mtDNA variants on which it arose. This is particularly clearly demonstrated for the mtDNA variant in *ND1* at nt 3394C, which causes amino acid substitution Y30H (Ji et al. 2012). When this variant arises on macrohaplogroup N mtDNAs, it reduces mitochondrial complex I activity by 15%–28% and markedly increases the penetrance of the milder LHON mutations (Brown et al. 1995; Liang et al. 2009). However, if the mutation arises on a macrohaplogroup M mtDNA and this mtDNA resides in high altitude, then this same variant is associated with maximum complex I activity and adaptation to high altitude (Ji et al. 2012).

Consistent with their functional importance, mtDNA haplogroups have been correlated with predisposition to a wide range of metabolic and degenerative diseases, various cancers, and longevity (Khusnutdinova et al. 2008; Wallace 2008; Gomez-Duran et al. 2010; Wallace et al. 2013). For example, Asian haplogroup F, which has been correlated with predilection to diabetes and obesity (Fuku et al. 2007), is associated with a 30% lower complex I activity relative to other macrohaplogroup N mtDNAs (Ji et al. 2012). In addition to metabolic and degenerative diseases, mtDNA haplogroups have been associated with the severity of sepsis (Baudouin et al. 2005) and the outcome of ischemia, strokes (Chinnery et al. 2010), and trauma (Gomez et al. 2009; Zhang et al. 2010a; Krysko et al. 2011).

Somatic mtDNA Mutations. Finally, additional clinically relevant mtDNA mutations accumulate over time in tissues. These somatic mtDNA mutations arise in tissues as well as in stem cell lineages with age and progressively erode mitochondrial function, generating the aging clock (Wallace 2005). De novo mtDNA mutations can accumulate at anytime throughout life from the oocyte to the cells of the elderly. The earlier in development the mutation occurs, the more broadly it will be distributed. Hence, mtDNA mutations that arise in the embryonic period can be dispersed throughout the body, while those that arise in an adult organ will be tissue specific (Holt et al. 1988; Coskun et al. 2010).

nDNA

Mitochondrial diseases can also result from mutations in any one of the hundreds of nDNA-coded mitochondrial genes. Most of the >200 pathogenic nDNA mitochondrial mutations that have been reported to date are highly deleterious and result in severe childhood disease (Koopman et al. 2012; Wallace et al. 2013).

nDNA–mtDNA Interaction

Mild nDNA mitochondrial gene variants can also become clinically relevant when combined with an incompatible mtDNA. Severe encephalomyopathy associated with a complete complex I deficiency was observed in the boys of one family. Genetic analysis revealed that their disease was the result of inheriting from their mother an X-linked *NDUFA1* gene mutation (G32R) that caused a 30% reduction in complex I activity. This occurred in the context of inheriting their mother's mtDNA, which harbored two additional complex I gene mutations, *ND1* (M21T) and *ND5* (M88T), and which also caused a 25% deficiency in complex I activity. In the mother, the NDUFA1 G32R mutation was masked by her normal X-chromosome gene. However, in her son, the hemizygous NDUFA1 G32R was unmasked and interacted directly with the mother's mtDNA *ND1* (M21T) and *ND5* (M88T) mtDNA variants to cause severe complex I deficiency and disease (Potluri et al. 2009). Similarly, the severity of the cardiomyopathy of members of a 13-generation Mennonite pedigree whose members are homozygous for a frameshift mutation in the heart muscle ANT1 isoform gene was found to be determined by their mtDNA haplogroup inherited from their mothers. Homozygous *ANT1* mutant individuals that harbored haplogroup H mtDNAs had mild cardiomyopathy, while those who harbored haplogroup U mtDNAs presented with severe cardiomyopathy leading to heart failure (Strauss et al. 2013).

Mitochondrial Pathophysiology of Complex Diseases

The complexity of the genetics and pathophysiology of common diseases can now be reinterpreted in the context of mitochondrial bioenergetic and genetic principles (Fig. 4) (Wallace 2011, 2013b). Assuming common diseases result from partial mitochondrial deficiencies, this could then perturb an array of physiological processes including energy production, REDOX state, Ca^{2+} homeostasis, ROS production, anabolic and catabolic metabolic pathways, etc. (Wallace et al. 2010). Alteration in mitochondrial bioenergetics would increase mtDNA damage and mutation rate, perturb mtDNA replication and mitophagy, and lead to the accumulation of somatic mtDNA mutations. This would result in the progressive decline in mitochondrial function with age. Cells that are sufficiently energetically impaired would malfunction and ultimately undergo apoptosis and necrosis. Thus, the accumulation of somatic mtDNA mutations becomes the aging clock in individuals born with a normal mitochondrial function. For individuals born with partial mitochondrial dysfunction, the accumulation of mtDNA mutations and mitochondrial damage could account for the delayed onset and progressive course of their diseases. The stochastic nature of this process could also explain variable expressivity and/or penetrance of disease.

Perturbation of mitochondrial bioenergetics can predispose to a wide range of "complex" diseases. Bioenergetic perturbations can result from genetic, epigenetic, and environmental factors. Alterations in nDNA-coded mitochondrial genes could impair energy metabolism by inactivating an OXPHOS polypeptide, perturbing antioxidant defenses, altering mtDNA replication and repair, or affecting mitochondrial quality control through alterations in mitochondrial fission and fusion or in mitophagy (Chen et al. 2010; Youle and van der Bliek 2012; Jokinen and Battersby 2013). Mitochondrial OXPHOS could also be perturbed by modulation of the expression of the nDNA-coded mitochondrial genes through variation in the epigenome (Wallace and Fan 2010).

Mitochondrial function could also be perturbed by mtDNA variation, either by recent deleterious mtDNA mutations or ancient adaptive mtDNA polymorphisms. Finally, mitochondrial energy production could be perturbed by the nature and availability of calories; the demands made on cellular energy for growth, maintenance, and reproduction; and the acute sensitivity of mitochondrial OXPHOS to a broad range of environmental toxins. As mitochondrial energetics declines, the organs with the highest energy requirements would be the first to show functional alterations. Even very subtle bioenergetic defects can adversely affect the central nervous system with its high mitochondrial energetic demand. Other sensitive organs include the heart, muscle, and kidney.

Alterations in the nature of available calories (carbohydrates, fats, proteins) would be differentially metabolized by individuals with different mtDNA haplogroups. Hence, mitochondrial alterations that perturb the flux of calories through the system could result in metabolic diseases such as diabetes, obesity, hypertension, and cardiovascular disease.

The mitochondria are also the most common bacterium in the human body, our bodies harboring on the order of 10^{17} mitochondria. Hence, damage to cells can release into the extracellular space and bloodstream mitochondrial *N*-formylmethionine-bearing polypeptides, mtDNA fragments, cardiolipin, and various other mitochondrial breakdown products, known as damage-associated molecular patterns (DAMPs), which can elicit an inflammatory re-

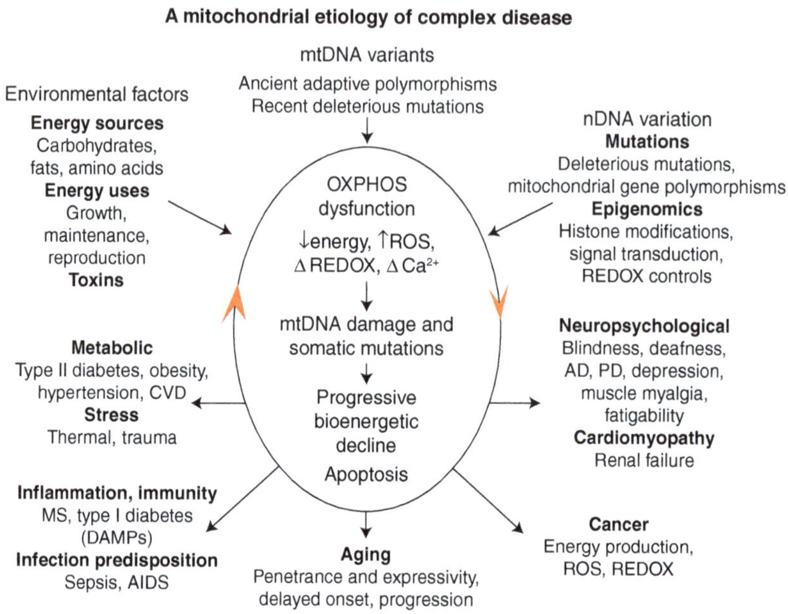

Figure 4. Integrated mitochondrial paradigm to explain the genetic and phenotypic complexities of metabolic and degenerative disease, aging, and cancer. The top three arrows show the three types of variation that impact on individual mitochondrial OXPHOS robustness and hence risk for developing disease symptoms. These include nuclear DNA (nDNA) variation encompassing DNA sequence changes and epigenomic modification of gene regulation and signal transduction pathways, mitochondrial DNA (mtDNA) variation including recent deleterious mutations and ancient adaptive polymorphisms, and environmental influences encompassing the availability and demand for calories and inhibition of mitochondrial function by environmental insults. The *central* oval encompasses the pathophysiological basis of disease processes and the basis of disease progression. The primary defect is reduction in the energy-transformation capacity of OXPHOS. This can result in reduced energy output, increased reactive oxygen species (ROS) production, altered REDOX status, and altered calcium homeostasis. The decline in OXPHOS efficiency can, in turn, perturb mitochondrial biogenesis, increase ROS production, impair mitophagy, etc., resulting in progressive increase in mtDNA damage and somatic mutations and further decline in mitochondrial function. Once mitochondrial function falls below the bioenergetic threshold of a tissue, symptoms ensue. Continued energetic failure can initiate cell destruction by apoptosis or necrosis. The *lower* five arrows summarize the disease categories and the phenotypic outcomes of perturbed mitochondrial energy transformation. The *bottom* arrow shows the effect of the stochastic accumulation of somatic mtDNA mutations resulting in delayed onset and a progressive course of diseases and aging. The *right* arrow indicates clinical problems that can result from reduced energy production in the most energetic tissues: the brain, heart, muscle, and kidney. The number and severity of symptoms in these organs reflect the degree and specific nature of the mitochondrial defect. The *left* arrow indicates the metabolic effects of mitochondrial dysfunction, which result in the perturbation of the body's energy balance. This results in the symptoms of the metabolic syndrome. The *lower right* arrow indicates that mitochondrial alterations are critical for cancer initiation, promotion, and metastasis. The *lower left* arrow outlines the hypothesized inflammatory and autoimmune responses that may result from the chronic introduction of the mitochondria's bacteria-like DNA and *N*-formylmethionine proteins into the bloodstream (Wallace 2011).

sponse (Zhang et al. 2010a,b; Krysko et al. 2011; Oka et al. 2012). Because apoptosis, which destroys the mitochondria before they are released into the bloodstream, is an energy-dependent process, the chronic energy deficiency of mitochondrial disease would foster the release of mitochondrial antigens. This, in turn, would result in local inflammation in degenerative diseases and systemic inflammation in autoimmune diseases.

 Cite this article as *Cold Spring Harb Perspect Biol* doi: 10.1101/cshperspect.a021220

Finally, the growth of cancer cells is directly limited by energetics. Hence, cancer cells must acquire modifications in their mitochondrial physiology to optimize energy production to their changing environments (Wallace 2012).

SEGREGATION OF HETEROPLASMIC GERMLINE mtDNA MUTATIONS

While the causes of mitochondrial dysfunction can be genetically complex because of the interaction of the large number of nDNA and mtDNA genes and variants, the most unpredictable aspect of mitochondrial genetics is the segregation of mtDNA heteroplasmy. Heteroplasmic mutants segregate along both the female germline and in somatic tissues. Therefore, understanding the quantitative genetics of mtDNA segregation is essential if we are to understand the mitochondrial etiology of complex diseases.

Germline Segregation of mtDNA Heteroplasmy

Human mtDNA Disease Segregation

Familial Transmission of Heteroplasmic mtDNA Mutations. Maternal transmission of heteroplasmic mtDNA mutations is now well documented for both tRNA and polypeptide mutations (Wallace et al. 2013). The first example of the interaction between mtDNA heteroplasmy variation and phenotype was the report that the $tRNA^{Lys}$ nt 8344A>G (Fig. 5A) mutation that causes myoclonic epilepsy and ragged red fiber (MERRF) disease was heteroplasmic (Wallace et al. 1988b; Shoffner et al. 1990). In this initial family, the phenotypic variability ranged from severe MERRF (III-1) to mild mitochondrial myopathy and electrophysiological aberrations (II-4, III-2, 3, and 4) (Fig. 5B). The severity of the clinical phenotypes varied with the percentage of mutant mtDNAs after the individuals were stratified by age (Fig. 5C).

The effect of mtDNA mutant heteroplasmy on phenotype is even more striking with the $tRNA^{Leu(UUR)}$ nt 3243A>G mutation associated with mitochondrial encephalomyopathy, lactic acidosis, and stroke-like episodes (MELAS) (Goto et al. 1990). When the heteroplasmy of this mutation is high, it can present as lethal childhood Leigh syndrome, MELAS, chronic progressive external ophthalmoplegia (CPEO), cardiomyopathy, migraines, diabetes mellitus, and deafness. In pedigrees with high heteroplasmy members, meiotic segregation of the mutant mtDNAs can result in the full range of phenotypes from asymptomatic to lethal disease (see example in Fig. 1 in Brown et al. 2001). Yet in other pedigrees, when the 3243A >G mutation is present in 10%–30% heteroplasmy, this same mutation results in only maternally inherited diabetes and deafness (see example in Fig. 1 in van den Ouweland et al. 1992).

Variability in clinical presentation as a result of heteroplasmic variation can also be observed in mtDNA polypeptide missense mutation pedigrees. The mtDNA ATP6 nt 8993T>G mutation, which causes the amino acid substitution L156R, is the most common example (Holt et al. 1990). This mutation was originally associated with the clinical designation of neurogenic muscle weakness, ataxia, and petinitus pigmentosum (NARP), but also can cause lethal childhood Leigh syndrome, olivopontocerebellar atrophy, cerebellar ataxia, and/or retinitis pigmentosa when present in ~70% to 100% of the mtDNAs (Tatuch et al. 1992; Ortiz et al. 1993).

However, heteroplasmic variation is not the only cause of clinical variation in mtDNA disease. In Leber hereditary optic neuropathy, pedigrees that are essentially homoplasmic for one of the common causal LHON mtDNA mutations, males are three to four times more likely to manifest mid-life subacute blindness than females. Other factors also affect the onset of blindness including mtDNA haplogroup background and environmental stressors such as smoking and alcohol abuse (Brown et al. 1997, 2002; Torroni et al. 1997; Sadun et al. 2003, 2011).

Variability of mtDNA Heteroplasmy in Maternal Oocytes. One reason for the high variability observed in the mtDNA heteroplasmy levels of maternal relatives is variation in the percentage of mutant mtDNAs in the oocytes of heteroplasmic women. Variability in oocyte heteroplasmy has been most intensively investigat-

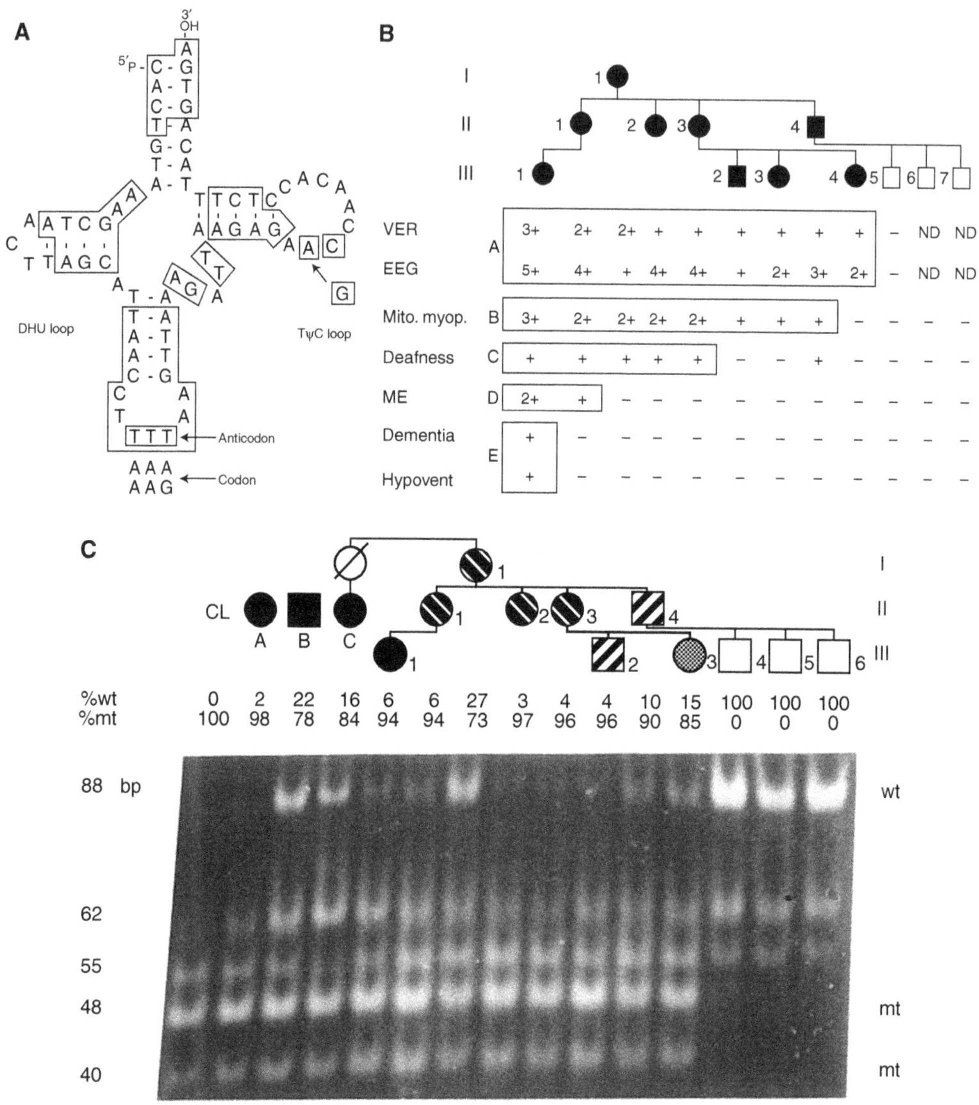

Figure 5. MERRF tRNA^{Lys} A8344G pedigree showing variable clinical expression in association with variable mtDNA mutant heteroplasmy modified by age. (*A*) Structure of tRNA^{Lys} showing position of A8344G transition in TΨC loop. (*B*) Pedigree of proband (III-1) showing that all maternal relatives had some clinical manifestations (filled symbols), though highly variable, while the three paternal sons were symptom free. VER, visual evoked response; EEG, electroencephalograph; Mito. myop., mitochondrial myopathy with ragged red fibers and abnormal mitochondria; deafness, sensory neural hearing loss; ME, myoclonic epilepsy; dementia, progressive cognitive decline; hypovent, hypoventilation. (*C*) Pedigree showing variable proportions of mutant-type (mt) and wild-type (wt) mtDNA along the maternal pedigree. A 183-nt PCR fragment was digested with CviJI. The wild-type (8344A) gave an 88-nt uncut fragment, whereas the mutant (8344G) created a new site cutting the 88-nt fragment into 48 and 40 nt fragments. "CL" is a cloned mutant fragment. Cases (*A*), (*B*), and (*C*) are independent pedigrees. Individual (*C*) is the maternal aunt of proband III-1 in (*A*), which manifested MERRF. All of the maternal relatives of the pedigree are heteroplasmic for the mutant mtDNA and the severity of the phenotype correlated with the percentage of heteroplasmy when corrected for age. (From Wallace et al. 1988b; reproduced, with permission, from the author and from Shoffner et al. 1990; reproduced, with permission, from the author and the National Academy of Sciences © 1990.)

 Cite this article as *Cold Spring Harb Perspect Biol* doi: 10.1101/cshperspect.a021220

ed in women harboring the $tRNA^{Leu(UUR)}$ 3243A>G and ATP6 8993T>G mutations. A 41-year-old mother who harbored the 3243A>G mutation in 18.1% of her quadriceps muscle and 7.2% of her leukocyte mtDNAs had two sons, one of whom was found at 15 years of age to harbor 11.7% 3243A>G mutant in his blood cells. The woman underwent a hysterectomy for endometriosis and 82 oocytes were recovered and analyzed for their 3243A>G mutation levels, which ranged from 0% to 45% with a mean heteroplasmy of 12.6% and a median of 8.2%. Eight of the oocytes (9.8%) lacked detectable mutant mtDNA and ~35 of the oocytes had heteroplasmy levels ranging from 1% to 45%. To model the distribution of oocyte genotypes, the researchers assumed that the distribution of oocyte heteroplasmy levels would approximate a binomial distribution, with the initial maternal mtDNA mutant allele frequency (p_0) being represented by the mean of the allele frequencies of the oocytes. Furthermore, it was hypothesized that the extent of the variation in mtDNA heteroplasmy levels was determined by a "bottleneck." This bottleneck was envisioned to reduce the number of mtDNA segregating units "N" within the maternal germline over "g" germline cell divisions, to be followed by expansion of the mtDNA copy number back to an infinite size. The variance (V) was calculated by the formula

$$V = p_0(1 - p_0)\{1 - [1 - (1/N)]^g\}.$$

From the mean oocyte heteroplasmy level of 12.6%, $p_0 = 0.126$, and with a variance of $V = 0.0143$, the authors estimated that the number of segregating units of the bottleneck (N) would be 173 if maintained over 24 germline cell divisions or eight if the bottleneck lasted for one cell generation (Brown et al. 2001).

Studies on the germline segregation of heteroplasmic 3243A>G mutant mtDNAs was extended to 38 preimplantation embryos from two women. The oral mucosa heteroplasmy of one woman was 30% and the mean heteroplasmy level of her embryos was 30% ± 15%. The mucosal heteroplasmy of the second woman was 27% and the mean of her embryos was 32% ± 23%. Six of the embryos had no detectable mutant <2%. Among the 35 embryos from the 27% heteroplasmy woman, 83% of the embryos harbored the 3243A>G mtDNA, with a heteroplasmy range of 5% to 77%, but none of her embryos were pure mutant.

Germline transmission of the heteroplasmy mtDNA ATP6 8993T>G gene missense mutation has also been found to result in oocytes with widely different hetroplasmy levels. In one case, an asymptomatic mother with 50% blood mutant mtDNAs had three boys; one died of Leigh syndrome with 98% mutant mtDNAs in muscle and fibroblasts, one died of sudden infant death syndrome (SIDS) with 92% mutant in blood, and one was affected with Leigh syndrome and harbored 87% mutant in blood. The mother was superovulated and seven oocytes could be recovered and genotyped. One of the oocytes had no detectable mutant mtDNA, whereas the remaining six oocytes had >95% mutant (Blok et al. 1997).

Low-Level Maternal Germline Heteroplasmy. While the maternal transmission of biallelic heteroplasmy of mtDNA disease has been extensively studied, the advent of next-generation sequencing (NGS) is now providing the capacity to determine whether the maternal germline might also harbor a spectrum of mtDNA variants each at a very low percentage of heteroplasmy. This is possible because NGS permits sequencing a region of the human mtDNA from a human sample more than a thousand times revealing rare variants. When the mtDNA sequence of the control region hypervariable region 2 (MT-HV2 from nt 162–455) and the COIII region (MT-CO3 from nt 9307–9591) were analyzed using the Roche 454 sequencing platform from blood and skeletal muscle samples of seven subjects, every subject was reported to harbor heteroplasmic variants in one or more bases at >0.2% heteroplasmy in both tissues. Overall, the number of variants per base position was greater in skeletal muscle than in blood and also was greater in MT-HV2 than in MT-CO3. Heteroplasmy levels >2% were also observed, but only in the muscle of three subjects within the MT-HV2; all other heteroplasmy levels were low (Payne et al. 2013).

Patients with mtDNA polymerase γ mutations, which decreased the fidelity of mtDNA replication, had elevated MT-HV2 mutations and these increased in frequency with age. When first-degree relatives of patients with mutations in the nDNA genes required for mtDNA maintenance were analyzed, 40% of the variants in a given individual were shared with their maternal relatives, as compared to only 12% of the variants shared with unrelated individuals. Seventy-one percent of the shared variants among relatives were primarily found in skeletal muscle as opposed to 13% shared muscle variants in unrelated individuals. Therefore, according to this study, low-level heteroplasmies (0.2%–2%) are present in all individuals and a significant proportion of these can be transmitted through the maternal lineage (Payne et al. 2013).

Using the Illumina platform, quality control criteria for mtDNA sequence validation and a heteroplasmy cutoff of 2%, a much lower frequency of heteroplasmic mutations was reported for blood and mucosal mtDNAs of nine individuals from three families. Four heteroplasmic mtDNA variants were reported, with one apparent germline mutation. Of the remaining variants one was prominent and two were low-heteroplasmy variants (Goto et al. 2011).

While these studies suggest that low-heteroplasmy variants are ubiquitous and can be maternally inherited, the rise of NGS to detect very low-level heteroplasmic mutant mtDNAs may be subject to artifacts. For example, most current protocols for making NGS libraries use PCR and PCR polymerases are error prone and thus could generate spurious mutations. One effort to overcome the potential of PCR artifacts is "duplex sequencing." This method requires that all sequence variants be confirmed by identification of the complementary nucleotide change on both DNA strands. The estimated error rate of the duplex sequencing method was estimated to be $1/10^9$. When this approach was applied to a human brain sample, the mtDNA mutation rate was found to be 3.5×10^{-5}, lower than that expected for the above studies. Still this is much higher than reported nDNA mutation rates (Schmitt et al. 2012). Given a mutant density of 3.5×10^{-5}, this tissue had about one mtDNA mutation per every two mtDNA molecules. Therefore, there is considerable genetic heterogeneity within the thousands of mtDNAs within a somatic cell. The frequency and heteroplasmy levels of potentially maternally transmitted low-heteroplasmy mutations merit further examination.

Bovine. Proof that low-heteroplasmy variants can be transmitted through the maternal germline would be if mtDNAs harboring one of two variant alleles were to alternately appear in successive generations. Such mtDNA allelic switching across maternal generations has been reported for bovine lineages. A synonymous sequence variant in the URF-5 (now *ND5*) gene at nt 12792C>T, detected as an HaeIII restriction fragment length polymorphism, was observed to switch from one allele to the other within two maternal generations in a 1982 study (Hauswirth and Laipis 1982). This variant was then linked to four mtDNA control region variants at nts 16074, 16079, 16231, and 16250, generating four different haplotypes. These also switched among generations over an eight-generation bovine pedigree (Olivo et al. 1983). While the observed haplotypes appeared to be homoplasmic in the animals studied, three offspring from one cow were found to be heteroplasmic, suggesting that genotypic switching was occurring by the germline transmission of low-heteroplasmic genotypes (Ashley et al. 1989).

A more extensive bovine survey of mtDNA variation revealed that a control region variant that changed a G to a C at the end of a homopolymeric run of Gs switched between the G and the C allele in 13 different mother–daughter pairs (Koehler et al. 1991). These early bovine observations were the first to lead to the hypothesis that there was an mtDNA copy-number bottleneck in the mammalian female germline.

Mouse. The Shoubridge laboratory has studied maternal germline segregation of mtDNAs in heteroplasmic mice. In their system, the mtDNAs of two different mouse lineages, NZB/BinJ (NZB) and BALB/cByJ (BALB), were combined by removing a bleb of cytoplasm from a one-cell embryo and fusing it to the one-cell embryo of the other mtDNA strain. The heteroplasmic embryos were then implanted into foster mothers at the two-cell stage. In the initial study, five founder females carrying 3.1% to 7.1% of the donor mtDNA were studied. These were crossed with BALB males and the progeny analyzed. The mean mtDNA heteroplasmy levels of the offspring were found to be similar to that of the founder mother, but the heteroplasmy levels of the individual offspring varied with the highest heteroplasmy levels being 29.6% in one pup (Jenuth et al. 1996).

In the mouse, the primary oocytes are thought to be derived from 50 primordial germ cells (PGCs) located at the base of the allantois of a 7.5-d postcoitum (dpc) mouse embryo. These cells are alkaline phosphatase (ALP) positive, permitting them to be identified and isolated. The PGCs migrate to the germinal ridge where they grow and differentiate into oogonia. The oogonia then proliferate by mitosis during embryonic development. In the mouse, the oogonia undergo 15 divisions to generate ~25,000 primary oocytes. In humans, about six to seven million primary oocytes are produced through roughly 24 cell divisions. The oogonia will either degenerate or differentiate into primary oocytes through asymmetric division, generating a primary oocyte and a daughter oogonium. By birth, most oogonia have either differentiated into primary oocytes or degenerated. The primary oocytes undergo oogenesis in which they enter meiosis and become arrested in prophase I where they remain until puberty begins and individual proto-oocytes complete differentiation, form follicles, and can be ovulated. Within the follicle, the oocyte completes the first meiotic division generating the first polar body and enters the second meiotic division where it becomes arrested at metaphase II. At fertilization, meiosis II is completed, the second polar body is extruded and the female and male pronuclei approach each other and fuse.

Based on ultrastructural analysis of mouse oogonia, the Shoubridge laboratory estimated that there were ~40 mitochondria per oogonium and assuming five mtDNAs per mitochondrion they concluded that an oogonium contained ~200 mtDNAs. Based on the assumption that the average number of mtDNAs per PGC and oogonium remained relatively constant throughout the PGC replication phase and on the observed distribution of oocyte and offspring NZB/BALB mtDNA heteroplasmy levels, the Shoubridge laboratory concluded that there must be ~185 (range 76–867) mtDNAs in an oogonium. The Shoubridge laboratory concluded that the rapid segregation of the mtDNA heteroplasmy in mammals could be explained by a drastic reduction in the number of mtDNA segregation units, a bottleneck, occurring in the PGCs of the female germline. Such a reduction in mtDNA segregating units would greatly increase the rate of genetic drift, leading to rapid segregation of different mtDNA types in different female germline cells. They estimated that the number of segregating mtDNA units in PGCs was ~200 and that the multiple cell divisions of the oogonia required to generate that large a number of primary oocytes were sufficient to account for the observed variance in heteroplasmy frequency of oocytes and offspring. They summarized: "Our study suggests that the probability of inheriting one of two mtDNA genotypes can be modeled as a binomial sampling process . . ." Thus, "It seems unlikely that strong positive or negative selection for pathogenic mtDNAs occurs in the oocyte in early embryogenesis . . ." (Jenuth et al. 1996). In short, germline heteroplasmy segregation is the product of the stochastic sampling process known as genetic drift.

Eleven years after the initial Shoubridge study reported an mtDNA PGC mtDNA bottleneck, Cao and associates published a paper reporting that the number of mtDNAs in PGCs was not as low as surmised by Shoubridge. Using ALP staining to identify PGCs in embryos between 7.5 and 13.5 dpc, Cao and associates determined by quantitative real-time polymerase chain reaction (PCR)(qRT-PCR) that the average mtDNA copy number was 1561 ± 161 (range 1350–1732), and that the minimum mtDNA copy number of the smallest PGCs was 953. They also estimated that there were ~100 mitochondria in a single PGC. Additional germ cell mtDNA copy-number estimates included PGCs at 13.5 dpc at 3.66×10^3, primary oocytes at 1.16×10^3, and mature oocytes at 1.57×10^5 mtDNAs per cell. By contrast, quantification of the mtDNAs in the somatic cells of 7.5-dpc embryos was reported as low, ranging from 57 to 3345 mtDNAs per cell. One complexity of these assessments discovered by this research group was that staining embryo cells with ALP partially inhibited mtDNA quantification. As an alternative approach to identifying germline cells, the authors used mice that expressed the enhanced green fluorescent protein (EGFP) driven by the 18-kb Oct-4 promoter (GOF-18/GFP). Oct-4 is transcribed in germ cells at d 9.5 to 13.5 dpc. Cao and colleagues then reported that the average mtDNA copy number for PCGs isolated using GOT-18/GFF was 1408 and 1294 in two experiments versus 673 and 736 for ALP-stained cells. Hence, they corrected their ALP-stained PGC mtDNA estimates by multiplying by 1.92. Based on these results, this group concluded that there was no constriction of mtDNA content in PGCs, and thus that the rapid mtDNA germline heteroplasmy segregation was not the product of a physical bottleneck in the number of mtDNA within the PGCs (Cao et al. 2007).

This report was followed a year later by a report from the Chinnery laboratory reaffirming that the mtDNA copy number in PGCs was on the order of 203 at 7.5 dpc, but that the mtDNA copy number in older PGCs increased to 1529 mtDNAs/cell by 14.5 dpc (Cree et al. 2008). To isolate PCGs without ALP staining, this group identified PGCs by the fluorescence of GFP transcribed from the *Stella (Dppa3)* promoter, which is specific for PGCs. Their studies revealed that the mouse oocyte contains 2.28×10^5 mtDNAs, that the mtDNAs do not replicate until the PGCs are formed, that the median mtDNA copy number in 5.5-dpc PGCs is 203, and the mean is 451. However, by 13.5 dpc when the number of primary oocytes is ~25,791, the median mtDNA copy number is 1529. Hence, the mtDNA copy number per cell increases from d 5.5 to 13.5. They also observed that by 14.5 dpc, female germline cells had a lower mtDNA content then male germline cells: 1376 ± 601 versus 2152 ± 951. From these observations, these authors built a mathematical model that encompassed both a severe mtDNA copy-number bottleneck in early mammalian PGCs as well as a subsequent mtDNA amplification phase, the combination of the two being able to account for the rapid germline segregation of mtDNA heteroplasmy (Cree et al. 2008).

At the end of 2008, the Shoubridge laboratory published another paper in which they quantified the mtDNA copy number in the germ cells of NZB/BALB heteroplasmic mice. In addition, they determined the proportion of the NZB and BALB mtDNAs in the germ cells at different stages. Based on the concern that ALP staining might result in spuriously low mtDNA copy numbers, this team used EGFP transcribed from the Oct4 (*Pou5fl*) promoter and then manually isolated the germ cells. They then quantified the mtDNA copy number in 8.5-dpc cells, observing a mean mtDNA copy number of ~280 with a median of 145. By 10.5 dpc, they found that the mtDNA copy number had increased to a mean of ~2800 and median of 2200. At 14.5 dpc when the PGCs have colonized the gonad, the mtDNA copy number had risen to ~6000 per cell. Thus, these data indicate that the mtDNA copy number decreases 700-fold from oocyte to PGC, but then increases 10- to 20-fold during expansion of the PCG population and the colonization of the gonad (Wai et al. 2008). While this was consistent with the observations of Cree et al. (2008), the Shoubridge laboratory then analyzed the proportion of NZB and BALB mtDNAs during embryonic development. This led to the surprising conclusion that the proportion of NZB and BALB mtDNAs did not change markedly during the mtDNA constriction and proliferation cycle of the PGCs and the oogonia. Hence, the Shoubridge laboratory concluded that "despite the severe reduction in mtDNA copy number, the mitochondrial genetic bottleneck does not occur during embryonic oogenesis." However, when they examined the mtDNA heteroplasmy variance in PGCs, oogonia, and primary oocytes in primordial follicles with <10,000 mtDNA per cell to that of postnatal mature ovulated oocytes and primary oocytes in secondary

 Cite this article as *Cold Spring Harb Perspect Biol* doi: 10.1101/cshperspect.a021220

follicles that harbor >10,000 mtDNAs per cell, they found that the mtDNA heteroplasmy variance had increased significantly. They then hypothesized that the "... genetic bottleneck must be the result of the selective replication of a random subset of mtDNA templates during the growth and maturation of the ovarian follicles," which must start in the primordial follicles. To identify this differential replication phase, the researchers pulse-labeled female pups with bromo-deoxyuridine (BrdU) injected at d 1 (P1) and (P4) postpartum. They then analyzed the mtDNA labeling of oocyte mtDNAs within mtDNA nucleoids, the nucleoids identified by their association with mtDNA-binding protein (Tfam), polymerase γ (PLOG), and single-strand binding protein (mt-SSB). This revealed that only a limited number of mtDNAs were replicating, replicating mtDNAs did not always correlate with Tfam reactivity, and the replicating mtDNAs were not consistently associated with the Balbiani body. The Balbiani body is an aggregate of Golgi elements surrounded by mitochondrial and endoplasmic reticulum. The Shoubridge group then concluded that the purpose of the reduced mtDNA copy number in early PGCs to ~200 mtDNAs is to permit selection against cells with high percentages of deleterious mtDNA mutations. However, "The genetic bottleneck for neutral (and less deleterious mtDNA sequence variants that are associated with most human disease) occurs during folliculogensis in early postnatal life ..." (Wai et al. 2008).

Although this Wai-Shoubridge study did not correlate the proposed differential mtDNA replication bottleneck with the Balbiani-body-associated mitochondrial cloud, comparative interspecific studies have been used to argue that the Balbiani body is important for regional mtDNA proliferation (Zhou et al. 2010).

The conclusion of the Wai-Shoubridge study that segregation did not occur during embryonic oogenesis contradicted their previous genetic-based conclusions (Jenuth et al. 1996) as well as the conclusions of the Chinnery group that a low mtDNA copy number in the early PGCs was one of two major factors in the rapid segregation of the mtDNA heteroplasmy along the female germline. In a follow-up report, the Chinnery laboratory argued that the discrepancy in the conclusions of the two Shoubridge laboratory studies was the result of the frequently extreme bias in the relative levels of the NZB versus the BALB mtDNAs, one mtDNA type always being present at a low percentage heteroplasmy. This is the product of the heteroplasmy being derived from a small bleb of cytoplasm from one oocyte fused to a much larger cytoplasm of the recipient oocyte. Because a greater variance and thus range of heteroplasmy levels can be generated from a starting mtDNA heteroplasmic ratio of 50:50 than can be generated from a starting ratio of 5:95, Chinnery argued that the variance levels observed by Shoubridge did not reflect the true extent of the heteroplasmy segregation (Samuels et al. 2010).

The question of whether or not there was a severe reduction in mtDNA copy number in PGCs was again raised by Cao and collaborators. They questioned the effectiveness of previous studies to identify true PGCs at 7.4 dpc. To increase the reliability of studying only PGCs, Cao and associates identified the PGCs using three different protein markers and also distinguished among PGCs isolated from early-bud (EB) and late-bud (LB) 7.5-d embryos. The first marker used was *Blimp1*, which is expressed in nascent PGCs and is PGCs specific. By introducing into mice a bacterial artificial chromosome harboring a monomeric red fluorescent protein gene (mRFP) expressed from the *Blimp1* promoter (*Blimp1*-mRFP), they were able to isolate positive cells manually. The appropriate cells were further validated by immunohistochenical staining for Stella (*PGC7*) and by ALP staining. Quantification of the mtDNA copy number of the *Blimp1*-mRFP-positive EB cells at 7.5 dpc gave a mean value of 1396 mtDNAs/cell and LB cells at 7.5 dpc of 1479. At 13.5 dpc, female germ cells gave a mean copy number of 1747 and male germ cells of 2039. Thus, Cao again concluded that mtDNA segregation does not occur because of a low mtDNA copy number in the early PGCs. Rather, segregation must occur later in female germ cell development without reduction in mtDNA copy number, presumably because of differential replication of mtDNAs during oocyte maturation. Consistent with this conclusion, they point out that mouse models heteroplasmic for highly deleterious mtDNA mutations including an mtDNA 4696-nt deletion (Sato et al. 2007) and an ND6 frameshift mutation (Fan et al. 2008) produced pups with lower percentages of mutant mtDNAs in successive litters. If the mtDNA variance were entirely determined by segregation and/or selection early in female germline development, then all litters should be generated from oocytes with the same distribution of heteroplasmy levels. Thus, the distribution of mtDNA heteroplasmy across female litters with advancing maternal age should be the same. If the mutant mtDNAs were segregating later in oocyte maturation, then the sicker oocytes would be preferentially lost from the ovary during the female reproductive lifespan, resulting in the successive decline in mutant mtDNAs over sequential litters. Why then might the PGCs retain high mtDNA copy numbers while adjacent somatic cells have lower mtDNA levels? These authors speculate that the high PGC mtDNA copy number permits the retention of heteroplasmic mutations that can be transmitted through the female germline and permit subsequent adaptation to changing environments (Cao et al. 2009).

Clearly, there is no consensus as to the cellular and molecular mechanism of rapid germline mtDNA heteroplasmic segregation. It could result from the rapid segregation of heteroplasmic mtDNAs because of genetic drift resulting from a physical bottleneck in the number of mtDNA segregating units in the PGCs (Jenuth et al. 1996; Cree et al. 2008), a combination of an initial severe reduction in the mtDNA population size followed by further segregation during subsequent replication (Cree et al. 2008; Khrapko 2008), an aggregation of a larger number of mtDNAs into homogeneous segregating units such as multiple mtDNA containing nucleoids (Carling et al. 2011), the replication of only a small proportion of the mtDNAs in the primordial follicle cells leading to biased transmission of a few mtDNAs (Wai et al. 2008; Carling et al. 2011; Jokinen and Battersby 2013), or to some as-yet unidentified factors.

Primate. Although the progeny of a heteroplasmic female mice can have an array of heteroplasmic ratios, the segregation rate does not seem as extreme as has been observed in human pedigrees harboring the tRNA$^{Leu(UUR)}$ nt 3243 A>G mutation or the ATP6 nt 8993T>G L126R mutations. To determine whether there are significant differences in mtDNA heteroplasmic segregation between rodents and primates, mtDNA segregation was studied in cytoplasmic mixing experiments of rhesus macaque oocytes.

The mtDNAs of macaque oocytes were mixed by karyoplast-cytoplast fusion. A karyoplast is a portion of the cytoplasm of a cell that contains the nucleus surrounded by the cellular plasma membrane. A cytoplast is a fragment of the cell that lacks the nucleus but contains most of the cytoplasm, mitochondria, and mtDNAs. In this macaque experiment, a micropipette was inserted under the zona pellucida of a metaphase stage-II embryo and 50% of the cytoplasm plus the nucleus was extracted. This karyoplast was then inserted under the zona pellucida of another oocyte from which the nucleus and half of the cytoplasm had been removed. The two cell fragments were then fused to generate a "reconstituted cell," which was fertilized by intracytoplasmic sperm injection (ICSI).

The karyoplast and cytoplast donors were derived from Indian and Chinese origin macaques, which differed in their mtDNA control region (D-loop) sequence, permitting the fate of the two mtDNA haplotypes to be monitored through development. The mean heteroplasmy in 15 analyzed reconstituted oocytes was 54.9% $\pm$ 10%. In two cell embryos, half of the embryos had ~50% of each haplotype in each blastomere, but the other half of the embryos exhibited significant differences among the blastomeres, the most divergent case being 36% and 70%. The divergence in percentage heteroplasmy increased in four- and eight-cell embryos, the coefficient of variance increasing from 17.7% to 25.0% to 30.9% and the range increasing from 13.3% to 25.3% to 43.2% in two-, four-, and eight-cell embryos, respectively. By the eight-cell stage, some embryo blastomeres differed by as much as 10% and 80% of the Indian and Chinese mtDNAs (Lee et al. 2012).

Three embryonic stem cell (ESC) lines were derived from the inner cell mass cells of heteroplasmic blastocytes and found to harbor heteroplasmy levels of the cytoplast mtDNA of 97.9%, 93%, and 5%. From nine clones of the 93% cell line, six were homoplasmic for the cytoplast mtDNA and three were heteroplasmic in the range of 90.7% to 92.9%, with the direction of segregation being independent of the nuclear origin (Lee et al. 2012).

Reconstituted oocyte-derived embryos were implanted in females and two fetuses were analyzed for their heteroplasmy levels. The male fetus harbored 26.3% of the cytoplast mtDNA, while the female fetus harbored 93.8% of the cytoplast mtDNA, with her tissue levels ranging from 91.1% in blood to 98.4% in kidneys. Recovery of the ovaries from the female fetus and analysis of 51 primordial oocytes revealed a continuous range of cytoplast donor mtDNAs from 3.7% to 99.2%. Hence, the somatic cell heteroplasmy level of the female fetus was essentially independent of the level of heteroplasmy in her germline cells (Lee et al. 2012).

A difference in heteroplasmy segregation in somatic versus germline cells was already apparent in the epiblast cell lineages, resulting in marked asymmetric segregation of the mtDNAs into the somatic tissues, even though the female germline remained capable of generating the full range of possible heteroplasmic levels. Hence, there appears to be two mtDNA segregation systems, one for somatic tissues that tends to move rapidly toward homoplasmy and is already functioning in the early embryo and the other that is acting in the female germline and generates a diverse array of oocyte heteroplasmy levels (Lee et al. 2012).

This difference between primate germline and somatic cell lineage heteroplasmy levels is reminiscent of the observation of Cao et al. (2007, 2009) who reported that the mtDNA copy number in the somatic cells of mouse embryos was markedly lower than that of the female germline cells. Such a somatic lineage mtDNA bottleneck would be conducive to the rapid segregation of mtDNA heteroplasmy in somatic cells.

Selection in mtDNA Germline Segregation

All of the above studies demonstrate that an intrinsic feature of heteroplasmic mtDNA mutations is their "imperfect transmission" (Chapman et al. 1982), which can be greatly accentuated along the maternal lineage by bottlenecks in the mtDNA copy number that foster intracellular heteroplasmic mtDNA genetic drift (Hauswirth and Laipis 1982; Charlesworth and Charlesworth 2010). Genetic drift is driven by fluctuations in gene pool size and, as the constriction of the gene pool size increases, the segregation rate of a heterogeneous pool of genetic variants toward homogeneity also increases (Takahata and Slatkin 1983). The rapid segregation of heteroplasmic mtDNA genotypes along the female germline by an extreme "bottleneck" counters the accumulation of large numbers of mtDNA mutations over many generations, which would progressively erode energetic function and ultimately compromise the viability of the species, a process known as "Muller's ratchet." In a sense, rapid mtDNA allelic drift acts like nuclear gene recombination in that it increases diversity while limiting the accumulation of deleterious mutations. By rapidly sorting out heteroplasmic alleles along the maternal lineage, mtDNA genetic drift permits the introduction of new mtDNA mutations into the population in the near homoplasmic state. These genotypes can then be acted on by selection with deleterious alleles being removed by purifying selection thus maintaining the integrity of the maternal germline. Nonpathogenic variants would accumulate along radiating maternal lineages at random, and rare advantageous alleles could become enriched within regional bioenergetic environments to found regional haplogroups. While severe mtDNA bottlenecks will increase the probability that a deleterious allele will be presented, resulting in genetic disease and in the rapid demise of that lineage (Bergstrom and Pritchard 1998), a severe mtDNA germline bottleneck would also increase the presentation of rare beneficial alleles, thus increasing species diversity and adaptability to changing environments. Thus, in contrast to nuclear genetics, in mtDNA genetics, genetic drift and selection are not two mutually exclusive or competing forces. Rather, they work together to inhibit the accumulation of deleterious mutations and facilitate the fixation of advantageous ones.

If we are to have a complete understanding of origin and biological importance of mtDNA genetic diversity, we need to understand the role that selection plays in shaping mtDNA genetic diversity. To date, only a few mouse models have been generated and studied that provide insight into the importance of selection in sorting out mtDNA heteroplasmic mtDNA mutations.

Mouse Germline Segregation of an mtDNA Deletion. In one mouse model system, an mtDNA deletion was introduced into the mouse female germline via somatic cell cytoplasmic hybrid (cybrid) fusions (Bunn et al. 1974; Wallace et al. 1975). A rearranged mtDNA was recovered from mouse brain by fusion of synaptosomes, neuronal synaptic bouton fragments containing mitochondria and mtDNAs, to cultured cells lacking mtDNA (ρ^0) (King and Attardi 1989; Chomyn 1996). Because ρ^0 cells cannot grow without uridine and pyruvate, ρ^0 cells that acquire partially functional mitochondria can be selected by growth in media lacking either pyruvate or uridine. The resulting synaptosome cybrids were screened for mtDNA mutations and one clone was identified that harbored a 4696-nt mtDNA deletion that removed six tRNAs and seven structural genes.

Next, enucleated cell cytoplasts from this mtDNA 4696-nt deletion cell line were fused to single-cell mouse embryos, the embryos implanted into foster mothers, and the resulting mice found to be heteroplasmic for the deleted mtDNA. Unlike human mtDNA deletion mutations, which are not generally maternally inherited, this deletion was transmitted through the female germline (Inoue et al. 2000). The heteroplasmic mice have COX-negative fibers in their heart and muscle, renal dysfunction, and male mice with $\geq$70% rearranged mtDNAs are infertile (Inoue et al. 2000; Nakada et al. 2006).

In humans, it has been observed that mtDNA duplications can be maternally inherited and that these duplicated molecules generate deletions by intramolecular recombination within postmitotic tissues (Wallace and Fan 2009). Hence, it is possible that the maternal transmission of the 4696-nt mtDNA deletion is the product of maternal transmission of a duplication, although this has not been determined.

Regardless of the molecular nature of the maternally transmitted rearranged mtDNA, analysis of the level of the deleted mtDNA in pups derived from heteroplasmic female mice revealed a striking directional loss of the rearranged mtDNA in successive litters. Among five females with tail deletion levels between 20% and 60%, the mean deletion heteroplasmy of all of the pups declined precipitously. By the third litter, all of the pups had less than 10% deleted mtDNA, with many having no detectable deleted mtDNA in their tail tissue. Interestingly, in two of the females, the mean mtDNA heteroplasmy of the pups of the first litter was higher than that of the tail of the mother. The same striking decline in the percentage of deleted mtDNA was seen in the oocytes of two females that were superovulated at d 44 and 117. In both cases, the oocyte heteroplasmy level in the first superovulation oocytes was less than that of the females and the reduction of heteroplasmy in the oocytes in the subsequent superovulation was even more marked. The same reduction in mtDNA deletion levels was observed in the litters of mice in which the ovaries of heteroplasmic deletion mice were grafted into ovarectimized C57BL/6 mice harboring *Mus spretus* mtDNAs (Sato et al. 2007). Apparently, then, oocytes with high mtDNA deletion levels can be ovulated while the mouse is young, but, as the female mouse ages, the high deletion oocytes are progressively lost. Hence, there must be selection against the formation or ovulation of the oocytes with highest deletion levels throughout the female mouse's lifespan.

Mouse Germline Segregation of a Frameshift mtDNA Mutation. An analogous finding has been reported for a mouse line that harbored a heteroplasmic mtDNA *ND6* frameshift mtDNA mutation. This mutation was the result of an insertion of a C at nt 13885 within the *mtDNA* (*ND6* 13885insC). The original mtDNA frameshift mutation was isolated in mouse L cells linked to an mtDNA homoplasmic *COI* T6589C missense mutation. When this cell line was enucleated and fused to a female mouse embryonic stem cell (mfESC), one of the cybrid clones was found to harbor mtDNAs that had reversed the *ND6* 13885insC mutation by deletion of an adjacent T, thus restoring the reading frame. Hence, this mfESC cybrid was homoplasmic for the *COI* T6589C missense mutation but heteroplasmic for the *ND6* 13885insC frameshift mutation.

Injection of this mfESC cybrid into mouse blastocysts resulted in the generation of one female mouse that was heteroplasmic for the *ND6* 13885insC frameshift mutation at 44% $\pm$ 3% in all of her tissues, with the percentage frameshift mutation being 47% in her tail. When this female was mated, all of the pups of her first and second litter had tail mtDNAs that were 14% frameshift, in the third litter all of the pups had 6% frameshift mtDNA, and in the fourth, fifth, and sixth litters the frameshift mtDNA was lost. When females that harbored 14% frameshift mtDNAs were mated, they produced pups with either 6% or 0% frameshift mtDNAs. When a female with 14% frameshift mtDNA was superovulated and 11 oocytes were genotyped, two oocytes were found to harbor 16% frameshift mtDNAs, one harbored 14%, one 12%, two 6%, and five 0% frameshift mtDNAs (Fan et al. 2008). Hence, there was a directional and concerted loss of the frameshift mtDNA over successive generations. Furthermore, the distribution of oocyte heteroplasmy levels was not Gaussian but was truncated such that there were oocytes with substantially less mutant mtDNA than the mother's tail genotype, but none had significantly more mutant mtDNAs than the mother. Therefore, germ cells or oocytes with a higher percentage of frameshift mtDNAs must have been selectively removed prior to ovulation. Thus, mammalian females have a prefertilization mtDNA selection that eliminates those oocytes harboring the most deleterious mtDNA mutations.

Independent evidence that a selection process exists in the female germline was obtained by analyzing the fate of mutant mtDNAs produced by the mouse polymerase γ (D257A *PolgA* exo$^-$) mutator gene. This revealed that while deleterious tRNA mutations were introduced into the germline, there was a dearth of deleterious polypeptide gene mutations (Stewart et al. 2008).

Germline Segregation of Single-Base tRNA Deletion Mutations. Using the mtDNA polymerase γ (D257A *PolgA* exo$^-$) mutator gene to generate germline mtDNA mutations, a mouse was isolated that was heteroplasmic for a tRNA^{Met} nt 3875 C deletion (tRNA^{Met} 3875 delC) mutation. The inheritance of the mtDNA harboring this mutation was monitored and found to be biased toward loss of the tRNA mutation. The tRNA^{Met} 3875delC mutation inhibits aminoacylation, but the biochemical defect was partially ameliorated by the up-regulation of the other mtDNA tRNAs and all of the mtDNA mRNAs except for that of *ND6* (Freyer et al. 2012).

Mice harboring the tRNA^{Met} 3875delC mutation had relatively similar heteroplasmy levels across tissues and these did not change with age. Yet, even with selective breeding, the levels of tRNA^{Met} 3875delC mutation heteroplasmy could not be increased above 86%. Hence, selection is clearly acting against this tRNA mutation at high levels of heteroplasmy. The mtDNA genotypes of 44 mothers and their 533 offspring were analyzed. When the heteroplasmy of the mothers was binned into groups of 40%–60%, 61%–70%, and >70%, the heteroplasmy levels of the pups of the 40%–60% and >70% mothers were found to have undergone nonrandom segregation. This conclusion was based on comparison of the offspring distribution of genotypes observed versus prediction made from the Kimura distribution that would be expected for a random genetic drift model. In contrast to the *ND6* 13885insC frameshift mutation and the nt 4696 deletion mtDNA mice in which the heteroplasmy levels of the pups of successive litters declined with age, the heteroplasmy distribution of the tRNA^{Met} nt 3875delC mutation mtDNAs did not significantly differ among generations (Freyer et al. 2012).

Analysis of female germ cell mtDNA tRNA^{Met} 3875delC mutation levels in 819 *Stel-*

la-GFP-isolated PGCs from 18 embryos at 13.5 d revealed a wide variation in heteroplasmy levels, with some PGCs cells being 100% mutant. Thus, germline cells could be homoplasmic for the mutant even though no homoplasmic animals were generated. This indicated that early in female germ cell development there was a stochastic distribution of heteroplasmic levels, but that the highest levels of mutant mtDNAs were lost as the oogonia or oocytes matured. The authors conclude that "... the variance in heteroplasmy among offspring is determined embryonically, consistent with a prenatal germline bottleneck governing the segregation of mutated mtDNA." Analysis of 340 oocytes from the ovaries of five 3.5-d-old neonate female mice revealed a relatively random distribution of mtDNA genotypes, consistent with a Kimura distribution prediction of random genetic drift. From this, the authors concluded that the distribution of mtDNA tRNA^{Met} 3875delC mutation genotypes of the oocytes was not truncated by selection. However, when the mean mutation level of the pups of a heteroplasmic female were plotted against the mother's tRNA^{Met} 3875delC heteroplasmy levels, the offspring of females with 45%–60% heteroplasmy had pups with an up to ~14% increased mutation load, while females with 65%–83% heteroplasmy levels exhibited a progressive decline in the average tRNA^{Met} nt 3875delC mutation levels of ~10%. The authors concluded "That selection is likely to have occurred after the oocyte stage ...," (Freyer et al. 2012), but it is noteworthy that none of the oocytes genotyped contained the tRNA^{Met} 3875delC mutation heteroplasmy at levels greater than ~85%, even though such genotypes were present in the PGCs. Therefore, the descendants of the PGCs with 100% mutant mtDNA must have been eliminated prior to the formation of the mature oocytes.

For the tRNA^{Met} 3875 delC mutation, two factors appear to be acting on mutant heteroplasmy levels. At 45%–60% mutant in the mother, there is a tendency for the mutant mtDNAs to increase in the offspring, while at heteroplasmy levels of 65%–83%, there was a tendency for their loss. Hence, the tRNA^{Met} 3875delC mutation levels seem to be maintained in the female germline by two opposing selective forces, one enriching for the mutant mtDNA at intermediate heteroplasmy levels and the other strongly selecting against the mutant at high levels of heteroplasmy.

Haplogroup Incompatibility in Heteroplasmic Mice. To determine whether mtDNA haplogroups might also show nonrandom segregation along the mouse female germline, NZB and 129S6 mtDNAs were combined within the female mouse germline by fusing cytoplasts from a culture mouse cell line harboring NZB mtDNAs with a mfESCs derived from a 129S6 mouse. The NZB-129 heteroplasmic mfESC cybrids were injected into blastocysts to generate a heteroplasmic mouse line. NZB and 129 mtDNAs differ in 91 nucleotide positions including 15 nonsynonymous amino acid substitutions, five tRNAvariants, seven rRNAvariants, and 11 control region variants. The heteroplasmic female mice were backcrossed for ~20 generations with C57BL6/J2 mice originally obtained from the Jackson Laboratory, and all subsequent females were mated with males from the same C57BL/6 J line maintained by brother/sister mating.

The C57BL6/J2 NZB-129 heteroplasmic mice were then allowed to segregate the two mtDNAs. This revealed a strong bias in segregation toward the loss of the NZB mtDNAs. Over 24 generations, 171 mice segregated to <3% NZB mtDNAs within one to two generations. Three out of seven mice that lost the NZB mtDNAs bred true as pure 129 mtDNAs. By contrast, more than 10 generations of selective breeding were required to generate 12 females with >97% NZB mtDNAs, and of three of these only one bred true for the NZB mtDNA.

In unrestricted segregation experiments, of 864 pups born to heteroplasmic mothers, 67% had a mean increase in 129 heteroplasmy, 28% had an increase in NZB heteroplasmy, and only 5% had a mean heteroplasmy similar to their mother. Moreover, the tendency to segregate NZB mtDNAs was highest when the mother's heteroplasmy level was in the 60%–80% range, with the tendency to lose the NZB mtDNAs declining when the percentage of heteroplasmy of the mother was strongly biased toward either

129 or NZB mtDNAs. "Therefore, for mice with approximately equal proportions of NZB and 129 mtDNAs, there was a strong bias to decrease NZB mtDNA among generations. The enrichment of the 129 mtDNAs was seen in the ovaries of the heteroplasmic females and to a lesser extent in the oocytes of the heteroplasmic females. The tendency to segregate the NZB mtDNAs was not simply the result of incompatibility between the C57BL/6 J nucleus and the NZB mtDNAs because the fertility and fecundity of the homoplasmic NZB mice was equal to or better than that of the 129 mice with the same nucleus" (Sharpley et al. 2012). Thus, the segregation pattern of the NZB-129 heteroplasmic mice was similar to that of the $tRNA^{Met}$ nt 3875delC mutation mice, although less directional or concerted than was seen for the mtDNA 4696-nt deletion and *ND6* 13885insC frameshift mutations.

This series of experiments unequivocally demonstrates that selection acts on mtDNA genotypes within the female germline, acting to weed out the most severe mtDNA mutations. Furthermore, the bias toward the elimination of a deleterious mtDNA is correlated with the severity of the mitochondrial defect associated with the mtDNA mutation.

Mathematical Models to Detect Selection in mtDNA Segregation. The striking inconsistencies observed among different estimates of the mtDNA copy numbers in mouse female germline cells, the variation in estimates of heteroplasmy level variances in progeny cells and offspring of different studies and model systems, and the role of selection in defining the transmission of heteroplasmic mtDNA genotypes mean that it is currently impossible to define the factors that determine the origin and inheritance of pathogenic mtDNA mutations. In an effort to more accurately compare different experiments and ultimately to better understand mtDNA mutant transmission, significant efforts have been made to model the transmission of heteroplasmic mtDNAs along the female germline.

Unfortunately, three factors have limited the investigator's ability to generalize the results from different studies: (1) development of methods for comparing differences in heteroplasmy variance among samples and experiments and to calculate the statistical significance of these differences, (2) development of appropriate mathematical models to describe the origin and transmission of mtDNA heteroplasmy, and (3) development of theory and methods to determine the role of selection in mtDNA heteroplasmy transmission.

Statistical Analysis of mtDNA Heteroplasmy Variance. A major limitation for defining the principles covering heteroplasmy transmission has been difficulty in comparing the heteroplasmy variance levels observed across different experimental systems. Difficulties arise at two levels, correcting for the effects of parental allele frequencies on the potential range of heteroplasmy variances of their descendants and difficulties in calculating the standard error of heteroplasmy variance estimates for testing the statistical significance of observed differences.

The heteroplasmy variance of offspring or derived cells is a function of the heteroplasmy level of the mothers for offspring or of the parental cells for derived cells, the parental heteroplasmy designated as p_0. This fact becomes particularly important when the heteroplasmic level of one of the mtDNA types in very low, $<10\%$. Under these conditions, the range of heteroplasmy values of the offspring or derived cells will be constrained by the frequency of the low parental allele, making the range of heteroplasmic variance levels aberrantly low. Heteroplasmic variance will then be maximum when $p_0 = 0.5$ and minimal when $p_0 < 0.1$ (Samuels et al. 2010). To correct for this systematic distortion in variance capacity, Samuels and collaborators have used Mendelian population genetics theory (Wright 1969) to normalize the heteroplasmy variances in proportion to p_0. This was accomplished by dividing the observed heteroplasmy variance by the term $p_0(1\text{-}p_0)$. This term is at maximum (0.25) when p_0 is 0.5, and becomes progressively smaller as p_0 gets smaller. Hence, dividing the observed variance by $p_0(1\text{-}p_0)$ proportionately increases population variance estimates for low values of p_0. With this correction, Samuels and associates discovered that a number of conclusions about changes in heteroplasmy variance drawn from various studies involving animals with low maternal p_0 could not be substantiated (Samuels et al. 2010).

Once the effects of extreme values of p_0 on variance estimates were corrected, it became necessary to calculate the standard error (SE) of the variance so that the statistical significance of differences in variance among different sample determinations could be calculated. At low levels of p_0, the required sample size to obtain statistical significance rapidly increases. Wonnapinij and colleagues (2010) pointed out that many experimental estimates of mtDNA heteroplasmy variance have been based on sample sizes ≤ 20. Their simulations indicate that these variance estimates have very large SEs. When Wonnapinij and colleagues estimated the SE of the variance for the original 1996 NZB-BALB mtDNA segregation of Jenuth and collaborators (1996), they concluded that once variance normalization was applied and variance SEs were calculated, the reported differences in heteroplasmy levels between primary and mature oocytes were not statistically significant. However, the heteroplasmy variance of the PGCs was significantly lower than that of primary and mature oocytes. This analysis supports the hypothesis that an mtDNA bottleneck occurs during the transition of PGCs to primary oocytes.

Similarly, Wonnapinij and colleagues reanalyzed the mouse mtDNA heteroplasmy variation of the Wai and associates (2008) study. Once the heteroplasmy variances were normalized and SE was estimated, Wonnapinij and colleagues concluded that the heteroplasmy variance values determined on postnatal d 11 and after relative to those of d 8 and earlier were not significantly different. Therefore, the conclusion of Wai and associates (2008) that an mtDNA heteroplasmy bottleneck occurs during postnatal folliculogenesis could not be substantiated (Wonnapinij et al. 2010). Wonnapinij and colleagues also pointed out that the mtDNA heteroplasmy variance levels in the early postnatal period differed between the Jenuth study (Jenuth et al. 1996) and Wai study (Wai et al. 2008). From these analyses, Wonnapinij and colleagues concluded that the major bottleneck responsible for mtDNA heteroplasmy variance occurs in the maternal germline between the PGCs and the oocytes.

Application of these analytical approaches to mouse versus human mtDNA heteroplasmic variance data revealed that human heteroplasmy variance was greater than mouse. Comparison of the 3243A>G mutation heteroplasmy levels of the 82 oocytes derived from the hysterectomy of the carrier mother (quadriceps = 18.11%; leukocytes = 7.25%) with the heteroplasmy variance of oocytes from the NZB-BALB mouse mtDNA studies revealed that the human normalized oocyte A3243G, mutation level variance was 0.13, while the mouse normalized oocyte mutation level variance was 0.02–0.04. Similarly, using an aggregate of human pedigrees harboring the NARP 8993T>G, LHON 11778G>A and 3460G>A, and MERRF 8344 A>G mutations encompassing 72 mother–child pairs with heteroplasmy levels in the range of 40%–60%, Wonnapinij and colleagues concluded that the normalized heteroplasmy mutation level variance of human offspring was ~0.37, while that for mouse offspring was 0.04–0.12 (Wonnapinij et al. 2010). This comparison assumes that the nature of the genetic differences between human mtDNA 3243A>G mutant and between NZB and BALB mtDNAs does not influence the heteroplasmy segregation rate.

A Neutralist Mathematical Model to Describe mtDNA Heteroplasmy Segregations. To determine whether mtDNA heteroplasmy variance has been affected by nonrandom factors such as selection, it was necessary to define the expectation for heteroplasmy variance if only stochastic factors were contributing to mtDNA segregation. The calculation of the expectations of random heteroplasmy transmission required a theoretical model. Generally, researchers have used the Sewall Wright variance formula, $V = p_0(1\text{-}p_0)(1\text{-}e^{-t/N_e})$ for determining the effect of random genetic drift on mtDNA heteroplasmy inheritance. Despite its simplicity, this approach ignores much of the information that is present in the total heteroplasmy distribution, especially when this is not symmetric. Additionally, a common assumption when comparing heteroplasmy distributions between cells or individuals is that the population follows a normal distribution. However, Wonnapinij and colleagues have argued that the normal distribution is not the optimal choice for representing the potential distribution of mtDNA heteroplasmic genotypes for two main reasons. First, the normal (or Gaussian) distribution is defined over a range of infinite minus and plus tails, yet genetic variation is constrained to allele frequencies (p_0) between zero and one. Second, the normal distribution is always symmetric, whereas mtDNA heteroplasmy distribution is often skewed. Therefore, Wonnapinij and colleagues (2008, 2010) argued that while the normal distribution can provide a reasonable model for the stochastic dispersion of heteroplasmy allele frequencies with p_0 values close to 0.5, it is less accurate at extreme allele frequencies.

To predict the entire heteroplasmy probability distribution including the probabilities of a single allele becoming fixed or lost, Wonnapinij and colleagues developed a tool based on Kimura's distribution, henceforth designated the Wonnapinij/Chinnery/Samuels-Kimura (WCS-K) tool. In the 1950s, Kimura (1955) developed a model of the possible allele frequency distribution for a finite population resulting from the random sampling of gametes. His continuous model includes a set of probability distribution functions that describe gene frequency distributions of populations under pure random genetic drift. These functions are represented by three equations: a probability $f(0, t)$ for losing an allele, a probability $f(1, t)$ for fixing that allele, and a probability distribution function $\varphi(x,t)$ that the allele is present at frequency x in the population. This model has three parameters: the initial gene frequency, p_0; the effective population size, N_e; and the number of generations, t. In the case of mtDNA heteroplasmy, p_0 is interpreted as the founder's heteroplasmy proportion or as a surrogate of the parental allele frequency calculated as the mean heteroplasmy in the derived cells or offspring of the progenitor. N_e is interpreted as a parameter related to the number of segregating mtDNA units. t is interpreted as the number of generations. The WCS-K tool avoids complications related to the definition of N_e by combining it with t into a single parameter called b, and treats b as one parameter ($b = e^{-t/Ne}$). Parameter b is determined from the heteroplasmy variance either of a number of offspring from a single mother or by pooling the heteroplasmy values of offspring from mothers with similar heteroplasmy levels ($b = 1\text{-}[V/p_0(1\text{-}p_0)]$ or $b = 1\text{-}V_{\text{normalized}}$). Wonnapinij and colleagues refer to parameter b as the bottleneck parameter, which then determines the width of the heteroplasmy distribution in the offspring (Wonnapinij et al. 2008; Samuels et al. 2013). To determine the statistical significance of the observed differences in the fit of experimental data with the Kimura distribution, the WCS-K tool uses the Kolmogorov-Smirmov (KS) test (Wonnapinij et al. 2008).

Using the Kimura model, Wonnapinij and colleagues reassessed the variance estimates and conclusions of a number of previous experimental studies on heteroplasmic segregation. When the authors applied the WCS-K tool to the heteroplasmy distribution of 82 oocytes recovered from the hysterectomy of the 3243A>G mutation woman (Brown et al. 2001), they concluded that the oocyte mtDNA distribution fit the expectations of the Kimura distribution and thus was by inference stochastic. Application of the tool to the Jenuth and collaborators data on the transmission to oocytes of the mouse NZB versus BALB mtDNAs from heteroplasmic mothers (Jenuth et al. 1996), the authors concluded that six of the eight mouse line datasets were consistent with the expectations of the Kimura distribution for random genetic drift, but that two datasets at the extreme ends of the heteroplasmic distribution failed to fit the Kimura distribution because of unexpectedly low numbers of oocytes that lacked one or the other mtDNA. Of course, the determination of whether an mtDNA is present or absent depends on the sensitivity of the method used to detect the minor allele.

Analysis of two datasets on heteroplasmic *Drosophila* eggs (Solignac et al. 1984; de Stordeur et al. 1989) resulted in five of six datasets fitting the Kimura distribution, the inference being that their distribution arose by random segregation. In one *D. simulans* dataset in which the mtDNA heteroplasmy was generated by cytoplasmic injection of *siIII* and *siII* mtDNA genomes (two naturally occurring *Drosophila* mtDNA sequences), the heteroplasmy distribution was found to fit the Kimura distribution and thus was random. However, in this case the founder female harbored 38.5% heteroplasmy and by the third generation the mean heteroplasmy of the offspring was $p_0 = 12.7\%$ ($V = 0.0096$). Therefore, although this striking directional shift could not have occurred by chance, the WCS-K tool was unable to detect deviation from randomness. Finally, in a *Drosophila* dataset derived from eggs 30 generations away from the founder, the results did not fit the expectations of the Kimura distribution, suggesting that factor(s) in addition to genetic drift had acted on this mtDNA segregation pattern (Wonnapinij et al. 2008).

Thus, by applying the WCS-K tool to heteroplasmy distributions, Wonnapinij and colleagues concluded that many were consistent with stochastic processes. Still, even this tool showed that some cases deviated significantly from the expectations of randomness.

In contrast to the seemingly modest role played by selection in determining transmission of heteroplasmy when analyzed using the WCS-K tool, the mouse experiments that analyzed the transmission of deleterious mutations including the mtDNA 4696 deletion (Sato et al. 2007), the *ND6* 13885insC mutation (Fan et al. 2008), the $tRNA^{Met}$ 3875delC mutation (Freyer et al. 2012), and the NZB mtDNA in 129-NZB haplogroup heteroplasmy (Sharpley et al. 2012) all unequivocally show evidence of mtDNA germline purifying selection. One set of potential limitations of using the Kimura distribution to model a stochastic distribution of mtDNA genotypes resides in the three parameters that relate to the neutralist basis of the distribution. These include: (1) a large population of N diploid parents, (2) no mutation, selection, or migration, and (3) no overlapping generations (Kimura 1955). One of the mathematical assumptions of the continuous Kimura model is that the effective size N must be sufficiently large so that terms of order $1/N^2$ and higher can be omitted without serious error. Studies of mammalian embryonic development suggest that the number of mitochondria or mtDNA molecules may be drastically reduced (Jansen and de Boer 1998; Krakauer and Mira 1999). If the number of segregating units remains ≥ 200 (Jenuth et al. 1996; Cree et al. 2008; Wai et al. 2008), then this assumption is still valid. If, however, the number of mtDNA segregating units were to approach $N = 8$, as suggested in one scenario for the 3243A>G genotype distribution of the 82 oocytes removed from a hysterectomy (Brown et al. 2001), then this term could introduce error into the conclusions of the Kimura model.

Another concern is whether the KS test for significance is appropriate and sufficiently sensitive to evaluate differences between the experimentally observed heteroplasmy distribution and the predictions of the Kimura model, especially when the heteroplasmy levels are ~0.5. By implementing the KS nonparametric test, the WCS-K tool avoids assuming that the data were sampled from Gaussian distributions, but there are drawbacks to using a nonparametric test. If the populations are approximately Gaussian (as in the case of ~0.5 heteroplasmy level), the nonparametric tests have less power (are less likely to give you a small p value), especially with small sample sizes (Stephens 1974). Additionally, the KS test is most sensitive when the distribution functions differ in a global fashion near the center of the distribution. But if there are repeated deviations between the distribution functions or the distribution functions have (or are adjusted to have) the same mean values, then the distribution functions cross each other multiple times and the maximum deviation between the distributions is reduced (Babu and Feigelson 2006).

Because the predictive ability of the WCS-K tool is based on the assumption of random genetic drift in a finite large population, the strength of the predictions of the WCS-K tool is strongly influenced by sample size. This makes it possible to change the statistical significance of a comparison simply by increasing the range of mtDNA heteroplasmy genotypes binned together. When the Kimura distribution was applied to analysis of heteroplasmy transmission of the $tRNA^{Met}$ 3875delC mutation, even for more central p_0 values where the effects on "*b*" are minimized, quite large sample sizes were required to show the directionality associated with selection (Freyer et al. 2012). While statistical evidence of selection was obtained when binning the data from mothers in the range of 40%–60% heteroplasmy (Freyer et al. 2012), it was not when binning females with heteroplasmy levels of 50%–60% (data not shown). This limitation is compounded because the currently available program can only analyze transmissions through one generation, thus limiting detection of cumulative effects of selection. Moreover, the effects of population size fluctuations on selection are also not considered and therefore not modeled (Wonnapinij et al. 2008, 2010).

While these technical considerations are pertinent, perhaps a more important reason why the WCS-K tool is relatively insensitive in identifying the effects of selection on mtDNA mutations is that the Kimura model was developed for changes in the allele frequency of Mendelian genes within finite populations. The Mendelian mode of inheritance was the only genetic model known in the 1950s when Kimura developed his model. However, the Mendelian rules of inheritance derive from the behavior of chromosomal genes and the biological unit upon which selection acts on Mendelian genes is the diploid organism. This unit of selection is fundamentally different from the situation with the mtDNA in which selection acts on cells and organisms with several thousand-fold ploidy.

This does not adversely affect the mathematics of allelic transmission if all alleles are neutral. However, it has a profound effect on whether the Kimura model can detect the effects of purifying selection during the transmission of a deleterious heteroplasmy mtDNA mutation. In diploid animals, selection acts on units of two gene copies for which there are only three allelic combinations: *a*/*a*, *a*/*b*, and *b*/*b* and two to three quantized phenoptypic states. For mtDNA genetics, selection acts on the cellular phenotype, which is a composite of the several thousand mtDNAs with different percentages of two or more allele systems.

For an nDNA allele at a population frequency of 1%, virtually every minor allele will be located in a heterozygous individual with a significant probability of being lost in subsequent generations of diploid individuals by drift. A cell with 1% mutant mtDNAs within a cell of 5000 mtDNAs would harbor 50 mutant mtDNAs. Because cytokinesis divides the cell in half, low-frequency mtDNA alleles are much less likely to be lost in successive generations than nDNA alleles. Finally, the mitochondria and mtDNAs within the cell are not autonomous units. Rather, they function as a collective population through repeated fission and fusion and sharing of all their gene products. Subtle alterations in the mtDNA sequence and heteroplasmy level can have significant effects on cellular energy production, REDOX regulation, ROS production, Ca^{2+} regulation, mtDNA copy number, mitochondrial fission and fusion, mtDNA replication and turnover, etc. For example, 20% of normal mtDNAs within a cell can mask the respiratory-deficient phenotypic of 80% 3243A>G mutant mtDNAs (Yoneda et al. 1994). Also, low-frequency deleterious mtDNAs can be retained in a cell lineage for many generations without adversely affecting the cellular phenotype. An *ND5* gene frameshift mutation, which was found to be homoplasmic in a human oncocytoma tumor, was present at low-heteroplasmy levels in the normal tissues of the patient and his two sisters. Thus, this deleterious mutation was silently transmitted through the maternal lineage (Gasparre et al. 2008). Hence, classical concepts of genotype–phenotype associations and their interaction with selection are violated by mtDNA genetics.

In conclusion, the formulations of Wonnapinij and associates (Wonnapinij et al. 2008, 2010) have provided a significant improvement for the mathematical analyses of mtDNA heteroplasmy variance in cells and offspring. However, the WCS-K tool is less powerful at detecting the role of selection on mtDNA mutation inheritance.

Heteroplasmic mtDNA Mutation Segregation in Somatic Cells and Tissues

Heteroplasmic mtDNAs also segregate in somatic cells and tissues and this greatly complicates phenotypic predictions in patients. Although it is standard practice to test for mutant nDNA genes in patient blood, the mtDNA heteroplasmy level in blood cells can be quite different from that of brain, heart, muscle, or kidney, the organs most susceptible to life-threatening mitochondrial disease.

The first reports of directional segregation of mtDNA heteroplasmy in mouse and human somatic cells appeared in the 1970s. In these experiments, the fate of the mtDNAs was monitored in replicating somatic cell hybrids and cybrids via the mtDNA chloramphenicol (CAP)-resistance (CAP^R) or -sensitivity (CAP^S) genetic markers. CAP^R is the product of mutations in the mtDNA 16S rRNA gene (Blanc et al. 1981a,b). When CAP^R and CAP^S cells of the same species and same cell lineage (mouse L cell to mouse L cell, human HeLa cells to human HeLa cells) were fused, the mixture of CAP^R and CAP^S mtDNAs was retained for extended periods and seemed to segregate stochastically. By contrast, when cell lines of the same species but different origin (mouse L cell to mouse RAG cells, human HeLa cells to human WiL2 [WAL2] or HT1080 cells) were fused, then one of the two mtDNAs was directionally and rapidly lost. In the case of cybrids prepared between enucleated CAP^R HeLa cells and WAL2A cells, the heteroplasmy could only be retained by continuous CAP selection (Wallace et al. 1976, 1977; Bunn et al. 1977; Wallace 1981, 1982, 1986). The directionality of mtDNA segregation was even more extreme when cells of different species were fused. When CAP^R human cells were fused to CAP^S mouse cells and the hybrids selected in CAP, the surviving hybrids retained the human mtDNA and all of the human chromosomes, and lost all of the mouse mtDNA and segregated the mouse chromosomes. This is the opposite of the well-established mouse–human hybrid segregation pattern (Wallace et al. 1976; Giles et al. 1980).

When clinically relevant mtDNA mutations became available, these mutants were also introduced into cultured cells. In six different cybrids lines heteroplasmic for the MELAS $tRNA^{Leu(UUR)}$ 3243A>G mutation, the mutant 3243A>G mtDNA progressively increased in frequency. Only cell lines that were predominantly homoplasmic were stable, with the exception of one cell line, which maintained a stable 30% 3243A>G mutant heteroplasmy (Yoneda et al. 1992).

Segregation of Human Somatic Cell mtDNA Mutations

Shortly after it was discovered that mtDNA deletions accumulate in muscle of patients with mitochondrial myopathy (Holt et al. 1988), it was found that mtDNA deletions were progressively lost from blood cells in CPEO and Kearn Sayre patients, though they become progressively enriched in muscle (Shoffner et al. 1989). Moreover, mtDNA deletions in humans were rarely transmitted through the female germline (Shoffner et al. 1989), though mtDNA duplications can be maternally transmitted (Ballinger et al. 1992, 1994).

The absence of mtDNA deletions in blood cells when they are prevalent in postmitotic tissues led to the hypothesis that severely deleterious mtDNA mutations might be inhibiting bone marrow stem cell replication. Those stem cells that spontaneously segregated the deleted mtDNA would then have a proliferative advantage and repopulate the bone marrow.

The MELAS $tRNA^{Leu(UUR)}$ 3243A>G mutation, like mtDNA deletions, commonly presents as CPEO and is also progressively lost from the blood cells, though not as rapidly as mtDNA deletions. Comparison of the 3243A>G heteroplasmy levels for 13 patients tested in white blood cells, oral mucosa cells, and urinary tract epithelial cells consistently showed that the blood cell mutant level was lowest and the urinary epithelial cells heteroplasmy was highest. For one patient the levels were 20%, 30%, and 65% and for another were 30%, 50%, and 80%. Because of their consistently high mutation heteroplasmy level, urinary epithelial cells are considered the best surrogate for postmitotic tissues when testing for 3243>G heteroplasmy levels of the body (Monnot et al. 2011).

The level of the 3243A>G mutation progressively declines in blood cells over the life of the individual. In a longitudinal analysis of 34 individuals carrying the 3243A>G mutation, 17 symptomatic and 17 carriers all showed an age-related decline in blood mutant heteroplasmy levels. For the symptomatic subjects the heteroplasmy level declined 0.534% per year, while for the asymptomatic carriers the heteroplasmy level declined 0.215% per year. At the same time, the symptomatic individual's phenotypes became worse while the asymptomatic individuals remained asymptomatic (Mehrazin et al. 2009). A mathematical model was developed for the decline of 3243A>G heteroplasmy in blood and compared to longitudinal data that has been collected on the blood cell heteroplasmy levels. The model and the data indicate that the loss of heteroplasmy is an exponential function of the form: $M_{\text{age-corrected}} = m(t)e^{St}$, where $m(t)$ is the average mutational level as a function of time and $S = 0.020 \pm 0.003$ (1/year) (Rajasimha et al. 2008).

While the percentage of the mtDNA $\text{tRNA}^{\text{Leu(UUR)}}$ 3243A>G mutation heteroplasmy declines with age in blood, the heteroplasmy level of the MERRF tRNA^{Lys} 8344A>G mutation does not (Rajasimha et al. 2008). Hence, different mtDNA mutations show different tissue-specific stabilities indicating that a variety of physiological processes must modulate the fate of heteroplasmic variants in tissues.

Segregation of Mouse Somatic Cell mtDNA Mutations

The original Jenuth et al. (1996) manuscript reported that germline segregation of mouse NZB-BALB heteroplasmy was random. However, in a subsequent paper these authors reported that NZB-BALB mtDNAs segregated directionally with age in different tissues (Jenuth et al. 1997). Over the life of an NZB-BALB heteroplasmic animal, the kidney and liver progressively increased the NZB mtDNA, while the blood and spleen cells accumulated the BALB mtDNA (Jenuth et al. 1997). The heteroplasmy of the tail, cerebral cortex, gastrocnemius muscle, heart ventricle, and lung were reported to remain relatively stable (Jenuth et al. 1997).

In a similar experiment involving mouse NZB and 129-mtDNA heteroplasmy, the directional segregation of liver and kidney toward NZB mtDNAs and of spleen toward 129 mtDNAs was confirmed. In addition, it was found that seminal vesicles, ovaries, and pancreas also segregated toward 129 mtDNAs. However, tail, brain, lung, and heart maintained relatively constant heteroplasmy (Sharpley et al. 2012). Hence, striking differences exist between germline and somatic cell mtDNA heteroplasmy segregation. Because the somatic tissue cells contain thousands of mtDNAs, this segregation is unlikely to be caused by a bottleneck involving a dramatic reduction in mtDNA copy number. Rather, this must involve a process by which either some mtDNAs are selectively replicated or eliminated or certain mitochondria are differentially propagated.

In contrast to the NZB/BALB mouse nuclei (Jenuth et al. 1997), it was discovered that the *Mus musculus castaneous* (CAST/Ei) mouse nucleus does not cause the tissue-specific segregation of the NZB and BALB mtDNAs. This discovery provided the opportunity to map genetic loci involved in the mtDNA sorting process. Heteroplasmic female *Mus musculus* mice were crossed with male *Mus musculus castaneous* (CAST/Ei) mice and the tissue-specific segregation of the NZB-BALB mtDNAs was found to be inherited as a simple Mendelian recessive trait. Using F2 intercross animals, Battersby, Shoubridge, and associates mapped three quantitative trait loci (QTLs). One locus on chromosome 5 at *D5mit25*, designated *Smdq1* (segregation of mitochondrial DNA QTL#1) accounted for 35% of the variance in the heteroplasmy segregation phenotype of liver. A second locus found on chromosome 2 (*Smdg2*) accounted for 16% of the heteroplasmy variance in the kidney. A third locus on chromosome 6 (*Smdq3*) accounted for 20% of the variance for the spleen (Battersby et al. 2003). Analysis of the maximal oxygen consumption rates of the liver mitochondria of NZB-BALB heteroplasmic animals with NZB mtDNAs levels between 0% and 91%–97% on the BALB background revealed no significant difference. In the liver, the rate of enrichment of the NZB mtDNAs was constant with age with a selective advantage over BALB of ~14% per replication cycle. No difference was found in the relative incorporation rates of BrdU of the two mtDNAs. Hence, the advantage of the NZB mtDNA did not appear to be caused by increased replication rate. Also, partial hepatectomy followed by liver lobe regeneration did not affect the NZB-BALB mtDNA ratio. Surprisingly, however, when hepatocytes were explanted from heteroplasmic mouse livers, the majority of the in vitro cultures reversed the segregation direction, losing the NZB mtDNA while enriching the BALB mtDNAs. Therefore, the authors surmised that direction and mechanism of segregation was not the result of differential mitochondrial physiology or mtDNA replication rate and thus might depend on mitochondrial maintenance or mitochondrial turnover (Battersby and Shoubridge 2001).

To determine whether the directional segregation of the NZB mtDNAs in hematopoietic tissues was the result of immune surveillance, NZB-BALB heteroplasmic females were crossed onto nuclear backgrounds that lacked the *Tap1*, β_2m, or *Rag1* genes. Absence of these genes should impair the presentation and recognition of mtDNA-coded peptides. However, the kinetics of selection for the BALB mtDNA was unaltered indicating that segregation was not the result of presentation of mitochondrially encoded peptides by the major histocompatibility locus (MHC) (Battersby et al. 2005).

Fine mapping of the chromosome 6 (*Smdq3*) locus led to the identification of the *Gimap3* locus (GTPase of immunity-associated protein 3), which proved to be mutant in CAST/Ei relative to BALB/c, with the CAST/Ei gene being differentially spliced. The *Gimap3* gene has two AUG start codons in exons 3 and 4 with a stop codon in exon 4 upstream of AUG start codon. When all five exons are spliced together as in BALB the second AUG is used to initiate the polypeptide. In CAST/Ei, a G to A transition in the splice acceptor site of exon 4 results in the deletion of exon 4. This results in the first AUG being used resulting in the addition of 58 amino acids to the amino terminus of the protein without affecting the carboxy-terminal transmembrane domain. Although the mtDNA copy number was not different between the two *Gimap3* alleles, the overexpression of the CAST/Ei allele in heteroplasmic mice slowed the mtDNA segregation rate (Jokinen et al. 2010).

Mitochondrial Physiological Control of Somatic and Germ Cell mtDNA Segregation

It has been suggested that in hematopoietic tissues the selection is age-dependent and proportional to the initial heteroplasmy level. This implies that selection occurs at the organelle level. Because Gimap3 is a mitochondrial outer membrane protein, it might participate in mitochondrial partitioning or sorting. *Gimap3* is part of a vertebrate-specific gene family and the protein is anchored to the mitochondrial outer membrane by its carboxyl terminus. It has been reported to be important in T-cell survival and development. One speculation is that "Gimap3 could act as a scaffold on mitochondrial membranes to remodel mitochondria in response to specific stimuli" (Jokinen et al. 2011; Jokinen and Battersby 2013).

Regulation of mtDNA Heteroplasmy by tRNA^{Arg} 9821 insA-Generated ROS

The discovery that *Gimap3* regulates mtDNA segregation in hematopoietic cells does not seem to be the basis for the enrichment of the NZB mtDNAs in either liver or kidney, each of which is modulated by a different genetic locus. It has been proposed that because the rate of mtDNA selection remains constant in the liver with age and is independent of the initial heteroplasmy level selection for the NZB mtDNA may act at the mtDNA level (Jokinen et al. 2011; Jokinen and Battersby 2013). This proposal corresponds to the discovery of an important mtDNA sequence difference between NZB versus BALB or 129 mtDNAs. NZB and NIH3T3 mtDNAs differ from BALB and 129 mtDNAs by having an additional A in the length of a homopolymer run of As in the DHU loop of tRNA^{Arg} gene at nt 9821 (tRNA^{Arg} 9821insA). This same variant has been found to be enriched in heteroplasmic cultured mouse L cells (Fan et al. 2012). Extensive biochemical analysis of mouse L929 ρ^0 cell cybrids that harbor either NIH3T3 or NZB mtDNAs revealed that they have the same mitochondrial respiration rate as cells with fewer As. However, the similarity in respiration rate proved to be the result of a compensatory 1.6- to 1.9-fold increase in mtDNA copy number in the tRNA^{Arg} 9821 insA cells. This increased mtDNA copy number was induced by increased mitochondrial ROS production associated with the tRNA^{Arg} 9821 insA allele. When the increased ROS production was neutralized by antioxidants, the mtDNA copy number declined and mitochondrial respiratory chain function was reduced. The increased ROS also partially inactivated the mitochondrial α-ketoglutarate dehydrogenase and aconitase enzymes thus reducing the efficiency of the tricarboxylic acid cycle and the tRNA^{Arg} 8921 insA cells had a reduced capacity to grow in galactose (Moreno-Loshuertos et al. 2006).

Because an alteration in a mitochondrial tRNA would be expected to inhibit mitochondrial protein synthesis, this could impede the synthesis of mitochondrial OXPHOS enzymes, inhibit electron transport, and result in increased ROS production. Hence, a major factor in the enrichment of the NZB mtDNAs in liver could be increased ROS production as a result of tRNA^{Arg} 9821 insA variant-induced partial mitochondrial protein synthesis defect (Fan et al. 2012). The increased ROS would increase the mtDNA copy number and this effect would become increasingly predominant as the proportion of NZB mtDNAs increased in the tissues.

Battersby and Shoubridge have argued that the tRNA^{Arg} 9821 insA could not be the basis of the NZB segregation because they were able to map two different chromosomal loci in the BALB/c CAST/Ei crosses that abolished NZB mtDNA directional segregation (Battersby and Shoubridge 2007). However, the gene(s) responsible for the liver and kidney replicative advantage of the NZB mtDNAs could be nDNA genes involved in regulating mitochondrial ROS production and/or REDOX regulation.

A ROS-driven mtDNA copy-number hypothesis could also explain the directional germline segregation of the NZB mtDNAs in NZB-129 heteroplasmic mice. Assuming that the female germline cells have mitochondria with a limited number of mtDNAs and fusion and fission are inhibited, then each mitochondrion would act as a semi-autonomous unit. Those PGCs and proto-oocytes with higher percentages of NZB mtDNAs would generate increased ROS that would damage the mitochondria and mtDNAs resulting in preferential loss of the germline cells with the higher NZB mtDNAs. This effect would increase until the percentage of NZB mtDNAs reached a sufficiently high level that the increased ROS signaled to the nucleus to increase the mtDNA copy number. The biochemical difference between the predominant 129 and NZB cells would then be negated permitting the very high NZB lineages to segregate toward homoplasmic NZB.

In contrast to germline cells, somatic tissue cells and cultured cells have mitochondria that contain multiple nucleoids and mitochondria that actively fuse within a cell mixing their contents and gene products. In this case, the mitochondrial polysomes of a heteroplasmic somatic cell can translate both mtDNAs and mRNAs and can complement each other in *trans*. This *trans* complementation of mtDNAs in heteroplasmic cells was first demonstrated in somatic cell hybrids and cybrids harboring CAP^R and CAP^S mtDNAs linked to different polypeptide electrophoretic forms of the *ND3* gene, MVI versus MVII. In heteroplasmic cells in which the mtDNA translation products were differentially labeled by growth in ^{35}S-methionine in the presence of cytosolic ribosome inhibitor, emetine, both MVI and MVII were equally labeled. When CAP was added, both MVI and MVII continued to be labeled. This means that the CAP^R ribosomes had to be translating the MV mRNA from the CAP^S mtDNA (Oliver and Wallace 1982; Oliver et al. 1983).

In the case of somatic tissues harboring NZB mtDNAs with the tRNA^{Arg} 9821insA allele mixed with either BALB or 129 mtDNAs, the ribosomes would use both mutant and normal tRNAs in proportion to the mtDNA heteroplasmy. Because inhibition of polysome elongation would be limited by the "slowest" ribosome, the presence of even a small proportion of the inhibitory NZB mtDNA $tRNA^{Arg}$ 9821 insA would impede protein synthesis, reduce mitochondrial OXPHOS, and increase ROS production. The increased ROS production would then drive an increase in mtDNA copy number. The resulting repeated cycles of amplification of the mtDNAs within NZB mtDNA-containing cells could result in the progressive enrichment of the NZB mtDNAs. Hence, the $tRNA^{Arg}$ 9821 insA allele could account for the progressive increase in NZB mtDNAs in liver and kidney of NZB and BALB or 129 heteroplasmic mice.

This hypothesis would also be consistent with the dramatic reversal of NZB segregation seen for NZB-BALB heteroplasmic hepatocytes when grown in culture in high glucose medium (Battersby and Shoubridge 2001). The high glucose would fully reduce the electron transport chain of the cells resulting in increased ROS production regardless of the haplotype combination. The increased mtDNA copy-number effect would be neutralized permitting the more energetically efficient BALB mtDNA to take over.

If mitochondrial ROS production is important in the replicative segregation of NZB versus BALB or 129 mtDNA heteroplasmy animals, then one might anticipate that antioxidant defenses might also be relevant in regulating their segregation. The degree of enrichment for the NZB mtDNA in liver and kidney and against NZB mtDNA in the germline could then be influenced by mouse strain-specific genetic variation. One potentially relevant nDNA genetic variant is the loss of the nicotinamide nucleotide transhydrogenase gene (*Nnt*) on chromosome 13 in the Jackson Laboratory strain C57BL/6 J (Toye et al. 2005; Huang et al. 2006; Fielder et al. 2012). NNT is a mitochondrial inner membrane protein, which uses the potential energy of the mitochondrial electrochemical gradient to transfer reducing equivalents from NADH to NADPH. The higher reducing potential of NADPH is required for the reduction of mitochondrial lipid peroxides and of H_2O_2 to H_2O via the glutathione peroxidases. NADPH is also essential for the regulation of multiple enzymes and transcription factors via oxidation-reduction of thiol-disulfides (Wallace 2012). Because the Jackson Laboratory strain of C57BL/6 J, which lacks the *Nnt* gene was used in the NZB-129 heteroplasmy studies, it is possible that mitochondrial ROS production caused by the NZB $tRNA^{Arg}$ 9821 insA allele was enhanced on this strain background resulting in enhanced selection against PGCs or primary or mature oocytes with significant levels of the NZB mtDNA and increased ROS production.

Regulation of mtDNA Heteroplasmy by Mitochondrial Fusion and Fission

Other mitochondrial quality assurance mechanisms such as mitochondrial dynamics and mitochondrial autophagy (mitophagy) might also be important in mtDNA heteroplasmy segregation (Youle and van der Bliek 2012; Jokinen and Battersby 2013). The mitochondria in somatic tissues and cultured cells are highly dynamic undergoing a frequent fusion and fission cycle. Perturbations of mitochondrial fission and fusion have been shown to affect the segregation of heteroplasmic mtDNAs. Mitochondrial fission is initiated by Drp1, a member of the dynamin protein family. Mitochondrial fusion is initiated by the Mfn1 and Mfn2 proteins that mediate the fusion of the outer mitochondrial membranes and OpaI, which mediates the fusion of the mitochondrial inner membranes (Youle and van der Bliek 2012). In a muscle-derived rhabdomyosarcoma cultured cell line carrying 80% $tRNA^{Leu(UUR)}$ 3243A>G mtDNAs, RNAi knockdown of the mitochondrial fusion Drp1 and hFis1 gene mRNAs resulted in an average 13% and 11% increase in the mutant mtDNA over the wild type, respectively. By contrast, RNAi knockdown of Opa1 mRNA did not affect the mutant heteroplasmy level. Hence, in this replicating cell line, increased mitochondrial fusion was permissive for retention of mutant mtDNAs (Malena et al. 2009).

In mice in which the mitochondrial fusion genes, *Mfn1* and *Mfn2*, were knocked down in skeletal muscle via floxed *Mfn1* and *Mfn2* genes and the muscle-specific *MLC1f* promoter-driven Cre, profound mtDNA depletion ensued. At 7 to 8 weeks, the double mutant mice exhibited hyperproliferation of muscle mitochondria, mtDNA depletion to ~250 mtDNAs/nucleus versus ~3500 mtDNA/nucleus in normal mouse muscle, a 5-fold increase in mtDNA point mutations, and a 14-fold increase in mtDNA deletions. Comparison of the mtDNA deletion levels at 8–13 weeks revealed an 80-fold increase in mtDNA deletions in $Mfn^{-/-}$, $Mfn2^{+/-}$ mice (2.3×10^{-5}/genome versus 2.8×10^{-7}/genome). When the homozygous mtDNA polymerase γ mutator locus, $PolgA^{D257A}$, was combined with the $Mfn^{-/-}$ mutation, the animals manifested a profound OXPHOS defect with complex I activity (Chen et al. 2010). Hence, both mtDNA fission and fusion are important in maintaining mtDNA integrity and the segregation of mtDNAs within cells and tissues.

Regulation of mtDNA Heteroplasmy by Inter-mtDNA Complementation

Mitochodnrial fission and fusion permit mtDNA mutant complementation within cells. The mtDNAs are contained in nucleoids and the distribution of mtDNAs within nucleoids and the partitioning of nucleoids into daughter mitochondria could be important in heteroplasmy segregation. In nucleoids, the mtDNAs are complexed with numerous copies of the mtDNA-packaging protein and transcription factor, TFAM. Knockdown of TFAM in HeLa cells perturbs the distribution of mtDNAs into daughter nucleoids and mitochondria (Kasashima et al. 2011). The number of mtDNAs per nucleoid has been estimated to be as low as one (~1.4) (Kukat et al. 2011) and between two and 10 (Gilkerson et al. 2008; Poe et al. 2010; Rebelo et al. 2011). If mtDNA nucleoids contain more than one mtDNA and nucleoids could exchange mtDNAs, then nucleoids could be heteroplasmic, which would significantly affect the dynamics of mtDNA segregation.

To determine whether nucleoids are heteroplasmic, two cell lines each harboring a homoplasmic, nonoverlapping mtDNA deletion that deleted one or more tRNAs (FLPΔ nt 7846-9748 and CWΔ nt 10155-15945) were fused into somatic cell hybrids. The homoplasmic mtDNA deletion cells each had fragmented mitochondria and were deficient in mitochondrial protein synthesis and respiration. However, in the hybrids containing the two deleted mtDNAs the mitochondrial were elongated and functional in both protein synthesis and respiration. By using hybridization probes that were internal to the two deletions and differentially fluorescently labeled, the presence of the two deleted mtDNAs were monitored. This revealed that the respiring hybrids harbored both mtDNAs in elongated mitochondria and most of the nucleoids hybridized to either one or the other probe, though in some cases the two hybridization probes seemed to overlap. When the hybrid cells were released from selection of respiratory sufficiency (uridine) within 12 d, 55% of the cells had segregated to pure CWΔ mtDNA and 42% had segregated to pure FLPΔ, with only 3% retaining both mtDNAs. As the mtDNAs segregated the nucleoids resolved into either one mtDNA or the other. This indicated that nucleoids do not exchange mtDNAs, and that the instances where the deleted mtDNAs colocalized in the hybrids was the result of two nucleoids being adjacent to each other and thus not resolved by light microscopy (Gilkerson and Schon 2008; Gilkerson et al. 2008).

Further evidence that the nucleoids are clones of a single mtDNA genotype have come from the analysis of cultured cells that were heteroplasmic for a 7522 nt deletion, ~80% deletion. By fragmenting the mitochondria, sorting the pico-green-stained mitochondria in a cell sorter, and performing differential PCR amplification for the deleted and normal mtDNA, it was again concluded that each nucleoid contained only one type of mtDNA (Poe et al. 2010). Therefore, the current evidence favors the conclusion that individual nucleoids harbor only one type of mtDNA.

Regulation of mtDNA Heteroplasmy by Mitophagy

The mtDNAs continually replicate within somatic cells including postmitotic cells, with the excess mtDNAs being removed by mitophagy. Thus, mitophagy provides another mechanism by which heteroplasmic mtDNA mutations can be segregated. Mitophagy is preceded by mitochondrial fission, which is thought to be asymmetric such that one daughter mitochondrion is more energetically competent than the other. Mitophagy is then envisioned to preferentially degrade the more respiration-compromised mitochondrion. This process is mediated by Parkin and PTEN-induced kinase 1 (PINK1), encoded by the *Park2* and *Park6* loci, respectively. In healthy mitochondria, PINK1 is incorporated into the mitochondrion but is rapidly degraded by the mitochondrial PARL rhomboid protease. However, if the mitochondrial membrane potential is significantly reduced, PINK1 becomes stabilized in the mitochondrial outer membrane where it phosphorylates serines on target proteins. PINK1 phosphorylation results in the attraction of the cytosolic protein Parkin to the mitochondrial outer membrane. Parkin is an E3 ubiquitin ligase that ubiquitinates mitochondrial outer membrane target proteins (Narendra et al. 2008, 2009, 2010; Guo 2010; Youle and van der Bliek 2012). Recent evidence has implied that Mfn1, Mfn2, and Miro1 (a protein involved in mitochondrial transport along microtubules) are important Parkin ubiquitination targets. The Parkin modification of mitochondrial proteins is followed by the attraction of the p62/SQSTM1 protein to the mitochondrion, which commits the mitochondrion to LC3-II encapsulation, fusion of the resulting autophagosome with a lysosome, and degradation (Narendra et al. 2012).

The relevance of mitophagy to mtDNA heteroplasmy segregation was first demonstrated in Parkin overexpression experiments. Cybrid cells that were stably heteroplasmic for a high percentage of an mtDNA COI mutation (COX-ICA65) had a 55% reduction in mitochondrial inner membrane potential and showed an increase in Parkin-bound mitochondria. Partial inhibition of mitochondrial fusion with vMIA and of the electron transport chain with azide to reduce mitochondrial electron transport, increased the Parkin bound to the mitochondria. Overexpressed Parkin in these cells over a 60-d period resulted in the selective loss of the COI mutant mtDNAs. Hence, mitophagy is capable of modulating mtDNA heteroplasmy.

However, in a cell line heteroplasmic for an mtDNA cytochrome *b* mutation with a 37% reduction in mitochondrial membrane potential overexpression of Parkin did not result in a reduction in the percentage of mutant mtDNAs (Suen et al. 2010). Mitophagy has been found to be dependent on both Parkin expression and mTORC1 inhibition in cells homoplasmic for deleterious mtDNA mutations (Gilkerson et al. 2012).

Because of the multiplicity of cellular processes that can impinge on mtDNA heteroplasmy, it is not surprising that different mtDNA mutations show different directionality and kinetics of segregation in different tissues.

IMPLICATIONS FOR MEDICINE

Complexities of Genetic Counseling

The absolute nature of maternal inheritance of the mtDNA creates an intractable dilemma for women that harbor high levels of a deleterious mtDNA mutation. They have a high probability that most if not all of their children will develop devastating diseases. Homoplasmic mtDNA mutations such as those commonly associated with LHON will be transmitted to all of a woman's offspring, though fortunately the penetrance of these milder mtDNA disease mutations is incomplete. For more deleterious mtDNA mutations, the outcome can be much worse. In one published pedigree where the mother was heteroplasmic for the mtDNA $tRNA^{Thr}$ 15923A>G mutation she conceived and lost seven pregnancies. Five of the conceptions were lost prenatally. One boy was born and died at 42 h and another girl was born and died at 56 h of severe OXPHOS defects (Fig. 6) (Yoon et al. 1993).

Even more common are women with subclinical levels of a pathogenic mutation such as the $tRNA^{Leu(UUR)}$ 3243A>G or ATP6 8993T>G mutations but repeatedly transmit a high percentage of the deleterious mutant mtDNA to their offspring. Unfortunately, pre-

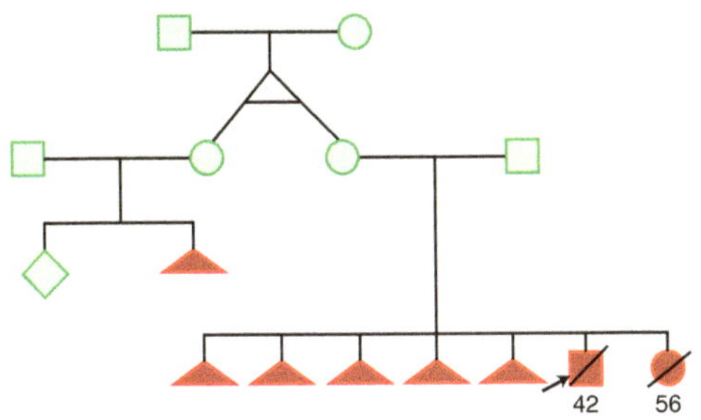

Figure 6. One published example of a germline lethal mtDNA mutation. A woman harboring a heteroplasmic mtDNA tRNA^{Thr} A15923G mutation lost five pregnancies in a row (triangles). The two surviving term infants died within 2 to 3 d. Postmortem mitochondrial analysis of one infant's skeletal muscle revealed that complex II + III and complex IV activities reduced to 6% and 5% of control and liver complex IV activity that was reduced 37% and complex II + III activity that was undetectable. The mother's twin sister had a similar reproductive history, yet both women were seemingly normal. (From Yoon et al. 1993; modified, with permission, from the authors.)

natal diagnostic procedures that have been effective in identifying fetuses with deleterious chromosomal gene mutations have proven to be relatively unreliable for diagnosing mtDNA diseases. These difficulties arise from the exclusively maternal inheritance of the mtDNA and the unpredictable nature of heteroplasmic mtDNA mutation segregation.

Obviously, all forms of prenatal diagnosis are useless for a woman harboring a homoplasmic mtDNA mutation because all of her fetuses will inherit her mutation. By contrast, heteroplasmic mtDNA mutations might be amenable to prenatal diagnosis if the woman was able to conceive a fetus with low levels of heteroplasmy. However, this can also be problematic because it is not certain that the fetal brain, heart, muscle, or kidney will have the same level of mtDNA mutation as found in the chorionic villus or amniocentesis sample. Furthermore, for most heteroplasmic mtDNA diseases, it is unclear what a "safe" level of heteroplasmy might be. Finally, some women might conceive multiple times and in every case the conceptus will have an unacceptably high level of mutant mtDNAs resulting in repeated fetal loss (Thorburn and Dahl 2001; Poulton and Bredenoord 2010).

Preimplantation Diagnosis. The ambiguities associated with chorionic villus sampling and amniocentesis diagnosis of mtDNA disease have led to investigation of the potential value of preimplantation genetic diagnosis (PGD). By determining the mtDNA heteroplasmic genotype of oocytes or embryos before fertilization or implantation, only those embryos with a desirable mtDNA genotype could be retained for implantation into the mother. Hence, many potential embryos could be screened with only the rare embryo with the optimal genotype permitted to go through gestation.

It would be particularly attractive if it were possible to remove the first polar body from an oocyte and determine its mtDNA genotype. Only those oocytes with the lowest polar body heteroplasmy level would then be utilized. Another possibility could be to fertilize the oocytes in vitro and then collect and genotype blastomeres or trophoblast cells to identify the low-heteroplasmy embryos.

The potential for PGD received significant initial support from a study of the mtDNA heteroplasmy genotypes of NZB-BALB mouse polar bodies, blastomeres, and embryos. In these studies, it was found that the mtDNA heteroplasmy level of the first polar body was within 0.1%–6.1% that of the rest of the embryo ($R^2 = 0.99$), as compared to the lack of correlation between the mother's mtDNA genotype and that of her embryos ($R^2 = 0.32$). Similarly, it was found that the mtDNA heteroplasmy of different blastomeres within the same embryo were also within 0%–6% of each other (Dean et al. 2003).

The positive results of the NZB-BALB heteroplasmy study were supported by studies on mice harboring the 4696 nt deleted mtDNA. Analysis of the second polar body following fertilization revealed that the polar body mtDNA genotype correlated with the percentage of deletion in the embryo with a coefficient of 0.95. Unfortunately, for mtDNA deletion mutations the benefits of PGD were compromised by the subsequent preferential replication of the deleted mtDNA during development. As mtDNA deletion-containing embryos developed, the percentage of deleted mtDNAs progressively increased, rising 17% during the 19 d after fertilization, 8% for the first 30 d after birth and before weaning, 8% for the first 100 d after weaning, and 6% for every subsequent 100 d. The maximum deletion level observed in any zygote was 78%, implying that the oogonia with higher levels of deletion did not yield oocytes. Ultimately, mtDNA deletion mice died of kidney failure, with the maximum deletion level observed in the kidneys being 90% (Sato et al. 2005). Because human mtDNA deletions have not been observed to be transmitted through the maternal lineage, the human female germ cells must have a lower tolerance for mitochondrial respiratory deficiency than the mouse (Wallace et al. 2013).

Based on the mouse polar body studies, sampling the mtDNA heteroplasmy levels of human meiosis I polar bodies would appear to be the most promising approach for PGD of heteroplasmic mtDNA diseases. Unfortunately, when human studies were conducted it was found that the mtDNA genotype of the first polar body might be very different from that of the embryo. The preimplantation embryos of three couples participating in assisted reproduction were studied, one each harboring the $tRNA^{Leu(UUR)}$ 3243A>G MELAS mutation, the $tRNA^{Lys}$ 8344A>G MERRF mutation, and the ATP6 9185T>C cerebellar ataxia and tubulopathy mutation. The maternal heteroplasmy levels were 25%, 75%–95%, and 14%–25%, respectively, in tissues. In only half of the embryos (27 of 51) were the polar body genotypes within ±10% of that of the embryo. In those cases where the polar body was homoplasmic for the normal mtDNA the embryo was also homoplasmic normal ($n = 7$). However, for half of the embryos the polar body genotype was significantly different from that of the embryo. In cases where the embryo's heteroplasmy levels were very high, the polar body genotype was significantly lower. For polar bodies with >60% mutant the polar body genotype was on the average 3.5% ± 5% higher than the embryo. For polar body genotypes that were <60%, the polar body heteroplasmy was 11.8% ± 7.8% lower than the embryo. Hence, it was concluded that in humans, polar body biopsy would not provide a reliable preimplantation test for human mtDNA diseases (Gigarel et al. 2011).

The second possibility could be sampling one of the blastomeres of an early cleavage stage embryo. Studies of rhesus macaque oocytes having a 50:50 heteroplasmy generated by karyoplast–cytoplast fusion, the heteroplasmy levels among primate blastomeres might be very different. In quantifying the divergence in percentage heteroplasmy, the coefficient of variance increased from 17.7% to 25.0% to 30.9% and the range increased from 13.3% to 25.3% to 43.2% in two-, four-, and eight-cell embryos, respectively. By the eight-cell stage, some embryo blastomeres differed by as much as 10% to 80%. Furthermore, the variance in heteroplasmy level among oocytes, blastomeres, and offspring was sufficiently great as to render absolute predictions on the transmission of heteroplasmy among cells and across generations unreliable (Lee et al. 2012).

In contrast to the macaque reconstituted cell studies, analysis of 3243A>G heteroplasmy levels in human embryos revealed that heteroplasmy levels of sister blastomeres at the two-cell stage were similar as were the heteroplasmy levels of embryos at 3 versus 5 d. Fetal extraembryonic or embryonic tissues of 12 fetuses from seven carrier mothers were also found to be relatively consistent from gestational d 120 to term. The heteroplasmy levels of the tissues of four fetuses were also found to be relatively uniform: 74% ± 0.7% (three tissues), 42% ± 0.8% (five tissues), 71% ± 2% (five tissues), and 78% ± 0.9% (seven tissues). Comparison of placental versus fetal tissue were also reported to harbor similar heteroplasmy levels: two cases encompassed tissue 3243A>G frequencies of 74%–75% and 78%, while a third ranged from 57% to 42%. While the pooled average of heteroplasmy levels between chorionic villi and amniotic fluid samples showed <10% variation, the heteroplasmy levels of individual trophoblast or amniocyte cells was quite variable. Thus, while there is marked variability in different embryos from the same woman, the tissues within an embryo can be relatively uniform even though the individual cells within a tissue can vary considerably (Monnot et al. 2011)

Initial studies have suggested that preimplantation embryo blastomeres have relatively similar mtDNA heteroplasmy levels, in contrast to the macaque report. A 30-year-old female carrier with a 35% 3243A>G mtDNA mutation heteroplasmy requested PGD after having a daughter diagnosed at 2 years of age with MELAs syndrome and 84% 3243A>G mutant mtDNAs. Eight unfertilized oocytes displayed mutation loads ranging from 9% to 90% and eight arrested embryos encompassing cleavage and morula stages had mutation loads ranging from 7% to 91%. Comparison of blastomere genotypes from five cleavage-stage embryos with mean heteroplasmy levels of 24.8% to 85.5% revealed standard deviations ranging from 0.5% to 2.9% (Treff et al. 2012).

Multiple additional studies have been conducted on the potential of PGD for women that are heteroplasmic for the MELAS $tRNA^{Leu(UUR)}$ 3243A>G and the NARP/Leigh syndrome ATP6 8993T>G mutations. In a study of three 8993T>G NARP and 18 3243A>G MELAS preimplantation embryos, the heteroplasmy levels were reported as stable among the different blastomeres. Another study of 10 8993T>G NARP and nine 3243A>G MELAS fetuses concluded that the various tissues had similar heteroplasmy levels and that the heteroplasmy levels did not change with gestational age. From this series, eight children carrying less than 30% mutant mtDNA in the prenatal period were born and appear healthy at 18 months to 9 years of age (Monnot et al. 2009b).

In a second series of 3243A>G and 8993T>G patients, 33 preimplantation embryos at 3 to 5 d and 25 fetuses at gestational age of 10 to 30 wk were analyzed from 16 unrelated carrier females. Twenty percent of the embryos lacked the mtDNA mutation. Of the 80% that had the mutation, the percentage of heteroplasmy did not differ significantly among blastomeres of the embryos. While the NARP embryos tended to have either <25% or >95% mutant mtDNAs, the MELAS embryos harbored a wide range of heteroplasmy levels from 0% to 80%. Among the fetuses, those harboring the NARP mutation tended to have either <30% or >65% mutant mtDNAs while the MELAS embryos again had a continuous distribution of heteroplasmy levels, though different tissue samples from the same embryo had similar heteroplasmy levels (Monnot et al. 2009a).

A third cohort of 8993T>G and 3243A>G patients concluded that for 8993T>G mutational loads fetuses with <60% mutant could have a "favorable" outcome, but for mutational loads of >90% there was a poor prognosis. By contrast the clinical correlation between genotype and phenotype for the 3243A>G mutation was significantly less clear. In general, it was stated that "assessment of mtDNA mutation load in chorionic villi, amniotic cells, or blastocytes ... show that the mutation percentage is stable at different times or in different cells of one embryo" (de Die-Smulders and Smeets 2009).

PGC was performed on 13 couples harboring the 8993T>G heteroplasmy mutation encompassing 20 embryos and 123 blastomeres. Of the 20 embryos, five had detectable 8993T>G heteroplasmy. The heteroplasmy levels of the blastomeres of each embryo were quite similar with ranges of 0%–8%, 4%–15%, 4%–15%, 9%–20%, 9%–20%, 10%–12%, 11%–22%, 18%–21%, 10%–12%, and 18%–21%. In contrast to the similarity of heteroplasmy levels of blastomeres within an embryo, the heteroplasmy levels among embryos varied considerably. Interestingly, the heteroplasmy levels of individual lymphocytes from a subject with a 44.3% whole blood heteroplasmy proved to be highly variable, ranging from 11% to 70% (Tajima et al. 2007).

For a healthy woman with 35% of the 8993T>G mutation in her blood and who had a daughter who died at 2 years with an mtDNA mutation load of >95%, chorionic villus sampling of two subsequent pregnancies revealed >95% 8993T>G mutation and were terminated. The woman then underwent two rounds of in vitro fertilization in association with PGD. Fifty-nine cells from 10 embryos were analyzed and the mutation load found to be consistent within the blastomeres of the individual embryos. Ten embryos and seven oocytes that were tested, 12 (70%) had very high mutation loads. Of the 10 embryos, six had high mutation loads, three had 30%–40% heteroplasmy, but one had a heteroplasmy level of 2.4%. This embryo was implanted and the pregnancy resulted in a healthy girl with a cord blood mutation load of 4% (Thorburn et al. 2009).

Therefore, in contrast to the macaque studies, the heteroplasmy levels of different blastomeres of human embryos appear to have relatively similar mtDNA heteroplasmy levels. Hence, blastomere biopsy and mtDNA heteroplasmy analysis from eight-cell embryos may provide a reasonable estimate of the heteroplasmy levels in the remaining blastomeres and reflect the heteroplasmy levels in the ensuing pregnancy amniocytes and cord blood samples (Poulton and Bredenoord 2010).

While mtDNA analysis of blastomere biopsies appears to be a promising approach for determining the heteroplasmy level of preimplantation embryos, blastomere biopsy removes a significant proportion of the embryo. An alternative could be to biopsy extraembryonic cells in the trophectoderm of the blastocyst. In one published case, a 30-year-old woman harboring 35% 3243A>G requested PGD following birth of a MELAS syndrome daughter. Six developmentally competent blastocysts with mean heteroplasmy values ranging from 24% to 91% were sampled. Three to four trophectoderm cells were obtained and compared to one to three inner cell mass samples. The standard deviations of the independent samples from the various trophoblasts ranged from 2.1% to 5.0% and the means of the trophoblast and inner cell mass samples were all within 3% of each other. Based on this data, one developmentally competent embryo with trophoblast heteroplasmy levels of 12% was chosen for implantation. This resulted in the successful delivery of a child for which a buccal cell DNA analysis at 1 mo revealed a heteroplasmy level of 15%. Subsequent tests at 5 and 12 mo of buccal swab and urine sediment DNA heteroplasmy by a commercial laboratory reported <10% mutation load (Treff et al. 2012). However, subsequent follow-up of this case has revealed significant problems in the placenta, infant, and child, and a mutation load in the child's blood and urine of 42%–52% (MJ Falk, pers. comm.).

Hence, there are still important ambiguities in the results on the effectiveness of human PGD in identifying and avoiding heteroplasmic mtDNA disease mutations. One recently recognized variable that might be relevant is that different pathogenic mtDNA mutations behave differently in both tissues and preimplantation embryos. Following in vitro fertilization embryos with mtDNA mutant genotypes of 70% 3243A>G, 90% 8344A>G, and 54% 9185T>G were not adversely affected in preimplantation development. However, the mtDNA copy number of 3243A>G embryos was found to increase from the germinal vesicle stage to the blastocyts ~2.7-fold, with the amount of the increase correlating with the percent heteroplasmy (R^2 = 0.47). Hence, for the 3243G mutation, the mtDNA copy number appears to partially compensate for the mtDNA defect. By contrast, preimplantation embryos harboring the 8344A>G mutation did not show any increase in mtDNA copy number, even though the mtDNA mutation load of the 8344A>G embryos was higher than that of the 3243A>G embryos: oocyte and embryo heteroplasmy levels for the 8344A>G were 81% ± 15% and 68% ± 17% versus 37% ± 29% and 36% ± 30% for the 3243A>G. Thus, the 3243A>G mutation must generate different physiological signals than the 8344A >G mutation (Monnot et al. 2013). How such mutation-specific effects have affected the reliability of PGD remains to be determined.

Germline Gene Therapy

While the PGD by blastomere sampling seems promising for determining the heteroplasmy levels of the 8993T>G NARP/Leigh syndrome mutation and possibly the 3243A>G MELAS mutation, there are still major limitations of this method. First, a woman may not produce oocytes that have low levels of mutant mtDNA, and without the appropriate oocyte her in vitro fertilization regime could be unproductive. Second, while the heteroplasmic levels of the ATP6 8993T>G L126R mutation may be consistent throughout the embryo and fetus, the potential for intertissue differences in heteroplasmy levels for the 3243A>G mutation may be significantly higher. Third, PGD is useless for homoplasmic women. Because of these limitations, there is increasing interest in developing methods to simply replace the mutant mtDNAs in preimplantation embryos by pronuclear or spindle transfer.

One early attempt at altering the mtDNA genotype in human oocytes was ooplasmic transfer. In this technique ~20% of the cytoplasm of an oocyte from a one woman is injected into an oocyte from another woman along with her husband's sperm (intracytoplasmic sperm injection, ICSI), and the embryos implanted into a woman's uterus. This technique was first used in an attempt to rejuvenate the oocytes of older women with the cytoplasm from the oocytes of younger women thus increasing their fertility. Multiple children have been born following application of this procedure and are alive today. Some of these children have been shown to be heteroplasmic for both the maternal oocyte mtDNA and the cytoplasmic donor mtDNA (Barritt et al. 2001a,b).

Given that it has been reported that the mixing of two different mouse mtDNAs within the same female germline can lead to offspring with neuro-psychiatric defects (Sharpley et al. 2012), concern has been raised that randomly mixing mtDNAs of different mtDNA lineages might be deleterious. Also, the addition of 20% normal mtDNAs would be unlikely to be sufficient to reduce the risk of the child's developing the mtDNA disease.

Rather than the transfer of mitochondria and mtDNAs, a more promising approach would be to isolate the nucleus from the oocyte or zygote of a woman harboring a deleterious mtDNA mutation and to transfer it into an enucleated oocyte or zygote from a woman with normal mtDNAs (Wallace 1987). As a model for pronuclear transfer, in mouse embryos for the heteroplasmic mtDNA 4696 nt deletion, the zygote nuclei were removed by micropipette from the mtDNA mutant zygotes following disaggregation of the cytoskeleton and the resulting karyoplast (plasma membrane bound pronuclei) fused to an enucleated oocyte by electroshock. About 6% of the mtDNA was carried along with the nucleus. Because the average mutant mtDNA in the karyoplast donor was 35%, this resulted in zygotes containing ~2% deleted mtDNAs. Eleven mice were born and at weaning their tail heteroplasmy levels were found to be 6%–12% mutant, mean 11%. Hence, zygote nuclear transfer resulted in a significant reduction in the transmission of the deleted mtDNAs. Unfortunately, in the case of the mtDNA deletion, the deletion levels progressively increased over the next 300 d to 5%–44% mutant, mean 23% (Sato et al. 2005).

Pronuclear transfer technology has also been extended to human preimplantation embryos. Using uni-pronuclear or tri-pronuclear human embryos that otherwise would have been destroyed, the cytoskeleton was disrupted with cytochalasin B and nocodazole, the pronuclei pinched off into a karyoplast, the karyoplast placed under the zona pellucida of an enucleated recipient zygote, and the two cells fused together with inactivated viral envelope proteins. About 22% of the resulting reconstituted embryos developed to the eight-cell stage and of those that received two pronuclei, 8.3% developed to the blastocyst stage, ~50% of the developmental potential of unmanipulated embryos. Using mtDNA control region polymorphisms to monitor the fate of the two mtDNAs, the mean mtDNA carryover along with the pronuclear karyoplast was 8.1% ± 7.6% ($n = 8$). Variation in the percentage of mutant mtDNAs in the different blastomeres within one eight-cell embryo was found to range from 5.1% to 30.2% and within another four-cell embryo to range from 8% to 39%. Further refinement of the karyoplast isolation technique resulted in nine embryos with an average carryover of karyoplast mtDNAs of 1.68% ± 1.81% ($n = 9$). Four of these embryos lacked detectable karyoplast mtDNA. Of those embryos that harbored karyoplast mtDNA the maximum donor mtDNA was found in a nine-cell embryo in which seven blastomeres lacked detectable donor mtDNAs, while the remaining two blastomeres harbored 6.1% and 11.4% donor mtDNAs (Craven et al. 2010).

Interestingly, the spread in blastomere heteroplasmy levels in the first round pronuclear transfer embryos (Craven et al. 2010) was reminiscent of the wide range of blastomere mtDNA heteroplasmy genotypes reported for the macaque embryo fusion experiments. This was much greater than the range of heteroplasmy levels observed in embryos generated from NZB/BALB heteroplasmic mice or in the blastomeres of heteroplasmic human preimplantation embryos. One possible explanation for this difference is that the oocyte cytoplasm may be relatively viscous resulting in the nonrandom partitioning of the mitochondria in different regions of the single-cell embryo. Depending on the angles of the initial cleavages, the mtDNAs of the karyoplast and the cytoplast could be asymmetrically distributed into daughter blastomeres resulting in differing heteroplasmy levels. By contrast, oocytes and embryos derived from heteroplasmic germline cells of the two mtDNAs could be more evenly distributed throughout the cytoplasm such that cleavages would generate blastomeres with similar mtDNA heteroplasmy levels.

One concern with the pronuclear transfer procedure is that it destroys an embryo. This concern has been overcome by transfer of the chromosomes from the pre-fertilization mtDNA mutant oocyte to an enucleated oocyte with normal mtDNA. In this case, however, the chromosomes of the mature oocyte are arrested in meiotic metaphase II. Hence, this method involves the physical transfer of the meiotic II spindle.

In the spindle transfer technique, the metaphase II (MII) spindle of the oocyte is visualized in the oocytes using the polarizing microscope and is removed by aspiration with a micropipette within a bleb of the ooycte cytoplasm and surrounded by the cell membrane. The spindle karyoplasts encompass ~1.5% of the volume of an enucleated oocyte cytoplast. Because the spindle is relatively free of surrounding cytoplasm and mitochondria very little mtDNA is transferred. The karyoplast is treated with a Sendai virus-derived fusigen and placed in the perivitelline space of the cytoplast opposite the first polar body where it fuses. The resulting reconstituted oocyte can then be fertilized by intracytoplasmic sperm injection (ICSI).

This technology was first developed and applied to the oocytes of rhesus macaque monkeys. It resulted in preimplantation embryos that developed normally to generate pluripotent stem cells and when blastocysts of four- to eight-cell cleavage embryos were transferred into the reproductive tract of females, one pair of twins and two singleton infants were born. These four monkeys had the maternal chromosomes of the spindle donor but the mtDNA of the cytoplast donor. No spindle donor oocyte mtDNA was detected at a detection sensitivity of ~3% (Tachibana et al. 2009). Three years after birth, the four-spindle transfer monkeys were phenotypically like control monkeys and showed no significant change in mtDNA carryover in blood and skin samples (Tachibana et al. 2013). Subsequent analysis revealed that the macaque karyoplasts carried 3538 ± 2310 mtDNAs while the cytoplast contributed 576,948 ± 209,069 mtDNA ($n = 9$) resulting in ~0.6% heteroplasmy. Of 102 spindle transfer oocytes 62% developed into blastocyts. Two female embryos generated from spindle transfer oocytes were permitted to develop in utero for 135 d and then analyzed for mtDNA heteroplasmy. Of 24 oocytes isolated from the ovaries of the female fetuses 22 had no detectable spindle donor mtDNA while one from each fetus was heteroplasmic with 16% and 14% heteroplasmy, respectively (Lee et al. 2012).

One of the technological limitations of spindle transfer is that both the spindle donor karyoplast and the mtDNA donor cytoplast must be ready at the same time, requiring paired induction of ovulation of two women, which can be quite difficult. Studies on macaque oocytes have shown that the spindle donor oocyte can be frozen using a vitrification procedure. These oocytes can then be thawed, the spindle removed, and transferred to a fresh enucleated oocyte without diminution of fertilization. Blastocyst formation efficiency was 88% and 68% of that of controls. Vitrification of the cytoplasm donor oocytes, by contrast, blocked development (Tachibana et al. 2013). Hence, spindle transfer has the potential of eliminating most of the mutant mtDNAs from the maternal lineage, but low levels of heteroplasmy can still resurface during germline transmission.

To assess the feasibility of moving this technology into the clinic, spindle transplantation has been applied to human oocytes. Of 64 spindle-transfer oocytes, 94% survived ICSI and 75% formed pronuclei. However, in contrast to macaque results only about half (48%) formed two normal pronuclei and two polar bodies. The remainders were abnormal. Only 13% of the nonspindle transfer embryos were abnormal. Of the normal spindle transfer embryos, 76% went on to form blastocysts. From the 16 blastocysts, nine formed ESCs (56%) and these were indistinguishable from the controls. The mean karyoplast derived mtDNA levels were very low, 0.5% ± 04% for spindle transfer oocytes and embryos and 0.6% ± 0.9% for human ESCs (Tachibana et al. 2013).

In an independent study using parthenogenetically activated embryos that have undergone endoreduplication to establish a diploid karyotype, spindle transfer was also analyzed for its capacity to replace the mtDNA of the spindle donor mother. As before, the spindle was removed from the oocyte in a bleb of cytoplasm, the resulting karyoplast placed under the zona pellucida of a comparably enucleated oocyte, and the karyoplast fused to the cytoplast by either Sendai viral fusigen or electric shock. Sendai fusion proved preferable because electroshock was prone to induce premature oocyte activation. This study determined that the number of mtDNAs carried by the karyoplast was 1129 ± 785 or 0.36% of the total mtDNA in a metaphase II oocyte, which contains 311,146 ± 206,521 mtDNAs. An important innovation of this study was the realization that partial disaggregation of the karyoplast spindle complex by a 2-h exposure to room temperature or prior vitrification followed by partial reaggregation of the spindle at 37°C significantly increased the efficiency of subsequent second polar body extrusion and embryo development. Several embryos were permitted to develop into blastocysts and from these ESC lines with normal karyotypes were derived. These were capable of differentiating into pancreatic cells, neurons, fibroblasts, and cardiomyocytes. Analysis of the mtDNA heteroplasmy level for all preimplantation embryos studied was 0.31% ± 0.27%. Of the three ESC lines that were generated, two lacked detectable karyoplast mtDNA while one had 2.79% ± 0.27% karyoplast mtDNA. The heteroplasmic ESC line subsequently lost the karyoplast mtDNA, the spindle donor mtDNAs becoming undetectable after passage 14 and remained so in 40 clones. Fibroblasts derived from one of the spindle transfer stem cell lines were converted back to induced pluripotent stem (iPS) cells, yet the mtDNAs from the karyoplast did not reappear. Finally, analysis of the mitochondrial respiratory complexes, respiration, and media acidification of one of the stem cell lines revealed that it was comparable to that of embryonic stem cells and iPS cells that had not undergone spindle transfer (Paull et al. 2013).

While both the pronuclear and spindle transfer techniques have been refined to transfer a minute amount of nuclear donor mtDNA, the possibility still remains that heteroplasmy can resurface in subsequent ESCs or maternal offspring (Paull et al. 2013). While low-level heteroplasmy of a pathogenic mutation would probably be masked by the predominance of normal mtDNAs, minimizing the risk of a clinical phenotype, there is a risk of incompatibility between the two mtDNA haplogroup lineages (Sharpley et al. 2012). This concern could be mitigated by screening potential oocyte mtDNA donor women for their mtDNA haplogroups. Then the mtDNA haplogroup of the woman harboring the deleterious mtDNA mutation could be matched with that of an oocyte donor; only matching mtDNA haplogroups would then be used.

Alternatives and Ethical Considerations

It has been argued that it may be unethical to manipulate human embryos in an effort to remove the risk of mtDNA disease. Certainly, all appropriate preclinical tests must be performed in an effort to reduce the risk for adverse outcomes in developing new human therapies. The question remains is human mtDNA germline gene therapy "ethical." To answer this question, we must factor into our cost-benefit analysis the implications not only for society but also for the family and their prospective children. Women harboring severely pathogenic mtDNA mutations want to have healthy children that are free of pain and can live productive lives. Currently children with high heteroplasmy levels of severely deleterious mtDNA mutations experience the progressive loss of mental and physical capabilities, often experiencing unremitting discomfort and pain and many will ultimately progress to premature death. For such families, this can mean repeated medical crises, numerous emergency room admissions, devastating medical bills, the loss of time, affection, and resources for other family members, ending only by the death of the child. Therefore, if the society is to rule on the morality of helping women have healthy children, than the families who have suffered the ravages of mtDNA disease and personally paid the cost should be given a major voice in the decision.

ACKNOWLEDGMENTS

This work was supported by NIH grants NS21328, NS070298, AG24373, and DK73691 and by Simons Foundation grant 205844 awarded to D.C.W.

REFERENCES

*Reference is also in this collection.

Ashley MV, Laipis PJ, Hauswirth WW. 1989. Rapid segregation of heteroplasmic bovine mitochondria. *Nucleic Acids Res* **17:** 7325–7331.

Babu GJ, Feigelson ED. 2006. Astrostatistics: Goodness-of-fit and all that! *Astronomical Data Analysis Software and Systems XVASP Conference Series* **351:** 127.

Ballinger SW, Shoffner JM, Hedaya EV, Trounce I, Polak MA, Koontz DA, Wallace DC. 1992. Maternally transmitted diabetes and deafness associated with a 10.4 kb mitochondrial DNA deletion. *Nat Genet* **1:** 11–15.

Ballinger SW, Shoffner JM, Gebhart S, Koontz DA, Wallace DC. 1994. Mitochondrial diabetes revisited. *Nat Genet* **7:** 458–459.

Balloux F, Handley LJ, Jombart T, Liu H, Manica A. 2009. Climate shaped the worldwide distribution of human mitochondrial DNA sequence variation. *Proc R Soc B* **276:** 3447–3455.

Barritt J, Willadsen S, Brenner C, Cohen J. 2001a. Cytoplasmic transfer in assisted reproduction. *Hum Reprod Update* **7:** 428–435.

Barritt JA, Brenner CA, Malter HE, Cohen J. 2001b. Mitochondria in human offspring derived from ooplasmic transplantation. *Hum Reprod* **16:** 513–516.

Battersby BJ, Shoubridge EA. 2001. Selection of a mtDNA sequence variant in hepatocytes of heteroplasmic mice is not due to differences in respiratory chain function or efficiency of replication. *Hum Mol Genet* **10:** 2469–2479.

Battersby BJ, Shoubridge EA. 2007. Reactive oxygen species and the segregation of mtDNA sequence variants. *Nat Genet* **39:** 571–572; author reply 572.

Battersby BJ, Loredo-Osti JC, Shoubridge EA. 2003. Nuclear genetic control of mitochondrial DNA segregation. *Nat Genet* **33:** 183–186.

Battersby BJ, Redpath ME, Shoubridge EA. 2005. Mitochondrial DNA segregation in hematopoietic lineages does not depend on MHC presentation of mitochondrially encoded peptides. *Hum Mol Genet* **14:** 2587–2594.

Baudouin SV, Saunders D, Tiangyou W, Elson JL, Poynter J, Pyle A, Keers S, Turnbull DM, Howell N, Chinnery PF. 2005. Mitochondrial DNA and survival after sepsis: A prospective study. *Lancet* **366:** 2118–2121.

Bergstrom CT, Pritchard J. 1998. Germline bottlenecks and the evolutionary maintenance of mitochondrial genomes. *Genetics* **149:** 2135–2146.

Blanc H, Adams CW, Wallace DC. 1981a. Different nucleotide changes in the large rRNA gene of the mitochondrial DNA confer chloramphenicol resistance on two human cell lines. *Nucleic Acids Res* **9:** 5785–5795.

Blanc H, Wright CT, Bibb MJ, Wallace DC, Clayton DA. 1981b. Mitochondrial DNA of chloramphenicol-resistant mouse cells contains a single nucleotide change in the region encoding the 3′ end of the large ribosomal RNA. *Proc Natl Acad Sci* **78:** 3789–3793.

Blok RB, Gook DA, Thorburn DR, Dahl HH. 1997. Skewed segregation of the mtDNA nt 8993 (T→G) mutation in human oocytes. *Am J Hum Genet* **60:** 1495–1501.

Brown WM, Prager EM, Wan A, Wilson AC. 1982. Mitochondrial DNA sequences in primates: Tempo and mode of evolution. *J Mol Evol* **18:** 225–239.

Brown MD, Torroni A, Reckord CL, Wallace DC. 1995. Phylogenetic analysis of Leber's hereditary optic neuropathy mitochondrial DNA's indicates multiple independent occurrences of the common mutations. *Hum Mutat* **6:** 311–325.

Brown MD, Sun F, Wallace DC. 1997. Clustering of Caucasian Leber hereditary optic neuropathy patients containing the 11778 or 14484 mutations on an mtDNA lineage. *Am J Hum Genet* **60:** 381–387.

Brown DT, Samuels DC, Michael EM, Turnbull DM, Chinnery PF. 2001. Random genetic drift determines the level of mutant mtDNA in human primary oocytes. *Am J Hum Genet* **68:** 533–536.

Brown MD, Starikovskaya E, Derbeneva O, Hosseini S, Allen JC, Mikhailovskaya IE, Sukernik RI, Wallace DC. 2002. The role of mtDNA background in disease expression: A new primary LHON mutation associated with Western Eurasian haplogroup J. *Hum Genet* **110:** 130–138.

Bunn CL, Wallace DC, Eisenstadt JM. 1974. Cytoplasmic inheritance of chloramphenicol resistance in mouse tissue culture cells. *Proc Natl Acad Sci* **71:** 1681–1685.

Bunn CL, Wallace DC, Eisenstadt JM. 1977. Mitotic segregation of cytoplasmic determinants for chloramphenicol resistance in mammalian cells. I: Fusions with mouse cell lines. *Somatic Cell Genet* **3:** 71–92.

Cao L, Shitara H, Horii T, Nagao Y, Imai H, Abe K, Hara T, Hayashi J, Yonekawa H. 2007. The mitochondrial bottleneck occurs without reduction of mtDNA content in female mouse germ cells. *Nat Genet* **39:** 386–390.

Cao L, Shitara H, Sugimoto M, Hayashi J, Abe K, Yonekawa H. 2009. New evidence confirms that the mitochondrial bottleneck is generated without reduction of mitochondrial DNA content in early primordial germ cells of mice. *PLoS Genet* **5:** e1000756.

Carling PJ, Cree LM, Chinnery PF. 2011. The implications of mitochondrial DNA copy number regulation during embryogenesis. *Mitochondrion* **11:** 686–692.

Chapman R, Stephens J, Lansman R, Avise J. 1982. Models of mitochondrial DNA transmission genetics and evolution in higher eucaryotes. *Genet Res* **40:** 41–57.

Charlesworth B, Charlesworth D. 2010. *Elements of evolutionary genetics*. Roberts and Company, Greenwood Village, CO.

Chen H, Vermulst M, Wang YE, Chomyn A, Prolla TA, McCaffery JM, Chan DC. 2010. Mitochondrial fusion is required for mtDNA stability in skeletal muscle and tolerance of mtDNA mutations. *Cell* **141:** 280–289.

Chinnery PF, Elliott HR, Syed A, Rothwell PM. 2010. Mitochondrial DNA haplogroups and risk of transient ischaemic attack and ischaemic stroke: A genetic association study. *Lancet Neurol* **9:** 498–503.

Chinnery PF, Elliott HR, Hudson G, Samuels DC, Relton CL. 2012. Epigenetics, epidemiology and mitochondrial DNA diseases. *Int J Epidemiol* **41:** 177–187.

Chomyn A. 1996. Platelet-mediated transformation of human mitochondrial DNA-less cells. *Methods Enzymol* **264:** 334–339.

Coskun PE, Wyrembak J, Derbereva O, Melkonian G, Doran E, Lott IT, Head E, Cotman CW, Wallace DC. 2010. Systemic mitochondrial dysfunction and the etiology of Alzheimer's disease and Down syndrome dementia. *J Alzheimers Dis* **20:** S293–S310.

Craven L, Tuppen HA, Greggains GD, Harbottle SJ, Murphy JL, Cree LM, Murdoch AP, Chinnery PF, Taylor RW, Lightowlers RN, et al. 2010. Pronuclear transfer in human embryos to prevent transmission of mitochondrial DNA disease. *Nature* **465:** 82–85.

Cree LM, Samuels DC, de Sousa Lopes SC, Rajasimha HK, Wonnapinij P, Mann JR, Dahl HH, Chinnery PF. 2008. A reduction of mitochondrial DNA molecules during embryogenesis explains the rapid segregation of genotypes. *Nat Genet* **40:** 249–254.

Dean NL, Battersby BJ, Ao A, Gosden RG, Tan SL, Shoubridge EA, Molnar MJ. 2003. Prospect of preimplantation genetic diagnosis for heritable mitochondrial DNA diseases. *Mol Hum Reprod* **9:** 631–638.

de Die-Smulders C, Smeets H. 2009. The challenge of prenatal and preimplantation genetic diagnosis of mitochondrial disorders. European Human Genetics Conference, Abstract S09.3, Vienna Austria, May 23–26, 2009. *Eur J Hum Genet* **17:** 9–10.

de Stordeur E, Solignac M, Monnerot M, Mounolou JC. 1989. The generation of transplasmic *Drosophila simulans* by cytoplasmic injection: Effects of segregation and selection on the perpetuation of mitochondrial DNA heteroplasmy. *Mol Gen Genet* **220:** 127–132.

Elliott HR, Samuels DC, Eden JA, Relton CL, Chinnery PF. 2008. Pathogenic mitochondrial DNA mutations are common in the general population. *Am J Hum Genet* **83:** 254–260.

Fan W, Waymire K, Narula N, Li P, Rocher C, Coskun PE, Vannan MA, Narula J, MacGregor GR, Wallace DC. 2008. A mouse model of mitochondrial disease reveals germline selection against severe mtDNA mutations. *Science* **319:** 958–962.

Fan W, Lin CS, Potluri P, Procaccio V, Wallace DC. 2012. MtDNA lineage analysis of mouse L cell lines reveals the accumulation of multiple mtDNA mutants and intermolecular recombination. *Genes Dev* **26:** 384–394.

Fielder TJ, Yi CS, Masumi J, Waymire KG, Chen HW, Wang S, Shi KX, Wallace DC, MacGregor GR. 2012. Comparison of male chimeric mice generated from microinjection of JM8.N4 embryonic stem cells into C57BL/6 J and C57BL/6NTac blastocysts. *Transgenic Res* **21:** 1149–1158.

Freyer C, Cree LM, Mourier A, Stewart JB, Koolmeister C, Milenkovic D, Wai T, Floros VI, Hagstrom E, Chatzidaki EE, et al. 2012. Variation in germline mtDNA heteroplasmy is determined prenatally but modified during subsequent transmission. *Nat Genet* **44:** 1282–1285.

Fuku N, Park KS, Yamada Y, Nishigaki Y, Cho YM, Matsuo H, Segawa T, Watanabe S, Kato K, Yokoi K, et al. 2007. Mitochondrial haplogroup N9a confers resistance against type 2 diabetes in Asians. *Am J Hum Genet* **80:** 407–415.

Gasparre G, Hervouet E, de Laplanche E, Demont J, Pennisi LF, Colombel M, Mege-Lechevallier F, Scoazec JY, Bonora E, Smeets R, et al. 2008. Clonal expansion of mutated mitochondrial DNA is associated with tumor formation and complex I deficiency in the benign renal oncocytoma. *Hum Mol Genet* **17:** 986–995.

Gigarel N, Hesters L, Samuels DC, Monnot S, Burlet P, Kerbrat V, Lamazou F, Benachi A, Frydman R, Feingold J, et al. 2011. Poor correlations in the levels of pathogenic mitochondrial DNA mutations in polar bodies versus oocytes and blastomeres in humans. *Am J Hum Genet* **88:** 494–498.

Giles RE, Stroynowski I, Wallace DC. 1980. Characterization of mitochondrial DNA in chloramphenicol-resistant interspecific hybrids and a cybrid. *Somatic Cell Genet* **6:** 543–554.

Gilkerson RW, Schon EA. 2008. Nucleoid autonomy: An underlying mechanism of mitochondrial genetics with therapeutic potential. *Commun Integr Biol* **1:** 34–36.

Gilkerson RW, Schon EA, Hernandez E, Davidson MM. 2008. Mitochondrial nucleoids maintain genetic autonomy but allow for functional complementation. *J Cell Biol* **181:** 1117–1128.

Gilkerson RW, De Vries RL, Lebot P, Wikstrom JD, Torgyekes E, Shirihai OS, Przedborski S, Schon EA. 2012. Mitochondrial autophagy in cells with mtDNA mutations results from synergistic loss of transmembrane potential and mTORC1 inhibition. *Hum Mol Genet* **21:** 978–990.

Gomez R, O'Keeffe T, Chang L, Huebinger R, Minei J, Barber R. 2009. Association of mitochondrial allele 4216C with increased risk for complicated sepsis and death after traumatic injury. *J Trauma* **66:** 850–857; discussion 857–858.

Gomez-Duran A, Pacheu-Grau D, Lopez-Gallardo E, Diez-Sanchez C, Montoya J, Lopez-Perez MJ, Ruiz-Pesini E. 2010. Unmasking the causes of multifactorial disorders: OXPHOS differences between mitochondrial haplogroups. *Hum Mol Genet* **19:** 3343–3353.

Goto Y, Nonaka I, Horai S. 1990. A mutation in the $tRNA^{Leu(UUR)}$ gene associated with the MELAS subgroup of mitochondrial encephalomyopathies. *Nature* **348:** 651–653.

Goto H, Dickins B, Afgan E, Paul IM, Taylor J, Makova KD, Nekrutenko A. 2011. Dynamics of mitochondrial heteroplasmy in three families investigated via a repeatable re-sequencing study. *Genome Biol* **12:** R59.

Guo M. 2010. What have we learned from *Drosophila* models of Parkinson's disease? *Prog Brain Res* **184:** 3–16.

Hauswirth WW, Laipis PJ. 1982. Mitochondrial DNA polymorphism in a maternal lineage of Holstein cows. *Proc Natl Acad Sci* **79:** 4686–4690.

Holt IJ, Harding AE, Morgan-Hughes JA. 1988. Deletions of muscle mitochondrial DNA in patients with mitochondrial myopathies. *Nature* **331:** 717–719.

Holt IJ, Harding AE, Petty RK, Morgan-Hughes JA. 1990. A new mitochondrial disease associated with mitochondrial DNA heteroplasmy. *Am J Hum Genet* **46:** 428–433.

Huang TT, Naeemuddin M, Elchuri S, Yamaguchi M, Kozy HM, Carlson EJ, Epstein CJ. 2006. Genetic modifiers of the phenotype of mice deficient in mitochondrial superoxide dismutase. *Hum Mol Genet* **15:** 1187–1194.

Inoue K, Nakada K, Ogura A, Isobe K, Goto Y, Nonaka I, Hayashi J-I. 2000. Generation of mice with mitochondrial dysfunction by introducing mouse mtDNA carrying a deletion into zygotes. *Nat Genet* **26:** 176–181.

Jansen RP, de Boer K. 1998. The bottleneck: Mitochondrial imperatives in oogenesis and ovarian follicular fate. *Mol Cell Endocrinol* **145:** 81–88.

Jenuth JP, Peterson AC, Fu K, Shoubridge EA. 1996. Random genetic drift in the female germline explains the rapid segregation of mammalian mitochondrial DNA. *Nat Genet* **14:** 146–151.

Jenuth JP, Peterson AC, Shoubridge EA. 1997. Tissue-specific selection for different mtDNA genotypes in heteroplasmic mice. *Nat Genet* **16:** 93–95.

Ji F, Sharpley MS, Derbeneva O, Alves LS, Qian P, Wang Y, Chalkia D, Lvova M, Xu J, Yao W, et al. 2012. Mitochondrial DNA variant associated with Leber hereditary optic neuropathy and high-altitude Tibetans. *Proc Natl Acad Sci* **109:** 7391–7396.

Jokinen R, Battersby BJ. 2013. Insight into mammalian mitochondrial DNA segregation. *Ann Med* **45:** 149–155.

Jokinen R, Marttinen P, Sandell HK, Manninen T, Teerenhovi H, Wai T, Teoli D, Loredo-Osti JC, Shoubridge EA, Battersby BJ. 2010. Gimap3 regulates tissue-specific mitochondrial DNA segregation. *PLoS Genet* **6:** e1001161.

Jokinen R, Junnila H, Battersby BJ. 2011. Gimap3: A foot-in-the-door to tissue-specific regulation of mitochondrial DNA genetics. *Small GTPases* **2:** 31–35.

Kasashima K, Sumitani M, Endo H. 2011. Human mitochondrial transcription factor A is required for the segregation of mitochondrial DNA in cultured cells. *Exp Cell Res* **317:** 210–220.

Kazuno AA, Munakata K, Nagai T, Shimozono S, Tanaka M, Yoneda M, Kato N, Miyawaki A, Kato T. 2006. Identification of mitochondrial DNA polymorphisms that alter mitochondrial matrix pH and intracellular calcium dynamics. *PLoS Genet* **2:** e128.

Khrapko K. 2008. Two ways to make an mtDNA bottleneck. *Nat Genet* **40:** 134–135.

Khusnutdinova E, Gilyazova I, Ruiz-Pesini E, Derbeneva O, Khusainova R, Khidiyatova I, Magzhanov R, Wallace DC. 2008. A mitochondrial etiology of neurodegenerative diseases: Evidence from Parkinson's disease. *Ann NY Acad Sci* **1147:** 1–20.

Kimura M. 1955. Solution of a process of random genetic drift with a continuous model. *Proc Natl Acad Sci* **41:** 144–150.

King MP, Attardi G. 1989. Human cells lacking mtDNA: Repopulation with exogenous mitochondria by complementation. *Science* **246:** 500–503.

Koehler CM, Lindberg GL, Brown DR, Beitz DC, Freeman AE, Mayfield JE, Myers AM. 1991. Replacement of bovine mitochondrial DNA by a sequence variant within one generation. *Genetics* **129:** 247–255.

Koopman WJ, Willems PH, Smeitink JA. 2012. Monogenic mitochondrial disorders. *N Engl J Med* **366:** 1132–1141.

Krakauer DC, Mira A. 1999. Mitochondria and germ-cell death. *Nature* **400:** 125–126.

Krysko DV, Agostinis P, Krysko O, Garg AD, Bachert C, Lambrecht BN, Vandenabeele P. 2011. Emerging role of damage-associated molecular patterns derived from mitochondria in inflammation. *Trends Immunol* **32:** 157–164.

Kukat C, Wurm CA, Spahr H, Falkenberg M, Larsson NG, Jakobs S. 2011. Super-resolution microscopy reveals that mammalian mitochondrial nucleoids have a uniform size and frequently contain a single copy of mtDNA. *Proc Natl Acad Sci* **108:** 13534–13539.

Lee HS, Ma H, Juanes RC, Tachibana M, Sparman M, Woodward J, Ramsey C, Xu J, Kang EJ, Amato P, et al. 2012. Rapid mitochondrial DNA segregation in primate preimplantation embryos precedes somatic and germline bottleneck. *Cell Rep* **1:** 506–515.

Leonard WR, Sorensen MV, Galloway VA, Spencer GJ, Mosher MJ, Osipova L, Spitsyn VA. 2002. Climatic influences on basal metabolic rates among circumpolar populations. *Am J Human Biol* **14:** 609–620.

Liang M, Guan M, Zhao F, Zhou X, Yuan M, Tong Y, Yang L, Wei QP, Sun YH, Lu F, et al. 2009. Leber's hereditary optic neuropathy is associated with mitochondrial ND1 T3394C mutation. *Biochem Biophys Res Commun* **383:** 286–292.

Malena A, Loro E, Di Re M, Holt IJ, Vergani L. 2009. Inhibition of mitochondrial fission favours mutant over wild-type mitochondrial DNA. *Hum Mol Genet* **18:** 3407–3416.

Mehrazin M, Shanske S, Kaufmann P, Wei Y, Coku J, Engelstad K, Naini A, De Vivo DC, DiMauro S. 2009. Longitudinal changes of mtDNA A3243G mutation load and level of functioning in MELAS. *Am J Med Genet A* **149A:** 584–587.

Mishmar D, Ruiz-Pesini E, Mondragon-Palomino M, Procaccio V, Gaut B, Wallace DC. 2006. Adaptive selection of mitochondrial complex I subunits during primate radiation. *Gene* **378:** 11–18.

Mitchell P. 1961. Coupling of phosphorylation to electron and hydrogen transfer by a chemi-osmotic type of mechanism. *Nature* **191:** 144–148.

MITOMAP. 2012. A human mitochondrial genome database. http://www.mitomap.org.

Monnot S, Gigarel N, Frydman N, Burlet P, Gobin S, Sinico M, Bonniere M, Rio M, Rotig A, Frydman R, et al. 2009a. NARP and MELAS mutations differentially impact mitochondrial DNA segregation throughout human embryofetal development. http://www.ashg.org/2009meeting/abstracts/fulltext/f21163htm.

Monnot S, Gigarel N, Hesters L, Burlet P, Benachi A, Dumez Y, Tachdjian G, Rotig A, Frydman R, Munnich A, et al. 2009b. The challenge of prenatal and preimplantation genetic diagnosis of mitochondrial DNA disorders. European Human Genetics Conference, Abstract C07.4, Vienna Austria, May 23–26, 2009. *Eur J Hum Genet* **17:** 24.

Monnot S, Gigarel N, Samuels DC, Burlet P, Hesters L, Frydman N, Frydman R, Kerbrat V, Funalot B, Martinovic J, et al. 2011. Segregation of mtDNA throughout human embryofetal development: m.3243A>G as a model system. *Hum Mutat* **32:** 116–125.

Monnot S, Samuels DC, Hesters L, Frydman N, Gigarel N, Burlet P, Kerbrat V, Lamazou F, Frydman R, Benachi A, et al. 2013. Mutation dependence of the mitochondrial DNA copy number in the first stages of human embryogenesis. *Hum Mol Genet* **22:** 1867–1872.

Moreno-Loshuertos R, Acin-Perez R, Fernandez-Silva P, Movilla N, Perez-Martos A, Rodriguez de Cordoba S, Gallardo ME, Enriquez JA. 2006. Differences in reactive oxygen species production explain the phenotypes associated with common mouse mitochondrial DNA variants. *Nat Genet* **38:** 1261–1268.

Nakada K, Sato A, Yoshida K, Morita T, Tanaka H, Inoue SI, Yonekawa H, Hayashi JI. 2006. Mitochondria-related male infertility. *Proc Natl Acad Sci* **103:** 15148–15153.

Narendra D, Tanaka A, Suen DF, Youle RJ. 2008. Parkin is recruited selectively to impaired mitochondria and promotes their autophagy. *J Cell Biol* **183:** 795–803.

Narendra D, Tanaka A, Suen DF, Youle RJ. 2009. Parkin-induced mitophagy in the pathogenesis of Parkinson disease. *Autophagy* **5:** 706–708.

Narendra DP, Jin SM, Tanaka A, Suen DF, Gautier CA, Shen J, Cookson MR, Youle RJ. 2010. PINK1 is selectively stabilized on impaired mitochondria to activate Parkin. *PLoS Biol* **8:** e1000298.

* Narendra D, Walker JE, Youle R. 2012. Mitochondrial quality control mediated by PINK1 and Parkin: Links to parkinsonism. *Cold Spring Harb Perspect Biol* **4:** a011338.

Neckelmann N, Li K, Wade RP, Shuster R, Wallace DC. 1987. cDNA sequence of a human skeletal muscle ADP/ATP translocator: lack of a leader peptide, divergence from a fibroblast translocator cDNA, and coevolution with mitochondrial DNA genes. *Proc Natl Acad Sci* **84:** 7580–7584.

Oka T, Hikoso S, Yamaguchi O, Taneike M, Takeda T, Tamai T, Oyabu J, Murakawa T, Nakayama H, Nishida K, et al. 2012. Mitochondrial DNA that escapes from autophagy causes inflammation and heart failure. *Nature* **485:** 251–255.

Oliver NA, Wallace DC. 1982. Assignment of two mitochondrially synthesized polypeptides to human mitochondrial DNA and their use in the study of intracellular mitochondrial interaction. *Mol Cell Biol* **2:** 30–41.

Oliver NA, Greenberg BD, Wallace DC. 1983. Assignment of a polymorphic polypeptide to the human mitochondrial DNA unidentified reading frame 3 gene by a new peptide mapping strategy. *J Biol Chem* **258:** 5834–5839.

Olivo PD, Van de Walle MJ, Laipis PJ, Hauswirth WW. 1983. Nucleotide sequence evidence for rapid genotypic shifts in the bovine mitochondrial DNA D-loop. *Nature* **306:** 400–402.

Ortiz RG, Newman NJ, Shoffner JM, Kaufman AE, Koontz DA, Wallace DC. 1993. Variable retinal and neurologic manifestations in patients harboring the mitochondrial DNA 8993 mutation. *Arch Ophthalmol* **111:** 1525–1530.

Paull D, Emmanuele V, Weiss KA, Treff N, Stewart L, Hua H, Zimmer M, Kahler DJ, Goland RS, Noggle SA, et al. 2013. Nuclear genome transfer in human oocytes eliminates mitochondrial DNA variants. *Nature* **493:** 632–637.

Payne BA, Wilson IJ, Yu-Wai-Man P, Coxhead J, Deehan D, Horvath R, Taylor RW, Samuels DC, Santibanez-Koref M, Chinnery PF. 2013. Universal heteroplasmy of human mitochondrial DNA. *Hum Mol Genet* **22:** 384–390.

Poe BG 3rd, Duffy CF, Greminger MA, Nelson BJ, Arriaga EA. 2010. Detection of heteroplasmy in individual mitochondrial particles. *Anal Bioanal Chem* **397:** 3397–3407.

Potluri P, Davila A, Ruiz-Pesini E, Mishmar D, O'Hearn S, Hancock S, Simon MC, Scheffler I, Wallace DC, Procaccio V. 2009. A novel NDUFA1 mutation leads to a progressive mitochondrial complex I-specific neurodegenerative disease. *Mol Genet Metab* **96:** 189–195.

Poulton J, Bredenoord AL. 2010. Applying pre-implantation genetic diagnosis to mtDNA diseases: Implications of scientific advances. 174th ENMC International Workshop, Naarden, The Netherlands, March 19–21, 2010. *Neuromuscul Disord* **20:** 559–563.

Rajasimha HK, Chinnery PF, Samuels DC. 2008. Selection against pathogenic mtDNA mutations in a stem cell population leads to the loss of the 3243A→G mutation in blood. *Am J Hum Genet* **82:** 333–343.

Rebelo AP, Dillon LM, Moraes CT. 2011. Mitochondrial DNA transcription regulation and nucleoid organization. *J Inherit Metab Dis* **34:** 941–951.

Rollins B, Martin MV, Sequeira PA, Moon EA, Morgan LZ, Watson SJ, Schatzberg A, Akil H, Myers RM, Jones EG, et al. 2009. Mitochondrial variants in schizophrenia, bipolar disorder, and major depressive disorder. *PLoS ONE* **4:** e4913.

Ruiz-Pesini E, Wallace DC. 2006. Evidence for adaptive selection acting on the tRNA and rRNA genes of the human mitochondrial DNA. *Hum Mutat* **27:** 1072–1081.

Ruiz-Pesini E, Mishmar D, Brandon M, Procaccio V, Wallace DC. 2004. Effects of purifying and adaptive selection on regional variation in human mtDNA. *Science* **303:** 223–226.

Sadun AA, Carelli V, Salomao SR, Berezovsky A, Quiros PA, Sadun F, DeNegri AM, Andrade R, Moraes M, Passos A, et al. 2003. Extensive investigation of a large Brazilian pedigree of 11778/haplogroup J Leber hereditary optic neuropathy. *Am J Ophthalmol* **136:** 231–238.

Sadun AA, La Morgia C, Carelli V. 2011. Leber's Hereditary Optic Neuropathy. *Curr Treat Options Neurol* **13:** 109–117.

Samuels DC, Wonnapinij P, Cree LM, Chinnery PF. 2010. Reassessing evidence for a postnatal mitochondrial genetic bottleneck. *Nat Genet* **42:** 471–473.

Samuels DC, Wonnapinij P, Chinnery PF. 2013. Preventing the transmission of pathogenic mitochondrial DNA mutations: Can we achieve long-term benefits from germline gene transfer? *Hum Reprod* **28:** 554–559.

Sato A, Kono T, Nakada K, Ishikawa K, Inoue S, Yonekawa H, Hayashi J. 2005. Gene therapy for progeny of mito-mice carrying pathogenic mtDNA by nuclear transplantation. *Proc Natl Acad Sci* **102:** 16765–16770.

Sato A, Nakada K, Shitara H, Kasahara A, Yonekawa H, Hayashi J. 2007. Deletion-mutant mtDNA increases in somatic tissues but decreases in female germ cells with age. *Genetics* **177:** 2031–2037.

Schaefer AM, Taylor RW, Turnbull DM, Chinnery PF. 2004. The epidemiology of mitochondrial disorders–past, present and future. *Biochim Biophys Acta* **1659:** 115–120.

Schaefer AM, McFarland R, Blakely EL, He L, Whittaker RG, Taylor RW, Chinnery PF, Turnbull DM. 2008. Prevalence of mitochondrial DNA disease in adults. *Ann Neurol* **63:** 35–39.

Schmitt MW, Kennedy SR, Salk JJ, Fox EJ, Hiatt JB, Loeb LA. 2012. Detection of ultra-rare mutations by next-generation sequencing. *Proc Natl Acad Sci* **109:** 14508–14513.

Sharpley MS, Marciniak C, Eckel-Mahan K, McManus MJ, Crimi M, Waymire K, Lin CS, Masubuchi S, Friend N, Koike M, et al. 2012. Heteroplasmy of mouse mtDNA is genetically unstable and results in altered behavior and cognition. *Cell* **151:** 333–343.

Shoffner JM, Lott MT, Voljavec AS, Soueidan SA, Costigan DA, Wallace DC. 1989. Spontaneous Kearns-Sayre/chronic external ophthalmoplegia plus syndrome associated with a mitochondrial DNA deletion: A slip-replication model and metabolic therapy. *Proc Natl Acad Sci* **86:** 7952–7956.

Shoffner JM, Lott MT, Lezza AM, Seibel P, Ballinger SW, Wallace DC. 1990. Myoclonic epilepsy and ragged-red fiber disease (MERRF) is associated with a mitochondrial DNA $tRNA^{Lys}$ mutation. *Cell* **61:** 931–937.

Snodgrass JJ, Leonard WR, Tarskaia LA, Alekseev VP, Krivoshapkin VG. 2005. Basal metabolic rate in the Yakut (Sakha) of Siberia. *Am J Human Biol* **17:** 155–172.

Snodgrass JJ, Leonard WR, Sorensen MV, Tarskaia LA, Mosher MJ. 2008. The influence of basal metabolic rate on blood pressure among indigenous Siberians. *Am J Phys Anthropol* **137:** 145–155.

Solignac M, Genermont J, Monnerot M, Mounolou JC. 1984. Genetics of mitochondria in Drosophila mtDNA inheritance in heteroplasmic strains of *Drosophila mauritiana*. *Mol Gen Genet* **197:** 183–188.

Stephens M. 1974. EDF statistics for goodness of fit and some comparisons. *J Am Stat Assoc* **69:** 730–737.

Stewart JB, Freyer C, Elson JL, Wredenberg A, Cansu Z, Trifunovic A, Larsson NG. 2008. Strong purifying selection in transmission of mammalian mitochondrial DNA. *PLoS Biol* **6:** e10.

Strauss KA, Dubiner L, Simon M, Zaragoza M, Sengupta PP, Li P, Narula N, Dreike S, Platt J, Procaccio V, et al. 2013. Severity of cardiomyopathy associated with adenine nucleotide translocator-1 deficiency correlates with mtDNA haplogroup. *Proc Natl Acad Sci* **110:** 3253–3458.

Suen DF, Narendra DP, Tanaka A, Manfredi G, Youle RJ. 2010. Parkin overexpression selects against a deleterious mtDNA mutation in heteroplasmic cybrid cells. *Proc Natl Acad Sci* **107:** 11835–11840.

Tachibana M, Sparman M, Sritanaudomchai H, Ma H, Clepper L, Woodward J, Li Y, Ramsey C, Kolotushkina O, Mitalipov S. 2009. Mitochondrial gene replacement in primate offspring and embryonic stem cells. *Nature* **461:** 367–372.

Tachibana M, Amato P, Sparman M, Woodward J, Sanchis DM, Ma H, Gutierrez NM, Tippner-Hedges R, Kang E, Lee HS, et al. 2013. Towards germline gene therapy of inherited mitochondrial diseases. *Nature* **493:** 627–631.

Tajima H, Sueoka K, Moon SY, Nakabayashi A, Sakurai T, Murakoshi Y, Watanabe H, Iwata S, Hashiba T, Kato S, et al. 2007. The development of novel quantification assay for mitochondrial DNA heteroplasmy aimed at preimplantation genetic diagnosis of Leigh encephalopathy. *J Assist Reprod Genet* **24:** 227–232.

Takahata N, Slatkin M. 1983. Evolutionary dynamics of extranuclear genes. *Genet Res* **42:** 257–268.

Tatuch Y, Christodoulou J, Feigenbaum A, Clarke JTR, Wherret J, Smith C, Rudd N, Petrova-Benedict R, Robinson BH. 1992. Heteroplasmic mtDNA mutation (T-G) at 8993 can cause Leigh disease when the percentage of abnormal mtDNA is high. *Am J Hum Genet* **50:** 852–858.

Thorburn DR, Dahl HH. 2001. Mitochondrial disorders: Genetics, counseling, prenatal diagnosis and reproductive options. *Am J Med Genet* **106:** 102–114.

Thorburn D, Wilton L, Stock-Myer S. 2009. Healthy baby girl born following preimplantation genetic diagnosis for mitochondrial DNA m.8993T>G mutation. *11th International Congress of Inborn Errors of Metabolism*, Abstract 117, Platform Presentation 3: Mitochondrial Disorders. *Mol Genet Metab* **98:** 5–6.

Torroni A, Petrozzi M, D'Urbano L, Sellitto D, Zeviani M, Carrara F, Carducci C, Leuzzi V, Carelli V, Barboni P, et al. 1997. Haplotype and phylogenetic analyses suggest that one European-specific mtDNA background plays a role in the expression of Leber hereditary optic neuropathy by increasing the penetrance of the primary mutations 11778 and 14484. *Am J Hum Genet* **60:** 1107–1121.

Toye AA, Lippiat JD, Proks P, Shimomura K, Bentley L, Hugill A, Mijat V, Goldsworthy M, Moir L, Haynes A, et al. 2005. A genetic and physiological study of impaired glucose homeostasis control in C57BL/6 J mice. *Diabetologia* **48:** 675–686.

Treff NR, Campos J, Tao X, Levy B, Ferry KM, Scott RT Jr. 2012. Blastocyst preimplantation genetic diagnosis (PGD) of a mitochondrial DNA disorder. *Fertil Steril* **98:** 1236–1240.

van den Ouweland JM, Lemkes HHP, Ruitenbeek W, Sandkjujl LA, deVijlder MF, Struyvenberg PAA, van de Kamp JJP, Maassen JA. 1992. Mutation in mitochondrial $tRNA^{Leu(UUR)}$ gene in a large pedigree with maternally transmitted type II diabetes mellitus and deafness. *Nat Genet* **1:** 368–371.

Wai T, Teoli D, Shoubridge EA. 2008. The mitochondrial DNA genetic bottleneck results from replication of a subpopulation of genomes. *Nat Genet* **40:** 1484–1488.

Wallace DC. 1981. Assignment of the chloramphenicol resistance gene to mitochondrial deoxyribonucelic acid and analysis of its expression in cultured human cells. *Mol Cell Biol* **1:** 697–710.

Wallace DC. 1982. Cytoplasmic inheritance of chloramphenicol resistance in mammalian cells. In *Techniques in somatic cell genetics* (ed. Shay JW), Chap. 12. pp. 159–187. Plenum, New York.

Wallace DC. 1986. Mitotic segregation of mitochondrial DNAs in human cell hybrids and expression of chloramphenicol resistance. *Somat Cell Mol Genet* **12:** 41–49.

Wallace DC. 1987. Maternal genes: Mitochondrial diseases. In *Medical and experimental mammalian genetics: A perspective* (ed. McKusick VA, Roderick TH, Mori J, Paul MW), pp. 137–190. A.R. Liss, New York.

Wallace DC. 2005. A mitochondrial paradigm of metabolic and degenerative diseases, aging, and cancer: A dawn for evolutionary medicine. *Annu Rev Genet* **39:** 359–407.

Wallace DC. 2007. Why do we have a maternally inherited mitochondrial DNA? Insights from Evolutionary Medicine. *Annu Rev Biochem* **76:** 781–821.

Wallace DC. 2008. Mitochondria as chi. *Genetics* **179:** 727–735.

Wallace DC. 2011. Bioenergetic origins of complexity and disease. *Cold Spring Harb Symp Quant Biol* **76:** 1–16.

Wallace DC. 2012. Mitochondria and cancer. *Nat Rev Cancer* **12:** 685–698.

Wallace DC. 2013a. Bioenergetics in human evolution and disease: Implications for the origins of biological complexity and the missing genetic variation of common diseases. *Philos Trans R Soc Lond B Biol Sci* doi: 10.1098/rstb.2012.0267.

Wallace DC. 2013b. Mitochondrial bioenergetic etiology of disease. *J Clin Invest* **123:** 1405–1412.

Wallace DC, Fan W. 2009. The pathophysiology of mitochondrial disease as modeled in the mouse. *Genes Dev* **23:** 1714–1736.

Wallace DC, Fan W. 2010. Energetics, epigenetics, mitochondrial genetics. *Mitochondrion* **10:** 12–31.

Wallace DC, Bunn CL, Eisenstadt JM. 1975. Cytoplasmic transfer of chloramphenicol resistance in human tissue culture cells. *J Cell Biol* **67:** 174–188.

Wallace DC, Pollack Y, Bunn CL, Eisenstadt JM. 1976. Cytoplasmic inheritance in mammalian tissue culture cells. *In Vitro* **12:** 758–776.

Wallace DC, Bunn CL, Eisenstadt JM. 1977. Mitotic segregation of cytoplasmic inherited genes for chloramphenicol resistance in mammalian cells. II: Fusions with human cell lines. *Somatic Cell Genet* **3:** 93–119.

Wallace DC, Ye JH, Neckelmann SN, Singh G, Webster KA, Greenberg BD. 1987. Sequence analysis of cDNAs for the human and bovine ATP synthase b-subunit: Mitochondrial DNA genes sustain seventeen times more mutations. *Curr Genet* **12:** 81–90.

Wallace DC, Singh G, Lott MT, Hodge JA, Schurr TG, Lezza AM, Elsas LJ, Nikoskelainen EK. 1988a. Mitochondrial DNA mutation associated with Leber's hereditary optic neuropathy. *Science* **242:** 1427–1430.

Wallace DC, Zheng X, Lott MT, Shoffner JM, Hodge JA, Kelley RI, Epstein CM, Hopkins LC. 1988b. Familial mitochondrial encephalomyopathy (MERRF): Genetic, pathophysiological, and biochemical characterization of a mitochondrial DNA disease. *Cell* **55:** 601–610.

Wallace DC, Brown MD, Lott MT. 1999. Mitochondrial DNA variation in human evolution and disease. *Gene* **238:** 211–230.

Wallace DC, Fan W, Procaccio V. 2010. Mitochondrial energetics and therapeutics. *Annu Rev Path* **5:** 297–348.

Wallace DC, Lott MT, Procaccio V. 2013. Mitochondrial medicine: The mitochondrial biology and genetics of metabolic and degenerative diseases, cancer, and aging. In *Emery and Rimoin's principles and practice of medical genetics* (ed. Rimoin DL, Pyeritz RE, Korf BR). Churchill Livingstone, Philadelphia.

Wonnapinij P, Chinnery PF, Samuels DC. 2008. The distribution of mitochondrial DNA heteroplasmy due to random genetic drift. *Am J Hum Genet* **83:** 582–593.

Wonnapinij P, Chinnery PF, Samuels DC. 2010. Previous estimates of mitochondrial DNA mutation level variance did not account for sampling error: comparing the mtDNA genetic bottleneck in mice and humans. *Am J Hum Genet* **86:** 540–550.

Wright S. 1969. Theory of gene frequencies. In *Evolution and the genetics of populations*, University of Chicago Press, Chicago.

Yoneda M, Miyatake TGA. 1994. Complementation of mutant and wild-type human mitochondrial DNAs coexisting since the mutation event and lack of complementation of DNAs introduced separately into a cell within distinct organelles. *Mol Cell Biol* **14:** 2699–2712.

Yoneda M, Chomyn A, Martinuzzi A, Hurko O, Attardi G. 1992. Marked replicative advantage of human mtDNA carrying a point mutation that causes the MELAS encephalomyopathy. *Proc Natl Acad Sci* **89:** 11164–11168.

Yoon KL, Ernst SG, Rasmussen C, Dooling EC, Aprille JR. 1993. Mitochondrial disorder associated with newborn cardiopulmonary arrest. *Pediatr Res* **33:** 433–440.

Youle RJ, van der Bliek AM. 2012. Mitochondrial fission, fusion, and stress. *Science* **337:** 1062–1105.

Zhang Q, Itagaki K, Hauser CJ. 2010a. Mitochondrial DNA is released by shock and activates neutrophils via p38 map kinase. *Shock* **34:** 55–559.

Zhang Q, Raoof M, Chen Y, Sumi Y, Sursal T, Junger W, Brohi K, Itagaki K, Hauser CJ. 2010b. Circulating mitochondrial DAMPs cause inflammatory responses to injury. *Nature* **464:** 104–107.

Zhou RR, Wang B, Wang J, Schatten H, Zhang YZ. 2010. Is the mitochondrial cloud the selection machinery for preferentially transmitting wild-type mtDNA between generations? Rewinding Muller's ratchet efficiently. *Curr Genet* **56:** 101–107.

Mechanisms of Protein Sorting in Mitochondria

Diana Stojanovski[1], Maria Bohnert[2,3], Nikolaus Pfanner[2,4], and Martin van der Laan[2,4]

[1]La Trobe Institute for Molecular Sciences, 3086 Melbourne, Australia

[2]Institut für Biochemie und Molekularbiologie, ZBMZ, Universität Freiburg, D-79104 Freiburg, Germany

[3]Fakultät für Biologie, Universität Freiburg, D-79104 Freiburg, Germany

[4]BIOSS Centre for Biological Signalling Studies, Universität Freiburg, D-79104 Freiburg, Germany

Correspondence: martin.van.der.laan@biochemie.uni-freiburg.de; maria.bohnert@biochemie.uni-freiburg.de

A protein's function is intimately linked to its correct subcellular location, yet the machinery required for protein synthesis is predominately cytosolic. How proteins are trafficked through the confines of the cell and integrated into the appropriate cellular compartments has puzzled and intrigued researchers for decades. Indeed, studies exploring this premise revealed elaborate cellular protein translocation and sorting systems, which ensure that all proteins are shuttled to the appropriate cellular destination, where they fulfill their specific functions. This holds true for mitochondria, where sophisticated molecular machines serve to recognize incoming precursor proteins and integrate them into the functional framework of the organelle. We summarize the recent progress in our understanding of mitochondrial protein sorting and the machineries and mechanisms that mediate and regulate this highly dynamic cellular process essential for survival of virtually all eukaryotic cells.

Mitochondria are multifunctional double-membrane-bound organelles that arose from a bacterial endosymbiont during the evolution of eukaryotic cells. Known as the powerhouses of the cell, mitochondria harbor the oxidative phosphorylation machinery for ATP synthesis, but also a large number of biosynthetic pathways. Moreover, they are intimately involved in complex cellular processes, like calcium homeostasis and programmed cell death. As a relic of their evolutionary origin, mitochondria contain their own genetic material and machineries to manufacture their own RNAs and proteins. However, the small circular mitochondrial genome encodes only a few proteins (8 and 13 polypeptides in yeast and humans, respectively). All remaining mitochondrial proteins (approximately 99%) are encoded by the nuclear genome and synthesized on cytosolic ribosomes in their precursor forms. To acquire their mature, functional state these precursor proteins need to be efficiently targeted and imported into mitochondria and sorted to the correct submitochondrial compartment: outer membrane, intermembrane space (IMS), inner membrane, and matrix. The inner mitochondrial membrane is further subdivided into the inner boundary membrane, which is closely opposed to the outer membrane, and large tubular invaginations, termed cristae membranes. Within the four mitochondrial compartments, sophisticated translocation, sorting, and assembly machineries

Cite this article as *Cold Spring Harb Perspect Biol* doi: 10.1101/cshperspect.a011320

serve to establish incoming precursors in a functional state within the context of their new environment. Advances in the last decade, particularly because of the application of proteomic approaches, have significantly extended the number of components and machineries known to be involved in mitochondrial protein import (Sickmann et al. 2003; Prokisch et al. 2004; Reinders et al. 2006; Pagliarini et al. 2008). These and previous discoveries have provided us with the current framework, which suggests the presence of at least six distinct translocation and assembly machineries within mitochondria (Fig. 1). In this article, we will summarize our current understanding of the machineries for mitochondrial protein import and describe the different molecular mechanisms that execute this essential task.

MITOCHONDRIAL PRECURSOR PROTEINS: SYNTHESIS AND TARGETING

It is widely accepted that the vast majority of mitochondrial precursor proteins are imported in a posttranslational manner. To this end, precursor proteins must be kept in an unfolded or loosely folded conformation to allow their passage

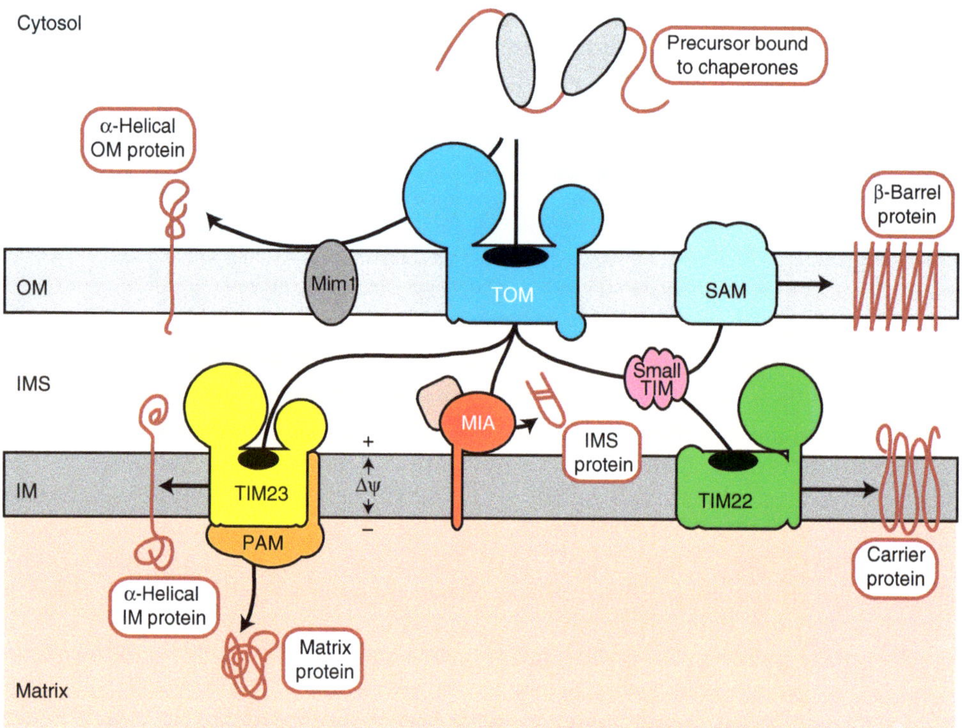

Figure 1. Overview of mitochondrial protein sorting pathways. Cytosolic chaperones deliver precursor proteins to the organelle in a translocation-competent state. Some α-helical proteins are inserted into the outer membrane with the help of Mim1. Virtually all other precursors initially traverse the outer membrane via the TOM complex and are subsequently routed to downstream sorting pathways. Biogenesis of outer membrane β-barrel proteins requires the small TIM chaperones of the IMS and the SAM complex. Cysteine-containing IMS proteins are imported via the MIA pathway. Metabolite carriers of the inner mitochondrial membrane are transferred by the small TIM chaperones to the TIM22 complex, which mediates their membrane integration. Presequence-containing precursors are directly taken over from the TOM complex by the TIM23 machinery that either inserts these proteins into the membrane or translocates them into the matrix in cooperation with the import motor PAM. OM, outer membrane; IMS, intermembrane space; IM, inner membrane, Δψ, membrane potential across the inner mitochondrial membrane.

Cite this article as *Cold Spring Harb Perspect Biol* doi: 10.1101/cshperspect.a011320

through tightly gated membrane pores. This is achieved by the binding of cytosolic factors to nascent precursors that stabilize them in a translocation-competent form and guide them to dedicated receptors on the mitochondrial surface (Fig. 1). The best characterized machineries that escort mitochondrial precursor proteins through the cytosol are the Hsp90/p23 and Hsc70/Hsp40 chaperone systems (Young et al. 2003; Bhangoo et al. 2007; Zara et al. 2009). Additional cytosolic proteins, like the arylhydrocarbon receptor-interacting protein (AIP), have been implicated in this process (Yano et al. 2003). However, there is evidence that at least some precursor proteins, like fumarase and Sod2, are imported cotranslationally into mitochondria in vivo (Luk et al. 2005; Yogev et al. 2007). Interestingly, approximately one-half of the mRNAs coding for mitochondrial proteins were reported to be localized to the mitochondrial surface (Marc et al. 2002; Garcia et al. 2007). Puf3, a member of the Pumilio/FBF family of RNA-binding proteins, has been shown to be involved in the targeting of mRNAs to mitochondria (Garcia-Rodriguez et al. 2007; Saint-Georges et al. 2008). A seemingly cotranslational import reaction may in some cases result from the fact that the initiation of precursor translocation is simply faster than the termination of translation for mitochondrial-associated mRNAs. From a physiological perspective, mRNA targeting to mitochondria may allow specific posttranscriptional processes to closely cooperate with the mitochondrial import machinery permitting rapid responses to metabolic changes and/or cellular stimuli (Devaux et al. 2010).

The final destination of a protein in the cell is determined by specific sequence elements within the precursor form that are referred to as signal sequences or targeting signals. These signals are recognized by companion receptors on the surface of the organelles. Mitochondrial targeting signals are categorized in two major classes: (i) "classical" amino-terminal cleavable presequences, and (ii) noncleavable internal import signals (Vögtle et al. 2009). Amino-terminal presequences are the most frequently found import signals and form positively charged amphipathic α-helices, which facilitate a chain of interactions between the precursors and defined import components by gradually increasing affinities (Moczko et al. 1997; Komiya et al. 1998; Marom et al. 2011). Presequences direct precursor proteins to the inner membrane or the matrix and are cleaved on import by the mitochondrial processing peptidase (MPP) (Taylor et al. 2001). Some precursors are further processed at their amino-termini by other proteases, like Oct1 or Icp55 (Naamati et al. 2009; Vögtle et al. 2009, 2011). Other precursor proteins, like cytochrome c_1 or cytochrome b_2, are initially inserted into the inner membrane by means of a hydrophobic stop-transfer signal downstream of the presequence and subsequently released into the IMS by inner membrane peptidase (IMP) cleavage (Glick et al. 1992; Nunnari et al. 1993). The presence of multiple internal targeting signals was first described for inner membrane proteins of the mitochondrial metabolite carrier family. These import signals largely overlap with the hydrophobic transmembrane segments of the mature carrier proteins (Wiedemann et al. 2001; Brandner et al. 2005). Only recently, membrane proteins that adopt a β-barrel conformation, like Tom40 or porin, were found to contain a specific import signal in the last β-strand of the precursors (β-signal) directing them into the outer mitochondrial membrane (Fig. 2A) (Kutik et al. 2008a). These rather short, evolutionary conserved β-signals are characterized by a bulky polar residue (predominantly lysine or glutamine), followed by an invariant glycine and two large hydrophobic residues. Shortly after the identification of the β-signal, a consensus sequence was identified that targets precursor proteins to the mitochondrial IMS (Milenkovic et al. 2009; Sideris et al. 2009). This mitochondrial IMS sorting signal (MISS) consists of a conserved leucine and valine residue together with a cysteine that forms a transient disulfide bond with the IMS receptor Mia40 (Fig. 3; see below).

THE TOM COMPLEX: A COMMON JUNCTION IN DISTINCT BIOGENESIS PATHWAYS

Independent of the type of import signal, the general translocase of the outer membrane (TOM

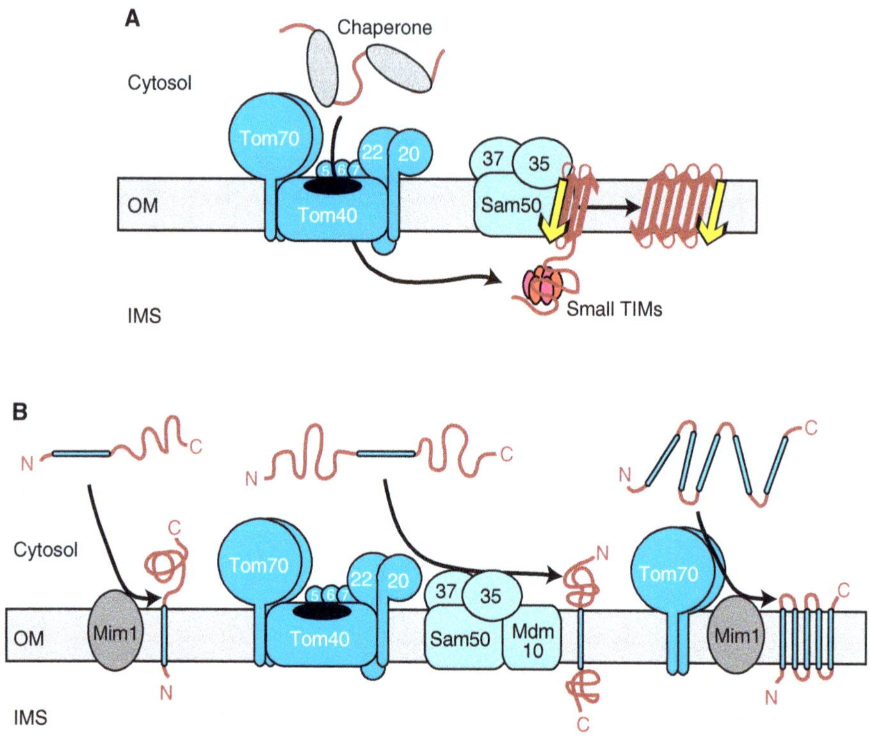

Figure 2. Biogenesis of outer membrane proteins. (*A*) Several mitochondrial outer membrane proteins are characterized by a β-barrel topology. On passage of the TOM complex, β-barrel precursors are guided through the aqueous IMS by hexameric small TIM complexes to the SAM complex. A specific import signal (β-signal), which is located within the last β-strand of the precursor, is recognized by the Sam35 receptor. Sam50 forms a central cavity in the SAM complex, where membrane insertion and folding of precursors takes place. (*B*) Multiple pathways mediate the insertion of α-helical proteins into the outer membrane. Signal-anchored precursors require Mim1 for membrane integration (*left*). The biogenesis of Tom22, which contains a single central transmembrane segment, depends on both the TOM and SAM–Mdm10 complexes (*middle*). Polytopic α-helical outer membrane proteins are recognized by the Tom70 receptor and transferred to Mim1 for membrane insertion (*right*). OM, outer membrane; IMS, intermembrane space; red arrow, β-strand; yellow arrow, β-signal; blue bar, transmembrane α-helix C, carboxy; N, amino.

complex) is the first line of contact at mitochondria for nearly all precursor proteins (Fig. 1). The central subunit of the TOM complex is the pore-forming β-barrel protein Tom40 (Hill et al. 1998; Künkele et al. 1998; Ahting et al. 1999; Becker et al. 2005). The primary receptors Tom20 and Tom70 are involved in precursor recognition; precursors with amino-terminal targeting signals generally bind to Tom20, whereas those with internal targeting signals preferentially interact with Tom70 (Brix et al. 1997; Abe et al. 2000; Wu and Sha 2006; Saitoh et al. 2007). Tom22 is crucial for the stability of the TOM complex and acts as a central receptor mediating the transfer of precursors from the primary receptor sites to the Tom40 channel (van Wilpe et al. 1999; Shiota et al. 2011). Moreover, the IMS domain of Tom22 contains a *trans* binding site for amino-terminal presequences (Moczko et al. 1997; Komiya et al. 1998). The small TOM proteins Tom5, Tom6, and Tom7 are involved in the biogenesis, stabilization, and dynamics of the TOM machinery (Dekker et al. 1998; Model et al.

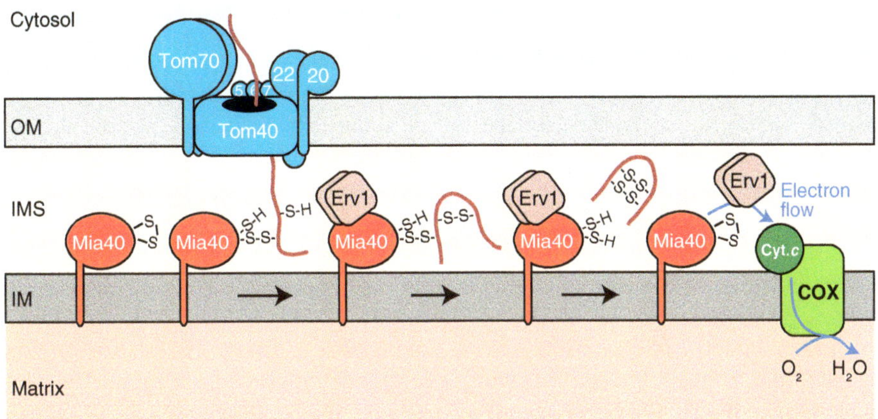

Figure 3. Import of cysteine-containing precursor proteins into the IMS. The Mia40 receptor forms transient intermolecular disulfide bonds with precursors emerging from the TOM complex. Mia40 functions not only as a receptor, but also as an oxidoreductase by catalyzing the formation of intramolecular disulfide bonds within the precursor together with the sulfhydryl oxidase Erv1. A ternary complex of Mia40, Erv1, and a bound precursor protein allows for the incorporation of multiple disulfide bonds. Erv1 re-oxidizes Mia40 for another round of import and transfers electrons via cytochrome *c* (Cyt.*c*) and cytochrome *c* oxidase (COX) to molecular oxygen. OM, outer membrane; IMS, intermembrane space; IM, inner membrane.

2001; Dukanovic et al. 2009; Becker et al. 2010, 2011a).

In a new twist, it has recently been reported that the function of the TOM complex is extensively regulated by cytosolic protein kinases (Schmidt et al. 2011). It was found that casein kinase 2 phosphorylates the central receptor Tom22 to facilitate its import and interaction with Tom20. Consequently, mutational inactivation of casein kinase 2 leads to defects in the TOM machinery. Additionally, phosphorylation of the Tom70 receptor by protein kinase A was shown to negatively regulate the import of metabolite carrier proteins in response to metabolic changes in the cell (Schmidt et al. 2011). These findings support the notion that the mitochondrial import machinery is not just a static entity, but rather is integrated into and regulated by cellular signaling networks that govern organelle biogenesis.

Once at the TOM complex, the majority of precursors are translocated across the outer membrane into the IMS, where they are transferred over to specialized protein sorting machineries of other submitochondrial compartments (Fig. 1). The mechanisms that allow for the initial translocation of several different classes of precursors through the TOM complex, and subsequent selective sorting into distinct downstream protein biogenesis pathways remain poorly understood.

DIVERSE PATHWAYS FOR TRAFFICKING TO THE MITOCHONDRIAL OUTER MEMBRANE

From a topological perspective, the outer membrane of mitochondria is an intriguing environment because it contains both β-barrel proteins and proteins that span the lipid bilayer with single or multiple α-helical transmembrane domains. Thus, the integrity of this subcompartment is upheld by the coexistence of multiple biogenesis pathways, which are likely acting in concert with each other. The presence of β-barrel proteins in the outer membranes of mitochondria and chloroplasts reflects the bacterial origin of these organelles (Dolezal et al. 2006). The basic mechanisms of β-barrel biogenesis appear to be conserved in evolution. BamA (Omp85, YaeT), the central component of the bacterial β-barrel assembly machinery (Knowles et al. 2009), is homologous to Sam50, the essential pore-forming component of the outer membrane sorting and assembly machinery (SAM complex; also termed TOB complex), which mediates the membrane insertion and folding of β-barrel proteins in mitochondria (Kozjak et al. 2003; Paschen et al. 2003; Gentle et al. 2004). Remarkably, the SAM complex is able to insert recombinantly expressed bacterial β-barrel proteins into the outer mitochondrial membrane (Walther et al. 2009). The SAM complex contains two additional peripheral membrane proteins, Sam35, which is essential for cell growth and the nonessential Sam37 (Wiedemann et al. 2003; Ishikawa et al. 2004; Milenkovic et al. 2004; Waizenegger et al. 2004). Mitochondrial β-barrel precursors are initially translocated across the outer membrane via the TOM complex (Fig. 2A). As β-barrel precursors emerge at the *trans* side of the TOM complex they are met by members of the small TIM chaperone family (Tim9/Tim10; Tim8/Tim13), which likely trap the β-barrel precursors in the IMS and shield the unfolded proteins from this aqueous environment on their way to the SAM machinery (Hoppins and Nargang 2004; Wiedemann et al. 2004). The molecular mechanism of precursor transfer from the TOM complex to the TIM chaperones and further to the SAM complex awaits to be elucidated. Either a concerted hand-over mechanism or the existence of soluble IMS intermediates appears conceivable. There is good evidence that on passage through the IMS, the aforementioned β-signal directs insertion of the precursor into the SAM complex by binding to the receptor Sam35 (Fig. 2A) (Kutik et al. 2008a). Complex formation of Sam35 and Sam50 with the β-signal is believed to initiate the entry of the precursor into a hydrophilic and proteinaceous membrane environment, which then accommodates β-barrel folding and insertion into the lipid phase (Kutik et al. 2008a). The physical nature of this membrane-embedded protein folding chamber is still under debate. Current models suggest that it may be formed mainly by a Sam50 oligomer (Paschen et al. 2003; Kutik et al. 2008a). Finally, both Sam37 and the amino-terminal IMS domain of Sam50 are involved in the release of the β-barrel precursor into the outer membrane (Chan and Lithgow 2008; Stroud et al. 2011a).

Several other outer membrane proteins participate in the assembly of β-barrel proteins. Mdm10 and Mim1 were found to interact with the SAM complex and are differentially required for the assembly of TOM complex components (Meisinger et al. 2004; Waizenegger et al. 2005; Becker et al. 2008). Mdm10 is also found in the endoplasmic reticulum-mitochondria encounter structure (ERMES), also known as mitochondrial distribution and morphology (MDM) complex, which contains four additional proteins: Mdm12, Mmm1, Mdm34, and the Miro GTPase Gem1 (Boldogh et al. 2003; Kornmann et al. 2011; Stroud et al. 2011b). ERMES was shown to tether the mitochondrial outer membrane to the endoplasmic reticulum (Kornmann et al. 2009). Anchoring of ERMES to the ER is mediated by the amino-terminal transmembrane segment of Mmm1 (Kornmann et al. 2009; Stroud et al. 2011b). ERMES was proposed to facilitate the exchange of phospholipids and calcium between the ER and mitochondria (Kornmann et al. 2009; Kornmann and Walter 2010; Gebert et al. 2011a). Interestingly, inactivation of ERMES proteins also affects β-barrel protein assembly (Meisinger et al. 2007; Wideman et al. 2010). The suggested role in lipid transport provides a possible explanation for the severe mitochondrial morphology defects observed with the loss of ERMES proteins (Boldogh et al. 2003; Meisinger et al. 2004, 2007). However, the true function of ERMES in the biogenesis of β-barrel proteins has not yet been clarified. The available data raise the intriguing possibility that lipid transfer from the ER to mitochondria and the biogenesis of mitochondrial β-barrel proteins may be coupled processes. It seems likely that the dually localized Mdm10 protein links the different roles of ERMES. Interestingly, the small TOM protein Tom7 was shown to regulate the dynamic interaction of Mdm10 with the SAM and ERMES complexes (Meisinger et al. 2006; Yamano et al. 2010; Becker et al. 2011a).

For proteins anchored to the outer membrane via α-helical regions, transmembrane domains are located at either the amino- (signal anchor) or carboxyl-terminus (tail anchor). Other α-helical outer membrane proteins span the lipid bilayer multiple times. There is a striking variety in the mechanisms adopted for their insertion (Fig. 2B). Membrane integration of signal-anchored proteins, such as Tom20 and Tom70, is facilitated by the outer membrane protein Mim1 (Becker et al. 2008; Hulett et al. 2008; Popov-Čeleketić et al. 2008a). However, a Mim1 counterpart for tail-anchored proteins has not yet been discovered, but rather the lipid composition of the outer membrane appears be crucial for specific insertion of these proteins (Setoguchi et al. 2006; Kemper et al. 2008). The precursor of Tom22, which contains one central transmembrane α-helix, requires both the TOM complex and the SAM–Mdm10 complex for its biogenesis (Fig. 2B) (Stojanovski et al. 2007; Thornton et al. 2010). This finding revealed that the substrate spectrum of the SAM complex is not limited to β-barrel precursors, but also includes specific α-helical outer membrane proteins. Finally, multi-spanning proteins, like the membrane fusion protein Ugo1, are assisted by the Tom70 receptor (but no other TOM components) and the Mim1 protein for their insertion into the outer membrane (Fig. 2B) (Otera et al. 2007; Becker et al. 2011b; Papic et al. 2011).

REDOX-DRIVEN PROTEIN IMPORT INTO THE MITOCHONDRIAL INTERMEMBRANE SPACE

Many IMS proteins are small in size (<20 kDa) and contain conserved cysteine residues in Cx_3C or Cx_9C motives that are involved in binding of cofactors and metals or the formation of disulfide bonds. The best characterized proteins of this kind are the small TIM chaperones that form hetero-oligomeric complexes with each subunit containing two disulfide bonds (Webb et al. 2006). Import of these proteins into mitochondria is mediated by the mitochondrial IMS import and assembly (MIA) machinery (Fig. 3). The MIA pathway involves two essential conserved core components, the import receptor Mia40 and the sulfhydryl oxidase Erv1 (ALR in humans) (Chacinska et al. 2004; Naoé et al. 2004; Mesecke et al. 2005). Mia40 is firmly attached to the inner mitochondrial membrane by an amino-terminal membrane anchor in yeast, whereas human Mia40 is a soluble IMS protein (Chacinska et al. 2008; Banci et al. 2009). Structural analysis has revealed that client proteins associate with Mia40 via a hydrophobic cleft on the surface and a redox-active CPC motif of Mia40 (Banci et al. 2009; Kawano et al. 2009). Polypeptide segments of both, Mia40 itself and Erv1/ALR, have been suggested to occupy the hydrophobic substrate binding site of Mia40 in the absence of a precursor (Banci et al. 2009; Kawano et al. 2009). Docking of substrate to Mia40 facilitates folding of the precursor and generation of a transient precursor-receptor disulfide intermediate with a cysteine residue in the precursor's import signal (Fig. 3) (Grumbt et al. 2007; Milenkovic et al. 2007, 2009; Sideris and Tokatlidis 2007; Sideris et al. 2009; Banci et al. 2010). Thus, Mia40 behaves as an oxidoreductase donating disulfide bonds to incoming precursors thereby trapping them in the IMS. This reaction involves the removal of electrons from the precursor and reduction of cysteine residues in Mia40. Analogous to other disulfide-generating systems, the sulfhydryl oxidase Erv1 reoxidizes Mia40 for further rounds of the reaction cycle (Fig. 3) (Mesecke et al. 2005; Stojanovski et al. 2008; Banci et al. 2011). Erv1 acts as a homo-dimer, and an intersubunit electron exchange mechanism is involved in the redox reaction (Bien et al. 2010). A role for the small redox-active peptide glutathione in proofreading of Mia40/Erv1-generated disulfide bonds has been suggested (Bien et al. 2010). For its own reoxidation, Erv1 transfers electrons via cytochrome *c* to the respiratory chain and, hence, molecular oxygen (Fig. 3) (Bihlmaier et al. 2007; Dabir et al. 2007). Interestingly, Erv1 was found in a ternary complex with Mia40 and a precursor protein. This direct association may facilitate the acquisition of multiple disulfides through a channeling mechanism, which does not require continual association and dissociation of the components (Stojanovski et al. 2008; Banci et al. 2011).

A recent study by von der Malsburg et al. (2011) revealed that mitofilin/Fcj1, an inner membrane protein required for the maintenance of inner membrane architecture, is involved in the MIA pathway. Mitofilin is part of the mitochondrial inner membrane organizing system (MINOS; also termed MICOS or MitOS), a large inner membrane protein complex that is crucial for the morphology of cristae membranes and involved in multiple contact sites between inner and outer mitochondrial membranes (Harner et al. 2011; Hoppins et al. 2011; von der Malsburg et al. 2011). Mitofilin associates with both the TOM machinery in the outer membrane and Mia40, thereby recruiting Mia40 into the proximity of incoming client proteins, as soon as they emerge from the Tom40 channel.

Insertion of Metabolite Carriers into the Inner Membrane via the TIM22 Complex

The large family of mitochondrial inner membrane metabolite carriers, such as the ADP/ATP carriers (AAC), functions in the transfer of metabolites between the mitochondrial matrix and IMS. Metabolite carriers as well as a few other integral membrane proteins with internal import signals, like the Tim23 protein, are inserted into the inner membrane by the translocase of the inner membrane 22 (TIM22 complex; carrier translocase) (Fig. 4). The TIM22 machinery consists of four membrane-integral components, Tim22, Tim54, Tim18, and Sdh3, and the peripherally associated IMS chaperones

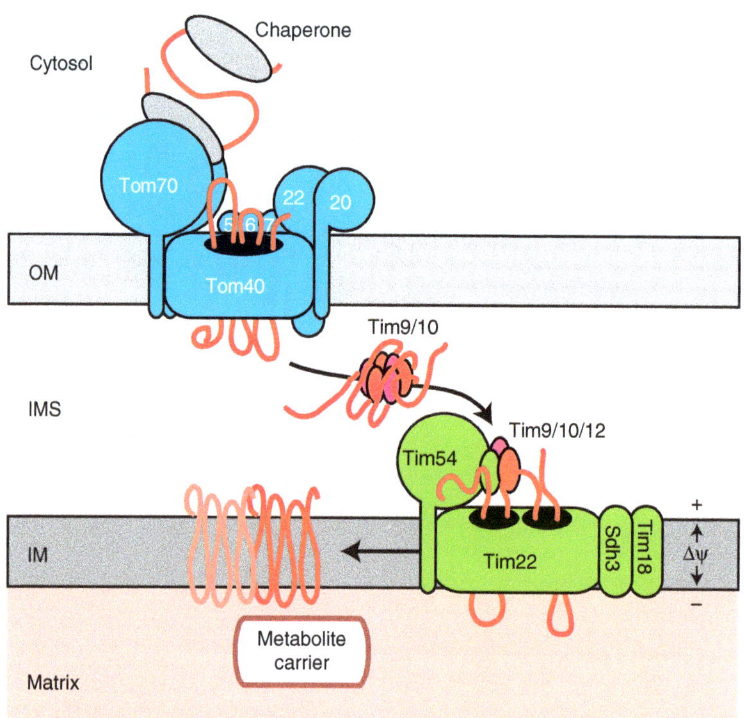

Figure 4. Insertion of metabolite carriers into the inner mitochondrial membrane. The Tom70 receptor takes over metabolite carrier precursors from cytosolic chaperones. The precursors pass the outer membrane in a loop conformation. In the IMS, the hydrophobic transmembrane segments of carriers are shielded by the Tim9/Tim10 chaperone complex, which delivers the precursor to the TIM22 machinery in the inner membrane. Tim54 and the adaptor protein Tim12 transfer the precursor to the protein-conducting channel formed by Tim22. Tim18, and Sdh3 are required for assembly and stability of the translocase. The membrane potential across the inner mitochondrial membrane ($\Delta\psi$) drives the insertion of metabolite carriers, which finally adopt their functional, dimeric state. OM, outer membrane; IMS, intermembrane space; IM, inner membrane.

 Cite this article as *Cold Spring Harb Perspect Biol* doi: 10.1101/cshperspect.a011320

Tim9, Tim10, and Tim12 (Sirrenberg et al. 1996, 1998; Kerscher et al. 1997, 2000; Koehler et al. 1998a,b, 2000; Gebert et al. 2011b). Tim22 forms a channel in the inner membrane for precursor insertion (Rehling et al. 2003), whereas Tim54 exposes a large domain to the IMS and is believed to provide a binding site for the Tim9–Tim10–Tim12 complex (Fig. 4) (Gebert et al. 2008; Wagner et al. 2008). Tim18 facilitates the assembly of Tim54 with Tim22 (Wagner et al. 2008). A recent study reported on the surprising finding that the integral inner membrane protein Sdh3, which has been known as a subunit of respiratory chain complex II (succinate dehydrogenase, SDH complex), is also a genuine subunit of the TIM22 complex. Sdh3 forms a subcomplex with Tim18, which is homologous to Sdh4, the binding partner of Sdh3 in the SDH complex (Gebert et al. 2011b). This finding is interesting from the evolutionary perspective, as it exemplifies how certain protein modules have been used to build different membrane protein machineries in mitochondria.

Cytosolic chaperones deliver carrier precursors to the Tom70 receptor in an ATP-dependent process (Wiedemann et al. 2001; Young et al. 2003; Bhangoo et al. 2007; Zara et al. 2009). They pass the TOM complex in a loop-like conformation and are bound in the IMS by the small TIM chaperones (Fig. 4). The hexameric Tim9/Tim10 and Tim8/Tim13 complexes transport these hydrophobic substrates across the aqueous IMS and deliver them to the inner membrane (Curran et al. 2002; Webb et al. 2006; Baker et al. 2009). Biochemical data suggest that the Tim9/Tim10 and Tim8/Tim13 complexes bind to the hydrophobic segments of the model substrates AAC and Tim23, respectively, concomitant with their chaperone-like role (Curran et al. 2002; Vergnolle et al. 2005; Davis et al. 2007). The crystal structure of the Tim9/Tim10 complex revealed an α-propeller topology with helical blades pointing away from a central opening that is lined by polar residues—a similar domain arrangement as for the chaperones Skp and prefoldin (Webb et al. 2006). With the help of the small TIM protein Tim12, the precursor-loaded chaperone complex docks to the membrane-embedded core of the TIM22 machinery, where the precursor is integrated into the inner membrane again by a looping mechanism (Fig. 4) (Rehling et al. 2003; Gebert et al. 2008). Membrane-inserted metabolite carriers subsequently assemble into their mature dimeric form. The only required energy source for the insertion process is the membrane potential ($\Delta\psi$) across the inner mitochondrial membrane.

INNER MEMBRANE INSERTION AND MATRIX TRANSLOCATION BY THE TIM23 MACHINERY

Precursor proteins with amino-terminal presequences (preproteins) are handed over from the TOM complex in the outer membrane to the translocase of the inner membrane 23 (TIM23 complex; presequence translocase). This process does not involve soluble IMS stages, but leads to the formation of two-membrane-spanning transport intermediates (TOM–TIM23 supercomplexes), in which the amino-terminal portion of the preprotein has been inserted into the TIM23 complex, whereas the carboxy-terminal domain is still inside the TOM complex (Dekker et al. 1997; Chacinska et al. 2003). The essential membrane-embedded core of the TIM23 complex ($TIM23^{CORE}$) is composed of three subunits (Fig. 5): the channel-forming Tim23 protein (Meinecke et al. 2006; van der Laan et al. 2007; Alder et al. 2008); Tim17, which is involved in gating of the Tim23 channel and membrane insertion of preproteins (Chacinska et al. 2005; Meier et al. 2005) and Tim50, which induces the closing of the Tim23 protein-conducting channel in the absence of preproteins (Meinecke et al. 2006). Tim50 exposes a large domain to the IMS, which is in close proximity to the outer membrane TOM complex, and together with the amino-terminal IMS domain of Tim23 constitutes the receptor domain of the TIM23 machinery (Gevorkyan-Airapetov et al. 2009; Mokranjac et al. 2009; Tamura et al. 2009a; Marom et al. 2011; Qian et al. 2011; Shiota et al. 2011). Two recent studies have indicated that Tim50 might have two distinct binding sites for preproteins in the conserved core domain and in the carboxy-terminal domain, respectively

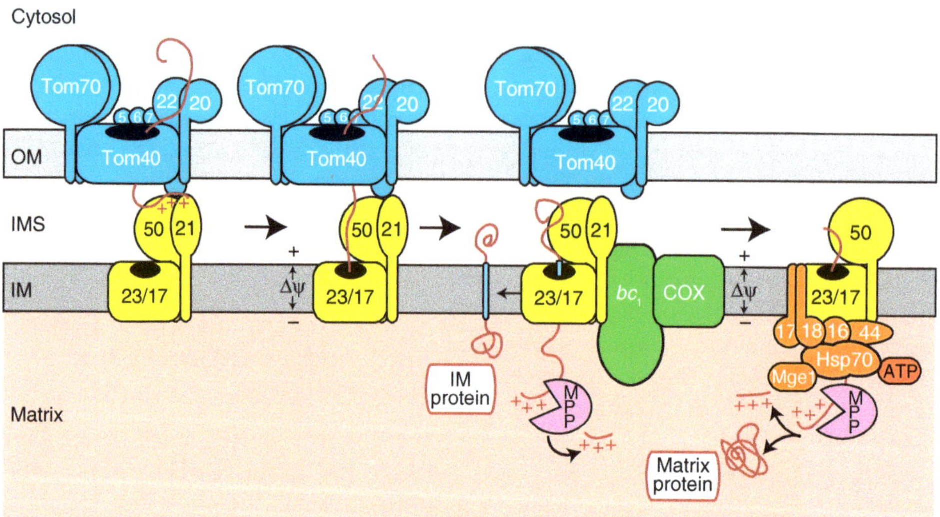

Figure 5. Import of presequence-containing precursor proteins (preproteins) into the inner membrane and matrix. Amino-terminal positively charged presequences are preferentially recognized by the Tom20 receptor on the mitochondrial surface. On the *trans* side of the protein-conducting Tom40 channel, presequences are bound by the IMS domain of Tom22. Membrane potential (Δψ)-dependent transfer of preproteins to the TIM23 machinery in the inner membrane involves Tim50 and Tim21. Preproteins with a hydrophobic stop-transfer signal (blue bar) are directly released from the protein-conducting channel formed by Tim23 into the inner membrane. This Δψ-dependent process is supported by the Tim21-dependent coupling of respiratory chain supercomplexes to TIM23. Stepwise translocation of soluble preproteins into the matrix requires the dynamic coupling of the ATP-driven mtHsp70-containing import motor, which leads to a loss of Tim21 from the TIM23 complex. Finally, the matrix processing peptidase (MPP) proteolytically removes presequences from both inner membrane and matrix preproteins. OM, outer membrane; IMS, intermembrane space; IM, inner membrane; bc_1, cytochrome bc_1 complex (complex III); COX, cytochrome *c* oxidase (complex IV).

(Qian et al. 2011; Schulz et al. 2011). The Tim21 protein connects with $TIM23^{CORE}$ and promotes preprotein transfer from the TOM to the TIM23 complex, probably by directly interacting with the IMS domain of Tom22 (Fig. 5) (Chacinska et al. 2005; Albrecht et al. 2006).

The $TIM23^{CORE}$-Tim21 complex has been termed $TIM23^{SORT}$, as this complex is sufficient to integrate preproteins with a hydrophobic stop-transfer signal adjacent to the amino-terminal presequence into the inner membrane in a Δψ-dependent manner (van der Laan et al. 2007). Remarkably, $TIM23^{SORT}$ is dynamically coupled to respiratory chain supercomplexes consisting of complex III (cytochrome bc_1 complex) and complex IV (cytochrome *c* oxidase), most likely via a direct interaction between Tim21 and the complex III subunit Qcr6 (Fig. 5) (van der Laan et al. 2006, 2007; Wiedemann et al. 2007; Saddar and Stuart 2008). Association of respiratory chain complexes to TIM23 was shown to stimulate the Δψ-dependent step of preprotein membrane insertion, which appears to be particularly important under conditions, where the overall Δψ across the inner membrane becomes limiting (van der Laan et al. 2006).

For complete import of precursors into the mitochondrial matrix the energy derived from Δψ in not sufficient. Therefore, the TIM23 complex has to be coupled to the ATP-driven presequence translocase-associated import motor (PAM) for matrix translocation (Fig. 5). The central subunit of PAM is the mitochondrial heat shock protein 70 (mtHsp70) that converts the energy from repeated cycles of ATP hydrolysis into a matrix-directed vectorial movement of the preprotein (Neupert and Herrmann 2007; Mapa et al. 2010). This activity of mtHsp70 is regulated by the multiple membrane-bound co-chaperones of PAM: The J-protein Pam18 stimulates the ATPase activity of mtHsp70 (D'Silva et al. 2003; Mokranjac et al. 2003; Truscott et al. 2003). The J-like protein Pam16 associates with Pam18 in the J-complex, recruits Pam18 to TIM23^{CORE}, and regulates its ATPase-stimulating function (Frazier et al. 2004; Kozany et al. 2004; Li et al. 2004; Mokranjac et al. 2006; D'Silva et al. 2008, Pais et al. 2011). Tim44 is crucial for the binding of both, mtHsp70 and J-complex to TIM23^{CORE} (D'Silva et al. 2004; Hutu et al. 2008; Schiller et al. 2008). The matrix-localized nucleotide exchange factor Mge1 is required for the completion of the mtHsp70 reaction cycle (Slutsky-Leiderman et al. 2007). Pam17 is thought to trigger the dynamic assembly of PAM (van der Laan et al. 2005; Hutu et al. 2008). The stepwise recruitment of PAM components goes along with a release of Tim21 from TIM23^{CORE} (Chacinska et al. 2005; van der Laan et al. 2005; Popov-Čeleketić et al. 2008b). Thus, the majority of TIM23^{CORE} complexes in mitochondria are either associated with Tim21 or PAM (Fig. 5) (Chacinska et al. 2005, 2010; van der Laan et al. 2005, 2006).

The picture of the TIM23 complex emerging from these findings is that of a highly dynamic molecular machine that switches between different functional and structural states depending on the signal information within an incoming preprotein. This view is best exemplified by the recently unraveled biogenesis pathway of the mitochondrial ATP-binding cassette (ABC) transporter Mdl1, a polytopic inner membrane protein with an amino-terminal presequence (Bohnert et al. 2010). Earlier studies suggested that polytopic inner membrane proteins are first fully translocated across the membrane in a PAM-dependent manner and afterwards integrated into the bilayer from the matrix side by the oxidase assembly (OXA) complex, a membrane protein insertase of bacterial origin that also mediates the biogenesis of mitochondrially encoded respiratory chain subunits (Neupert and Herrmann 2007). However, membrane insertion of Mdl1 was found to be a modular process, in which certain transmembrane segment are laterally released from the TIM23 complex independently of PAM, whereas other transmembrane segments are routed to the inner membrane via the matrix and the OXA machinery (Bohnert et al. 2010). Such a complex import mechanism requires multiple functional switches and modular rearrangements within the TIM23 complex that must be tightly controlled by specific import signals and partner proteins of TIM23.

CONCLUSIONS AND PERSPECTIVES

Protein trafficking to and within mitochondria is possibly the most intricate in the context of a eukaryotic cell (the same would apply to chloroplasts). Indeed, it is fascinating how nature has answered the difficult question of protein sorting within mitochondria by providing elaborate transport systems to ensure that precursor proteins acquire their mature and functional state. Future perspectives for the field include addressing the role of phospholipids in protein transport processes. Recent studies have indicated that cardiolipin is important for the stability and functionality of not only respiratory chain complexes, but also protein translocation devices, like the TIM23/PAM machinery and the TOM and SAM complexes (Kutik et al. 2008b; Gebert et al. 2009; Tamura et al. 2009b). Certainly, high-resolution structures of the membrane-embedded core domains of translocation complexes will be invaluable for learning more about the organization and operation of these machineries. More information is also required on how the import systems are regulated and how the effectiveness of import is promoted through specific coupling of translocation machineries. Recent insights into the role of cytosolic kinases for the assembly and activity of the TOM complex and the cooperation of the TOM machinery with downstream protein sorting pathways have highlighted the central importance of this aspect (Chacinska et al. 2010; Schmidt et al. 2011; von der Malsburg et al. 2011). Despite many efforts, we still do not fully understand how specific import signals in precursor proteins are recognized by receptor domains and protein-conducting channels and how these signals are transmitted to induce large-scale rearrangements and functional switches in the mitochondrial protein import systems. Finally, an exciting link between protein import and machineries implicated in the regulation of mitochondrial morphology has emerged, which needs to be further explored (Meisinger et al. 2007; Kornmann et al. 2010; von der Malsburg et al. 2011).

In this article, we have highlighted our current understanding of mitochondrial import based on research that mainly utilized fungal models. It is undeniable that this field will continue to rely on such models; however, the crucial importance of understanding these processes in mammalian cells should be emphasized. Despite a sound knowledge of the protein import machinery in mammalian cells, there has so far been little focus on how problems in mitochondrial protein import can contribute to human disease, although evidence supporting such connections is mounting. For instance, Mohr-Tranebjaerg syndrome (MTS) is caused by mutations in the small TIM chaperone deafness dystonia polypeptide 1 (DDP1/TIMM8/Tim8) (Koehler et al. 1999). The underlying severity of the disease may be attributable to impaired biogenesis of the TIM23 complex as a result of dysfunctional Tim8. Defects in Pam18 (DNAJC19) have been linked to dilated cardiomyopathy with ataxia (DCMA) (Davey et al. 2006; Sinha et al. 2010). TIMM17A mRNA expression was recently shown to be significantly higher in breast cancer compared with normal or benign breast tissue (Salhab et al. 2010). Moreover, TIM17 was identified as a multicopy suppressor of mtDNA instability in specific mutant backgrounds in yeast, which may be relevant for certain forms of mtDNA depletion syndrome (Iacovino et al. 2009). Hsp60, a central player in protein folding and assembly of mitochondria, has been implicated in spastic paraplegia-13 (SPG13), an autosomally dominant form of pure hereditary spastic paraplegia (Fontaine et al. 2000; Hansen et al. 2002). These observations reiterate the significance of a functional mitochondrial protein biogenesis network for health and disease and the importance of gaining a greater understanding of the import machinery in higher eukaryotes.

ACKNOWLEDGMENTS

This work is supported by the Deutsche Forschungsgemeinschaft, Sonderforschungsbereich 746, Excellence Initiative of the German Federal and State Governments (EXC 294), Gottfried Wilhelm Leibniz Program, the Fonds der Chemischen Industrie, the Australian Research Council, and the Clive and Vera Ramaciotti Foundation.

REFERENCES

Abe Y, Shodai T, Muto T, Mihara K, Torii H, Nishikawa S, Endo T, Kohda D. 2000. Structural basis of presequence recognition by the mitochondrial protein import receptor Tom20. *Cell* **100:** 551–560.

Ahting U, Thun C, Hegerl R, Typke D, Nargang FE, Neupert W, Nussberger S. 1999. The TOM core complex: The general protein import pore of the outer membrane of mitochondria. *J Cell Biol* **147:** 959–968.

Albrecht R, Rehling P, Chacinska A, Brix J, Cadamuro SA, Volkmer R, Guiard B, Pfanner N, Zeth K. 2006. The Tim21 binding domain connects the preprotein translocases of both mitochondrial membranes. *EMBO Rep* **7:** 1233–1238.

Alder NN, Jensen RE, Johnson AE. 2008. Fluorescence mapping of mitochondrial TIM23 complex reveals a water-facing, substrate-interacting helix surface. *Cell* **134:** 439–450.

Baker MJ, Webb CT, Stroud DA, Palmer CS, Frazier AE, Guiard B, Chacinska A, Gulbis JM, Ryan MT. 2009. Structural and functional requirements for activity of the Tim9-Tim10 complex in mitochondrial protein import. *Mol Biol Cell* **20:** 769–779.

Banci L, Bertini I, Cefaro C, Ciofi-Baffoni S, Gallo A, Martinelli M, Sideris DP, Katrakili N, Tokatlidis K. 2009. MIA40 is an oxidoreductase that catalyzes oxidative protein folding in mitochondria. *Nat Struct Mol Biol* **16:** 198–206.

Banci L, Bertini I, Cefaro C, Cenacchi L, Ciofi-Baffoni S, Felli IC, Gallo A, Gonnelli L, Luchinat E, Sideris D, et al. 2010. Molecular chaperone function of Mia40 triggers consecutive induced folding steps of the substrate in mitochondrial protein import. *Proc Natl Acad Sci* **107:** 20190–20195.

Banci L, Bertini I, Calderone V, Cefaro C, Ciofi-Baffoni S, Gallo A, Kallergi E, Lionaki E, Pozidis C, Tokatlidis K. 2011. Molecular recognition and substrate mimicry drive the electron-transfer process between MIA40 and ALR. *Proc Natl Acad Sci* **108:** 4811–4816.

Becker L, Bannwarth M, Meisinger C, Hill K, Model K, Krimmer T, Casadio R, Truscott KN, Schulz GE, Pfanner N, et al. 2005. Preprotein translocase of the outer mitochondrial membrane: Reconstituted Tom40 forms a characteristic TOM pore. *J Mol Biol* **353:** 1011–1020.

Becker T, Pfannschmidt S, Guiard B, Stojanovski D, Milenkovic D, Kutik S, Pfanner N, Meisinger C, Wiedemann N. 2008. Biogenesis of the mitochondrial TOM complex: Mim1 promotes insertion and assembly of signal-anchored receptors. *J Biol Chem* **283:** 120–127.

Becker T, Guiard B, Thornton N, Zufall N, Stroud DA, Wiedemann N, Pfanner N. 2010. Assembly of the mitochondrial protein import channel: Role of Tom5 in two-stage interaction of Tom40 with the SAM complex. *Mol Biol Cell* **21:** 3106–3113.

Becker T, Wenz LS, Thornton N, Stroud D, Meisinger C, Wiedemann N, Pfanner N. 2011a. Biogenesis of mitochondria: Dual role of Tom7 in modulating assembly of the preprotein translocase of the outer membrane. *J Mol Biol* **405:** 113–124.

Becker T, Wenz LS, Krüger V, Lehmann W, Müller JM, Goroncy L, Zufall N, Lithgow T, Guiard B, Chacinska A, et al. 2011b. The mitochondrial import protein Mim1 promotes biogenesis of multispanning outer membrane proteins. *J Cell Biol* **194:** 387–395.

Bhangoo MK, Tzankov S, Fan AC, Dejgaard K, Thomas DY, Young JC. 2007. Multiple 40-kDa heat-shock protein chaperones function in Tom70-dependent mitochondrial import. *Mol Biol Cell* **18:** 3414–3428.

Bien M, Longen S, Wagener N, Chwalla I, Herrmann JM, Riemer J. 2010. Mitochondrial disulfide bond formation is driven by intersubunit electron transfer in Erv1 and proofread by glutathione. *Mol Cell* **37:** 516–528.

Bihlmaier K, Mesecke N, Terziyska N, Bien M, Hell K, Herrmann JM. 2007. The disulfide relay system of mitochondria is connected to the respiratory chain. *J Cell Biol* **179:** 389–395.

Bohnert M, Rehling P, Guiard B, Herrmann JM, Pfanner N, van der Laan M. 2010. Cooperation of stop-transfer and conservative sorting mechanisms in mitochondrial protein transport. *Curr Biol* **20:** 1227–1232.

Boldogh IR, Nowakowski DW, Yang HC, Chung H, Karmon S, Royes P, Pon LA. 2003. A protein complex containing Mdm10p, Mdm12p, and Mmm1p links mitochondrial membranes and DNA to the cytoskeleton-based segregation machinery. *Mol Biol Cell* **14:** 4618–4627.

Brandner K, Rehling P, Truscott KN. 2005. The carboxyl-terminal third of the dicarboxylate carrier is crucial for productive association with the inner membrane twin-pore translocase. *J Biol Chem* **280:** 6215–6221.

Brix J, Dietmeier K, Pfanner N. 1997. Differential recognition of preproteins by the purified cytosolic domains of the mitochondrial import receptors Tom20, Tom22, and Tom70. *J Biol Chem* **272:** 20730–20735.

Chacinska A, Rehling P, Guiard B, Frazier AE, Schulze-Specking A, Pfanner N, Voos W, Meisinger C. 2003. Mitochondrial translocation contact sites: Separation of dynamic and stabilizing elements in formation of a TOM-TIM-preprotein supercomplex. *EMBO J* **22:** 5370–5381.

Chacinska A, Pfannschmidt S, Wiedemann N, Kozjak V, Sanjuán Szklarz LK, Schulze-Specking A, Truscott KN, Guiard B, Meisinger C, Pfanner N. 2004. Essential role of Mia40 in import and assembly of mitochondrial intermembrane space proteins. *EMBO J* **23:** 3735–3746.

Chacinska A, Lind M, Frazier AE, Dudek J, Meisinger C, Geissler A, Sickmann A, Meyer HE, Truscott KN, Guiard B, et al. 2005. Mitochondrial presequence translocase: Switching between TOM tethering and motor recruitment involves Tim21 and Tim17. *Cell* **120:** 817–829.

Chacinska A, Guiard B, Müller JM, Schulze-Specking A, Gabriel K, Kutik S, Pfanner N. 2008. Mitochondrial biogenesis, switching the sorting pathway of the intermembrane space receptor Mia40. *J Biol Chem* **283:** 29723–29729.

Chacinska A, van der Laan M, Mehnert CS, Guiard B, Mick DU, Hutu DP, Truscott KN, Wiedemann N, Meisinger C, Pfanner N, et al. 2010. Distinct forms of mitochondrial TOM-TIM supercomplexes define signal-dependent states of preprotein sorting. *Mol Cell Biol* **30:** 307–318.

Chan NC, Lithgow T. 2008. The peripheral membrane subunits of the SAM complex function codependently in mitochondrial outer membrane biogenesis. *Mol Biol Cell* **19:** 126–136.

Curran SP, Leuenberger D, Oppliger W, Koehler CM. 2002. The Tim9p-Tim10p complex binds to the transmembrane domains of the ADP/ATP carrier. *EMBO J* **21:** 942–953.

Dabir DV, Leverich EP, Kim SK, Tsai FD, Hirasawa M, Knaff DB, Koehler CM. 2007. A role for cytochrome *c* and cytochrome *c* peroxidase in electron shuttling from Erv1. *EMBO J* **26:** 4801–4811.

Davey KM, Parboosingh JS, McLeod DR, Chan A, Casey R, Ferreira P, Snyder FF, Bridge PJ, Bernier FP. 2006. Mutation of DNAJC19, a human homologue of yeast inner mitochondrial membrane co-chaperones, causes DCMA syndrome, a novel autosomal recessive Barth syndrome-like condition. *J Med Genet* **43:** 385–393.

Davis AJ, Alder NN, Jensen RE, Johnson AE. 2007. The Tim9p/10p and Tim8p/13p complexes bind to specific sites on Tim23p during mitochondrial protein import. *Mol Biol Cell* **18:** 475–486.

Dekker PJ, Martin F, Maarse AC, Bömer U, Müller H, Guiard B, Meijer M, Rassow J, Pfanner N. 1997. The Tim core complex defines the number of mitochondrial translocation sites and can hold arrested preproteins in the absence of matrix Hsp70-Tim44. *EMBO J* **16:** 5408–5419.

Dekker PJ, Ryan MT, Brix J, Müller H, Hönlinger A, Pfanner N. 1998. Preprotein translocase of the outer mitochondrial membrane: molecular dissection and assembly of the general import pore complex. *Mol Cell Biol* **18:** 6515–6524.

Devaux F, Lelandais G, Garcia M, Goussard S, Jacq C. 2010. Posttranscriptional control of mitochondrial biogenesis: Spatio-temporal regulation of the protein import process. *FEBS Lett* **584:** 4273–4279.

Dolezal P, Likic V, Tachezy J, Lithgow T. 2006. Evolution of the molecular machines for protein import into mitochondria. *Science* **313:** 314–318.

D'Silva PD, Schilke B, Walter W, Andrew A, Craig EA. 2003. J protein cochaperone of the mitochondrial inner membrane required for protein import into the mitochondrial matrix. *Proc Natl Acad Sci* **100:** 13839–13844.

D'Silva P, Liu Q, Walter W, Craig EA. 2004. Regulated interactions of mtHsp70 with Tim44 at the translocon in the mitochondrial inner membrane. *Nat Struct Mol Biol* **11:** 1084–1091.

D'Silva PR, Schilke B, Hayashi M, Craig EA. 2008. Interaction of the J-protein heterodimer Pam18/Pam16 of the mitochondrial import motor with the translocon of the inner membrane. *Mol Biol Cell* **19:** 424–432.

Dukanovic J, Dimmer KS, Bonnefoy N, Krumpe K, Rapaport D. 2009. Genetic and functional interactions between the mitochondrial outer membrane proteins Tom6 and Sam37. *Mol Cell Biol* **29:** 5975–5988.

Fontaine B, Davoine CS, Durr A, Paternotte C, Feki I, Weissenbach J, Hazan J, Brice A. 2000. A new locus for autosomal dominant pure spastic paraplegia, on chromosome 2q24-q34. *Am J Hum Genet* **66:** 702–707.

Frazier AE, Dudek J, Guiard B, Voos W, Li Y, Lind M, Meisinger C, Geissler A, Sickmann A, Meyer HE, et al. 2004. Pam16 has an essential role in the mitochondrial protein import motor. *Nat Struct Mol Biol* **11:** 226–233.

Garcia M, Darzacq X, Delaveau T, Jourdren L, Singer RH, Jacq C. 2007. Mitochondria-associated yeast mRNAs and the biogenesis of molecular complexes. *Mol Biol Cell* **18:** 362–368.

Garcia-Rodriguez LJ, Gay AC, Pon LA. 2007. Puf3p, a Pumilio family RNA binding protein, localizes to mitochondria and regulates mitochondrial biogenesis and motility in budding yeast. *J Cell Biol* **176:** 197–207.

Gebert N, Chacinska A, Wagner K, Guiard B, Koehler CM, Rehling P, Pfanner N, Wiedemann N. 2008. Assembly of the three small Tim proteins precedes docking to the mitochondrial carrier translocase. *EMBO Rep* **9:** 548–554.

Gebert N, Joshi AS, Kutik S, Becker T, McKenzie M, Guan XL, Mooga VP, Stroud DA, Kulkarni G, Wenk MR, et al. 2009. Mitochondrial cardiolipin involved in outer-membrane protein biogenesis: Implications for Barth syndrome. *Curr Biol* **19:** 2133–2139.

Gebert N, Ryan MT, Pfanner N, Wiedemann N, Stojanovski D. 2011a. Mitochondrial protein import machineries and lipids: A functional connection. *Biochim Biophys Acta* **1808:** 1002–1011.

Gebert N, Gebert M, Oeljeklaus S, von der Malsburg K, Stroud DA, Kulawiak B, Wirth C, Zahedi RP, Dolezal P, Wiese S, et al. 2011b. Dual function of Sdh3 in the respiratory chain and TIM22 protein translocase of the mitochondrial inner membrane. *Mol Cell* **44:** 811–818.

Gentle I, Gabriel K, Beech P, Waller R, Lithgow T. 2004. The Omp85 family of proteins is essential for outer membrane biogenesis in mitochondria and bacteria. *J Cell Biol* **164:** 19–24.

Gevorkyan-Airapetov L, Zohary K, Popov-Čeleketić D, Mapa K, Hell K, Neupert W, Azem A, Mokranjac D. 2009. Interaction of Tim23 with Tim50 is essential for protein translocation by the mitochondrial TIM23 complex. *J Biol Chem* **284:** 4865–4872.

Glick BS, Brandt A, Cunningham K, Müller S, Hallberg RL, Schatz G. 1992. Cytochromes c_1 and b_2 are sorted to the intermembrane space of yeast mitochondria by a stop-transfer mechanism. *Cell* **69:** 809–822.

Grumbt B, Stroobant V, Terziyska N, Israel L, Hell K. 2007. Functional characterization of Mia40p, the central component of the disulfide relay system of the mitochondrial intermembrane space. *J Biol Chem* **282:** 37461–37470.

Hansen JJ, Durr A, Cournu-Rebeix I, Georgopoulos C, Ang D, Nielsen MN, Davoine CS, Brice A, Fontaine B, Gregersen N, et al. 2002. Hereditary spastic paraplegia SPG13 is associated with a mutation in the gene encoding the mitochondrial chaperonin Hsp60. *Am J Hum Genet* **70:** 1328–1332.

Harner M, Körner C, Walther D, Mokranjac D, Kaesmacher J, Welsch U, Griffith J, Mann M, Reggiori F, Neupert W. 2011. The mitochondrial contact site complex, a determinant of mitochondrial ultrastructure. *EMBO J* **30:** 4356–4370.

Hill K, Model K, Ryan MT, Dietmeier K, Martin F, Wagner R, Pfanner N. 1998. Tom40 forms the hydrophilic channel of the mitochondrial import pore for preproteins. *Nature* **395:** 516–521.

Hoppins SC, Nargang FE. 2004. The Tim8-Tim13 complex of *Neurospora crassa* functions in the assembly of proteins into both mitochondrial membranes. *J Biol Chem* **279:** 12396–12405.

Hoppins S, Collins SR, Cassidy-Stone A, Hummel E, Devay RM, Lackner LL, Westermann B, Schuldiner M, Weissman JS, Nunnari J. 2011. A mitochondrial-focused genetic interaction map reveals a scaffold-like complex required for inner membrane organization in mitochondria. *J Cell Biol* **195:** 323–340.

Hulett JM, Lueder F, Chan NC, Perry AJ, Wolynec P, Likic VA, Gooley PR, Lithgow T. 2008. The transmembrane segment of Tom20 is recognized by Mim1 for docking to the mitochondrial TOM complex. *J Mol Biol* **376:** 694–704.

Hutu DP, Guiard B, Chacinska A, Becker D, Pfanner N, Rehling P, van der Laan M. 2008. Mitochondrial protein import motor: Differential role of Tim44 in the recruitment of Pam17 and J-complex to the presequence translocase. *Mol Biol Cell* **19:** 2642–2649.

Iacovino M, Granycome C, Sembongi H, Bokori-Brown M, Butow RA, Holt IJ, Bateman JM. 2009. The conserved translocase Tim17 prevents mitochondrial DNA loss. *Hum Mol Genet* **18:** 65–74.

Ishikawa D, Yamamoto H, Tamura Y, Moritoh K, Endo T. 2004. Two novel proteins in the mitochondrial outer membrane mediate β-barrel protein assembly. *J Cell Biol* **166:** 621–627.

Kawano S, Yamano K, Naoé M, Momose T, Terao K, Nishikawa S, Watanabe N, Endo T. 2009. Structural basis of yeast Tim40/Mia40 as an oxidative translocator in the mitochondrial intermembrane space. *Proc Natl Acad Sci* **106:** 14403–14407.

Kemper C, Habib SJ, Engl G, Heckmeyer P, Dimmer KS, Rapaport D. 2008. Integration of tail-anchored proteins into the mitochondrial outer membrane does not require any known import components. *J Cell Sci* **121:** 1990–1998.

Kerscher O, Holder J, Srinivasan M, Leung RS, Jensen RE. 1997. The Tim54p-Tim22p complex mediates insertion of proteins into the mitochondrial inner membrane. *J Cell Biol* **139:** 1663–1675.

Kerscher O, Sepuri NB, Jensen RE. 2000. Tim18p is a new component of the Tim54p-Tim22p translocon in the mitochondrial inner membrane. *Mol Biol Cell* **11:** 103–116.

Knowles TJ, Scott-Tucker A, Overduin M, Henderson IR. 2009. Membrane protein architecture: the role of the BAM complex in outer membrane protein assembly. *Nat Rev Microbiol* **7:** 206–214.

Koehler CM, Jarosch E, Tokatlidis K, Schmid K, Schweyen RJ, Schatz G. 1998a. Import of mitochondrial carriers mediated by essential proteins of the intermembrane space. *Science* **279:** 369–373.

Koehler CM, Merchant S, Oppliger W, Schmid K, Jarosch E, Dolfini L, Junne T, Schatz G, Tokatlidis K. 1998b. Tim9p, an essential partner subunit of Tim10p for the import of mitochondrial carrier proteins. *EMBO J* **17:** 6477–6486.

Koehler CM, Leuenberger D, Merchant S, Renold A, Junne T, Schatz G. 1999. Human deafness dystonia syndrome is a mitochondrial disease. *Proc Natl Acad Sci* **96:** 2141–2146.

Koehler CM, Murphy MP, Bally NA, Leuenberger D, Oppliger W, Dolfini L, Junne T, Schatz G, Or E. 2000. Tim18p, a new subunit of the TIM22 complex that mediates insertion of imported proteins into the yeast mitochondrial inner membrane. *Mol Cell Biol* **20:** 1187–1193.

Komiya T, Rospert S, Koehler C, Looser R, Schatz G, Mihara K. 1998. Interaction of mitochondrial targeting signals with acidic receptor domains along the protein import pathway: Evidence for the "acid chain" hypothesis. *EMBO J* **17:** 3886–3898.

Kornmann B, Walter P. 2010. ERMES-mediated ER-mitochondria contacts: Molecular hubs for the regulation of mitochondrial biology. *J Cell Sci* **123:** 1389–1393.

Kornmann B, Currie E, Collins SR, Schuldiner M, Nunnari J, Weissman JS, Walter P. 2009. An ER-mitochondria tethering complex revealed by a synthetic biology screen. *Science* **325:** 477–481.

Kornmann B, Osman C, Walter P. 2011. The conserved GTPase Gem1 regulates endoplasmic reticulum-mitochondria connections. *Proc Natl Acad Sci* **108:** 14151–14156.

Kozany C, Mokranjac D, Sichting M, Neupert W, Hell K. 2004. The J domain-related cochaperone Tim16 is a constituent of the mitochondrial TIM23 preprotein translocase. *Nat Struct Mol Biol* **11:** 234–241.

Kozjak V, Wiedemann N, Milenkovic D, Lohaus C, Meyer HE, Guiard B, Meisinger C, Pfanner N. 2003. An essential role of Sam50 in the protein sorting and assembly machinery of the mitochondrial outer membrane. *J Biol Chem* **278:** 48520–48523.

Künkele KP, Heins S, Dembowski M, Nargang FE, Benz R, Thieffry M, Walz J, Lill R, Nussberger S, Neupert W. 1998. The preprotein translocation channel of the outer membrane of mitochondria. *Cell* **93:** 1009–1019.

Kutik S, Stojanovski D, Becker L, Becker T, Meinecke M, Krüger V, Prinz C, Meisinger C, Guiard B, Wagner R, et al. 2008a. Dissecting membrane insertion of mitochondrial β-barrel proteins. *Cell* **132:** 1011–1024.

Kutik S, Rissler M, Guan XL, Guiard B, Shui G, Gebert N, Heacock PN, Rehling P, Dowhan W, Wenk MR, et al. 2008b. The translocator maintenance protein Tam41 is required for mitochondrial cardiolipin biosynthesis. *J Cell Biol* **183:** 1213–1221.

Li Y, Dudek J, Guiard B, Pfanner N, Rehling P, Voos W. 2004. The presequence translocase-associated protein import motor of mitochondria. Pam16 functions in an antagonistic manner to Pam18. *J Biol Chem* **279:** 38047–38054.

Luk E, Yang M, Jensen LT, Bourbonnais Y, Culotta VC. 2005. Manganese activation of superoxide dismutase 2 in the mitochondria of *Saccharomyces cerevisiae*. *J Biol Chem* **280:** 22715–22720.

Mapa K, Sikor M, Kudryavtsev, Waegemann K, Kalinin S, Seidel CA, Neupert W, Lamb DC, Mokranjac D. 2010. The conformational dynamics of the mitochondrial Hsp70 chaperone. *Mol Cell* **38:** 89–100.

Marc P, Margeot A, Devaux F, Blugeon C, Corral-Debrinski M, Jacq C. 2002. Genome-wide analysis of mRNAs targeted to yeast mitochondria. *EMBO Rep* **3:** 159–164.

Marom M, Dayan D, Demishtein-Zohary K, Mokranjac D, Neupert W, Azem A. 2011. Direct interaction of mitochondrial targeting presequences with purified components of the TIM23 complex. *J Biol Chem* **286:** 43809–43815.

Meier S, Neupert W, Herrmann JM. 2005. Conserved N-terminal negative charges in the Tim17 subunit of the TIM23 translocase play a critical role in the import of preproteins into mitochondria. *J Biol Chem* **280:** 7777–7785.

Meinecke M, Wagner R, Kovermann P, Guiard B, Mick DU, Hutu DP, Voos W, Truscott KN, Chacinska A, Pfanner N, et al. 2006. Tim50 maintains the permeability barrier of the mitochondrial inner membrane. *Science* **312:** 1523–1526.

Meisinger C, Rissler M, Chacinska A, Sanjuán Szklarz LK, Milenkovic D, Kozjak V, Schönfisch B, Lohaus C, Meyer HE, Yaffe MP, et al. 2004. The mitochondrial morphology protein Mdm10 functions in assembly of the preprotein translocase of the outer membrane. *Dev Cell* **7:** 61–71.

Meisinger C, Wiedemann N, Rissler M, Strub A, Milenkovic D, Schönfisch B, Müller H, Kozjak V, Pfanner N. 2006. Mitochondrial protein sorting: Differentiation of β-barrel assembly by Tom7-mediated segregation of Mdm10. *J Biol Chem* **281:** 22819–22826.

Meisinger C, Pfannschmidt S, Rissler M, Milenkovic D, Becker T, Stojanovski D, Youngman MJ, Jensen RE, Chacinska A, Guiard B, et al. 2007. The morphology proteins Mdm12/Mmm1 function in the major β-barrel assembly pathway of mitochondria. *EMBO J* **26:** 2229–2239.

Mesecke N, Terziyska N, Kozany C, Baumann F, Neupert W, Hell K, Herrmann JM. 2005. A disulfide relay system in the intermembrane space of mitochondria that mediates protein import. *Cell* **121:** 1059–1069.

Milenkovic D, Kozjak V, Wiedemann N, Lohaus C, Meyer HE, Guiard B, Pfanner N, Meisinger C. 2004. Sam35 of the mitochondrial protein sorting and assembly machinery is a peripheral outer membrane protein essential for cell viability. *J Biol Chem* **279:** 22781–22785.

Milenkovic D, Gabriel K, Guiard B, Schulze-Specking A, Pfanner N, Chacinska A. 2007. Biogenesis of the essential Tim9-Tim10 chaperone complex of mitochondria: Site-specific recognition of cysteine residues by the intermembrane space receptor Mia40. *J Biol Chem* **282:** 22472–22480.

Milenkovic D, Ramming T, Müller JM, Wenz LS, Gebert N, Schulze-Specking A, Stojanovski D, Rospert S, Chacinska

A. 2009. Identification of the signal directing Tim9 and Tim10 into the intermembrane space of mitochondria. *Mol Biol Cell* **20:** 2530–2539.

Moczko M, Bömer U, Kübrich M, Zufall N, Hönlinger A, Pfanner N. 1997. The intermembrane space domain of mitchondrial Tom22 functions as a trans binding site for preproteins with N-terminal targeting sequences. *Mol Cell Biol* **17:** 6574–6584.

Model K, Meisinger C, Prinz T, Wiedemann N, Truscott KN, Pfanner N, Ryan MT. 2001. Multistep assembly of the protein import channel of the mitochondrial outer membrane. *Nat Struct Biol* **8:** 361–370.

Mokranjac D, Sichting M, Neupert W, Hell K. 2003. Tim14, a novel key component of the import motor of the TIM23 protein translocase of mitochondria. *EMBO J* **22:** 4945–4956.

Mokranjac D, Bourenkov G, Hell K, Neupert W, Groll M. 2006. Structure and function of Tim14 and Tim16, the J and J-like components of the mitochondrial protein import motor. *EMBO J* **25:** 4675–4685.

Mokranjac D, Sichting M, Popov-Čeleketić D, Mapa K, Gevorkyan-Airapetov L, Zohary K, Hell K, Azem A, Neupert W. 2009. Role of Tim50 in the transfer of precursor proteins from the outer to the inner membrane of mitochondria. *Mol Biol Cell* **20:** 1400–1407.

Naamati A, Regey-Rudzki N, Galperin S, Lill R, Pines O. 2009. Dual targeting of Nfs1 and discovery of its novel processing enzyme, Icp55. *J Biol Chem* **284:** 30200–30208.

Naoé M, Ohwa Y, Ishikawa D, Ohshima C, Nishikawa S, Yamamoto H, Endo T. 2004. Identification of Tim40 that mediates protein sorting to the mitochondrial intermembrane space. *J Biol Chem* **279:** 47815–47821.

Neupert W, Herrmann JM. 2007. Translocation of proteins into mitochondria. *Annu Rev Biochem* **76:** 723–749.

Nunnari J, Fox TD, Walter P. 1993. A mitochondrial protease with two catalytic subunits of nonoverlapping specificities. *Science* **262:** 1997–2004.

Otera H, Taira Y, Horie C, Suzuki Y, Suzuki H, Setoguchi K, Kato H, Oka T, Mihara K. 2007. A novel insertion pathway of mitochondrial outer membrane proteins with multiple transmembrane segments. *J Cell Biol* **179:** 1355–1363.

Pagliarini DJ, Calvo SE, Chang B, Sheth SA, Vafai SB, Ong SE, Walford GA, Sugiana C, Boneh A, Chen WK, et al. 2008. A mitochondrial protein compendium elucidates complex I disease biology. *Cell* **134:** 112–123.

Pais JE, Schilke B, Craig EA. 2011. Re-evaluation of the role of the Pam18:Pam16 interaction in translocation of proteins by the mitochondrial Hsp70-based import motor. *Mol Biol Cell* **22:** 4740–4749.

Papic D, Krumpe K, Dukanovic J, Dimmer KS, Rapaport D. 2011. Multispan mitochondrial outer membrane protein Ugo1 follows a unique Mim1-dependent import pathway. *J Cell Biol* **194:** 397–405.

Paschen SA, Waizenegger T, Stan T, Preuss M, Cyrklaff M, Hell K, Rapaport D, Neupert W. 2003. Evolutionary conservation of biogenesis of β-barrel membrane proteins. *Nature* **426:** 862–866.

Popov-Čeleketić J, Waizenegger T, Rapaport D. 2008a. Mim1 functions in an oligomeric form to facilitate the integration of Tom20 into the mitochondrial outer membrane. *J Mol Biol* **376:** 671–680.

Popov-Čeleketić D, Mapa K, Neupert W, Mokranjac D. 2008b. Active remodelling of the TIM23 complex during translocation of preproteins into mitochondria. *EMBO J* **27:** 1469–1480.

Prokisch H, Scharfe C, Camp DG 2nd, Xiao W, David L, Andreoli C, Monroe ME, Moore RJ, Gritsenko MA, Kozany C, et al. 2004. Integrative analysis of the mitochondrial proteome in yeast. *PLoS Biol* **2:** e160.

Qian X, Gebert M, Höpker J, Yan M, Li J, Wiedemann N, van der Laan M, Pfanner N, Sha B. 2011. Structural basis for the function of Tim50 in the mitochondrial presequence translocase. *J Mol Biol* **411:** 513–519.

Rehling P, Model K, Brandner K, Kovermann P, Sickmann A, Meyer HE, Kühlbrandt W, Wagner R, Truscott KN, Pfanner N. 2003. Protein insertion into the mitochondrial inner membrane by a twin-pore translocase. *Science* **299:** 1747–1751.

Reinders J, Zahedi RP, Pfanner N, Meisinger C, Sickmann A. 2006. Toward the complete yeast mitochondrial proteome: multidimensional separation techniques for mitochondrial proteomics. *J Proteome Res* **5:** 1543–1554.

Saddar S, Dienhart MK, Stuart RA. 2008. F_1F_0-ATP synthase complex influences the assembly state of the cytochrome *bc*$_1$-cytochrome oxidase supercomplex and its association with the TIM23 machinery. *J Biol Chem* **282:** 6677–6686.

Saint-Georges Y, Garcia M, Delaveau T, Jourdren L, Le Crom S, Lemoine S, Tanty V, Devaux F, Jacq C. 2008. Yeast mitochondrial biogenesis: A role for the PUF RNA-binding protein Puf3p in mRNA localization. *PLoS ONE* **4:** e2293.

Saitoh T, Igura M, Obita T, Ose T, Kojima R, Maenaka K, Endo T, Kohda D. 2007. Tom20 recognizes mitochondrial presequences through dynamic equilibrium among multiple bound states. *EMBO J* **26:** 4777–4787.

Salhab M, Patani N, Jiang W, Mokbel K. 2010. High TIMM17A expression is associated with adverse pathological and clinical outcomes in human breast cancer. *Breast Cancer* **19:** 153–160.

Schiller D, Cheng YC, Liu Q, Walter W, Craig EA. 2008. Residues of Tim44 involved in both association with the translocon of the inner mitochondrial membrane and regulation of mitochondrial Hsp70 tethering. *Mol Cell Biol* **28:** 4424–4433.

Schmidt O, Harbauer AB, Rao S, Eyrich B, Zahedi RP, Stojanovski D, Schönfisch B, Guiard B, Sickmann A, Pfanner N, et al. 2011. Regulation of mitochondrial protein import by cytosolic kinases. *Cell* **144:** 227–239.

Schulz C, Lytovchenko O, Melin J, Chacinska A, Guiard B, Neumann P, Ficner R, Jahn O, Schmidt B, Rehling P. 2011. Tim50's presequence receptor domain is essential for signal driven transport across the TIM23 complex. *J Cell Biol* **195:** 643–656.

Setoguchi K, Otera H, Mihara K. 2006. Cytosolic factor- and TOM-independent import of C-tail-anchored mitochondrial outer membrane proteins. *EMBO J* **25:** 5635–5647.

Shiota T, Mabuchi H, Tanaka-Yamano S, Yamano K, Endo T. 2011. In vivo protein-interaction mapping of a mitochondrial translocator protein Tom22 at work. *Proc Natl Acad Sci* **108:** 15179–15183.

Sickmann A, Reinders J, Wagner Y, Joppich C, Zahedi R, Meyer HE, Schönfisch B, Perschil I, Chacinska A, Guiard B, et al. 2003. The proteome of *Saccharomyces cerevisiae* mitochondria. *Proc Natl Acad Sci* **100:** 13207–13212.

Sideris DP, Tokatlidis K. 2007. Oxidative folding of small Tims is mediated by site-specific docking onto Mia40 in the mitochondrial intermembrane space. *Mol Microbiol* **65:** 1360–1373.

Sideris DP, Petrakis N, Katrakili N, Mikropoulou D, Gallo A, Ciofi-Baffoni S, Banci L, Bertini I, Tokatlidis K. 2009. A novel intermembrane space-targeting signal docks cysteines onto Mia40 during mitochondrial oxidative folding. *J Cell Biol* **187:** 1007–1022.

Sinha D, Joshi N, Chittoor B, Samji P, D'Silva P. 2010. Role of magmas in protein transport and human mitochondria biogenesis. *Hum Mol Genet* **19:** 1248–1262.

Sirrenberg C, Bauer MF, Guiard B, Neupert W, Brunner M. 1996. Import of carrier proteins into the mitochondrial inner membrane mediated by Tim22. *Nature* **384:** 582–585.

Sirrenberg C, Endres M, Folsch H, Stuart RA, Neupert W, Brunner M. 1998. Carrier protein import into mitochondria mediated by the intermembrane proteins Tim10/Mrs11 and Tim12/Mrs5. *Nature* **391:** 912–915.

Slutsky-Leiderman O, Marom M, Iosefson O, Levy R, Maoz S, Azem A. 2007. The interplay between components of the mitochondrial protein translocation motor studied using purified components. *J Biol Chem* **282:** 33935–33942.

Stojanovski D, Guiard B, Kozjak-Pavlovic V, Pfanner N, Meisinger C. 2007. Alternative function for the mitochondrial SAM complex in biogenesis of α-helical TOM proteins. *J Cell Biol* **179:** 881–893.

Stojanovski D, Milenkovic D, Müller JM, Gabriel K, Schulze-Specking A, Baker MJ, Ryan MT, Guiard B, Pfanner N, Chacinska A. 2008. Mitochondrial protein import: Precursor oxidation in a ternary complex with disulfide carrier and sulfhydryl oxidase. *J Cell Biol* **183:** 195–202.

Stroud DA, Becker T, Qiu J, Stojanovski D, Pfannschmidt S, Wirth C, Hunte C, Guiard B, Meisinger C, Pfanner N, et al. 2011a. Biogenesis of mitochondrial β-barrel proteins: The POTRA domain is involved in precursor release from the SAM complex. *Mol Biol Cell* **22:** 2823–2833.

Stroud DA, Oeljeklaus S, Wiese S, Bohnert M, Lewandrowski U, Sickmann A, Guiard B, van der Laan M, Warscheid B, Wiedemann N. 2011b. Composition and topology of the endoplasmic reticulum-mitochondria encounter structure. *J Mol Biol* **413:** 743–750.

Tamura Y, Harada Y, Shiota T, Yamano K, Watanabe K, Yokota M, Yamamoto H, Sesaki H, Endo T. 2009a. Tim23-Tim50 pair coordinates functions of translocators and motor proteins in mitochondrial protein import. *J Cell Biol* **184:** 129–141.

Tamura Y, Endo T, Iijima M, Sesaki H. 2009b. Ups1p and Ups2p antagonistically regulare cardiolipin metabolism in mitochondria. *J Cell Biol* **185:** 1029–1045.

Taylor AB, Smith BS, Kitada S, Kojima K, Miyaura H, Otwinowski Z, Ito A, Deisenhofer J. 2001. Crystal structures of mitochondrial processing peptidase reveal the mode for specific cleavage of import signal sequences. *Structure* **9:** 615–625.

Thornton N, Stroud DA, Milenkovic D, Guiard B, Pfanner N, Becker T. 2010. Two modular forms of the mitochondrial sorting and assembly machinery are involved in biogenesis of α-helical outer membrane proteins. *J Mol Biol* **396:** 540–549.

Truscott KN, Voos W, Frazier AE, Lind M, Li Y, Geissler A, Dudek J, Müller H, Sickmann A, Meyer HE, et al. 2003. A J-protein is an essential subunit of the presequence translocase-associated protein import motor of mitochondria. *J Cell Biol* **163:** 707–713.

van der Laan M, Chacinska A, Lind M, Perschil I, Sickmann A, Meyer HE, Guiard B, Meisinger C, Pfanner N, Rehling P. 2005. Pam17 is required for architecture and translocation activity of the mitochondrial protein import motor. *Mol Cell Biol* **25:** 7449–7458.

van der Laan M, Wiedemann N, Mick DU, Guiard B, Rehling P, Pfanner N. 2006. A role for Tim21 in membrane-potential-dependent preprotein sorting in mitochondria. *Curr Biol* **16:** 2271–2276.

van der Laan M, Meinecke M, Dudek J, Hutu DP, Lind M, Perschil I, Guiard B, Wagner R, Pfanner N, Rehling P. 2007. Motor-free mitochondrial presequence translocase drives membrane integration of preproteins. *Nat Cell Biol* **9:** 1152–1159.

van Wilpe S, Ryan MT, Hill K, Maarse AC, Meisinger C, Brix J, Dekker PJ, Moczko M, Wagner R, Meijer M, et al. 1999. Tom22 is a multifunctional organizer of the mitochondrial preprotein translocase. *Nature* **401:** 485–489.

Vergnolle MA, Baud C, Golovanov AP, Alcock F, Luciano P, Lian LY, Tokatlidis K. 2005. Distinct domains of small Tims involved in subunit interaction and substrate recognition. *J Mol Biol* **351:** 839–849.

Vögtle FN, Wortelkamp S, Zahedi RP, Becker D, Leidhold C, Gevaert K, Kellermann J, Voos W, Sickmann A, Pfanner N, et al. 2009. Global analysis of the mitochondrial N-proteome identifies a processing peptidase critical for protein stability. *Cell* **139:** 428–439.

Vögtle FN, Prinz C, Kellermann J, Lottspeich F, Pfanner N, Meisinger C. 2011. Mitochondrial protein turnover: Role of the precursor intermediate peptidase Oct1 in protein stabilization. *Mol Biol Cell* **22:** 2135–2143.

von der Malsburg K, Müller JM, Bohnert M, Oeljeklaus S, Kwiatkowska P, Becker T, Loniewska-Lwowska A, Wiese S, Rao S, Milenkovic D, et al. 2011. Dual role of mitofilin in mitochondrial membrane organization and protein biogenesis. *Dev Cell* **21:** 694–707.

Wagner K, Gebert N, Guiard B, Brandner K, Truscott KN, Wiedemann N, Pfanner N, Rehling P. 2008. The assembly pathway of the mitochondrial carrier translocase involves four preprotein translocases. *Mol Cell Biol* **28:** 4251–4260.

Waizenegger T, Habib SJ, Lech M, Mokranjac D, Paschen SA, Hell K, Neupert W, Rapaport D. 2004. Tob38, a novel essential component in the biogenesis of β-barrel proteins of mitochondria. *EMBO Rep* **5:** 704–709.

Waizenegger T, Schmitt S, Zivkovic J, Neupert W, Rapaport D. 2005. Mim1, a protein required for the assembly of the TOM complex of mitochondria. *EMBO Rep* **6:** 57–62.

Walther DM, Papic D, Bos MP, Tommassen J, Rapaport D. 2009. Signals in bacterial β-barrel proteins are functional in eukaryotic cells for targeting to and assembly in mitochondria. *Proc Natl Acad Sci* **106:** 2531–2536.

Webb CT, Gorman MA, Lazarou M, Ryan MT, Gulbis JM. 2006. Crystal structure of the mitochondrial chaperone TIM9.10 reveals a six-bladed α-propeller. *Mol Cell* **21:** 123–133.

Wideman JG, Go NE, Klein A, Redmond E, Lackey SW, Tao T, Kalbacher H, Rapaport D, Neupert W, Nargang FE. 2010. Roles of the Mdm10, Tom7, Mdm12, and Mmm1 proteins in the assembly of mitochondrial outer membrane proteins in *Neurospora crassa*. *Mol Biol Cell* **21:** 1725–1736.

Wiedemann N, Pfanner N, Ryan MT. 2001. The three modules of ADP/ATP carrier cooperate in receptor recruitment and translocation into mitochondria. *EMBO J* **20:** 951–960.

Wiedemann N, Kozjak V, Chacinska A, Schönfisch B, Rospert S, Ryan MT, Pfanner N, Meisinger C. 2003. Machinery for protein sorting and assembly in the mitochondrial outer membrane. *Nature* **424:** 565–571.

Wiedemann N, Truscott KN, Pfannschmidt S, Guiard B, Meisinger C, Pfanner N. 2004. Biogenesis of the protein import channel Tom40 of the mitochondrial outer membrane: Intermembrane space components are involved in an early stage of the assembly pathway. *J Biol Chem* **279:** 18188–18194.

Wiedemann N, van der Laan M, Hutu DP, Rehling P, Pfanner N. 2007. Sorting switch of mitochondrial presequence translocase involves coupling of motor module to respiratory chain. *J Cell Biol* **179:** 1115–1122.

Wu Y, Sha B. 2006. Crystal structure of yeast mitochondrial outer membrane translocon member Tom70p. *Nat Struct Mol Biol* **13:** 589–593.

Yamano K, Tanaka-Yamano S, Endo T. 2010. Tom7 regulates Mdm10-mediated assembly of the mitochondrial import channel protein Tom40. *J Biol Chem* **285:** 41222–41231.

Yano M, Terada K, Mori M. 2003. AIP is a mitochondrial import mediator that binds to both import receptor Tom20 and preproteins. *J Cell Biol* **163:** 45–56.

Yogev O, Karniely S, Pines O. 2007. Translation-coupled translocation of yeast fumarase into mitochondria *in vivo*. *J Biol Chem* **282:** 29222–29229.

Young JC, Hoogenraad NJ, Hartl FU. 2003. Molecular chaperones Hsp90 and Hsp70 deliver preproteins to the mitochondrial import receptor Tom70. *Cell* **112:** 41–50.

Zara V, Ferramosca A, Robitaille-Foucher P, Palmieri F, Young JC. 2009. Mitochondrial carrier protein biogenesis: role of the chaperones Hsc70 and Hsp90. *Biochem J* **419:** 369–375.

The Mitochondrial Nucleoid: Integrating Mitochondrial DNA into Cellular Homeostasis

Robert Gilkerson[1,2], Liliana Bravo[1], Iraselia Garcia[1], Norma Gaytan[1], Alan Herrera[1], Alicia Maldonado[2], and Brandi Quintanilla[1]

[1]Department of Biology, University of Texas-Pan American, Edinburg, Texas 78539-2999

[2]Department of Clinical Laboratory Sciences, University of Texas-Pan American, Edinburg, Texas 78539-2999

Correspondence: gilkersonrw@utpa.edu

The packaging of mitochondrial DNA (mtDNA) into DNA-protein assemblies called nucleoids provides an efficient segregating unit of mtDNA, coordinating mtDNA's involvement in cellular metabolism. From the early discovery of mtDNA as "extranuclear" genetic material, its organization into nucleoids and integration into both the mitochondrial organellar network and the cell at large via a variety of signal transduction pathways, mtDNA is a crucial component of the cell's homeostatic network. The mitochondrial nucleoid is composed of a set of DNA-binding core proteins involved in mtDNA maintenance and transcription, and a range of peripheral factors, which are components of signaling pathways controlling mitochondrial biogenesis, metabolism, apoptosis, and retrograde mitochondria-to-nucleus signaling. The molecular interactions of nucleoid components with the organellar network and cellular signaling pathways provide exciting clues to the dynamic integration of mtDNA into cellular metabolic homeostasis.

ORIGINS: "EXTRANUCLEAR DNA" AND BIOENERGETICS

Historically, it has been clear that mitochondria play by their own rules. Beginning with the observations that mitochondria independently synthesize protein (McLean et al. 1958) and RNA (Wintersberger 1964), followed by the discovery of DNA within the organelle (Nass et al. 1965; Nass 1969), mitochondria have emerged as a highly distinct organellar system for providing ATP to the cell. The then-controversial endosymbiont theory (Sagan 1967), proposing that mitochondria are descended from proteobacteria engulfed by eukaryotic cells, was later confirmed by the similarity of mtDNA to bacterial, rather than human nuclear, DNA. This "extranuclear" DNA is in fact essential for mitochondrial ATP production (DiMauro and Schon 2003).

The 16,569 bp circular human mtDNA encodes two tRNAs, 22 rRNAs, and 13 polypeptides necessary for the proper assembly and function of the mitochondrial complexes of oxidative phosphorylation (OXPHOS) (Anderson et al. 1981), which are located in the mitochondrial inner membrane. Complexes I, II, III, and IV transfer electrons, ultimately donating them to molecular oxygen, to create the chemiosmotic proton-motive gradient ($\Delta\psi_m$) which is used

Cite this article as *Cold Spring Harb Perspect Biol* doi: 10.1101/cshperspect.a011080

by complex V, the F_1F_0 ATP synthase, to synthesize ATP from ADP and P_i. These complexes are large, multisubunit macromolecular assemblies, in which all but a small handful of polypeptides are encoded by nuclear genes. The assembly of these complexes requires the coordination of both nuclear and mitochondrial transcription and translation, followed by complex assembly in the mitochondria and insertion into the inner membrane. The necessity of producing and assembling complexes of polypeptides derived from both chromosomal and organellar genomes means that mitochondria are susceptible to mutations in either.

Defects of both nuclear DNA and mtDNA have been shown to have devastating consequences for bioenergetics in human health. Mutations in nuclear-encoded factors such as mitochondrial DNA polymerase γ (POLG) and thymidine kinase 2 (TK2) cause depletion of mtDNA and increased mtDNA mutation, frequently manifesting as neuromuscular disease (Oskoui et al. 2006; Akman et al. 2008). Mutations or depletion of mtDNA cause a range of neuromuscular disorders, such as the classical mitochondrial diseases Kearns-Sayre syndrome (KSS), which is caused by large-scale deletions of mtDNA (Δ-mtDNAs), and myoclonus epilepsy with lactic acidosis and stroke-like episodes (MELAS), which is caused by point mutations in mtDNA-encoded tRNAs (reviewed in DiMauro and Schon 2003; Greaves et al. 2012). Although each of the mitochondrial neuromuscular disorders is individually rare, pathogenic mtDNA mutations as a whole occur at a high frequency of 1:200 individuals (Elliott et al. 2008; Greaves et al. 2012). The study of mitochondrial neuromuscular disorders has uncovered fundamental concepts of genotype-phenotype relationships for mtDNA mutations and mitochondrial function. Typically, wild-type and mutant mtDNAs exist side by side within the same cell and tissue, a situation called heteroplasmy. The relative proportions of the two mtDNA variants is a crucial determinant of mitochondrial function in cells carrying mutant mtDNAs: Cells can typically withstand a very high overall mtDNA mutation load (often ~80%–90%, depending on the nature of the mutation), but when the mutation load exceeds the threshold of function, the cell can no longer maintain mitochondrial function (DiMauro and Schon 2003).

In addition to the classical mitochondrial diseases, mtDNA mutations and defects are emerging in a variety of prevalent human diseases in disparate tissues, such as Parkinson disease (Bender et al. 2006; Kraytsberg et al. 2006), multiple sclerosis (Campbell et al. 2012), heart failure (Karamanlidis et al. 2010), diabetes (Ritov et al. 2010), and aging (Bender et al. 2006). The crucial role of mtDNA in cellular metabolism and bioenergetics, as well as its profound importance to human health, motivates a better understanding of the elegant complexity of mtDNA's macromolecular organization and integration into both the organellar network and cell at large. Despite the bioenergetic importance and medical relevance of mtDNA, a number of misconceptions regarding its organization have persisted. MtDNA has often been thought to exist in a "naked" or unprotected state. However, mtDNA is efficiently packaged into a macromolecular assembly that allows for dynamic genetic maintenance and integration into cellular signaling.

THE NUCLEOID: A MACROMOLECULAR ASSEMBLY FOR MITOCHONDRIAL GENETICS

Given their endosymbiotic bacterial origins, it is not surprising that the organization of mtDNA is similar to that of bacterial DNA. The bacterial genome undergoes a 10^4-fold compaction of its volume to form the bacterial nucleoid (de Vries 2010). Similarly, the mitochondrial nucleoid typically has a diameter of ~100 nm (Kukat et al. 2011), whereas a completely relaxed 16.6 kb circular DNA would have a length on the order of ~2 μm within the mitochondrial matrix. Beginning with the original fluorescence micrographs of human mtDNA within the cell (Satoh and Kuroiwa 1991), nucleoids have been clearly evident by microscopy as discrete punctae distributed throughout the mitochondrial network (Alam et al. 2003; Legros et al. 2004). This punctate appearance reveals the unique organization of mtDNA within the organellar network. The mitochondrial nucleoid is composed of a range of proteins with a diverse array of functions, from DNA packaging and transcription to factors with broader signaling functions, facilitating mtDNA's integration into crucial cell-wide metabolic and proliferative signaling networks. Nucleoid packaging provides an efficient means of ensuring that mitochondrial genetic material is distributed throughout the cell's mitochondria and is responsive to cellular metabolic needs.

The packaging of mtDNA into nucleoids is mediated by the nucleoid's protein components. As the different individual protein factors of the nucleoid have been cataloged, the macromolecular structure of the nucleoid has been suggested to consist of a core domain comprised of key mtDNA-associated proteins, and a peripheral zone, consisting of factors which are less constitutively bound, many of which have important signaling roles with the rest of the cell (Fig. 1). The core nucleoid components center around transcription and replication factors such as transcription factor A, mitochondrial (TFAM), mitochondrial single-strand binding protein (mtSSB), POLG, and mtRNA polymerase (POLRMT), which were found in both cross-linked and native nucleoid fractions of HeLa cells (Bogenhagen et al. 2008). TFAM is a 24-kDa protein with two high-mobility group (HMG)-box domains. HMG-family proteins are able to bend, wrap, and unwind DNA. Although it was early identified as a transcription factor for mtDNA (Fisher and Clayton 1985), it has become clear that TFAM plays an equally important role as a packaging protein for mtDNA (Alam et al. 2003). TFAM packages DNA in vitro (Kaufman et al. 2007) and colocalizes with mtDNA in nucleoids by immunomicroscopy (Iborra et al. 2004; Legros et al. 2004). Crystallographic studies have shown that TFAM induces U-turn bends in DNA (Ngo et al. 2011; Rubio-Cosials et al. 2011). Nucleoids tightly package mtDNA, to the extent that freely diffusible matrix protein is excluded from the nucleoid (Brown et al. 2011). Collectively, these studies show that TFAM is a crucial packaging factor for mtDNA, in addition to its role as a key transcription factor for mtDNA. Although

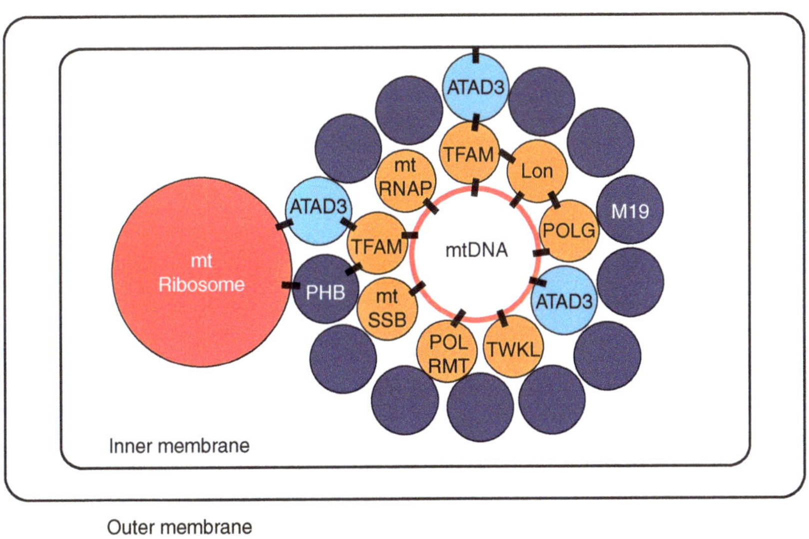

Figure 1. Schematic of interactions at the mitochondrial nucleoid. Core nucleoid factors (gold) have known mtDNA-binding activity. Peripheral factors (blue) do not directly interact with mtDNA, but associate with the nucleoid via protein–protein interactions. ATAD3 (light blue) associates with mtDNA, the mitochondrial ribosome, and the mitochondrial inner membrane. Shown protein–protein and DNA–protein interactions are designated by black rectangles.

TFAM is crucial to mtDNA maintenance, overexpression of TFAM can actually result in excessive binding of mtDNA, effectively choking off transcription of mtDNA (Maniura-Weber et al. 2004). Thus, although TFAM appears to be the major mtDNA packaging protein, the other core constituents of the nucleoid appear to function primarily in replication, transcription, and translation. POLG, POLG2, mtSSB, and the DNA helicase Twinkle (all nucleoid core factors), are the minimal factors necessary for mitochondrial replication (reviewed in Pohjoismaki and Goffart 2011), whereas POLRMT and mtTFB are the minimal factors necessary for mitochondrial transcription. Intriguingly, Shadel and coworkers have found that TFAM is in fact dispensable for transcription in vitro (Shutt et al. 2011), further indicating that TFAM's primary importance is in mtDNA packaging. The Lon protease has been found to be an integral nucleoid core factor (Lu et al. 2007; Bogenhagen et al. 2008). Similarly, the POLGb subunit has been shown to play a role in the organization of the nucleoid (Di Re et al. 2009), as has the newly discovered CSB factor (associated with Cockayne syndrome) (Berquist et al. 2012). The nucleoid's core factors thus both package mtDNA to bring together the factors necessary for replication and transcription.

Although a great deal of research has explored the protein composition of the nucleoid, its DNA content has been a crucial question of nucleoid biology. The number of mtDNAs contained within a single nucleoid is a fundamental question of this genetic unit. The studies surrounding this issue have recently come full circle, suggesting that nucleoids carry either one or two mtDNA molecules in a single nucleoid. The early studies of Satoh and Kuroiwa (1991) quantitated the fluorescence of individual nucleoids by ethidium bromide and calculating the quantity of mtDNA contained within it, calibrated against the fluorescence of known quantities of phage DNA. Using this direct, elegant method, they found that a single mitochondrion, on average, carried 4.6 mtDNAs. Further, each mitochondrion carried an average of 3.2 nucleoids, with most mitochondria carrying a single nucleoid. Thus, an individual nucleoid contains 1.4 mtDNAs per nucleoid, suggesting that nucleoids contain either one or two mtDNAs. Subsesquent studies have taken approaches in which the number of nucleoid foci within a cell was counted, followed by quantitation of mtDNA copy number per cell, then obtaining an average value of mtDNAs per nucleoid. Using these estimates, a number of researchers found that rapidly dividing cells, such as HeLa and osteosarcoma cell lines, maintain 2–10 mtDNAs per nucleoid (Iborra et al. 2004; Legros et al. 2004; Gilkerson et al. 2008). More recently, however, investigators have used next-generation imaging methods (Kukat et al. 2011), allowing higher-resolution imaging of the mitochondrial nucleoid in situ. Strikingly, these studies reveal that nucleoids are smaller than previously thought (and often aggregate). These small nucleoids contain an average of 1.4 mtDNAs per nucleoid (Kukat et al. 2011). Intriguingly, this value is identical to that first found by Satoh and Kuroiwa (above), suggesting that this issue has come "full circle" and that a single nucleoid likely consists of only one or two mtDNA molecules.

Integration and Dynamics within the Organellar Network

The highly dynamic, responsive nature of mitochondria as an organellar network has been a revolutionary new frontier in mitochondrial biology. The dynamics of the mitochondrial network involve a complex set of interactions that modulate mitochondrial morphology and shape the organellar network (Nunnari and Suomalainen 2012). Although mitochondria have traditionally been regarded as a population of individual football-shaped organelles, it is clear that mitochondria actually balance between fragmented and networked states of organization in response to both fusion and fission events. Mitochondrial fusion is mediated by mitofusins 1 and 2 (MFN1 and MFN2) and OPA1, whereas fission events are mediated by FIS1, DRP1, and MFF (Twig et al. 2008; Otera et al. 2010). These two pathways allow for modulation of mitochondrial organization in response to a variety of cellular cues. Mitochondrial morphology is linked to function, as a loss of mitochondrial bioenergetic capacity results in an inability to maintain the fused, networked mitochondrial organization (Santra et al. 2004); this loss of interconnectivity is likely owing to the collapse of the mitochondrial transmembrane potential across the inner membrane (Legros et al. 2004). Thus, mitochondrial morphology can provide a visual indication of the relative "health" of a cell's mitochondria, as cells with genetic or pharmacologic mitochondrial dysfunction show an inability to maintain a united mitochondrial network (Legros et al. 2004; Gilkerson et al. 2008). Defects in mitochondrial organization and trafficking are appearing as a common manifestation of mitochondrial dysfunction in diseases such as Parkinson's disease and Huntington's disease (Schon and Przedborski 2011).

The molecular interactions that integrate the nucleoid into this dynamic organellar network are beginning to emerge. Within the organelle, the nucleoid is located within the protein-dense matrix compartment. The factors associated with the nucleoid's peripheral zone, such as the ATPase AAA domain-containing protein 3 (ATAD3) and prohibitins 1 and 2 (PHB1 and PHB2), do not appear to directly bind mtDNA, but provide molecular scaffolding, help facilitate mitochondrial translation, and interact with important cell-wide signaling pathways. The mitochondrial M19 protein (mitochondrial nucleoid factor 1 [MNF1]) has been identified as a nucleoid-associated protein, and appears to reside in the peripheral zone of the nucleoid (Sumitani et al. 2009). M19 appears to play a role in modulating mitochondrial ATP production, providing a crucial link between mtDNA and mitochondrial bioenergetic function (Cambier et al. 2012). Prohibitin appears to play an important role in stabilizing TFAM at the mitochondrial nucleoid (Kasashima et al. 2008), and is strongly implicated in mitochondrial protein synthesis (He et al. 2012). Prohibitin forms oligomeric rings in the inner mitochondrial membrane that are necessary for the maintenance of mitochondrial morphology (Merkwirth and Langer 2009). Researchers have long postulated that mitochondrial DNA may be "tethered" to the inner membrane (Albring et al. 1977; Van Tuyle and McPherson 1979); however, the actual tethering interactions have remained elusive. More recently, the ATAD3 protein has been shown to associate with both the mitochondrial nucleoid and the inner membrane (He et al. 2007; Bogenhagen et al. 2008). He et al. (2012) propose that ATAD3 may actually provide a scaffold to which mitochondrial nucleoids and ribosomes may bind to within the matrix (He et al. 2012), tethering both within close proximity to each other at the inner membrane. ATAD3 may play a unique central role in nucleoid organization, as it associates with both the inner membrane and the mitochondrial ribosome, and also binds to D-loop sequences of mtDNA (He et al. 2007, 2012; Bogenhagen et al. 2008). ATAD3 and prohibitin are both required for mitochondrial protein synthesis, and appear to bring the mitochondrial nucleoid and translational machinery into close physical proximity (He et al. 2012), allowing for an efficient coordination of mtDNA-encoded mitochondrial protein synthesis and physical attachment to the mitochondrial inner membrane (Fig. 1). Accordingly, nucleoids do not appear to freely diffuse within the matrix compartment of the mitochondrion. Although it has been shown that fusions of wild-type cells with cells lacking mtDNA (ρ^0 cells) result in repopulation of the mtDNA-depleted cells, nucleoids were found to migrate more slowly than GFP targeted to the mitochondrial outer membrane (Legros et al. 2004). This finding indicates that nucleoid movement is limited by their attachment to the mitochondrial inner membrane, rather than freely diffusing through the matrix compartment. Iborra et al. (2004) found that nucleoids move in a "constrained walk," consistent with more recent findings via superresolution fluorescence microscopy (Brown et al. 2011).

An immediately striking feature of nucleoid organization in fluorescence microscopy is the regularity at which nucleoids appear along mitochondrial filaments, in which mtDNA is distributed evenly throughout the cell by appearing at discrete intervals throughout mitochondria (Iborra et al. 2004; Legros et al. 2004). More striking yet was the observation that, when in a fragmented morphology, all mitochondria within a cell will contain at least one nucleoid (Margineantu et al. 2002)! These observations strongly suggest that mitochondrial nucleoid maintenance is a regulated process, coordinated with the mitochondrial fission and fusion machinery to ensure the efficient distribution of mtDNA throughout the cell, while simultaneously protecting the nucleoid from fission events at the nucleoid itself. Genetic inhibition of mitochondrial fission, via RNAi of DRP1, results in loss of nucleoid content and mtDNA copy number (Parone et al. 2008), whereas disruption of mitochondrial fusion through knockout of MFN1 and MFN2 similarly results in loss of mtDNA copy number (Chen et al. 2011). Live cell imaging of nucleoids revealed that mitochondrial division typically occurs just to the side of nucleoids (Iborra et al. 2004). Thus, it appears very likely that nucleoid factors are intricately connected to the fission and fusion machinery to ensure nucleoid propagation.

Conversations beyond the Organelle: MtDNA and Cellular Signaling

As the organizing unit of mtDNA, the nucleoid is likely to play a crucial role in coordinating organellar response to cellular signaling cues. As mitochondria are increasingly found to participate in crucial cellular signaling networks, a number of nucleoid-associated factors are found to have signaling functions that may determine mitochondrial response to nuclear-encoded signaling factors, particularly those involving metabolism and apoptosis. In addition, there is growing evidence that retrograde mitochondria-to-nucleus cross talk may be mediated by mtDNA and nucleoid-associated factors. Collectively, the nucleoid integrates mtDNA into cell-wide signaling networks, playing a direct role in mitochondrial participation in cellular signaling.

The regulation of mitochondrial content by metabolic signaling pathways has revealed a vital mechanism of nuclear-mitochondrial cross talk. Transcription of the nuclear-encoded mitochondrial OXPHOS complex subunits is controlled by the nuclear respiratory factors 1 and 2 (NRF-1 [Evans and Scarpulla 1990] and NRF-2 [Virbasius et al. 1993]). These NRFs are modulated by the peroxisome proliferation γ-related coactivator (PGC)-1α master regulator (Puigserver et al. 1998) and its related factor, PRC (Scarpulla 2008). Increased PGC-1α signaling leads to elevated levels of nuclear-encoded, mitochondrially localized proteins and thus increased mitochondrial content within the cell. Interestingly, a number of nucleoid-associated factors play direct roles in the modulation of mtDNA copy number, providing direct biogenesis signaling from within the nucleoid. TFAM plays a key role in modulating mtDNA copy number, and has in fact been shown to accurately reflect the overall level of mtDNA within the cell (Ekstrand et al. 2004; Kanki et al. 2004). TFAM, Lon, mtSSB, and NRF-1 and -2 have all been found to be coordinately up-regulated in response to myogenic differentiation (Wagatsuma et al. 2011). Importantly, however, it must be noted that this is a delicate coordination, as overexpression of TFAM can actually inhibit transcription of mtDNA (Matsushima et al. 2010a) and cause mitochondrial dysfunction (Ylikallio et al. 2010), though overexpression of TFAM has been shown to ameliorate mitochondrial dysfunction in neuronal and cardiac settings (Ikeuchi et al. 2005; Hokari et al. 2010). It was shown previously that cells will accurately regulate mtDNA content on the basis of mass of mtDNA, rather than number of copies per se (Tang et al. 2000); this seems consistent with a role for TFAM as the principal means by which the cell regulates mtDNA content. Interestingly, Lon-mediated proteolytic degradation of TFAM is emerging as a major mechanism of mtDNA copy control, in which TFAM levels are regulated by the Lon protease to modulate the overall copy number of mtDNA within the cell (Matsushima et al. 2010b; Ambro et al. 2012; Matsushima and Kaguni 2012). The Lon protease has been identified as a core nucleoid component (Bogenhagen et al. 2008) with the ability to bind mtDNA (Liu et al. 2004). The dynamics of TFAM's association with mtDNA are still being elucidated, but it is clear that TFAM plays a very elegant role in maintaining mtDNA in nucleoids. Thus, the presence of known metabolic

 Cite this article as *Cold Spring Harb Perspect Biol* doi: 10.1101/cshperspect.a011080

signaling molecules at the nucleoid allows for the efficient, precise modulation of mtDNA content, transcription, translation (via the near-by mitochondrial ribosomes), and thus mito-chondrial function, in response to cellular sig-naling cues.

Nucleoid "Give-and-Take" in Cellular Signaling

Although it is clear that nuclear signaling net-works, particularly those involving mitochon-drial biogenesis via PGC-1α and related factors, can impact mitochondrial content function, there is also an accumulation of evidence sug-gesting that mitochondria can communicate with the rest of the cell via retrograde signaling to the nucleus. Experiments in *Drosophila mel-anogaster* have shown that genetically induced ablation of mitochondrial genes leads to cell-cycle arrest via either reactive oxygen species (ROS)-induced JNK signaling, or activation of AMP kinase via low ATP levels (Owusu-Ansah et al. 2008). Studies in mammalian systems indi-cate that either genetically or pharmacologically induced mitochondrial dysfunction results in altered expression of nuclear-encoded signaling factors (Gilkerson et al. 2012). Accordingly, al-though the loss of mitochondrial function in mtDNA-depleted cells is certainly detrimental to cellular homeostasis, the resultant down-reg-ulation of nuclear transcription of prosurvival signaling molecules, even those not associated with mitochondrial bioenergetics per se, is po-tentially an even more devastating mechanism of cellular dysfunction. This suggests that nu-cleoid-associated factors may alter their locali-zation to participate in crucial metabolic signal-ing networks, consistent with findings from yeast, in which the mitochondrial nucleoid un-dergoes remodeling in response to metabolic stress (Kucej et al. 2008). Thus, there are likely a range of nucleoid factors that undergo dy-namic changes in localization based on cellular signaling cues. Taken as a whole, these interac-tions reveal the dynamic interplay and integral involvement of the nucleoid in metabolic and apoptotic signaling. Although a range of impor-tant signaling functions have been assigned to nucleoid-associated factors, the dynamics of the interactions mediating the nucleoid's role in crucial signaling networks is only beginning to be understood. mtDNA and the nucleoid may thus be at the front line of environmental inter-actions.

ACKNOWLEDGMENTS

We thank Dr. Eric A. Schon for comments on the manuscript. This work is supported by a University of Texas-Pan American Faculty Re-search Council Award (R.G.) and the National Science Foundation Undergraduate Research Mentoring program DBI 0934013 (N.G. and B.Q.).

REFERENCES

Akman HO, Dorado B, Lopez LC, Garcia-Cazorla A, Vila MR, Tanabe LM, Dauer WT, Bonilla E, Tanji K, Hirano M. 2008. Thymidine kinase 2 (H126N) knockin mice show the essential role of balanced deoxynucleotide pools for mitochondrial DNA maintenance. *Hum Mol Genet* **17:** 2433–2440.

Alam TI, Kanki T, Muta T, Ukaji K, Abe Y, Nakayama H, Takio K, Hamasaki N, Kang D. 2003. Human mitochon-drial DNA is packaged with TFAM. *Nucleic Acids Res* **31:** 1640–1645.

Albring M, Griffith J, Attardi G. 1977. Association of a pro-tein structure of probable membrane derivation with HeLa cell mitochondrial DNA near its origin of replica-tion. *Proc Natl Acad Sci* **74:** 1348–1352.

Ambro L, Pevala V, Bauer J, Kutejova E. 2012. The influence of ATP-dependent proteases on a variety of nucleoid-as-sociated processes. *J Struct Biol* **179:** 181–192.

Anderson S, Bankier AT, Barrell BG, de Bruijn MH, Coul-son AR, Drouin J, Eperon IC, Nierlich DP, Roe BA, Sanger F, et al. 1981. Sequence and organization of the human mitochondrial genome. *Nature* **290:** 457–465.

Bender A, Krishnan KJ, Morris CM, Taylor GA, Reeve AK, Perry RH, Jaros E, Hersheson JS, Betts J, Klopstock T, et al. 2006. High levels of mitochondrial DNA deletions in substantia nigra neurons in aging and Parkinson dis-ease. *Nat Genet* **38:** 515–517.

Berquist BR, Canugovi C, Sykora P, Wilson DM III, Bohr VA. 2012. Human Cockayne syndrome B protein reciprocally communicates with mitochondrial proteins and pro-motes transcriptional elongation. *Nucl Acids Res* **40:** 8392–8405.

Bogenhagen DF, Rousseau D, Burke S. 2008. The layered structure of human mitochondrial DNA nucleoids. *J Biol Chem* **283:** 3665–3675.

Brown TA, Tkachuk AN, Shtengel G, Kopek BG, Bog-enhagen DF, Hess HF, Clayton DA. 2011. Superresolution fluorescence imaging of mitochondrial nucleoids reveals their spatial range, limits, and membrane interaction. *Mol Cell Biol* **31:** 4994–5010.

Cambier L, Rassam P, Chabi B, Mezghenna K, Gross R, Eveno E, Auffray C, Wrutniak-Cabello C, Lajoix AD, Pomies P. 2012. M19 modulates skeletal muscle differentiation and insulin secretion in pancreatic β-cells through modulation of respiratory chain activity. *PLoS ONE* **7:** e31815.

Campbell GR, Kraytsberg Y, Krishnan KJ, Ohno N, Ziabreva I, Reeve A, Trapp BD, Newcombe J, Reynolds R, Lassmann H, et al. 2012. Clonally expanded mitochondrial DNA deletions within the choroid plexus in multiple sclerosis. *Acta Neuropath* **124:** 209–220.

Chen Y, Liu Y, Dorn GW II. 2011. Mitochondrial fusion is essential for organelle function and cardiac homeostasis. *Circ Res* **109:** 1327–1331.

de Vries R. 2010. DNA condensation in bacteria: Interplay between macromolecular crowding and nucleoid proteins. *Biochimie* **92:** 1715–1721.

DiMauro S, Schon EA. 2003. Mitochondrial respiratory-chain diseases. *N Engl J Med* **348:** 2656–2668.

Di Re M, Sembongi H, He J, Reyes A, Yasukawa T, Martinsson P, Bailey LJ, Goffart S, Boyd-Kirkup JD, Wong TS, et al. 2009. The accessory subunit of mitochondrial DNA polymerase γ determines the DNA content of mitochondrial nucleoids in human cultured cells. *Nucl Acids Res* **37:** 5701–5713.

Ekstrand MI, Falkenberg M, Rantanen A, Park CB, Gaspari M, Hultenby K, Rustin P, Gustafsson CM, Larsson NG. 2004. Mitochondrial transcription factor A regulates mtDNA copy number in mammals. *Hum Mol Gen* **13:** 935–944.

Elliott HR, Samuels DC, Eden JA, Relton CL, Chinnery PF. 2008. Pathogenic mitochondrial DNA mutations are common in the general population. *Am J Hum Genet* **83:** 254–260.

Evans MJ, Scarpulla RC. 1990. NRF-1: A *trans*-activator of nuclear-encoded respiratory genes in animal cells. *Genes Dev* **4:** 1023–1034.

Fisher RP, Clayton DA. 1985. A transcription factor required for promoter recognition by human mitochondrial RNA polymerase. Accurate initiation at the heavy- and light-strand promoters dissected and reconstituted in vitro. *J Biol Chem* **260:** 11330–11338.

Gilkerson RW, Schon EA, Hernandez E, Davidson MM. 2008. Mitochondrial nucleoids maintain genetic autonomy but allow for functional complementation. *J Cell Biol* **181:** 1117–1128.

Gilkerson RW, De Vries RL, Lebot P, Wikstrom JD, Torgyekes E, Shirihai OS, Przedborski S, Schon EA. 2012. Mitochondrial autophagy in cells with mtDNA mutations results from synergistic loss of transmembrane potential and mTORC1 inhibition. *Hum Mol Genet* **21:** 978–990.

Greaves LC, Reeve AK, Taylor RW, Turnbull DM. 2012. Mitochondrial DNA and disease. *J Pathol* **226:** 274–286.

He J, Mao CC, Reyes A, Sembongi H, Di Re M, Granycome C, Clippingdale AB, Fearnley IM, Harbour M, Robinson AJ, et al. 2007. The AAA$^+$ protein ATAD3 has displacement loop binding properties and is involved in mitochondrial nucleoid organization. *J Cell Biol* **176:** 141–146.

He J, Cooper HM, Reyes A, Di Re M, Sembongi H, Litwin TR, Gao J, Neuman KC, Fearnley IM, Spinazzola A, et al. 2012. Mitochondrial nucleoid interacting proteins support mitochondrial protein synthesis. *Nucl Acids Res* **40:** 6109–6121.

Hokari M, Kuroda S, Kinugawa S, Ide T, Tsutsui H, Iwasaki Y. 2010. Overexpression of mitochondrial transcription factor A (TFAM) ameliorates delayed neuronal death due to transient forebrain ischemia in mice. *Neuropathol* **30:** 401–407.

Iborra FJ, Kimura H, Cook PR. 2004. The functional organization of mitochondrial genomes in human cells. *BMC Biol* **2:** 9.

Ikeuchi M, Matsusaka H, Kang D, Matsushima S, Ide T, Kubota T, Fujiwara T, Hamasaki N, Takeshita A, Sunagawa K, et al. 2005. Overexpression of mitochondrial transcription factor A ameliorates mitochondrial deficiencies and cardiac failure after myocardial infarction. *Circulation* **112:** 683–690.

Kanki T, Ohgaki K, Gaspari M, Gustafsson CM, Fukuoh A, Sasaki N, Hamasaki N, Kang D. 2004. Architectural role of mitochondrial transcription factor A in maintenance of human mitochondrial DNA. *Mol Cell Biol* **24:** 9823–9834.

Karamanlidis G, Nascimben L, Couper GS, Shekar PS, del Monte F, Tian R. 2010. Defective DNA replication impairs mitochondrial biogenesis in human failing hearts. *Circ Res* **106:** 1541–1548.

Kasashima K, Sumitani M, Satoh M, Endo H. 2008. Human prohibitin 1 maintains the organization and stability of the mitochondrial nucleoids. *Exp Cell Res* **314:** 988–996.

Kaufman BA, Durisic N, Mativetsky JM, Costantino S, Hancock MA, Grutter P, Shoubridge EA. 2007. The mitochondrial transcription factor TFAM coordinates the assembly of multiple DNA molecules into nucleoid-like structures. *Mol Biol Cell* **18:** 3225–3236.

Kraytsberg Y, Kudryavtseva E, McKee AC, Geula C, Kowall NW, Khrapko K. 2006. Mitochondrial DNA deletions are abundant and cause functional impairment in aged human substantia nigra neurons. *Nat Genet* **38:** 518–520.

Kucej M, Kucejova B, Subramanian R, Chen XJ, Butow RA. 2008. Mitochondrial nucleoids undergo remodeling in response to metabolic cues. *J Cell Sci* **121:** 1861–1868.

Kukat C, Wurm CA, Spahr H, Falkenberg M, Larsson NG, Jakobs S. 2011. Super-resolution microscopy reveals that mammalian mitochondrial nucleoids have a uniform size and frequently contain a single copy of mtDNA. *Proc Natl Acad Sci* **108:** 13534–13539.

Legros F, Malka F, Frachon P, Lombes A, Rojo M. 2004. Organization and dynamics of human mitochondrial DNA. *J Cell Sci* **117:** 2653–2662.

Liu T, Lu B, Lee I, Ondrovicova G, Kutejova E, Suzuki CK. 2004. DNA and RNA binding by the mitochondrial lon protease is regulated by nucleotide and protein substrate. *J Biol Chem* **279:** 13902–13910.

Lu B, Yadav S, Shah PG, Liu T, Tian B, Pukszta S, Villaluna N, Kutejova E, Newlon CS, Santos JH, et al. 2007. Roles for the human ATP-dependent Lon protease in mitochondrial DNA maintenance. *J Biol Chem* **282:** 17363–17374.

Maniura-Weber K, Goffart S, Garstka HL, Montoya J, Wiesner RJ. 2004. Transient overexpression of mitochondrial transcription factor A (TFAM) is sufficient to stimulate mitochondrial DNA transcription, but not sufficient to increase mtDNA copy number in cultured cells. *Nucl Acids Res* **32:** 6015–6027.

Margineantu DH, Gregory Cox W, Sundell L, Sherwood SW, Beechem JM, Capaldi RA. 2002. Cell cycle dependent morphology changes and associated mitochondrial DNA redistribution in mitochondria of human cell lines. *Mitochondrion* **1:** 425–435.

Matsushima Y, Kaguni LS. 2012. Matrix proteases in mitochondrial DNA function. *Biochim Biophys Acta* **1819:** 1080–1087.

Matsushima Y, Goto Y, Kaguni LS. 2010a. Mitochondrial Lon protease regulates mitochondrial DNA copy number and transcription by selective degradation of mitochondrial transcription factor A (TFAM). *Proc Natl Acad Sci* **107:** 18410–18415.

Matsushima Y, Goto Y, Kaguni LS. 2010b. Mitochondrial Lon protease regulates mitochondrial DNA copy number and transcription by selective degradation of mitochondrial transcription factor A (TFAM). *Proc Natl Acad Sci* **107:** 18410–18415.

McLean JR, Cohn GL, Brandt IK, Simpson MV. 1958. Incorporation of labeled amino acids into the protein of muscle and liver mitochondria. *J Biol Chem* **233:** 657–663.

Merkwirth C, Langer T. 2009. Prohibitin function within mitochondria: Essential roles for cell proliferation and cristae morphogenesis. *Biochim Biophys Acta* **1793:** 27–32.

Nass S. 1969. The significance of the structural and functional similarities of bacteria and mitochondria. *Int Rev Cytol* **25:** 55–129.

Nass S, Nass MM, Hennix U. 1965. Deoxyribonucleic acid in isolated rat-liver mitochondria. *Biochim Biophys Acta* **95:** 426–435.

Ngo HB, Kaiser JT, Chan DC. 2011. The mitochondrial transcription and packaging factor Tfam imposes a U-turn on mitochondrial DNA. *Nat Struct Mol Biol* **18:** 1290–1296.

Nunnari J, Suomalainen A. 2012. Mitochondria: In sickness and in health. *Cell* **148:** 1145–1159.

Oskoui M, Davidzon G, Pascual J, Erazo R, Gurgel-Giannetti J, Krishna S, Bonilla E, De Vivo DC, Shanske S, DiMauro S. 2006. Clinical spectrum of mitochondrial DNA depletion due to mutations in the thymidine kinase 2 gene. *Archiv Neurol* **63:** 1122–1126.

Otera H, Wang C, Cleland MM, Setoguchi K, Yokota S, Youle RJ, Mihara K. 2010. Mff is an essential factor for mitochondrial recruitment of Drp1 during mitochondrial fission in mammalian cells. *J Cell Biol* **191:** 1141–1158.

Owusu-Ansah E, Yavari A, Mandal S, Banerjee U. 2008. Distinct mitochondrial retrograde signals control the G1-S cell cycle checkpoint. *Nat Genet* **40:** 356–361.

Parone PA, Da Cruz S, Tondera D, Mattenberger Y, James DI, Maechler P, Barja F, Martinou JC. 2008. Preventing mitochondrial fission impairs mitochondrial function and leads to loss of mitochondrial DNA. *PLoS ONE* **3:** e3257.

Pohjoismaki JL, Goffart S. 2011. Of circles, forks and humanity: Topological organisation and replication of mammalian mitochondrial DNA. *Bioessays* **33:** 290–299.

Puigserver P, Wu Z, Park CW, Graves R, Wright M, Spiegelman BM. 1998. A cold-inducible coactivator of nuclear receptors linked to adaptive thermogenesis. *Cell* **92:** 829–839.

Ritov VB, Menshikova EV, Azuma K, Wood R, Toledo FG, Goodpaster BH, Ruderman NB, Kelley DE. 2010. Deficiency of electron transport chain in human skeletal muscle mitochondria in type 2 diabetes mellitus and obesity. *Amer J Physiol* **298:** E49–E58.

Rubio-Cosials A, Sidow JF, Jimenez-Menendez N, Fernandez-Millan P, Montoya J, Jacobs HT, Coll M, Bernado P, Sola M. 2011. Human mitochondrial transcription factor A induces a U-turn structure in the light strand promoter. *Nat Struct Mol Biol* **18:** 1281–1289.

Sagan L. 1967. On the origin of mitosing cells. *J Theor Biol* **14:** 255–274.

Santra S, Gilkerson RW, Davidson M, Schon EA. 2004. Ketogenic treatment reduces deleted mitochondrial DNAs in cultured human cells. *Ann Neurol* **56:** 662–669.

Satoh M, Kuroiwa T. 1991. Organization of multiple nucleoids and DNA molecules in mitochondria of a human cell. *Exp Cell Res* **196:** 137–140.

Scarpulla RC. 2008. Nuclear control of respiratory chain expression by nuclear respiratory factors and PGC-1-related coactivator. *Ann NY Acad Sci* **1147:** 321–334.

Schon EA, Przedborski S. 2011. Mitochondria: The next (neurode)generation. *Neuron* **70:** 1033–1053.

Shutt TE, Bestwick M, Shadel GS. 2011. The core human mitochondrial transcription initiation complex: It only takes two to tango. *Transcription* **2:** 55–59.

Sumitani M, Kasashima K, Ohta E, Kang D, Endo H. 2009. Association of a novel mitochondrial protein M19 with mitochondrial nucleoids. *J Biochem* **146:** 725–732.

Tang Y, Schon EA, Wilichowski E, Vazquez-Memije ME, Davidson E, King MP. 2000. Rearrangements of human mitochondrial DNA (mtDNA): New insights into the regulation of mtDNA copy number and gene expression. *Mol Biol Cell* **11:** 1471–1485.

Twig G, Elorza A, Molina AJ, Mohamed H, Wikstrom JD, Walzer G, Stiles L, Haigh SE, Katz S, Las G, et al. 2008. Fission and selective fusion govern mitochondrial segregation and elimination by autophagy. *EMBO J* **27:** 433–446.

Van Tuyle GC, McPherson ML. 1979. A compact form of rat liver mitochondrial DNA stabilized by bound proteins. *J Biol Chem* **254:** 6044–6053.

Virbasius JV, Virbasius CA, Scarpulla RC. 1993. Identity of GABP with NRF-2, a multisubunit activator of cytochrome oxidase expression, reveals a cellular role for an ETS domain activator of viral promoters. *Genes Dev* **7:** 380–392.

Wagatsuma A, Kotake N, Yamada S. 2011. Muscle regeneration occurs to coincide with mitochondrial biogenesis. *Mol Cell Biochem* **349:** 139–147.

Wintersberger E. 1964. [DNA-dependent RNA synthesis in rat liver mitochondria]. *Hoppe-Seyler's Zeit Physiol Chemie* **336:** 285–288.

Ylikallio E, Tyynismaa H, Tsutsui H, Ide T, Suomalainen A. 2010. High mitochondrial DNA copy number has detrimental effects in mice. *Hum Mol Genet* **19:** 2695–2705.

Clinical and Molecular Features of *POLG*-Related Mitochondrial Disease

Jeffrey D. Stumpf[1], Russell P. Saneto[2], and William C. Copeland[1]

[1]Laboratory of Molecular Genetics, National Institute of Environmental Health Sciences, National Institutes of Health, Research Triangle Park, North Carolina 27709

[2]Division of Pediatric Neurology, Seattle Children's Hospital/University of Washington, Seattle, Washington 98105

Correspondence: copelan1@niehs.nih.gov

The inability to replicate mitochondrial genomes (mtDNA) by the mitochondrial DNA polymerase (pol γ) leads to a subset of mitochondrial diseases. Many mutations in *POLG*, the gene that encodes pol γ, have been associated with mitochondrial diseases such as myocerebrohepatopathy spectrum (MCHS) disorders, Alpers-Huttenlocher syndrome, myoclonic epilepsy myopathy sensory ataxia (MEMSA), ataxia neuropathy spectrum (ANS), and progressive external ophthalmoplegia (PEO). This chapter explores five important topics in *POLG*-related disease: (1) clinical symptoms that identify and distinguish *POLG*-related diseases, (2) molecular characterization of defects in polymerase activity by *POLG* disease variants, (3) the importance of holoenzyme formation in disease presentation, (4) the role of pol γ exonuclease activity and mutagenesis in disease and aging, and (5) novel approaches to therapy and avoidance of toxicity based on primary research in pol γ replication.

While many polymerases are responsible for replicating billions of nucleotides in the nucleus, DNA polymerase γ replicates thousands of copies of the 16-kilobase mitochondrial genome (mtDNA) in each human cell. The holoenzyme of DNA polymerase γ (pol γ) consists of a catalytic subunit (encoded by *POLG* at chromosomal locus 15q25) and a dimeric form of its accessory subunit (encoded by *POLG2* at chromosomal locus 17q24.1) (Longley et al. 1998b; Lim et al. 1999). MtDNA encodes 13 proteins that are essential for the electron transport chain that provides most of the ATP in the cell. Therefore, mtDNA replication is essential for life as demonstrated by the embryonic lethality of *POLG* knockout mice (Hance et al. 2005). Studies over the last decade have identified over 200 mutations in *POLG* that are associated with certain mitochondrial diseases (http://tools.niehs.nih.gov/polg/) (Longley et al. 2005: Copeland 2008; Longley et al. 2010; Walter et al. 2010; Stumpf and Copeland 2011; Tang et al. 2011). *POLG*-related disorders are currently defined by at least five major phenotypes of neurodegenerative disease that include: (1) Alpers-Huttenlocher syndrome (AHS), (2) childhood myocerebrohepatopathy spectrum (MCHS), (3) myoclonic epilepsy myopathy sensory ataxia (MEMSA),

Cite this article as *Cold Spring Harb Perspect Biol* doi: 10.1101/cshperspect.a011395

(4) the ataxia neuropathy spectrum (ANS), and (5) progressive external ophthalmoplegia (PEO) with or without sensory ataxic neuropathy and dysarthria (SANDO) (Wong et al. 2008; Cohen et al. 2010; Saneto and Naviaux 2010; Saneto et al. 2010). These diseases associate with mutations that have been shown to decrease the activity of the polymerase, the fidelity of replication, and the formation of the holoenzyme. This work reviews the important conclusions regarding *POLG*-related diseases and discusses some strategies for disease treatment options.

CLINICAL SYMPTOMS THAT IDENTIFY AND DISTINGUISH *POLG*-RELATED DISEASES

The clinical identification of mitochondrial diseases associated with *POLG* mutations is difficult because *POLG*-related diseases evolve over time and have an overlapping range of symptoms with multiple organ system involvement and with different degrees of severity and timing of onset. However, several important clues to the diagnosis include the tempo of disease progression, the presenting organ system, the combination of symptoms (but not necessarily the sequence of presentation or severity of organ involvement), and increasing involvement of neurological signs and symptoms during illness (especially viral infections producing fevers, or other physiological stressors) (Nguyen et al. 2006; Cohen et al. 2010). While no single test identifies *POLG*-related diseases, proper diagnosis depends on the age of presentation and associated biochemical and neuroimaging findings. By definition, *POLG* mutations are necessary for *POLG*-related disease, and identification of mutations known to facilitate disease (discussed later) is necessary for use as a diagnostic tool. The correct combination of laboratory testing and clinical observations (Table 1) identifies specific diseases and allows the monitoring of the course of the disease. Although the broad clinical expression of *POLG*-related diseases can make diagnoses difficult, there are phenotypic similarities and ages of onset that reflect distinct syndromes (Table 2) (Wong et al. 2008; Saneto and Naviaux 2010).

Table 1. Evaluation of patients with *POLG* disease[a]

Organ system	Test
CNS	EEG (with possible Video-EEG required for subtle seizures)
	MRI of the brain (T1/T2/FLAIR/Diffusion)
	MRS of the brain (TE 30 and either TE 130 (1.5 T) or TE 288 (3.0 T)
	Neurological examination
Vision	Formal visual exam (including eye movements)
	Electroretinogram and visual evoked potential if clinically indicated
Hearing	Brain stem auditory evoked potentials if clinically indicated
Cardiac	EKG (may also consider ECHO if clinically indicated)
Liver	Liver biopsy (for pathological evaluation and mitochondrial DNA content)
	Liver enzyme and function studies (ALT, AST, GGT, fasting glucose, albumin, prothrombin time and INR, ammonia, bilirubin, cholesterol)
Gastrointestinal	Gastric emptying and bowel motility if clinically indicated
	Swallowing study if clinically indicated
Respiratory	Pulmonary function testing
	Sleep polysomnography to evaluate hypercapnia during sleep
Systemic	Blood acylcarnitines, total and free carnitine, plasma amino acids

[a]Adapted from Haas et al. (2007, 2008) and Cohen and Naviaux (2010).

Even with a multitude of tests for mitochondrial disease, there are many examples where the results may lead to a difficult or improper diagnosis. For instance, extracting muscle to assay electron transport chain complex activity is useful in identifying mitochondrial cytopathy, but not necessarily *POLG*-related diseases (Gonzalez-Vioque et al. 2006; de Vries et al. 2007; Saneto et al. 2010). In addition, the use of mitochondrial DNA (mtDNA) depletion assays can be misleading for two reasons: normal mtDNA values would not exclude *POLG* disease, especially early in the course of the disease, and

Table 2. Major clinical syndromes associated with *POLG* mutations[a]

Age of onset	Syndrome	References
Neonatal/infancy	Myocerebrohepatopathy spectrum (MCHS)	Ferrari et al. 2005; de Vries et al. 2007
Infancy/childhood	Alpers-Huttenlocher syndrome (AHS)	Naviaux and Nguyen 2004
Adolescent/young adult	Ataxia neuropathy spectrum (ANS)[b]	Hakonen et al. 2005
	Myoclonus, epilepsy, myopathy, sensory ataxia (MEMSA)	Van Goethem et al. 2004
	Progressive external ophthalmoplegia (PEO)[c,d]	Van Goethem et al. 2001

[a]Table data adapted from Saneto and Naviaux (2010).

[b]ANS has also been labeled as mitochondrial recessive ataxia syndrome (MIRAS), sensory ataxic neuropathy with dysarthria and ophthalmoplegia (SANDO), and spino-cerebellar ataxia epilepsy syndrome (SCAE) (Tzoulis et al. 2006).

[c]PEO is labeled as PEO+ when associated with additional symptoms that may include peripheral neuropathy, ataxia, hearing loss, depression, premature menopause, hypogonadism, Parkinsonism, cataracts, or GI dysmotility.

[d]Dominant and recessive mutations in *POLG* have been associated with PEO.

mtDNA copy number is variable among different tissues (Dimmock et al. 2010). Finally, Van Goetham et al. (2003b) described several patients with mitochondrial neurogastrointestinal encephalopathy (MNGIE)-like syndrome, but without the cardinal finding of leukoencephalopathy as described in true MNGIE (Hirano et al. 1994). The MNGIE-like patients do not have deficiency in thymidine phosphorylase activity and would not respond to allogenic stem cell transplantation. Thus, proper treatment may be delayed or incorrectly given. To assist in the proper diagnosis of *POLG* related disease, the following section lists clinical features of the five most characterized *POLG*-related disorders.

Myocerebrohepatopathy Spectrum (MCHS) Disorders

MCHS disorders present in the first two to three years of life and result in early death. The clinical features of MCHS disorders include encephalopathy/developmental delay, liver dysfunction, myopathy, and hypotonia. Like other mitochondrial disorders, gastrointestinal (GI) dysmotility and failure to thrive occurs in MCHS disorders, albeit at earlier ages. Seizures may occur in these patients, but are not a necessary or dominant feature. Liver involvement can be the most dramatic feature with early failure leading to death. For instance, one compound heterozygous MCHS patient with G848S and R1096C mutations in *POLG* presented at 5 months of age with encephalopathy, hypotonia, and liver failure (RP Saneto, unpubl.). The biopsy showed that the liver pathology was distinct from what is commonly seen in another infantile onset *POLG*-related disease, Alpers-Huttenlocher (Nguyen et al. 2006). This patient died at 11 months of age because of liver failure.

Alpers-Huttenlocher Syndrome (AHS)

Also referred to as Alpers syndrome, AHS presents most frequently during infancy (similar to MCHS) and is fatal. Epilepsy is usually the predominant feature in AHS with an early predilection of epileptiform discharges over the occipital region of the brain, and generalized discharges as the disease progresses (Tulinius and Hagne 1991; Engelsen et al. 2008; Wolf et al. 2009; Saneto et al. 2010). Early seizures have the semiology of occipital lobe seizures with vomiting, headache, and myoclonus. Most patients have multiple episodes of epilepsia partialis continua and/or convulsive status epilepticus. Early in the course of epilepsy, seizures are often controlled with conventional medications. However, as the disease progresses, the seizures become increasingly resistant to valproic acid and other medications that trigger liver failure (Bicknese et al. 1992; Engelsen et al. 2008; Uusimaa et al. 2008; Wolf et al. 2009; Saneto et al. 2010). Even without valproic acid dosing, liver failure occurs in AHS patients, and liver pathology cannot distinguish the samples that were treated with valproic acid (Harding 1990). Characteristic features of liver pathology in AHS require at least three of the following histological findings: (1) microvesicular steatosis, (2) bile ductular proliferation, (3) hepatocyte dropout, (4) bridging fibrosis or cirrhosis, (5) collapse of liver cell plates, (6) parenchymal disarray or disorganization of normal lobular architecture, (7) regenerative nodules, and (8) oncocytic change in scattered hepatocytes not affected by steatosis (Nguyen et al. 2006).

In addition to liver pathology, other common clinical features of AHS include headaches that are migrainous in nature, cortical blindness usually in fully developed AHS, and myoclonus often in the later phases of the disease (Nguyen et al. 2006; Cohen et al. 2010). Myoclonus may become so rapid that they become indistinguishable from myoclonic seizures. The onset and tempo of cognitive impairment is variable and often triggered by infectious illness.

Myoclonic Epilepsy Myopathy Sensory Ataxia (MEMSA)

MEMSA has the dominant features of epilepsy, myopathy, and ataxia, but without ophthalmoplegia. A subclinical sensory polyneuropathy leading to ataxia is usually the first sign of the disease. Usually seen later in the disease course, myoclonic seizures are often focal, frequently involving an arm, and can become generalized over time and comorbid with a progressive encephalopathy. Two features distinguish MEMSA from other similar diseases. First, the lack of ophthalmoplegia and prominence of myopathy sets these patients apart from other *POLG* syndromes. Second, histochemical assays of the muscle reveal the absence of ragged-red fibers distinguishing this syndrome from myoclonus epilepsy with ragged-red fibers (Van Goethem et al. 2003b).

Ataxia Neuropathy Spectrum (ANS)

ANS presents from early adolescence to the late third decade, and may be the most difficult to diagnose. While the predominant features of ANS are neuropathy and ataxia without myopathy, patients may also express progressive external ophthalmoplegia (Van Goethem et al. 2004). A wide variety of symptoms have been associated with ANS, including mild cognitive impairment, involuntary movements, psychiatric symptoms, myoclonus, blindness, hearing loss, and liver involvement (Van Goethem et al. 2004; Hakonen et al. 2005; Winterthun et al. 2005; Tzoulis et al. 2006). Commonly, ANS patients also present with white matter brain lesions as shown by MRI (Van Goethem et al. 2004).

Progressive External Ophthalmoplegia (PEO)

PEO (or PEO+, which indicates multiple organ involvement) is an adult onset disorder that occurs with dominant or recessive *POLG* mutations (Van Goethem et al. 2001; Lamantea et al. 2002). Although the major clinical finding is a progressive weakness of the extraocular eye muscles resulting in ptosis and loss of eye movements in the horizontal and vertical directions (Van Goethem et al. 2001), a generalized myopathy is present in most patients. Other symptoms of PEO+ include sensory ataxia, neuropathy, dysarthria, myopathy, and restless leg syndrome (Fadic et al. 1997; Van Goethem et al. 2001; Milone et al. 2008; Aitken et al. 2009; Blok et al. 2009). Cosegregating with PEO families are features of Parkinsonism (Luoma et al. 2004), premature ovarian failure, sensory ataxia (Pagnamenta et al. 2006), and cataracts (Luoma et al. 2004; Blok et al. 2009).

MOLECULAR CHARACTERIZATION OF DEFECTS IN POLYMERASE ACTIVITY BY *POLG* DISEASE VARIANTS

Identification of *POLG* mutations is the unifying characteristic of *POLG*-related disease. Ever since the landmark paper by Van Goethem et al. (2001) identified the cosegregation of PEO with *POLG* mutations, searches for other disease-associated mutations have resulted in an exponentially increasing number of mutations found in disease patients (http://tools.niehs.nih.gov/polg). Most mutations are found in compound heterozygotes containing multiple mutations in *cis* and in *trans*. However, some mutations do not affect enzyme function in known biochemical assays or in yeast model systems. The consideration that as few as 43 mutations have been described in multiple publications describing nonrelated families raises doubts as to whether all of the hundreds of mutations that have only shown to cosegregate with disease in a single publication are in fact causative of disease. In fact, at least 32 mutations have been identified only in *cis* with another *POLG* mutation, and the relative contribution of each mutation on the same chromosome is difficult to interpret without further analysis. Although it is useful to have a comprehensive record of *POLG*-associated mutations, it is imperative to characterize mutations that are likely to cause disease by recording adequate genotypic family histories, determining mitochondrial dysfunction in model systems, or measuring deficiency of enzymatic activity (summarized in Table 3).

The catalytic subunit of pol γ is a 140-kD protein containing DNA polymerase, 3′-5′ exonuclease and 5′ dRP lyase activities (Gray and Wong 1992; Longley et al. 1998a,b). Defects in DNA polymerase activity have been detected in many *POLG* mutations studied, suggesting that the reduced polymerase activity facilitates disease progression. Although 5′ dRP lyase is an important step in mtDNA base excision repair, the critical residues for lyase activity are unknown, making it difficult to ascertain the importance of lyase activity in maintenance of mitochondrial genomes.

Mutations in *POLG* were first identified in 2001 as a disease locus for PEO (Van Goethem et al. 2001). With the exception of one mutation, all dominant *POLG* mutations responsible for developing PEO are mapped to the polymerase domain of pol γ. Three of the substitutions, H932Y, R943H, and Y955C, change side chains that interact directly with the incoming dNTP (Graziewicz et al. 2004; Stumpf et al. 2010). These enzymes retain less than 1% of the wild-type polymerase activity and display a severe decrease in processivity (Graziewicz et al. 2004), characteristics that likely cause the severe clinical presentation in R943H and Y955C heterozygotes (Graziewicz et al. 2004). In addition, in vitro assays showed that the Y955C substitution also increases nucleotide misinsertion errors 10- to 100-fold in the absence of exonucleolytic proofreading (Ponamarev et al. 2002). In vivo evidence in yeast showed that alteration of the conserved tyrosine in the yeast mitochondrial polymerase gene, *mip1*, enhanced mtDNA damage, increased mutagenesis, and resulted in cells with dysfunctional mitochondria (Baruffini et al. 2006; Stuart et al. 2006). Similar to dominance seen in human patients, heterozygous yeast that alter the conserved tyrosine show a dominant increase in mitochondrial dysfunction, suggesting that the mutant enzymes compete with the wild-type polymerases for binding at the replication fork (Stuart et al. 2006). Additionally, a mouse transgenic model with the Y955C *POLG* allele targeted to the heart resulted in cardiomyopathy, loss of mtDNA, and enlarged hearts (Lewis et al. 2007). These models strongly suggest that large reductions in pol γ polymerase activity are sufficient to cause mitochondrial dysfunction that is central to *POLG*-related disease.

Other *POLG* disease-associated mutations have been assayed for polymerase activity to determine the potential to cause disease. Polymerase active site mutations G923D and A957S are found in PEO patients and have been shown to exhibit 21% and 23% polymerase activity, respectively (Graziewicz et al. 2004). Analysis of a cluster of Alpers mutations in the thumb domain has shown a striking correlation with the severity of the defect and the degree of conservation of amino acid sequences among various eukaryotes (Kasiviswanathan et al. 2009). Mutations in the most conserved sites represented by G848S, T851A, R852C, and R853Q exhibited less than 1% WT enzyme activity (Kasiviswanathan et al. 2009). Mutations in codons for less conserved amino acids (Gln879 and Thr885) resulted in only moderate reduction in activity (Kasiviswanathan et al. 2009). Haplotype analysis of the Finnish population demonstrates a carrier frequency of 1:125 for a common polymerase domain mutation, W748S (Hakonen et al. 2005). It has been shown that the W748S mutation caused the polymerase to have a low

Table 3. Genetic and biochemical evidence for *POLG* variants causing disease

Human mutation	Yeast mutation	Human disease	Biochemical data	Genetic data	References
L244P	L211P	Alpers		Mild increase in petite frequency, 6- to 10-fold increase in mutagenesis	Stumpf et al. 2010
G268A	G244A	PEO		2.5-Fold increase in petite frequency in yeast 10-Fold increase in point mutations (haploid yeast)	Baruffini et al. 2006
G303R	G259R	Alpers		100% Petite in haploid yeast	(Baruffini et al. 2011)
L304R	L260R	PEO		40–100% Petites in haploid yeast, 7.9% in diploid	Stumpf et al. 2010
S305R	C261R	Alpers, ANS		Petite frequency 40-fold higher than the C261S "humanized" wild-type	Baruffini et al. 2011
Q308H	Q264H	PEO		100% Petites in haploid yeast	Stumpf et al. 2010
R309 L	R265 L	PEO		35% Petites in haploid yeast	Stumpf et al. 2010
A467T	I416T	Alpers syndrome, ANS, PEO	Defect in accessory subunit binding 4% Catalytic activity because of low DNA binding, low k_{cat}, high K_m	Causes mtDNA and nuclear DNA damage in yeast, 9% petites as haploid yeast	Chan et al. 2005a; Luoma et al. 2005; Stuart et al. 2006
W748S	NA	Alpers syndrome, ANS, PEO	Defect in catalysis, DNA binding, and enzyme stability		Chan et al. 2006
R807C	R607C	SANDO		100% Petites in haploid yeast	Stumpf et al. 2010
R807P	R607P	PEO		67% Petites in haploid yeast and elevated mutation frequency	Stumpf et al. 2010
G848S	G651S	Alpers, PEO	<1% Polymerase activity, reduced gap-filling activity	46% Petites in haploid yeast and elevated mutation frequency	Baruffini et al. 2007; Kasiviswanathan et al. 2009; Stumpf et al. 2010
T851A	T654A	Alpers	<1% Polymerase activity	100% Petites in haploid yeast	Kasiviswanathan et al. 2009; Stumpf et al. 2010)
R852C	R655C	Alpers	<1% Polymerase activity		Kasiviswanathan et al. 2009; Stumpf et al. 2010

Cite this article as *Cold Spring Harb Perspect Biol* doi: 10.1101/cshperspect.a011395

R853W	R656W	PEO with ptosis		100% Petites in haploid yeast	Stumpf et al. 2010
R853Q	R656Q	Alpers	<1% Polymerase activity	100% Petites in haploid yeast	Kasiviswanathan et al. 2009; Stumpf et al. 2010
N864S	N667S	PEO		100% Petites in haploid yeast	Stumpf et al. 2010
E873X	E676	Alpers syndrome		Stop codon leads to nonsense-mediated decay of message	Chan et al. 2005b
Q879H	NA	Alpers	50% Polymerase activity		Kasiviswanathan et al. 2009
T885S	NA	Alpers	70% Polymerase activity		Kasiviswanathan et al. 2009
G888S	G691S	Alpers		Mild increase in petite frequency and mutagenesis	Stumpf et al. 2010
A889T	A692T	PEO	Reduced gap-filling activity	17% Petite frequency in haploid yeast	Stumpf et al. 2010
G923D	G725D	PEO	21% Polymerase activity	100% Petites in haploid yeast, only 7.6% in diploid, dominant in humans	Graziewicz et al. 2004; Stuart et al. 2006; Stumpf et al. 2010
D930N	D732N	Alpers		100% Petites in haploid yeast	Baruffini et al. 2011
H932Y	H734Y	PEO, ataxia	Reduced gap-filling activity	100% Petites in haploid yeast	Stumpf et al. 2010
R943H	R745H	PEO	0.2% Polymerase activity because of high K_m(dNTP) and lower k_{cat}	100% Petites in haploid yeast, 37% in diploid yeast, loss of mtDNA in yeast, dominant in humans	Graziewicz et al. 2004; Stuart et al. 2006; Stumpf et al. 2010
K947R	K749R	PEO		100% Petites in haploid yeast	Baruffini et al. 2011
Y955C	Y757C	PEO, Parkinsonism	0.03% Polymerase catalytic activity, mutator polymerase	Loss of mtDNA in transgenic mice, oxidative stress, cardiomyopathy, high petite frequency in yeast (100% in haploid, 80% in diploid), dominant in humans	Ponamarev et al. 2002; Graziewicz et al. 2004; Baruffini et al. 2006; Stuart et al. 2006
A957S	A759S	PEO	23% Polymerase activity	35% Petites in haploid yeast, only 2.5% in diploid, dominant in humans	Graziewicz et al. 2004; Stuart et al. 2006; Stumpf et al. 2010
A957P	A759P	Alpers		100% Petites in haploid yeast	Stumpf et al. 2010

Continued

Table 3. *Continued*

Human mutation	Yeast mutation	Human disease	Biochemical data	Genetic data	References
R964C	NA	NRTI toxicity	Reduced catalytic activity, low k_{cat}	Two- to threefold increase in petite frequency upon exposure to stavudine in haploid yeast	Bailey et al. 2009a; Baruffini and Lodi 2010
G1051R	G807R	PEO, ataxia	Slightly reduced gap-filling activity	Mild increase of petites in haploid yeast	Stumpf et al. 2010
P1073 L	P829 L	Alpers		35-Fold increase in petites in haploid yeast	Baruffini et al. 2011
G1076V	G832V	PEO		100% Petites in haploid yeast	Stumpf et al. 2010
R1096C	R853C	PEO, Alpers		100% Petites in haploid yeast	Stumpf et al. 2010
R1096H	R853H	Alpers		100% Petites in haploid yeast	Stumpf et al. 2010
S1104C	S861C	PEO		100% Petites in haploid yeast	Stumpf et al. 2010
V1106I	V863I	PEO		100% Petites in haploid yeast	Stumpf et al. 2010
E1143G	E900G	Polymorphism in 4% in population	1.4-Fold more active than WT	Temperature-sensitive mutant in yeast, high petite frequency at 36°C	Chan et al. 2006; Baruffini et al. 2007
D1184N	D941N	PEO		33% Increase in petites in haploid yeast and mild mutagenesis	Stumpf et al. 2010

NA, not applicable (because there is no homolog in yeast); ANS, ataxia neuropathy syndromes.

catalytic activity and a severe DNA-binding defect (Chan et al. 2006). The W748S mutation is often found in combination with E1143G in ataxia patients (Hakonen et al. 2005; Winterthun et al. 2005) and was originally identified as an SNP in 4% of the general population (Di Fonzo et al. 2003; Van Goethem et al. 2004). However, biochemical evidence shows that the E1143G substitution partially reduces the deleterious effects of the W748S mutation, suggesting that E1143G may modulate the effects of disease mutations in *POLG* (Chan et al. 2006).

Yeast has been a useful model system to determine the functionality of *POLG* mutants by utilizing changes in residues that are in conserved stretches of amino acids (Baruffini et al. 2006, 2007, 2011; Stuart et al. 2006; Stumpf et al. 2010). Most of the mutations in conserved regions are in the polymerase domain, and many are completely unable to retain mitochondrial function because of disruption of mtDNA replication. These include PEO/SANDO-associated mutations R807C, R807P, R853W, N864S, G923D, H932Y, K947R, G1076V, R1096C, S1104C, and V1106I, and Alpers-associated mutations G848S, T851A, R853Q, D930N, A957P, P1073A, and R1096H (Stuart et al. 2006; Stumpf et al. 2010; Baruffini et al. 2011). By measuring a base substitution in mtDNA that confers erythromycin resistance, it was shown that mtDNA mutant frequency is increased by some disease-associated mutations (Baruffini et al. 2006, 2007, 2011; Stumpf et al. 2010). However, biochemical evidence of several mutants showed that the incorporation of an incorrect nucleotide is not more kinetically favorable (Kasiviswanathan et al. 2009; Szczepanowska and Foury 2010). Therefore, it is still unclear what causes the increase in mutagenesis in these mutant strains. Finally, information from the yeast studies allowed the comparison of multiple substitutions of the same residue that were associated with different diseases (Stumpf et al. 2010). The mutations associated with Alpers were more severe than those associated with PEO, which agrees with the in vitro studies that suggest that the biochemical defects correlate with the severity and age of onset found in patients (Chan and Copeland 2009).

POLG LINKER REGION IS IMPORTANT FOR BINDING TO THE ACCESSORY SUBUNIT

The linker (also called spacer) region is between the polymerase and exonuclease domain and was named prior to biochemical characterizations of pol γ that determined its functions. Structural studies identified critical contacts between the catalytic subunit of the holoenzyme and the homodimeric accessory subunit (encoded by *POLG2*) that are required for processivity (Lee et al. 2009). In particular, Glu540 and a helix comprised of amino acids 540–558 are crucial to holoenzyme assembly (Lee et al. 2009; Lee et al. 2010a). Ala467 is also important for subunit interaction, and substitution to threonine (A467T) is a common *POLG* disease mutation. Not only is the A467T mutant enzyme compromised for its interaction with the accessory subunit, it also retains only 4% polymerase activity compared to WT enzyme (Chan et al. 2005a).

The A467T mutation is estimated to occur in 36% of all alleles associated with *POLG* disease. (Ferrari et al. 2005; Horvath et al. 2006; Nguyen et al. 2006; de Vries et al. 2007; Wong et al. 2008). In the general population, the frequency of the A467T mutation has been found to exist in between 0.2% to 1% of asymptomatic European populations (Van Goethem et al. 2001; Luoma et al. 2005; Winterthun et al. 2005; Horvath et al. 2006). Ataxic patients who are homozygous for the A467T mutation present with symptoms in their early to late teens (Van Goethem et al. 2004; Winterthun et al. 2005; Tzoulis et al. 2006); however, patients have often been reported to be compound heterozygous with A467T and at least one other *POLG* disease-associated mutation. Because of the severity of the A467T mutation on processivity and catalysis (Chan et al. 2005a), it is assumed that the other mutation determines the disease progression. For instance, Naviaux et al. (1999) reported an Alpers patient with reduced electron transport chain function, dicarboxylic aciduria, fulminant hepatic failure, refractory epilepsy, and lactic acidosis, which resulted in death at 42 months. Analysis of mtDNA and pol γ activity from skeletal muscle biopsy indicated a reduction of mitochondrial DNA content to 30% of wild-type levels and no detectable pol γ activity (Naviaux et al. 1999). Later sequencing of the *POLG* gene revealed compound heterozygosity with an A467T substitution in *trans* with a G to T nonsense mutation that converts Glu873 (GAG) to a stop codon (Naviaux and Nguyen 2004). *POLG* mRNAs containing the Glu873 nonsense mutation are degraded from the pool of mRNAs by nonsense mediated decay resulting in mono-allelic expression of *POLG* containing only the A467T mutation (Chan et al. 2005b).

Other examples of clinical presentations of compound heterozygotes include SANDO patients with A467T in one *POLG* allele and either R3P, L304R, or R627W in the other (Van Goethem et al. 2003a). One patient with ataxia-myopathy syndrome was shown to have A467T in one *POLG* allele and R627Q mutation in *cis* with the Q1236H single nucleotide polymorphism in the other *POLG* allele (Luoma et al. 2005). Other patients with ataxia were found to be heterozygous with the A467T mutation on one allele and W748S in *cis* with the E1143G mutation in the other allele (Van Goethem et al. 2004).

Very little is known about other linker domain mutations. Information about one linker domain mutation has demonstrated the difficulty of assigning disease causation to mutations, despite biochemical assays, structural data, and model systems. The G517V substitution was identified in simple and compound heterozygotes in many different disease presentations including Leigh syndrome, neuropathy, myopathy, sideroblastic anemia, and Kearns Sayre syndrome (Sarzi et al. 2007; Wong et al. 2008). However, the G517V substitution in recombinant pol γ exhibits no major changes in catalysis, DNA binding, or binding to the accessory subunit (Kasiviswanathan and Copeland 2011). There are several possibilities to explain the contradiction between the association to mitochondrial disease and the biochemical data. It is possible that other pol γ activities that are not routinely assayed, such as dRP lyase activity, are important for mtDNA replication. Alternatively, the assays may require a less minimalistic replisome and substrate to dissect the difference in activity between the enzymes, suggesting the need for proper reconstitution of mtDNA replication in vitro. More importantly, it is clear that the preponderance of evidence of association between a particular mutation and disease is neither sufficient nor a logical argument to propose a causal relationship. In the absence of genetic or biochemical data, the proper characterization of disease-associated mutations requires adequate genotypic and clinical family history.

POL γ EXONUCLEASE ACTIVITY, DISEASE, AND MTDNA MUTAGENESIS

Identification of mutations associated with mitochondrial disease in the exonuclease domain resulted in an obvious and reasonable hypothesis: exonuclease domain mutations reduce the proofreading function of pol γ, resulting in an mtDNA mutator phenotype and ultimately leading to mitochondrial dysfunction and disease. Because specific mtDNA mutations at high enough cellular concentrations (or heteroplasmy) cause maternally inherited mitochondrial disease, general mtDNA mutagenesis caused by decreased proofreading activity could be a possible contributor to human disease.

Surprisingly, mutations that have been studied in the exonuclease domain, which are most conserved from humans to yeast, have not caused increases in mutagenesis in vivo or decreases in exonuclease activity in vitro (Baruffini et al. 2006; Stumpf et al. 2010; Szczepanowska and Foury 2010). In fact, kinetic data showed that disease-associated mutations increase exonuclease activity for both correct and mismatched primer-template termini (Lee et al. 2010b; Szczepanowska and Foury 2010). Because not all disease-associated mutations have been assayed for exonuclease activity, it is still possible that lack of proofreading can cause *POLG*-related diseases. However, proofreading activity may not factor into diseases that clinicians identify with *POLG* mutations. It is unknown if exonuclease-deficient pol γ variants are found in the unaffected population, or in people affected by noncanonical *POLG*-associated diseases. It is also formally possible that mutations that eliminate exonuclease function of pol γ are embryonic lethal. This seems unlikely considering that mice with exonuclease-deficient pol γ are viable (Trifunovic et al. 2004; Kujoth et al. 2005), and exonuclease-deficient pol γ efficiently replicates DNA in vitro.

To test the potential effects of eliminating mitochondrial proofreading function on disease, several groups used mouse models that disrupted pol γ exonuclease activity. In the first mouse model that eliminated pol γ exonuclease activity, the pol γ variant was transgenically targeted to the heart, resulting in severe cardiomyopathy accompanied by mtDNA mutations and deletions (Zhang et al. 2000). Several years later, two independent groups created mice homozygous for mutations that disrupted exonuclease function (Trifunovic et al. 2004; Kujoth et al. 2005). These mice exhibited premature aging between 6 and 9 months, characterized by graying hair, loss of hair and hearing, curvature of the spine, enlarged hearts, and decreased body weight and bone density (Trifunovic et al. 2004; Kujoth et al. 2005). These observations have not only shown that exonuclease-deficient pol γ does not cause embryonic lethality, but also have created discussion about the role of mtDNA mutations in aging and mitochondrial disease.

The degree of increase of mtDNA mutagenesis in *POLG* exonuclease-deficient mice was originally unclear. The increase in mtDNA mutagenesis reported in the mutants (three- to eightfold) is similar to the accumulation of mutations detected in 2- to 3-year-old mice (three- to 11-fold) (Trifunovic et al. 2004, 2005; Kujoth et al. 2005; Vermulst et al. 2007). However, mutation frequencies in young, wild-type mice is at or below the limit of detection using the method of PCR cloning and sequencing, which introduces mutations at 1.3×10^{-4} mutations per base pair (Kujoth et al. 2005). This limitation was alleviated by an alternative method of quantifying mutation frequencies called the "random capture method," where the frequency of mutations that cause resistance to restriction endonuclease digestion is enriched, allowing more accurate estimations of mutation frequency (7.1×10^{-7} mutations per base pair in wild-type mice and 1.6×10^{-4} mutations per base pair in young heterozygotes) (Vermulst et al. 2007). Mutation frequency in homozygous mutant mice was confirmed using next-generation sequencing technology (Williams et al. 2010). The mutation frequency of heterozygotes, which were asymptomatic, was much higher than aged wild-type mice (5.4×10^{-6} mutations per base pair) (Vermulst et al. 2007). Therefore, the increase in mutation frequency that occurs in older mice is not sufficient to cause phenotypes associated with aging. However, it is still possible that the extremely high mutation rate that occurs in homozygous *POLG* exonuclease-deficient mice is sufficient to cause or expedite the aging process. Both the relationship of mtDNA mutagenesis to disease and causes of increased mutagenesis are critical topics that need further exploration.

In addition to detecting point mutations, the random capture assay detected a 90-fold increase in mtDNA deletions in homozygous *POLG* exonuclease-deficient mice as compared to age-matched wild-type or heterozygotes (Vermulst et al. 2008). While mtDNA deletions in wild-type and heterozygotes mostly occur among direct repeats of six or more nucleotides, mtDNA deletions in homozygous mutants occur independently of direct repeats (Vermulst et al. 2008). The mechanism for deletions between direct repeats is often attributed to strand-slippage, where a mispriming event occurs downstream of the correct target, a process that appears to be prevented by the proofreading exonuclease function of a polymerase. Interestingly, the lack of increase of deletions in the heterozygote suggests that the wild-type pol γ protects against deletions that are caused by the exonuclease-deficient variant, suggesting an interplay between separate domains of both enzymes similar to the idea of extrinsic proofreading (Nick McElhinny et al. 2006). The role of mtDNA deletions in aging is controversial because there is disagreement about the absolute number of mtDNA deletions in the mitochondria. Because the random capture assay only detects relative numbers of deletions among various samples, deletion quantification has depended on Southern blots and long-range PCR techniques that are helpful when the majority of molecules have a particular deletion (Edgar et al.

2009), but not when the size and distribution of deletions are expected to be random (Vermulst et al. 2009). Next-generation sequencing techniques detected the presence of deletions or tandem duplications only in the control region (the technique cannot distinguish between the two) in homozygous mutant mice (Williams et al. 2010). Other studies have found relatively abundant levels of linear nonreplicating molecules with large mtDNA deletions, probably because of pausing at the control and OriL regions during mtDNA replication (Bailey et al. 2009b; Ameur et al. 2011). However, amplification of large sections of single mtDNA molecules extracted from 2- to 3-year-old mice showed that only 0.07%–0.2% of molecules contained mtDNA deletions, arguing against the model that mtDNA deletions drive normal aging (Guo et al. 2010). Similar to the story of mtDNA mutations, mtDNA deletions may not contribute to normal aging, and their role in premature aging in mice with exonuclease-deficient pol γ is still unclear.

An important point that has confounded the study of mtDNA mutagenesis is the complicated measurement of mtDNA mutation rates, which in nuclear DNA studies are expressed as mutations per base pair per genome sequenced. To know the number of genomes sequenced, many approximations need to be made about the number and rate of genome replications per cell and mitochondrion, the rate of mitophagy (the targeted destruction of mitochondria) per cell, the number of mtDNA genomes that are destroyed during each mitophagy event, and whether mitophagy occurs more frequently with mutant genomes. These variables make all of the previous mutation frequencies difficult to interpret. For instance, if mtDNA deletions are preferentially biased for mitophagy, then the absolute number of deleted mtDNA molecules may underrepresent the true number of deletion events. Also, if mutant mtDNA tend to segregate and be amplified in particular cells, then mutant frequency may not reflect the number of cells with suboptimal ATP production. Mutation frequencies are normally determined in a population of cells; however, determining the mtDNA dynamics of a mutation requires measurements of individual cells. For instance, in heterozygous *POLG* exonuclease-deficient 15-month-old mice, 20% of cells in the duodenum were respiratory deficient as visualized by staining for loss of cytochrome oxidase (COX) activity (Vermulst et al. 2008). Even though the vast majority of mtDNA molecules contained a point mutation in these cells (but not deletions), the heterozygotes were asymptomatic presumably because there were enough cells with mitochondrial function. In the prematurely aging homozygous mutants, there were much higher percentages of COX-negative cells in the brain, liver, and duodenum (Vermulst et al. 2008). It is still undetermined whether the COX-negative cells in the homozygous mutants contain mostly mtDNA mutations or deletions, a finding that should resolve the controversy.

GENETIC AND ENVIRONMENTAL EFFECTORS OF MITOCHONDRIAL DYSFUNCTION: SUGGESTIONS FOR THERAPEUTIC APPROACHES?

Unfortunately, treatment of *POLG*-related diseases has been limited to symptom management and supportive care. The predominance of liver involvement in the younger child has led to liver transplantation in some patients, but the outcome has been poor and is not recommended for *POLG*-related disease because of neurological involvement (Tzoulis et al. 2006; Wong et al. 2008). Even though seizures in *POLG* patients can be resistant to anti-seizure medications, the use of valproic acid should be avoided, because of induction of liver failure (Horvath et al. 2006; Tzoulis et al. 2006; Chinnery and Zeviani 2008). With a paucity of options known to treat *POLG*-related mitochondrial diseases, it is important to use basic research to understand genetic and environmental changes that can suppress the molecular phenotype related to *POLG* mutations.

Targeting Oxidative Stress for Treatment of *POLG*-Related Disease

Therapies that target oxidative stress caused by dysfunctional mitochondria show promise.

Recently, an early open-label study has suggested that a novel drug, α-tocotrienol quinone (ATQ3), may have therapeutic value in treating mitochondrial disease (Shrader et al. 2011). This drug is a strong antioxidant, which is hypothesized to block free radicals and thereby prevent oxidative injury. While this medication is currently making its way through the clinical trial pathway, the true therapeutic value needs to be validated.

Studies using basic model systems may help understand the potential effectiveness of antioxidant therapy. A yeast study with homologous disease-associated mutations in the mitochondrial polymerase showed that high frequencies of mitochondrial dysfunction could be suppressed by treatment with the antioxidants MitoQ and dihydrolipoic acid (Baruffini et al. 2006, 2011). However, the effect was specific to certain mutations for reasons that still need to be explored. In addition, the transgenic mouse model with the Y955C *POLG* allele targeted to the heart developed cardiomyopathy, loss of mtDNA, enlarged heart, and increased levels of 8-oxo-dG in its mtDNA, suggesting a role for oxidative damage in disrupting mtDNA replication and a benefit from antioxidant therapy (Lewis et al. 2007). Further support for the potential of antioxidant therapy is demonstrated by the attenuation of age-dependent cardiomyopathy in mice with mutations that inactivate pol γ exonuclease function by overexpression of catalase, which reduces oxidative damage (Dai et al. 2010).

Availability of Nucleotides Modifies Mitochondrial Dysfunction

Mitochondrial toxicity is a common and dangerous side effect of HIV patients treated with nucleotide reverse transcription inhibitors (NRTIs). While the NRTIs were designed to inhibit HIV reverse transcriptase and not cellular polymerases, NRTI incorporation inhibits pol γ replication (Kakuda 2000), and may cause mitochondrial toxicity in some cases. In support of this hypothesis, a *POLG* disease-associated mutation changing Arg964 to cysteine (R964C) was suggested to facilitate stavudine-induced toxicity (Yamanaka et al. 2007). Biochemical characterization of the R964C variant demonstrated a threefold decrease in discrimination against stavudine (Bailey et al. 2009a). A yeast system that allowed stavudine uptake showed a very mild increase in frequency of mitochondrial dysfunction in the R964C yeast homolog (Baruffini and Lodi 2010). More advanced experimental models will be required to determine whether this decrease in stavudine discrimination is sufficient to cause mitochondrial toxicity.

Nucleotide concentrations may be crucial in proper mtDNA replication as evidenced by the disease-associated mutations in the small subunit of ribonucleotide reductase that catalyzes the limiting step of dNTP production (Bourdon et al. 2007). Several disease-associated mutations show reduced binding affinities for nucleotides and would be sensitive to low nucleotide concentrations. One such example is the H932Y variant, which is able to incorporate nucleotides at near normal rates when the nucleotide concentration is raised 100-fold over the concentration needed for the wild-type enzyme (Stumpf et al. 2010). The increased frequency of mitochondrial dysfunction by the homologous mutation in yeast is suppressed by overexpressing ribonucleotide reductase, which increases nucleotide concentration (Stumpf et al. 2010). The suppression of disease-associated mutations by ribonucleotide reductase overexpression occurs for other nucleotide-binding defective mutants (Lecrenier and Foury 1995; Baruffini et al. 2006), suggesting a mechanism that could be targeted for mitochondrial therapies.

Exercise Attenuates Premature Aging in *POLG* Mutant Mice

A recent provocative study showed that premature aging in the *POLG* exonuclease-deficient mice can be at least delayed for several months by endurance exercise (Safdar et al. 2011). Not only are the mice that undergo exercise endurance phenotypically indistinguishable from wild-type, they also have similar frequency of mutant mtDNA and COX activity as wild-type. Interestingly, they exhibit about 50% higher mtDNA copy number. However, the mechanism for attenuating premature aging is unclear. It has been suggested that exercise promotes mitochondrial biogenesis via PGC-1α facilitating a process that either dilutes mutant mtDNA or targets dysfunctional mtDNA for destruction (Safdar et al. 2011). According to this model, mitochondrial dynamics caused by fusion and fission is sufficient to offset the mutation load that is created by the lack of exonuclease activity. It is interesting to note that endurance exercise can only be undertaken by a fully developed mouse after many mitochondrial divisions and presumably the majority of mtDNA mutations have occurred. This suggests that the suppressive effects of endurance exercise may occur even if training starts later in life. Also, it would be interesting to know how temporary the effect of endurance training is on the accumulation of mtDNA mutations. This work should inspire future studies on the therapeutic effects of exercise for *POLG*-related diseases.

CONCLUSIONS

From the initial discovery of the first *POLG* disease-associated mutation a decade ago, research on *POLG* and mtDNA replication has spawned a wealth of information. Clinical reports have vastly improved and broadened the diagnoses and documented the mutations in many *POLG*-related diseases. The search for the polymerase defects in disease-associated mutants has resulted in the biochemical, genetic, and crystallographic dissection of the important activities of the polymerase. The development of the mitochondrial "mutator" mouse has inspired many labs to understand the role of mitochondrial mutations in pathology. Finally, the use of these tools allows more mechanistic approaches to therapeutic strategies. These developments in conjunction with new technology will hopefully provide new research tools that are needed to fully understand some of the mysteries of *POLG*-related diseases. For instance, improved sequencing technologies will allow broader studies of the entire spectrum of mutations in nuclear genes that affect mtDNA replication in disease patients. Also, reconstitution of mtDNA replication from tissue extracts and recombinant proteins will provide improved methods for determining mutations and environmental toxicants that cause mtDNA depletion. The better understanding of mtDNA dynamics will help determine how cells deal with mtDNA depletion and mutagenesis during the course of the disease. Finally, appropriate model systems will be developed to understand tissue specificity in mitochondrial diseases. This information will be crucial for improving both diagnosis and treatment of *POLG*-related diseases.

ACKNOWLEDGMENTS

We and the rest of the *POLG* field are indebted to the hard work of Margaret Humble for maintaining and updating the pol γ mutation database (http://tools.niehs.nih.gov/polg). We also thank Drs. Richard Gradman and Tammy Collins for comments and suggestions from their critical reading of this manuscript. This work is supported by the Intramural Research Program of the National Institutes of Health, the National Institute of Environmental Health Sciences (ES 065078), and the Mitochondrial Research Guild at Seattle Children's Hospital.

REFERENCES

Aitken H, Gorman G, McFarland R, Roberts M, Taylor RW, Turnbull DM. 2009. Clinical reasoning: Blurred vision and dancing feet: Restless legs syndrome presenting in mitochondrial disease. *Neurology* **72:** e86–90.

Ameur A, Stewart JB, Freyer C, Hagstrom E, Ingman M, Larsson NG, Gyllensten U. 2011. Ultra-deep sequencing of mouse mitochondrial DNA: mutational patterns and their origins. *PLoS Genet* **7:** e1002028.

Bailey CM, Kasiviswanathan R, Copeland WC, Anderson KS. 2009a. R964C mutation of DNA polymerase γ imparts increased stavudine toxicity by decreasing nucleoside analog discrimination and impairing polymerase activity. *Antimicrob Agents Chemother* **53:** 2610–2612.

Bailey LJ, Cluett TJ, Reyes A, Prolla TA, Poulton J, Leeuwenburgh C, Holt IJ. 2009b. Mice expressing an error-prone DNA polymerase in mitochondria display elevated replication pausing and chromosomal breakage at fragile sites of mitochondrial DNA. *Nucleic Acids Res* **37:** 2327–2335.

Baruffini E, Lodi T. 2010. Construction and validation of a yeast model system for studying in vivo the susceptibility to nucleoside analogues of DNA polymerase γ allelic variants. *Mitochondrion* **10:** 183–187.

Baruffini E, Lodi T, Dallabona C, Puglisi A, Zeviani M, Ferrero I. 2006. Genetic and chemical rescue of the Saccharomyces cerevisiae phenotype induced by mitochondrial DNA polymerase mutations associated with progressive external ophthalmoplegia in humans. *Hum Mol Genet* **15:** 2846–2855.

Baruffini E, Ferrero I, Foury F. 2007. Mitochondrial DNA defects in Saccharomyces cerevisiae caused by functional interactions between DNA polymerase γ mutations associated with disease in human. *Biochim Biophys Acta* **1772:** 1225–1235.

Baruffini E, Horvath R, Dallabona C, Czermin B, Lamantea E, Bindoff L, Invernizzi F, Ferrero I, Zeviani M, Lodi T. 2011. Predicting the contribution of novel *POLG* mutations to human disease through analysis in yeast model. *Mitochondrion* **11:** 182–190.

Bicknese AR, May W, Hickey WF, Dodson WE. 1992. Early childhood hepatocerebral degeneration misdiagnosed as valproate hepatotoxicity. *Ann Neurol* **32:** 767–775.

Blok MJ, Van den Bosch BJ, Jongen E, Hendrickx A, de Die-Smulders CE, Hoogendijk JE, Brusse E, de Visser M, Poll-The BT, Bierau J, et al. 2009. The unfolding clinical spectrum of *POLG* mutations. *J Med Genet* **46:** 776–785.

Bourdon A, Minai L, Serre V, Jais JP, Sarzi E, Aubert S, Chretien D, de Lonlay P, Paquis-Flucklinger V, Arakawa H, et al. 2007. Mutation of RRM2B, encoding p53–controlled ribonucleotide reductase (p53R2), causes severe mitochondrial DNA depletion. *Nat Genet* **39:** 776–780.

Chan SS, Copeland WC. 2009. DNA polymerase γ and mitochondrial disease: Understanding the consequence of *POLG* mutations. *Biochim Biophys Acta* **1787:** 312–319.

Chan SSL, Longley MJ, Copeland WC. 2005a. The common A467T mutation in the human mitochondrial DNA polymerase (*POLG*) compromises catalytic efficiency and interaction with the accessory subunit. *J Biol Chem* **280:** 31341–31346.

Chan SSL, Longley MJ, Naviaux RK, Copeland WC. 2005b. Mono-allelic *POLG* expression resulting from nonsense-mediated decay and alternative splicing in a patient with Alpers syndrome. *DNA Repair* **4:** 1381–1389.

Chan SSL, Longley MJ, Copeland WC. 2006. Modulation of the W748S mutation in DNA polymerase γ by the E1143G polymorphism in mitochondrial disorders. *Hum Mol Genet* **15:** 3473–3483.

Chinnery PF, Zeviani M. 2008. 155th ENMC workshop: Polymerase γ and disorders of mitochondrial DNA synthesis, 21–23 September 2007, Naarden, The Netherlands. *Neuromuscul Disord* **18:** 259–267.

Cohen BH, Naviaux RK. 2010. The clinical diagnosis of *POLG* disease and other mitochondrial DNA depletion disorders. *Methods* **51:** 364–373.

Cohen BH, Chinnery PF, Copeland WC. 2010. *POLG*-related disorders. In *Gene reviews at gene tests: Medical genetics information resource* (http://wwwgenetestsorg), University of Washington, Seattle, WA.

Copeland WC. 2008. Inherited mitochondrial diseases of DNA replication. *Annu Rev Med* **59:** 131–146.

Dai DF, Chen T, Wanagat J, Laflamme M, Marcinek DJ, Emond MJ, Ngo CP, Prolla TA, Rabinovitch PS. 2010. Age-dependent cardiomyopathy in mitochondrial mutator mice is attenuated by overexpression of catalase targeted to mitochondria. *Aging Cell* **9:** 536–544.

de Vries MC, Rodenburg RJ, Morava E, van Kaauwen EP, Ter Laak H, Mullaart RA, Snoeck IN, van Hasselt PM, Harding P, van den Heuvel LP, et al. 2007. Multiple oxidative phosphorylation deficiencies in severe childhood multi-system disorders due to polymerase γ (*POLG1*) mutations. *Eur J Pediatr* **166:** 229–234.

Di Fonzo A, Bordoni A, Crimi M, Sara G, Bo RD, Bresolin N, Comi GP. 2003. *POLG* mutations in sporadic mitochondrial disorders with multiple mtDNA deletions. *Hum Mutat* **22:** 498–499.

Dimmock D, Tang LY, Schmitt ES, Wong LJ. 2010. Quantitative evaluation of the mitochondrial DNA depletion syndrome. *Clin Chem* **56:** 1119–1127.

Edgar D, Shabalina I, Camara Y, Wredenberg A, Calvaruso MA, Nijtmans L, Nedergaard J, Cannon B, Larsson NG, Trifunovic A. 2009. Random point mutations with major effects on protein-coding genes are the driving force behind premature aging in mtDNA mutator mice. *Cell Metab* **10:** 131–138.

Engelsen BA, Tzoulis C, Karlsen B, Lillebo A, Laegreid LM, Aasly J, Zeviani M, Bindoff LA. 2008. *POLG1* mutations cause a syndromic epilepsy with occipital lobe predilection. *Brain* **131:** 818–828.

Fadic R, Russell JA, Vedanarayanan VV, Lehar M, Kuncl RW, Johns DR. 1997. Sensory ataxic neuropathy as the presenting feature of a novel mitochondrial disease. *Neurology* **49:** 239–245.

Ferrari G, Lamantea E, Donati A, Filosto M, Briem E, Carrara F, Parini R, Simonati A, Santer R, Zeviani M. 2005. Infantile hepatocerebral syndromes associated with mutations in the mitochondrial DNA polymerase-γ A. *Brain* **128:** 723–731.

Gonzalez-Vioque E, Blazquez A, Fernandez-Moreira D, Bornstein B, Bautista J, Arpa J, Navarro C, Campos Y, Fernandez-Moreno MA, Garesse R, et al. 2006. Association of novel *POLG* mutations and multiple mitochondrial DNA deletions with variable clinical phenotypes in a Spanish population. *Arch Neurol* **63:** 107–111.

Gray H, Wong TW. 1992. Purification and identification of subunit structure of the human mitochondrial DNA polymerase. *J Biol Chem* **267:** 5835–5841.

Graziewicz MA, Longley MJ, Bienstock RJ, Zeviani M, Copeland WC. 2004. Structure-function defects of human mitochondrial DNA polymerase in autosomal dominant progressive external ophthalmoplegia. *Nat Struct Mol Biol* **11:** 770–776.

Guo X, Kudryavtseva E, Bodyak N, Nicholas A, Dombrovsky I, Yang D, Kraytsberg Y, Simon DK, Khrapko K. 2010. Mitochondrial DNA deletions in mice in men: Substantia nigra is much less affected in the mouse. *Biochim Biophys Acta* **1797:** 1159–1162.

Haas RH, Parikh S, Falk MJ, Saneto RP, Wolf NI, Darin N, Cohen BH. 2007. Mitochondrial disease: A practical approach for primary care physicians. *Pediatrics* **120:** 1326–1333.

Haas RH, Parikh S, Falk MJ, Saneto RP, Wolf NI, Darin N, Wong LJ, Cohen BH, Naviaux RK. 2008. The in-depth evaluation of suspected mitochondrial disease. *Mol Genet Metab* **94:** 16–37.

Hakonen AH, Heiskanen S, Juvonen V, Lappalainen I, Luoma PT, Rantamaki M, Goethem GV, Lofgren A, Hackman P, Paetau A, et al. 2005. Mitochondrial DNA polymerase W748S mutation: A common cause of autosomal recessive ataxia with ancient European origin. *Am J Hum Genet* **77:** 430–441.

Hance N, Ekstrand MI, Trifunovic A. 2005. Mitochondrial DNA polymerase γ is essential for mammalian embryogenesis. *Hum Mol Genet* **14:** 1775–1783.

Harding BN. 1990. Progressive neuronal degeneration of childhood with liver disease (Alpers-Huttenlocher syndrome): A personal review. *J Child Neurol* **5:** 273–287.

Hirano M, Silvestri G, Blake DM, Lombes A, Minetti C, Bonilla E, Hays AP, Lovelace RE, Butler I, Bertorini TE, et al. 1994. Mitochondrial neurogastrointestinal encephalomyopathy (MNGIE): Clinical, biochemical, and genetic features of an autosomal recessive mitochondrial disorder. *Neurology* **44:** 721–727.

Horvath R, Hudson G, Ferrari G, Futterer N, Ahola S, Lamantea E, Prokisch H, Lochmuller H, McFarland R, Ramesh V, et al. 2006. Phenotypic spectrum associated with mutations of the mitochondrial polymerase γ gene. *Brain* **129:** 1674–1684.

Kakuda TN. 2000. Pharmacology of nucleoside and nucleotide reverse transcriptase inhibitor-induced mitochondrial toxicity. *Clin Ther* **22:** 685–708.

Kasiviswanathan R, Copeland WC. 2011. Biochemical analysis of the G517V *POLG* variant reveals wild-type like activity. *Mitochondrion* **11:** 929–934.

Kasiviswanathan R, Longley MJ, Chan SS, Copeland WC. 2009. Disease mutations in the human mitochondrial DNA polymerase thumb subdomain impart severe defects in MtDNA replication. *J Biol Chem* **284:** 19501–19510.

Kujoth GC, Hiona A, Pugh TD, Someya S, Panzer K, Wohlgemuth SE, Hofer T, Seo AY, Sullivan R, Jobling WA, et al. 2005. Mitochondrial DNA mutations, oxidative stress, and apoptosis in mammalian aging. *Science* **309:** 481–484.

Lamantea E, Tiranti V, Bordoni A, Toscano A, Bono F, Servidei S, Papadimitriou A, Spelbrink H, Silvestri L, Casari G, et al. 2002. Mutations of mitochondrial DNA polymerase γ are a frequent cause of autosomal dominant or recessive progressive external ophthalmoplegia. *Ann Neurol* **52:** 211–219.

Lecrenier N, Foury F. 1995. Overexpression of the RNR1 gene rescues *Saccharomyces cerevisiae* mutants in the mitochondrial DNA polymerase-encoding MIP1 gene. *Mol Gen Genet* **249:** 1–7.

Lee YS, Kennedy WD, Yin YW. 2009. Structural insight into processive human mitochondrial DNA synthesis and disease-related polymerase mutations. *Cell* **139:** 312–324.

Lee YS, Lee S, Demeler B, Molineux IJ, Johnson KA, Yin YW. 2010a. Each monomer of the dimeric accessory protein for human mitochondrial DNA polymerase has a distinct role in conferring processivity. *J Biol Chem* **285:** 1490–1499.

Lee YS, Molineux IJ, Yin YW. 2010b. A single mutation in human mitochondrial DNA polymerase pol γ A affects both polymerization and proofreading activities, but only as a holoenzyme. *J Biol Chem* **285:** 28105–28116.

Lewis W, Day BJ, Kohler JJ, Hosseini SH, Chan SSL, Green E, Haase CP, Keebaugh E, Long R, Ludaway T, et al. 2007. MtDNA depletion, oxidative stress, cardiomyopathy, and death from transgenic cardiac targeted human mutant polymerase γ. *Lab Invest* **87:** 326–335.

Lim SE, Longley MJ, Copeland WC. 1999. The mitochondrial p55 accessory subunit of human DNA polymerase γ enhances DNA binding, promotes processive DNA synthesis, and confers *N*-ethylmaleimide resistance. *J Biol Chem* **274:** 38197–38203.

Longley MJ, Prasad R, Srivastava DK, Wilson SH, Copeland WC. 1998a. Identification of 5′-deoxyribose phosphate lyase activity in human DNA polymerase γ and its role in mitochondrial base excision repair in vitro. *Proc Natl Acad Sci* **95:** 12244–12248.

Longley MJ, Ropp PA, Lim SE, Copeland WC. 1998b. Characterization of the native and recombinant catalytic subunit of human DNA polymerase γ: Identification of residues critical for exonuclease activity and dideoxynucleotide sensitivity. *Biochemistry* **37:** 10529–10539.

Longley MJ, Graziewicz MA, Bienstock RJ, Copeland WC. 2005. Consequences of mutations in human DNA polymerase γ. *Gene* **354:** 125–131.

Longley MJ, Humble MM, Sharief FS, Copeland WC. 2010. Disease variants of the human mitochondrial DNA helicase encoded by C10orf2 differentially alter protein stability, nucleotide hydrolysis and helicase activity. *J Biol Chem* **285:** 29690–29702.

Luoma P, Melberg A, Rinne JO, Kaukonen JA, Nupponen NN, Chalmers RM, Oldfors PA, Rautakorpi I, Peltonen PL, Majamaa PK, et al. 2004. Parkinsonism, premature menopause, and mitochondrial DNA polymerase γ mutations: Clinical and molecular genetic study. *Lancet* **364:** 875–882.

Luoma PT, Luo N, Loscher WN, Farr CL, Horvath R, Wanschitz J, Kiechl S, Kaguni LS, Suomalainen A. 2005. Functional defects due to spacer-region mutations of human mitochondrial DNA polymerase in a family with an ataxia-myopathy syndrome. *Hum Mol Genet* **14:** 1907–1920.

Milone M, Brunetti-Pierri N, Tang LY, Kumar N, Mezei MM, Josephs K, Powell S, Simpson E, Wong LJ. 2008. Sensory ataxic neuropathy with ophthalmoparesis caused by *POLG* mutations. *Neuromuscul Disord* **18:** 626–632.

Naviaux RK, Nguyen KV. 2004. POLG Mutations associated with Alpers' syndrome and mitochondrial DNA depletion. *Ann Neurol* **55:** 706–712.

Naviaux RK, Nyhan WL, Barshop BA, Poulton J, Markusic D, Karpinski NC, Haas RH. 1999. Mitochondrial DNA polymerase γ deficiency and mtDNA depletion in a child with Alpers' syndrome. *Ann Neurol* **45:** 54–58.

Nguyen KV, Sharief F, Chan SSL, Copeland WC, Naviaux RK. 2006. Molecular diagnosis of Alpers syndrome. *J Hepatology* **45:** 108–116.

Nick McElhinny SA, Pavlov YI, Kunkel TA. 2006. Evidence for extrinsic exonucleolytic proofreading. *Cell Cycle* **5:** 958–962.

Pagnamenta AT, Taanman JW, Wilson CJ, Anderson NE, Marotta R, Duncan AJ, Bitner-Glindzicz M, Taylor RW, Laskowski A, Thorburn DR, et al. 2006. Dominant inheritance of premature ovarian failure associated with mutant mitochondrial DNA polymerase γ. *Hum Reprod* **21:** 2467–2473.

Ponamarev MV, Longley MJ, Nguyen D, Kunkel TA, Copeland WC. 2002. Active site mutation in DNA polymerase γ associated with progressive external ophthalmoplegia causes error-prone DNA synthesis. *J Biol Chem* **277:** 15225–15228.

Safdar A, Bourgeois JM, Ogborn DI, Little JP, Hettinga BP, Akhtar M, Thompson JE, Melov S, Mocellin NJ, Kujoth GC, et al. 2011. Endurance exercise rescues progeroid aging and induces systemic mitochondrial rejuvenation in mtDNA mutator mice. *Proc Natl Acad Sci* **108:** 4135–4140.

Saneto RP, Naviaux RK. 2010. Polymerase γ disease through the ages. *Dev Disabil Res Rev* **16:** 163–174.

Saneto RP, Lee IC, Koenig MK, Bao X, Weng SW, Naviaux RK, Wong LJ. 2010. *POLG* DNA testing as an emerging standard of care before instituting valproic acid therapy for pediatric seizure disorders. *Seizure* **19:** 140–146.

Sarzi E, Bourdon A, Chretien D, Zarhrate M, Corcos J, Slama A, Cormier-Daire V, de Lonlay P, Munnich A, Rotig A. 2007. Mitochondrial DNA depletion is a prevalent cause of multiple respiratory chain deficiency in childhood. *J Pediatr* **150:** 531–534, e531–e536.

Shrader WD, Amagata A, Barnes A, Enns GM, Hinman A, Jankowski O, Kheifets V, Komatsuzaki R, Lee E, Mollard P, et al. 2011. α-Tocotrienol quinone modulates oxidative stress response and the biochemistry of aging. *Bioorg Med Chem Lett* **21:** 3693–3698.

Stuart GR, Santos JH, Strand MK, Van Houten B, Copeland WC. 2006. Mitochondrial DNA defects in S. cerevisiae with mutations in DNA polymerase γ associated with progressive external ophthalmolplegia. *Hum Mol Genet* **15:** 363–374.

Stumpf JD, Copeland WC. 2011. Mitochondrial DNA replication and disease: Insights from DNA polymerase γ mutations. *Cell Mol Life Sci* **68:** 219–233.

Stumpf JD, Bailey CM, Spell D, Stillwagon M, Anderson KS, Copeland WC. 2010. mip1 Containing mutations associated with mitochondrial disease causes mutagenesis and depletion of mtDNA in *Saccharomyces cerevisiae*. *Hum Mol Genet* **19:** 2123–2133.

Szczepanowska K, Foury F. 2010. A cluster of pathogenic mutations in the 3′-5′ exonuclease domain of DNA polymerase γ defines a novel module coupling DNA synthesis and degradation. *Hum Mol Genet* **19:** 3516–3529.

Tang S, Wang J, Lee NC, Milone M, Halberg MC, Schmitt ES, Craigen WJ, Zhang W, Wong LJ. 2011. Mitochondrial DNA polymerase γ mutations: An ever expanding molecular and clinical spectrum. *J Med Genet* **48:** 669–681.

Trifunovic A, Wredenberg A, Falkenberg M, Spelbrink JN, Rovio AT, Bruder CE, Bohlooly YM, Gidlof S, Oldfors A, Wibom R, et al. 2004. Premature ageing in mice expressing defective mitochondrial DNA polymerase. *Nature* **429:** 417–423.

Trifunovic A, Hansson A, Wredenberg A, Rovio AT, Dufour E, Khvorostov I, Spelbrink JN, Wibom R, Jacobs HT, Larsson NG. 2005. Somatic mtDNA mutations cause aging phenotypes without affecting reactive oxygen species production. *Proc Natl Acad Sci* **102:** 17993–17998.

Tulinius MH, Hagne I. 1991. EEG findings in children and adolescents with mitochondrial encephalomyopathies: A study of 25 cases. *Brain Dev* **13:** 167–173.

Tzoulis C, Engelsen BA, Telstad W, Aasly J, Zeviani M, Winterthun S, Ferrari G, Aarseth JH, Bindoff LA. 2006. The spectrum of clinical disease caused by the A467T and W748S *POLG* mutations: A study of 26 cases. *Brain* **129:** 1685–1692.

Uusimaa J, Hinttala R, Rantala H, Paivarinta M, Herva R, Roytta M, Soini H, Moilanen JS, Remes AM, Hassinen IE, et al. 2008. Homozygous W748S mutation in the *POLG1* gene in patients with juvenile-onset Alpers syndrome and status epilepticus. *Epilepsia* **49:** 1038–1045.

Van Goethem G, Dermaut B, Lofgren A, Martin JJ, Van Broeckhoven C. 2001. Mutation of *POLG* is associated with progressive external ophthalmoplegia characterized by mtDNA deletions. *Nature Genet* **28:** 211–212.

Van Goethem G, Martin JJ, Dermaut B, Lofgren A, Wibail A, Ververken D, Tack P, Dehaene I, Van Zandijcke M, Moonen M, et al. 2003a. Recessive *POLG* mutations presenting with sensory and ataxic neuropathy in compound heterozygote patients with progressive external ophthalmoplegia. *Neuromuscul Disord* **13:** 133–142.

Van Goethem G, Mercelis R, Lofgren A, Seneca S, Ceuterick C, Martin JJ, Van Broeckhoven C. 2003b. Patient homozygous for a recessive *POLG* mutation presents with features of MERRF. *Neurology* **61:** 1811–1813.

Van Goethem G, Luoma P, Rantamaki M, Al Memar A, Kaakkola S, Hackman P, Krahe R, Lofgren A, Martin JJ, De Jonghe P, et al. 2004. *POLG* mutations in neurodegenerative disorders with ataxia but no muscle involvement. *Neurology* **63:** 1251–1257.

Vermulst M, Bielas JH, Kujoth GC, Ladiges WC, Rabinovitch PS, Prolla TA, Loeb LA. 2007. Mitochondrial point mutations do not limit the natural lifespan of mice. *Nature Genet* **39:** 540–543.

Vermulst M, Wanagat J, Kujoth GC, Bielas JH, Rabinovitch PS, Prolla TA, Loeb LA. 2008. DNA deletions and clonal mutations drive premature aging in mitochondrial mutator mice. *Nature Genet* **40:** 392–394.

Vermulst M, Wanagat J, Loeb LA. 2009. On mitochondria, mutations, and methodology. *Cell Metab* **10:** 437.

Walter MC, Czermin B, Muller-Ziermann S, Bulst S, Stewart JD, Hudson G, Schneiderat P, Abicht A, Holinski-Feder E, Lochmuller H, et al. 2010. Late-onset ptosis and myopathy in a patient with a heterozygous insertion in *POLG2*. *J Neurol* **257:** 1517–1523.

Williams SL, Huang J, Edwards YJ, Ulloa RH, Dillon LM, Prolla TA, Vance JM, Moraes CT, Zuchner S. 2010. The mtDNA mutation spectrum of the progeroid *Polg* mutator mouse includes abundant control region multimers. *Cell Metab* **12:** 675–682.

Winterthun S, Ferrari G, He L, Taylor RW, Zeviani M, Turnbull DM, Engelsen BA, Moen G, Bindoff LA. 2005. Autosomal recessive mitochondrial ataxic syndrome due to mitochondrial polymerase γ mutations. *Neurology* **64:** 1204–1208.

Wolf NI, Rahman S, Schmitt B, Taanman JW, Duncan AJ, Harting I, Wohlrab G, Ebinger F, Rating D, Bast T. 2009. Status epilepticus in children with Alpers' disease caused by *POLG1* mutations: EEG and MRI features. *Epilepsia* **50:** 1596–1607.

Wong LJ, Naviaux RK, Brunetti-Pierri N, Zhang Q, Schmitt ES, Truong C, Milone M, Cohen BH, Wical B, Ganesh J, et al. 2008. Molecular and clinical genetics of mitochondrial diseases due to *POLG* mutations. *Hum Mutat* **29:** E150–E172.

Yamanaka H, Gatanaga H, Kosalaraksa P, Matsuoka-Aizawa S, Takahashi T, Kimura S, Oka S. 2007. Novel mutation of human DNA polymerase γ associated with mitochondrial toxicity induced by anti-HIV treatment. *J Infect Dis* **195:** 1419–1425.

Zhang D, Mott JL, Chang SW, Denniger G, Feng Z, Zassenhaus HP. 2000. Construction of transgenic mice with tissue-specific acceleration of mitochondrial DNA mutagenesis. *Genomics* **69:** 151–161.

Mitochondrial Biogenesis through Activation of Nuclear Signaling Proteins

John E. Dominy and Pere Puigserver

Department of Cancer Biology, Dana-Farber Cancer Institute and Department of Cell Biology, Harvard Medical School, Boston, Massachusetts 02215

Correspondence: pere_puigserver@dfci.harvard.edu

The dynamics of mitochondrial biogenesis and function is a complex interplay of cellular and molecular processes that ultimately shape bioenergetics capacity. Mitochondrial mass, by itself, represents the net balance between rates of biogenesis and degradation. Mitochondrial biogenesis is dependent on different signaling cascades and transcriptional complexes that promote the formation and assembly of mitochondria—a process that is heavily dependent on timely and coordinated transcriptional control of genes encoding for mitochondrial proteins. In this article, we discuss the major signals and transcriptional complexes, programming mitochondrial biogenesis, and bioenergetic activity. This regulatory network represents a new therapeutic window into the treatment of the wide spectrum of mitochondrial and neurodegenerative diseases characterized by dysregulation of mitochondrial dynamics and bioenergetic deficiencies.

Mitochondria are dense, double membrane-enclosed organelles that are present in all mammalian cells except mature red blood corpuscles. They uniquely possess their own genome in a circular DNA molecule. Although this genome encodes for only a small fraction of the total genes needed for mitochondrial organelle assembly and function, it is thus necessarily complemented by nuclear-encoded genes. Mitochondria provide a wide variety of biochemical services to the cell. Although it is conventionally accepted that energy production through the aerobic oxidation of carbon substrates and the generation of ATP is the principal function of mitochondria, this narrow view overlooks the many other impressive biosynthetic and regulatory capacities of this versatile organelle. Mitochondria, for instance, are essential for the synthesis of pyrimidines and purines, contribute to the synthesis of heme, regulate nitrogen balance through the urea cycle, produce ketone bodies, are necessary for sex hormone production, are involved in the processing of xenobiotics, play a critical role in redox balance, and are key regulators of the apoptotic program in mammals. Given the centrality of mitochondria to the maintenance of so many pathways in mammalian cells, it is not surprising that there exist multiple layers of control that enable a cell to coordinate its net mitochondrial activity with fuel sources, biosynthetic demands, proliferation rates, and external stimuli. One important layer in this signaling control is the expression of mitochondrial nuclear-derived genes

Cite this article as *Cold Spring Harb Perspect Biol* doi: 10.1101/cshperspect.a015008

that initiate the cellular program of mitochondrial biogenesis. Defects and failure in the integrity of this signaling system can cause serious perturbations to cellular metabolism and give rise to a broad range of tissue specific as well as systemic pathologies in humans. The same signaling system, on the other hand, also opens many therapeutic opportunities for the correction of mitochondrial defects. In this article, we discuss some of the major signaling factors that serve to regulate mitochondrial biogenesis and their implications for the mitigation of mitochondrial disorders.

REGULATION OF MITOCHONDRIAL BIOGENESIS AND FUNCTION THROUGH NUCLEAR-BASED GENE EXPRESSION

Although mitochondria contain in excess of 1000 proteins, its tiny circular genome encodes for only 13 proteins, 22 tRNAs, and two rRNAs; the balance of this genetic shortage is made up by the nuclear genome (Calvo and Mootha 2010). Because the vast majority of mitochondrial genes are situated in the nucleus, transcription complexes at promoters of these genes control their expression. This affords the cell the opportunity to coordinately regulate the expression of both mitochondrial and non-mitochondrial gene sets to affect a unified specific cellular response. Below we highlight some of the major nuclear transcriptional complexes regulating mitochondrial gene expression and the diverse signaling pathways to which they respond (Fig. 1).

Nuclear Respiratory Factors

Sequence analysis of the electron transport gene cytochrome *c* promoter initially identified nuclear respiratory factor 1 (NRF-1) as an important regulator of gene expression (Evans and Scarpulla 1989). NRF-1 controls the expression of a significant number of the proteins that make up the five respiratory complexes, as well as proteins integral to mitochondrial import and heme biosynthesis (Scarpulla 2008). NRF-1 is also able to integrate nuclear control of the transcriptional and replicative activity of the mitochondrial genome through the direct modulation of transcription factor A mitochondrial (TFAM) and transcription factor B proteins (TFBs) gene expression; TFAM and TFBs are major regulators of mitochondrial DNA transcription and replication (Gleyzer et al. 2005). NRF-1 may play an integral role in coupling mitochondrial replication with nuclear replication over the course of the cell cycle through regulation of E2F transcription factor targets and mitochondrial genes, indicating a link between cell division and mitochondrial biogenesis (Cam et al. 2004).

Studies involving the promoter region of cytochrome *c* oxidase complex subunit IV (COXIV) revealed the presence of another nuclear respiratory factor, NRF-2, which was able to regulate the expression of proteins in the electron transport chain (Carter et al. 1992). NRF-2 promotes the expression of genes that encode for mitochondrial complex IV cytochrome *c* oxidase (Gugneja et al. 1995). Like NRF1, NRF2 also controls the expression of TFAM and TFBs. The differential regulation of NRF1 and NRF2 is not completely understood but phosphorylation of these factors can alter their transcriptional activities (Scarpulla 2008). In addition, specific coactivators of the PGC-1 family sculpt NRFs-dependent control of mitochondrial gene expression in response to different signaling pathways (Wu et al. 1999).

Nuclear Hormone Receptors

Within the superfamily of nuclear hormone receptors, there are several subclasses that control mitochondrial function and biogenesis. The PPAR group of receptors is one such set. PPARα, which can be activated by long-chain fatty acids and eicosanoids, can potently promote the expression of genes involved in mitochondrial β-oxidation (Gulick et al. 1994). The activation of PPARγ activation, on the other hand, triggers mitochondrial biogenesis in white adipose tissue (Wilson-Fritch et al. 2004); PPARγ activation is also necessary to induce the expression of mitochondrial genes involved in the brown fat-mediated thermogenic program (Puigserver et al. 1998). PPARδ, in

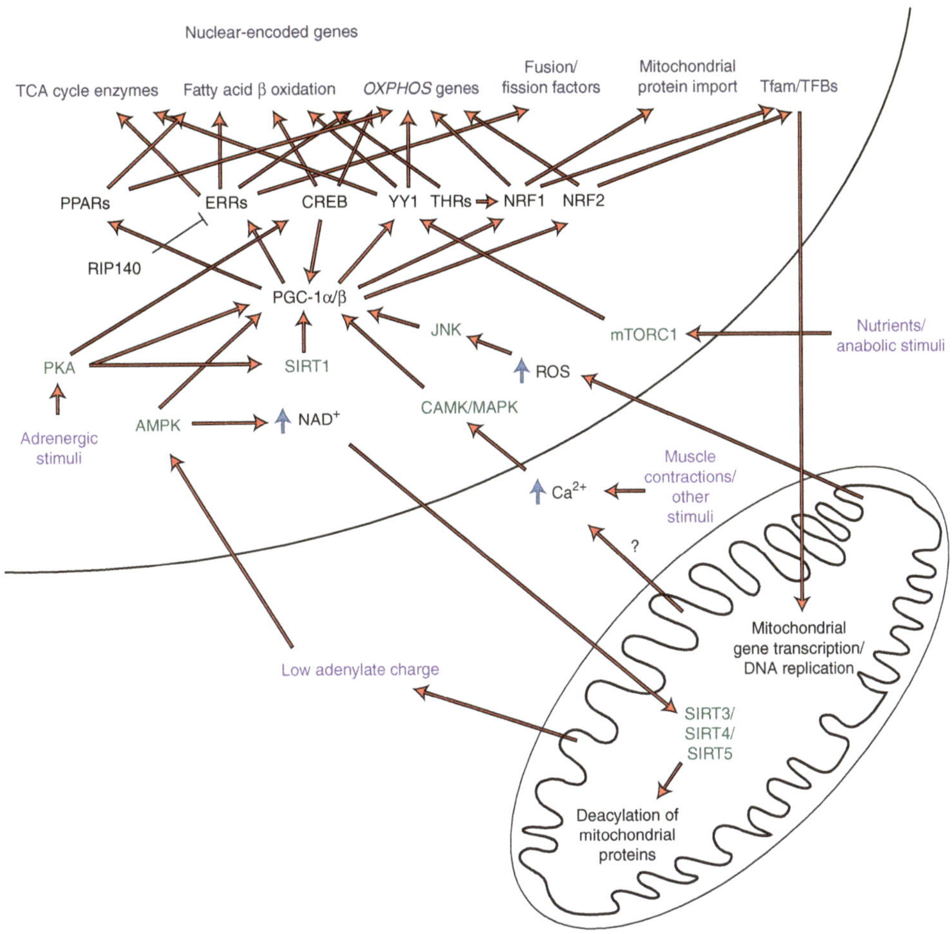

Figure 1. An overview of the transcriptional complexes regulating mitochdondrial gene expression and the signaling pathways converging upon them. Illustrated are the major transcriptional complexes/chromatin remodeling factors directly associated with changes in mitochondrial gene expression (see text for additional information).

skeletal muscle, is able to up-regulate mitochondrial genes associated with fatty acid oxidation and specify slow twitch fiber type (Narkar et al. 2008; Kleiner et al. 2009).

Thyroid hormone receptors (THRs) can also promote mitochondrial biogenesis and tissue-specific function. This includes mitochondrial-driven thermogenesis that occurs in brown fat during the adaptation to lower temperatures (Silva 1995). In some instances, THRs directly drive the transcription of nuclear-encoded genes, whereas, in others, the effects can occur indirectly through the thyroid hormone-mediated up-regulation of NRF-1 (Venditti et al. 2009). A truncated form of thyroid hormone receptor α has been reported to be localized to the mitochondrion and may serve to directly activate the transcription of genes in the mitochondrial genome (Casas et al. 1999).

Estrogen-related receptors ERR-α, ERR-β, and ERR-γ are yet another set of nuclear hormone receptors capable of promoting mitochondrial biogenesis. These receptors have no known endogenous ligands and are primary expressed in tissues with high oxidative metabolism capacities (Eichner and Giguere 2011). DNA-binding sites for ERR-α have been mapped in a large number of nuclear-encoded mitochondrial genes, including those involved in oxidative phosphorylation, fatty acid oxidation, TCA cycle, and factors regulating mitochondrial fusion/fission. Although ERRs are considered orphan nuclear receptors, their activity can certainly be modulated through recruitment of coactivators/corepressor complexes. Indeed, the transcriptional coactivators PGC-1α/β are efficacious activators of ERRs and promote the expression of mitochondrial genes (Mootha et al. 2004; Schreiber et al. 2004). The transcriptional activity of this transcription factor/activator module is controlled through deacetylation by sirtuin 1, which allows for an important regulatory mechanism that connects nutrient/hormonal signaling to the control of mitochondrial function (Wilson et al. 2010). From the standpoint of repression of ERR activity, the RIP140 nuclear hormone corepressor seems to be important (Wilson et al. 2010).

Other Transcription Factors: CREB and YY1

CREB, a cAMP activated transcription factor, can promote the expression of several mitochondrial genes, including the proteins which make up complex IV and enzymes involved in the β-oxidation pathway (Gopalakrishnan and Scarpulla 1994). Many mitochondrial genes contain YY1 binding sites within their promoter regions and this transcription factor has been found to work in conjunction with PGC-1α to regulate their expression (Cunningham et al. 2007). Loss of YY1 expression in skeletal muscle causes exercise intolerance and morphological signs of mitochondrial myopathy in mice (Blattler et al. 2012b).

Transcriptional Coactivators

Transcriptional coactivators do not bind to DNA but can nonetheless promote gene transcription by enhancing the activity of true DNA-binding transcription factors. Biologically, these proteins can set into motion genome-wide transcriptional changes by coactivating many different transcription factors. In mammalian cells, the PGC-1 family of coactivators (PGC-1α, PGC-1β, and PRC) potentiate the activity of several transcription factors involved in the basic functions of the mitochondrion as well as its rate of biogenesis (Puigserver and Spiegelman 2003; Kelly and Scarpulla 2004). Indeed, PGC-1α and PGC-1β are sufficient to increase total mitochondrial mass, reactive oxygen species scavenging enzymes, oxidative phosphorylation components, mitochondrial metabolic pathways, protein import complexes, proteins involved in fission and fusion, and the levels of mitochondrial sirtuins (Mootha et al. 2003; Uldry et al. 2006; Cunningham et al. 2007; Rasbach et al. 2010; Handschin and Spiegelman 2011). This process occurs in part through the ability of PGC-1 proteins to potentiate the function of the NRFs, ERRs, and YY1.

In skeletal muscle, the increased mitochondrial biogenesis caused by transgenic overexpression of PGC-1α can attenuate the development of mitochondrial disease in mouse models of respiratory complex IV/cytochrome *c* oxidase deficiencies (Lin et al. 2002; Wenz et al. 2008; Viscomi et al. 2011). It has also been shown that PGC-1α/β expression improves mitochondrial respiration capacity in cells from human patients with a complex III or IV deficiency (Srivastava et al. 2009). It should also be noted that age-related declines in mitochondrial function are also mitigated by enhanced activity of PGC-1α. Skeletal muscle specific PGC-1α transgenic mice are less susceptible to exercise-induced fatigue and are protected against age-related diseases, such as sarcopenia and metabolic diseases, and as a consequence, these mice show an increased life span (Wenz et al. 2009). Nevertheless, it has yet to be determined whether the beneficial health effects of PGC-1α in skeletal muscle are attributable to its promotion of mitochondrial function or its effects on nonmitochondrial gene expression.

Sirtuin Proteins

Sirtuins are NAD^+-dependent protein deacylases that are homologous to yeast Sir2p, which is one of the genes involved in transcriptional silencing in that organism (Klar et al. 1979). In mammals there are seven sirtuin paralogs. SIRT1, SIRT6, and SIRT7 are nuclear proteins,

SIRT3, SIRT4, and SIRT5 are imported into mitochondria, and SIRT2 is principally cytoplasmic (Finkel et al. 2009; Haigis and Sinclair 2010; Verdin et al. 2010; Guarente 2011). Through their deacylation activities, sirtuins modulate the biological activities of a wide array of target pathways. From the standpoint of mitochondrial biology, SIRT1, SIRT3, SIRT4, and SIRT5 have profound effects on the function of this organelle. SIRT1 deacetylates several key transcription factors that result in the up-regulation of numerous genes involved in mitochondrial respiration (Rodgers et al. 2005; Purushotham et al. 2009) and may (Price et al. 2012) or may not (Park et al. 2012) be necessary for the promitochondrial effects of polyphenolic stilbenoid compounds like resveratrol. In addition, sirtuins deacetylate numerous metabolic enzymes to govern their specific activity. Changes in SIRT3 activity have been shown to be an important determinant in the acetylation state of mitochondrial in response to nutrient availability (Hirschey et al. 2011; Hebert et al. 2013). The acetylation of many mitochondrial proteins, such as isocitrate dehydrogenase 2 (Yu et al. 2012), can alter their catalytic/biological function. As such, loss of SIRT3 activity results in profound aberrations in mitochondrial function and exacerbates the metabolic pathologies associated with chronic nutrient excess. Although no consistent deacylase activities have been shown for SIRT4, it has been shown that SIRT4 antagonizes mitochondrial capacity for fatty acid oxidation in hepatic and skeletal muscle cells (Nasrin et al. 2010). SIRT5, which has been shown to display desuccinylase, demalonylase, and deacetylase activities, targets carbamoyl phosphate synthetase 1, and can regulate mitochondrial urea cycle flux (Nakagawa et al. 2009; Du et al. 2011).

SIGNALING MODULES THAT CONTROL MITOCHONDRIAL BIOGENESIS AND FUNCTION

Cells are constantly engaged in both temporal and spatial control of mitochondrial biogenesis in response to nutrient availability, hormonal cues, and changes in temperature. Integration of these variables frequently involves the use of signaling pathways that ultimately converge upon the transcriptional regulators described above. In this section, we discuss the key signaling modules that control mitochondrial genes and lead to increased mitochondrial mass.

AMP-Activated Protein Kinase (AMPK)

AMPK is a phylogenetically ancient cellular sensor that is triggered by high cellular energy demands (Hardie 2007). As such, it is not surprising that AMPK is strongly connected to mitochondrial bioenergetics. When ATP synthesis is impaired or ATP is being consumed and generating AMP at a high rate, AMPK becomes active, which, in turn, phosphorylates several enzymes involved in the stimulation of catabolic pathways, such as glucose transport and fatty acid oxidation and the inhibition of anabolic pathways, such as glycogen synthesis and lipogenesis (Kahn et al. 2005). Chronic activation of AMPK is also able to induce mitochondrial biogenesis through the control of nuclear transcription (Fig. 2). Essential to this process are the AMPK-mediated activation of PGC-1α and SIRT1. AMPK can directly phosphorylate and activate PGC-1α (Jager et al. 2007); other pathways that activate AMPK can result in a subsequent increase in PGC-1α protein expression and mitochondrial biogenesis (Birkenfeld et al. 2011). AMPK through Nampt-dependent and -independent mechanisms is also able to increase NAD^+ levels, which promotes SIRT1 activity and the deacetylation/activation of PGC-1α (Canto et al. 2009). Thus, the AMPK/SIRT1/PGC1α signaling axis is thought to play a key role in the adaptive metabolic programming that occurs during caloric restriction and exercise (Canto and Auwerx 2009). Nevertheless, it has been alternatively suggested that caloric restriction-mediated activation of AMPK serves to maintain mitochondrial protein biosynthesis independently of transcription (Miller et al. 2012).

Calcium Signaling

Intracellular calcium dynamics are affected by a wide array of signaling pathways and calcium, as

Mitochondrial dynamic process	Nuclear transcription control
• Mitochondrial unfolded protein response	- Zinc finger ZC376.7 (*C. elegans*) - CHOP/C/EBPβ - PGC-1α/HSF1?
• Mitochondrial fusion/fission	- PGC-1α/PGC-1β - ERRα; YY1
• Mitophagy	- PARIS/PGC-1α
• Mitochondrial motility	- Unknown
• Mitochondrial protein import	- PGC-1α/PGC-1β - ERRα; NRF-1; YY1
• Mitochondrial transcription	- PGC-1α/PGC-1β - ERRα; NRF-1; YY1

Figure 2. Transcriptional control of processes involved in the regulation of mitochondrial dynamics. Nuclear transcription factors regulate the transcription of specific proteins that are basic components of specific mitochondrial processes associated with mitochondrial dynamics (see text for further information).

a second messenger molecule, and constitutes an important mechanism for initiating changes in mitochondrial gene expression. This phenomenon is most evident in skeletal muscle, a tissue that displays coordinated changes in calcium homeostasis as part of its basic contractile mechanism. In this tissue, calcium-sensitive protein networks, such as calcium/calmodulin-dependent kinase and p38 mitogen-activated kinase, are critical actors in the mobilization of the transcriptional machinery that promotes mitochondrial proliferation and fatty acid oxidation (Wu et al. 2002; Wright et al. 2007).

cAMP Pathway

Yet another well-conserved second messenger system capable of altering mitochondrial function is the cAMP-PKA pathway. The principal transcriptional effector of this pathway is the CREB transcription factor, whose phosphorylation by PKA promotes CREB activity and target gene transcription (Mayr and Montminy 2001). PKA activation can also enhance the expression of PGC-1α, through the intermediacy of CREB (Herzig et al. 2001), as well as directly activate the catalytic function of SIRT1 (Gerhart-Hines et al. 2011). Biologically, one of the most salient examples of the power of cAMP-mediated effects on mitochondrial function is found in the adaptive nonshivering thermogenic response to lower temperatures (Cannon and Nedergaard 2004). This heat-generating process is highly dependent on induction of uncoupled mitochondrial respiration in adipose depots—primarily brown fat tissue—through the up-regulation of uncoupling proteins such as UCP1 and the up-regulation of oxidative pathways (Cannon and Nedergaard 2004). Sympathetic stimulation of brown and "beige" fat tissues increases cAMP-PKA tone in these tissues and precipitates the initial uncoupling response. Whereas the specific determination of the brown fat cell lineage is transcriptionally specified through PRDM16 (Seale et al. 2007), the mitochondrial thermogenic response is defined by the PGC-1 coactivators, which account for approximately 30% of the cAMP response in brown fat cells (Uldry et al. 2006).

mTOR Pathway

The mechanistic target of rapamycin (mTOR) pathway is a logic gate for integrating cell growth and cell size with the availability of growth factors, energy levels, and certain nutrients, such as the branched-chain amino acids. These processes require a net increase in cellular anabolic/biosynthetic capacities, which will ultimately depend on enhanced mitochondrial biogenesis and bioenergetics. The effects of mTOR on mitochondrial biology occur through several mechanisms. One of the mechanisms seems to be independent of transcription through direct effects on mitochondrial organelles and stimulation of respiration (Schieke et al. 2006). In a second mechanism, mTOR modulates mitochondrial gene expression through the regulation of different transcription factors. In liver, mTOR affects genes of fatty acid oxidation and ketone body production through its ability to activate NCoR and inactivate PPARα (Sengupta et al. 2010). In skeletal muscle, mTOR binds to the transcription factor YY1 and recruits PGC1α, increasing the ability of YY1 to activate mitochondrial gene expression. mTOR inhibition by using rapamycin, for example, results in a dissociation of PGC1α and decreases mitochondrial gene expression (Cunningham et al. 2007; Blattler et al. 2012a).

Retrograde Signaling from Mitochondria to the Nucleus

The retrograde transmission of regulatory signals from the mitochondria to the nucleus, from a teleological standpoint, makes sense in that deficits or excesses in mitochondrial activity would need to be ultimately corrected by adjustments in nuclear gene expression. This phenomenon has been actively explored in yeast (Liu and Butow 2006). In mammalian cells, however, the nature of these signals and the mechanisms that they use are less well understood. Nevertheless, evidence of this regulatory phenomenon can be seen in the skeletal muscle of patients with mitochondrial diseases, such as those with genetic mutations of oxidative respiration proteins, which display increased mitochondrial proliferation—perhaps as an effort to compensate for impaired mitochondrial activity (DiMauro and Hirano 2009; Wallace and Fan 2009). The nature of this communication is unknown, but calcium and ROS signaling have been connected to this retrograde control.

Calcium Signaling

Much like the endoplasmic reticulum, mitochondria are quantitatively significant stores of calcium in the cell. As such, they can provide a potential signaling avenue for the regulation of cellular processes—indeed, overloading the mitochondrial calcium pool is a signal for the induction of apoptosis. At this point, although mechanisms of calcium uptake by the mitochondria have been characterized (Baughman et al. 2011; De Stefani et al. 2011), there is no clearly delineated pathway by which mitochondrial calcium can be released to regulate the calcium sensitive transcription/chromatin factor pathways discussed above. Nonetheless, PGC-1α increases mitochondrial calcium release, which establishes a positive feedback loop for the maintenance of a mutual connection between mitochondrial and nuclear gene expression (Bianchi et al. 2006).

Reactive Oxygen Species (ROS)

ROS are produced at respiratory chain complexes located in the inner mitochondrial membrane. In *Drosophila*, disruption of respiratory complex 1 signals to components of the G1 to S cell cycle machinery trough JNK and FoxO (Owusu-Ansah et al. 2008). In mammalian cells, a similar retrograde signaling has been uncovered through a systematic analysis of different mitochondrial dysfunction mutations; this pathway increases *OXPHOS* genes and involves ROS signaling through JNK and the RXRα/PGC1α pathway (Chae et al. 2013).

CONTROL OF MITOCHONDRIAL PROCESSES THAT CONTRIBUTE TO THE BIOGENESIS OF MITOCHONDRIA

Mitochondrial Unfolded Protein Response and Quality Control

The mitochondria contains an internal protease system that monitors accumulation of unfolded or misfolded proteins (encoded by the nuclear or mitochondrial genomes) in the different compartments, including the outer membrane, intermembrane space, inner membrane, and matrix. Thus, during mitochondrial biogenesis or stress, chaperones and quality-control proteases recognize hydrophobic amino acids that are in the interior structure of normal folded proteins (Baker and Haynes 2011). In addition to this organelle's internal quality control, there is a retrograde signaling to the nucleus via translocation of several transcription factors. In *Caenorhabditis elegans,* mitochondrial chaperones such as mtHSP70 proteins bind to unfolded mitochondrial proteins; however, if the amount of unfolded proteins exceeds the chaperone capacity, a protease called ClpXP degrades these proteins into peptides. Mitochondrial efflux of these peptides activates a transcription factor, ZC376.7, which translocates to the nucleus to activate expression of mitochondrial chaperone genes (Haynes et al. 2010). In mammalian cells, accumulation of unfolded proteins in the mitochondrial matrix increases CHOP/C/EBPβ transcriptional activities on promoters encoding for mitochondrial chaperone genes, including chaperonins 10 and 60, mtDNAJ, and ClpP (Zhao et al. 2002). In hepatocyte cells, PGC-1α interacts with HSF1; however, it is unknown whether this pathway controls the mitochondrial unfolded protein response (Charos et al. 2012).

Mitochondrial Fusion/Fission

Mitochondria are dynamic organelles that undergo constant division or fission and fusion (Twig et al. 2008; Westermann 2010; Corrado et al. 2012). The mitochondrial mass undergoes different shapes depending on the cellular state. Fusion of the outer mitochondrial membrane is executed by the activity of two dynamin-related GTPases, MFN1 and MFN2, that form homo- or heterodimers. In addition, OPA1—another dynamin-related GTPase—controls mitochondrial fusion and has also been involved in cytochrome release and apoptosis. The mitochondrial fission machinery is made up of DRP1, FIS1, MFF, and MIEF1. DRP1 is a GTPase protein that is localized in the cytoplasm, but after calcineurin-mediated dephosphorylation is recruited to the mitochondria. Once at the mitochondria DRP1 oligomerizes and forms ring structures that will constrict and produce fission. DRP1 has adaptor proteins that either activate fission (MFF) or suppress fission (MIEF) modulating the DRP1 GTPase activity. FIS1 is the DRP1 receptor located in the outer mitochondrial membrane and also interacts and sequesters MIEF1. In cells, fusion makeup of interconnected mitochondria is essential to maintain a highly metabolic and energetic state. In contrast, quiescent or resting cells have fragmented mitochondria with small spherical shapes with low metabolic activities (Chen and Chan 2010).

Thus, it seems plausible that the mechanisms that control mitochondrial biogenesis are tightly associated with the fusion machinery. Along these lines, the transcriptional coactivators PGC-1α and PGC1-β are positive activators of MFN2 (Cartoni et al. 2005; Soriano et al. 2006). Whether there is independent transcriptional control of the fission machinery, or if their regulation is simply bundled with the rest of the nuclear-encoded mitochondrial proteins is currently unknown.

Mitophagy

Mitochondrial mass is defined by the balances of mitochondrial formation and degradation. This latter process largely occurs through a particular type of cellular autophagy, termed mitophagy, that clears defective mitochondria. Mitophagy is a cellular degradation process that maintains "healthy" mitochondrial mass (Ashrafi and Schwarz 2013). Elongated and fused mitochondrial are resistant to mitophagy, whereas fragmented mitochondria generated by fission are more accessible to mitophagy. The elimination of damaged mitochondria is mediated by the activities of PINK1 (PTEN-induced putative protein kinase 1) and the E3 ubiquitin ligase, Parkin. PINK1 and Parkin accumulate at the mitochondrial surface of damaged organelles through the loss of mitochondrial membrane potential (Youle and van der Bliek 2012). In turn, this promotes separation from the functional mitochondrial network followed by targeting defective mitochondria for degradation. This latter process requires Parkin-dependent ubiquitination of mitochondrial proteins, but the critical Parkin substrates necessary for mitophagy are unknown. MFN1 and MFN2 are ubiquitinated by Parkin, although they are not required for mitophagy. Mitochondria are selectively removed upon treatment with chemical uncouplers. In this process, opening of the mitochondrial permeability transition pore (PTP) also triggers selective mitophagy. In addition, starvation signals such as glucagon can also induce mitophagy but require opening of the PTP. Similar to mitochondrial fusion/fission, little is known about the transcriptional control of mitophagy. Interestingly, mutations that occur in Parkinson's disease and inactivate the E3 ligase activity of Parkin, lead to the accumulation of Parkin substrates (Narendra et al. 2012). Among them is PARIS, a zinc finger corepressor, which can suppress the transcriptional coactivator PGC1α (Shin et al. 2011). Consistent with this, PGC-1α responsive genes are underexpressed in microdissected dopaminergic neurons of Parkinson's disease mouse models, suggesting that this pathway could be causative in this disease (Beal 2009; Zheng et al. 2010). Based on these data, it is possible to speculate that during mitophagy Parkin activation can signal to the nucleus through Paris degradation, resulting in an increase in PGC-1α and will subsequently promote mitochondrial biogenesis.

Mitochondrial Motility

Depending on the cell type, an increase in mitochondrial mass needs to be coordinated with a precise intracellular localization of mitochondrial organelles to meet local energetic cellular demands. Mitochondrial motility is influenced by the organelle's size, which is in turn controlled by fusion/fission processes. Failure in these processes prevents efficient intracellular movement of mitochondria, as is most evident in neurons, which critically depend on this pathway to localize mitochondria close to synaptic sites (Chen and Chan 2009).

Mitochondrial transport is based on microtubule dynamics. Kinesin-1 type motors are recruited to mitochondria through outer mitochondrial membrane proteins Miro1 and Miro2, involving an adaptor protein termed Milton. This is a calcium-dependent process: High concentrations of calcium halt mitochondria through the EF hands of Miro, anchoring mitochondria to sites of high ATP requirements. It is unclear whether calcium is the only regulator of this movement or there are additional sites of control including establishment of sensing gradients or transcriptional/translational regulation (Stowers et al. 2002; Wang et al. 2011).

Mitochondrial Protein Import

Biogenesis of functional and competent mitochondria requires the import and assembly of proteins synthesized in the cytoplasm. The mitochondria contain more than 1000 proteins that are mostly imported from the cytoplasm through the translocase of the outer mitochondrial membrane (TOM complex) (Endo et al. 2011; Gebert et al. 2011).

Once the mitochondrial protein has passed the TOM channel, it is sorted by interactions with different machineries located in the intermembrane space and inner membrane. Proteins that contain a presequence target signaling to the matrix are imported by the translocase of the inner membrane (TIM23) complex that is associated with the presequence translocase-associated motor (PAM). The TIM23 complex is also involved in laterally releasing proteins to the inner mitochondrial membrane. Several complexes in the intermembrane space participate in sorting mitochondrial proteins. For example, the MIA (mitochondrial intermembrane assembly) complex promotes oxidative folding of intermembrane proteins. The Tim9–Tim10 chaperone complex transfers hydrophobic proteins to the mitochondrial outer membrane through the SAM complex (e.g., β-barrel proteins) or the mitochondrial inner membrane through the TIM22 carrier complex (e.g., inner membrane carrier proteins). Additional import pathways include the oxidase assembly (OXA) machinery that transfers mitochondrial proteins from the matrix ribosomes to the inner membrane, or the mitochondrial import

1 (Mim1) that imports α-helical outer membrane proteins (Endo and Yamano 2009; Schmidt et al. 2010).

The dynamics of this complex mitochondrial protein import process will depend not only on the number of existing import complexes but also on the amounts of mitochondrial protein translated in the cytoplasm. Little is known about the protein stability or activities of these mitochondrial proteins, for example, control by ubiquitination or other posttranslational modifications (phosphorylation, acetylation, etc.) that control the import processes. However, an important regulatory control is the coordinated transcription of genes encoding for these proteins that will determine the rate of synthesis and cellular concentrations that will promote and contribute to mitochondrial mass. In mammalian cells, this transcriptional control involves the PGC-1 pathway and its interactions with transcription factors including NRFs and YY1 that are bound to promoters of mitochondrial import genes (Blattler et al. 2012b; Scarpulla et al. 2012). Whether there is selectivity or specificity among the different import pathways described here is currently unknown.

Mitochondrial Transcription

Balanced and efficient mitochondrial biogenesis requires the activation of mitochondrial transcription (Peralta et al. 2012). The mammalian mitochondrial DNA is a circular double-stranded genome (~16.5 kb) that encodes 37 genes. Transcription of the mitochondrial DNA (mtDNA) is tightly coordinated with nuclear genes (particularly involved in OXPHOS) to achieve competent bioeenergetic mitochondrial organelles. The mtDNA contains two strands: the heavy (H) and the light (L) strands. The H strand contains most of the genes. In addition, there are two noncoding regions designated as the D-loop that harbors the promoter of both strands and the origin of replication. Mitochondrial transcription is bidirectional and starts in the D-loop region through the assembly of a complex composed by TFAM, TFB2M, and the RNA polymerase POLRMT. MTERF proteins also bind to promoters and control mitochondrial transcription (Rebelo et al. 2011). How the transcription of mitochondrial-encoded genes is precisely controlled is not completely understood. Several of the proteins that control mitochondrial DNA transcription have posttranslational modifications, including phosphorylation and acetylation. For example, TFAM is specifically phosphorylated (Lu et al. 2013) and acetylated (Dinardo et al. 2003), suggesting that its activity could be modified through cycles of acetylation/deacetylation in response to energetic signals. Another level of control of this mitochondrial transcriptional machinery is by inducing amounts of its components. Along these lines elements essential to mitochondrial transcription, such as TFAM, TFB2M, MTERF, and POLRMT, are regulated through the transcriptional coactivators PGC-1α and PGC-1β through their interaction with NRFs, ERRs, and YY1 (Mootha et al. 2004; Blattler et al. 2012b; Scarpulla et al. 2012).

THERAPEUTIC POTENTIAL OF SIGNALING PATHWAYS FOR THE ATTENUATION OF MITOCHONDRIAL-RELATED PATHOLOGIES

Mitochondrial dysfunction, whether it is attributable to deleterious mutations in nuclear or mtDNA genes (DiMauro and Hirano 2009; Wallace and Fan 2009) or to secondary systemic changes caused by factors such as aging/senescence, significantly impairs cell function and can result in a spectrum of pathologies. In certain circumstances, such as aging, the mechanisms for why or how mitochondrial function and dynamics become abnormal are unknown (Petersen et al. 2003; Reznick et al. 2007). Regardless of the mechanism, however, data gleaned from signaling studies suggest that certain elements of these pathways can be leveraged to either prevent or attenuate a wide array of mitochondrial dysfunctions.

Genetically Based Mitochondrial Disorders

Despite the fact that significant advances have been made in understanding the causative molecular mechanisms underlying many genetically based human mitochondrial disorders, there is a frightening paucity of demonstrable treatments (Hirano et al. 2012). In theory, however, interventions that boost the net mitochondrial activity in these patients to restore energy production and antioxidant capacities to some minimum threshold should be sufficient to attenuate afflicted organ dysfunction and the subsequent pathologies. As such, there are several druggable signaling avenues to pursue.

The first is the PGC-1α signaling axis. As mentioned earlier, transgenic overexpression of PGC-1α has been shown to significantly blunt the pathological effects of cytochrome *c* oxidase (Viscomi et al. 2011), COXIV deficiency in mice (Lin et al. 2002; Wenz et al. 2008), and COXIII/IV deficiency in human patient-derived cell lines. Thus, small molecule therapies that potentiate PGC-1α function or the function of one of its cognate transcription factors could yield tangible benefits. To date, efforts to specifically exploit this pathway have been largely limited to the testing of PPAR panagonists such as the fibrate family of compounds, which are already in clinical use as antihyperlipidemic drugs. Studies with bezofibrate suggest that it is of moderate use in humans suffering from mutations in carnitine palmitoyltransferase II deficiency (Bonnefont et al. 2009). Data on bezofibrate's ability to correct the phenotypes associated with mutations in respiratory complex proteins, in contrast, have been confined largely to mouse models and have yielded very mixed results (Wenz et al. 2008; Viscomi et al. 2011; Yatsuga and Suomalainen 2012). The major limitation of fibrates is that they target only a single cognate transcription factor of PGC-1α, namely, PPARα, which is involved in up-regulation of the mitochondrial proteome. Small molecules that specifically up-regulate PGC-1α expression or promote its inherent biological activity, such as through deacetylation, are no doubt needed for a full effect.

The second signaling axis is that mediated by AMPK activation. High throughput screenings for compounds that improve the growth and ATP levels of fibroblasts from humans with a complex I deficiency revealed the AMPK activator 5-aminoimidazole-4-carboxamide ribonucleoside (AICAR) as one of the most efficacious molecules (Golubitzky et al. 2011). In several mouse models of cytochrome *c* oxidase deficiency, AICAR treatment was also shown to lead to a partial rescue of COXIV activity and some measure of improvement in motor performance (Viscomi et al. 2011).

The third most therapeutically attractive targets for mitochondrial diseases are the sirtuin family of proteins. Within this family, SIRT1 and SIRT3 offer the strongest connection with improvements in mitochondrial health. Small molecule activation of SIRT1 has been shown in vivo to improve mitochondrial respiratory capacity in mice on a high-fat diet (Minor et al. 2011). In healthy human volunteers, a next-generation SIRT1 activator (SRT2104) has been shown to decrease serum lipid levels and increase the recovery of adenosine diphosphate and phosphocreatine after exercise, which is suggestive of an increase in mitochondrial oxidative phosphorylation (Libri et al. 2012). Although there are no human in vivo studies of the effects of SIRT1 activation on mitochondrial diseases, in vitro studies with the putative SIRT1 activator resveratrol have shown improvements in mitochondrial fatty acid oxidation in fibroblasts derived from patients with carnitine palmitoyltransferase II or very long-chain acyl-CoA dehydrogenase deficiencies (Bastin et al. 2011). In the case of SIRT3, up-regulation of activity has been shown to improve mitochondrial homeostasis in mice (Brown et al. 2013), although no data are available in humans. Ideally, therapies that simultaneously increase SIRT1 and SIRT3 activities, such as supplementation with the precursor nicotinamide riboside to increase the levels of their NAD^+ cosubstrate (Canto et al. 2012), would be the most desirable for promoting mitochondrial function.

Prevention of Mitochondrial Dysregulation Associated with Aging and Metabolic Disorders

Many of the therapeutic strategies described for the amelioration of genetically based mitochondrial disorders are applicable to the decline in mitochondrial function that accompanies age-related senescence or certain metabolic disease states such as Type-2 diabetes. Other therapeutic options are available to the nongenetic mitochondrial disorders, because of their generally reduced phenotypic severity. These options include dietary manipulation and physical exercise. Restriction of caloric intake can result in profound alterations in mitochondrial physiology, including increased proliferation, enhanced bioenergetic efficiency, reduced membrane potential, and reduced production of oxidative species in mice (Lopez-Lluch et al. 2006) and in humans (Civitarese et al. 2007). These mitochondrial changes are thought to be a major contributor to the ability of caloric restriction to prevent the development of diseases linked to aging. Indeed, calorie restriction extends life span in different organisms but the molecular mechanisms are not completely understood (Sohal and Weindruch 1996; Spindler 2001). Proposed mechanisms for the longevity effects include changes in the signaling pathways associated with nutrient availability and insulin levels that converge to regulate mTOR (Kaeberlein et al. 2005; Zoncu et al. 2011) and FoxOs (Salih and Brunet 2008) protein activities. In terms of mitochondrial function, however, other regulatory pathways appear to be connected with the effects of caloric restriction, namely, the AMPK, sirtuins, and PGC1α regulatory network.

Exercise is an extremely powerful inducer of mitochondrial activity and proliferation—particularly in skeletal muscle (Yan et al. 2011). Coincident with this change, exercise is also able to strongly protect against many types of age-related metabolic disorders. Less clear, however, is whether exercise can mimic the longevity effects of caloric restriction. An important point, however, is whether the beneficial health effects of exercise are to an increase in the mass of mitochondrial and/or the myriad other cellular and systemic changes that occur with exercise.

CONCLUDING REMARKS

Researchers across multiple disciplines have made considerable progress by uncovering many important signaling pathways that control mitochondrial function at the level of nuclear transcription and discerning how these pathways are targeted through extrinsic signals to ultimately govern mitochondrial activities. Collectively, these signaling pathways can effectively boost the cell's mitochondrial mass, improve the mitochondrial bioenergetics profile, and protect it against oxidative stress. As such, they represent a very fertile ground for the development of therapies for the rectification of diseases characterized by mitochondrial dysfunction.

ACKNOWLEDGMENTS

Mitochondrial-related research in the Puigserver laboratory is supported from the Dana-Farber Cancer Institute funds and grants from the NIH/NIDDK, American Diabetes Association, American Heart Association, and Muscle Dystrophy Association.

REFERENCES

*Reference is also in this collection.

Ashrafi G, Schwarz TL. 2013. The pathways of mitophagy for quality control and clearance of mitochondria. *Cell Death Differ* **20:** 31–42.

Baker BM, Haynes CM. 2011. Mitochondrial protein quality control during biogenesis and aging. *Trends Biochem Sci* **36:** 254–261.

Bastin J, Lopes-Costa A, Djouadi F. 2011. Exposure to resveratrol triggers pharmacological correction of fatty acid utilization in human fatty acid oxidation-deficient fibroblasts. *Hum Mol Genet* **20:** 2048–2057.

Baughman JM, Perocchi F, Girgis HS, Plovanich M, Belcher-Timme CA, Sancak Y, Bao XR, Strittmatter L, Goldberger O, Bogorad RL, et al. 2011. Integrative genomics identifies MCU as an essential component of the mitochondrial calcium uniporter. *Nature* **476:** 341–345.

Beal MF. 2009. Therapeutic approaches to mitochondrial dysfunction in Parkinson's disease. *Parkinsonism Relat Disord* **15** (Suppl 3)**:** S189–S194.

Bianchi K, Vandecasteele G, Carli C, Romagnoli A, Szabadkai G, Rizzuto R. 2006. Regulation of Ca^{2+} signalling and Ca^{2+}-mediated cell death by the transcriptional coactivator PGC-1α. *Cell Death Differ* **13:** 586–596.

Birkenfeld AL, Lee HY, Guebre-Egziabher F, Alves TC, Jurczak MJ, Jornayvaz FR, Zhang D, Hsiao JJ, Martin-Montalvo A, Fischer-Rosinsky A, et al. 2011. Deletion of the mammalian INDY homolog mimics aspects of dietary restriction and protects against adiposity and insulin resistance in mice. *Cell Metab* **14:** 184–195.

Blattler SM, Cunningham JT, Verdeguer F, Chim H, Haas W, Liu H, Romanino K, Ruegg MA, Gygi S, Shi Y, et al. 2012a. Yin Yang 1 deficiency in skeletal muscle protects against rapamycin-induced diabetic-like symptoms through activation of insulin/IGF signaling. *Cell Metab* **15:** 505–517.

Blattler SM, Verdeguer F, Liesa M, Cunningham JT, Vogel RO, Chim H, Liu H, Romanino K, Shirihai OS, Vazquez F, et al. 2012b. Defective mitochondrial morphology and bioenergetic function in mice lacking the transcription factor Yin Yang 1 in skeletal muscle. *Mol Cell Biol* **32:** 3333–3346.

Bonnefont JP, Bastin J, Behin A, Djouadi F. 2009. Bezafibrate for an inborn mitochondrial β-oxidation defect. *N Engl J Med* **360:** 838–840.

Brown K, Xie S, Qiu X, Mohrin M, Shin J, Liu Y, Zhang D, Scadden DT, Chen D. 2013. SIRT3 reverses aging-associated degeneration. *Cell Rep* **3:** 319–327.

Calvo SE, Mootha VK. 2010. The mitochondrial proteome and human disease. *Annu Rev Genomics Hum Genet* **11:** 25–44.

Cam H, Balciunaite E, Blais A, Spektor A, Scarpulla RC, Young R, Kluger Y, Dynlacht BD. 2004. A common set of gene regulatory networks links metabolism and growth inhibition. *Mol Cell* **16:** 399–411.

Cannon B, Nedergaard J. 2004. Brown adipose tissue: Function and physiological significance. *Physiol Rev* **84:** 277–359.

Canto C, Auwerx J. 2009. PGC-1α, SIRT1 and AMPK, an energy sensing network that controls energy expenditure. *Curr Opin Lipidol* **20:** 98–105.

Canto C, Gerhart-Hines Z, Feige JN, Lagouge M, Noriega L, Milne JC, Elliott PJ, Puigserver P, Auwerx J. 2009. AMPK regulates energy expenditure by modulating NAD^+ metabolism and SIRT1 activity. *Nature* **458:** 1056–1060.

Canto C, Houtkooper RH, Pirinen E, Youn DY, Oosterveer MH, Cen Y, Fernandez-Marcos PH, Yamamoto H, Andreux PA, Cettour-Rose P, et al. 2012. The NAD^+ precursor nicotinamide riboside enhances oxidative metabolism and protects against high-fat diet-induced obesity. *Cell Metab* **15:** 838–847.

Carter RS, Bhat NK, Basu A, Avadhani NG. 1992. The basal promoter elements of murine cytochrome *c* oxidase subunit IV gene consist of tandemly duplicated ets motifs that bind to GABP-related transcription factors. *J Biol Chem* **267:** 23418–23426.

Cartoni R, Leger B, Hock MB, Praz M, Crettenand A, Pich S, Ziltener JL, Luthi F, Deriaz O, Zorzano A, et al. 2005. Mitofusins 1/2 and ERRα expression are increased in human skeletal muscle after physical exercise. *J Physiol* **567:** 349–358.

Casas F, Rochard P, Rodier A, Cassar-Malek I, Marchal-Victorion S, Wiesner RJ, Cabello G, Wrutniak C. 1999. A variant form of the nuclear triiodothyronine receptor c-ErbAα1 plays a direct role in regulation of mitochondrial RNA synthesis. *Mol Cell Biol* **19:** 7913–7924.

Chae S, Ahn BY, Byun K, Cho YM, Yu MH, Lee B, Hwang D, Park KS. 2013. A systems approach for decoding mitochondrial retrograde signaling pathways. *Science Signal* **6:** rs4.

Charos AE, Reed BD, Raha D, Szekely AM, Weissman SM, Snyder M. 2012. A highly integrated and complex PPARGC1A transcription factor binding network in HepG2 cells. *Genome Res* **22:** 1668–1679.

Chen H, Chan DC. 2009. Mitochondrial dynamics—fusion, fission, movement, and mitophagy—in neurodegenerative diseases. *Hum Mol Genet* **18:** R169–R176.

Chen H, Chan DC. 2010. Physiological functions of mitochondrial fusion. *Ann NY Acad Sci* **1201:** 21–25.

Civitarese AE, Carling S, Heilbronn LK, Hulver MH, Ukropcova B, Deutsch WA, Smith SR, Ravussin E. 2007. Calorie restriction increases muscle mitochondrial biogenesis in healthy humans. *PLoS Med* **4:** e76.

Corrado M, Scorrano L, Campello S. 2012. Mitochondrial dynamics in cancer and neurodegenerative and neuroinflammatory diseases. *Int J Cell Biol* **2012:** 729290.

Cunningham JT, Rodgers JT, Arlow DH, Vazquez F, Mootha VK, Puigserver P. 2007. mTOR controls mitochondrial oxidative function through a YY1-PGC-1α transcriptional complex. *Nature* **450:** 736–740.

De Stefani D, Raffaello A, Teardo E, Szabo I, Rizzuto R. 2011. A forty-kilodalton protein of the inner membrane is the mitochondrial calcium uniporter. *Nature* **476:** 336–340.

DiMauro S, Hirano M. 2009. Pathogenesis and treatment of mitochondrial disorders. *Adv Exp Med Biol* **652:** 139–170.

Dinardo MM, Musicco C, Fracasso F, Milella F, Gadaleta MN, Gadaleta G, Cantatore P. 2003. Acetylation and level of mitochondrial transcription factor A in several organs of young and old rats. *Biochem Biophys Res Commun* **301:** 187–191.

Du J, Zhou Y, Su X, Yu JJ, Khan S, Jiang H, Kim J, Woo J, Kim JH, Choi BH, et al. 2011. Sirt5 is a NAD-dependent protein lysine demalonylase and desuccinylase. *Science* **334:** 806–809.

Eichner LJ, Giguere V. 2011. Estrogen related receptors (ERRs): A new dawn in transcriptional control of mitochondrial gene networks. *Mitochondrion* **11:** 544–552.

Endo T, Yamano K. 2009. Multiple pathways for mitochondrial protein traffic. *Biol Chem* **390:** 723–730.

Endo T, Yamano K, Kawano S. 2011. Structural insight into the mitochondrial protein import system. *Biochim Biophys Acta* **1808:** 955–970.

Evans MJ, Scarpulla RC. 1989. Interaction of nuclear factors with multiple sites in the somatic cytochrome *c* promoter. Characterization of upstream NRF-1, ATF, and intron Sp1 recognition sequences. *J Biol Chem* **264:** 14361–14368.

Finkel T, Deng CX, Mostoslavsky R. 2009. Recent progress in the biology and physiology of sirtuins. *Nature* **460:** 587–591.

Gebert N, Ryan MT, Pfanner N, Wiedemann N, Stojanovski D. 2011. Mitochondrial protein import machineries and lipids: A functional connection. *Biochim Biophys Acta* **1808:** 1002–1011.

Gerhart-Hines Z, Dominy JE Jr, Blattler SM, Jedrychowski MP, Banks AS, Lim JH, Chim H, Gygi SP, Puigserver P. 2011. The cAMP/PKA pathway rapidly activates SIRT1 to promote fatty acid oxidation independently of changes in NAD^+. *Mol Cell* **44:** 851–863.

Gleyzer N, Vercauteren K, Scarpulla RC. 2005. Control of mitochondrial transcription specificity factors (TFB1M and TFB2M) by nuclear respiratory factors (NRF-1 and

NRF-2) and PGC-1 family coactivators. *Mol Cell Biol* **25:** 1354–1366.

Golubitzky A, Dan P, Weissman S, Link G, Wikstrom JD, Saada A. 2011. Screening for active small molecules in mitochondrial complex I deficient patient's fibroblasts, reveals AICAR as the most beneficial compound. *PloS ONE* **6:** e26883.

Gopalakrishnan L, Scarpulla RC. 1994. Differential regulation of respiratory chain subunits by a CREB-dependent signal transduction pathway. Role of cyclic AMP in cytochrome *c* and COXIV gene expression. *J Biol Chem* **269:** 105–113.

Guarente L. 2011. Franklin H. Epstein Lecture: Sirtuins, aging, and medicine. *N Engl J Med* **364:** 2235–2244.

Gugneja S, Virbasius JV, Scarpulla RC. 1995. Four structurally distinct, non-DNA-binding subunits of human nuclear respiratory factor 2 share a conserved transcriptional activation domain. *Mol Cell Biol* **15:** 102–111.

Gulick T, Cresci S, Caira T, Moore DD, Kelly DP. 1994. The peroxisome proliferator-activated receptor regulates mitochondrial fatty acid oxidative enzyme gene expression. *Proc Natl Acad Sci* **91:** 11012–11016.

Haigis MC, Sinclair DA. 2010. Mammalian sirtuins: Biological insights and disease relevance. *Annu Rev Pathol* **5:** 253–295.

Handschin C, Spiegelman BM. 2011. PGC-1 coactivators and the regulation of skeletal muscle fiber-type determination. *Cell Metab* **13:** 351; author reply 352.

Hardie DG. 2007. AMP-activated/SNF1 protein kinases: Conserved guardians of cellular energy. *Nat Rev Mol Cell Biol* **8:** 774–785.

Haynes CM, Yang Y, Blais SP, Neubert TA, Ron D. 2010. The matrix peptide exporter HAF-1 signals a mitochondrial UPR by activating the transcription factor ZC376.7 in *C. elegans*. *Mol Cell* **37:** 529–540.

Hebert AS, Dittenhafer-Reed KE, Yu W, Bailey DJ, Selen ES, Boersma MD, Carson JJ, Tonelli M, Balloon AJ, Higbee AJ, et al. 2013. Calorie restriction and SIRT3 trigger global reprogramming of the mitochondrial protein acetylome. *Mol Cell* **49:** 186–199.

Herzig S, Long F, Jhala US, Hedrick S, Quinn F, Bauer A, Rudolph D, Schutz G, Yoon C, Puigserver P, et al. 2001. CREB regulates hepatic gluconeogenesis through the coactivator PGC-1. *Nature* **413:** 179–183.

Hirano M, Garone C, Quinzii CM. 2012. CoQ_{10} deficiencies and MNGIE: Two treatable mitochondrial disorders. *Biochim Biophys Acta* **1820:** 625–631.

Hirschey MD, Shimazu T, Jing E, Grueter CA, Collins AM, Aouizerat B, Stancakova A, Goetzman E, Lam MM, Schwer B, et al. 2011. SIRT3 deficiency and mitochondrial protein hyperacetylation accelerate the development of the metabolic syndrome. *Mol Cell* **44:** 177–190.

Jager S, Handschin C, St-Pierre J, Spiegelman BM. 2007. AMP-activated protein kinase (AMPK) action in skeletal muscle via direct phosphorylation of PGC-1α. *Proc Natl Acad Sci* **104:** 12017–12022.

Kaeberlein M, Powers RW III, Steffen KK, Westman EA, Hu D, Dang N, Kerr EO, Kirkland KT, Fields S, Kennedy BK. 2005. Regulation of yeast replicative life span by TOR and Sch9 in response to nutrients. *Science* **310:** 1193–1196.

Kahn BB, Alquier T, Carling D, Hardie DG. 2005. AMP-activated protein kinase: Ancient energy gauge provides clues to modern understanding of metabolism. *Cell Metab* **1:** 15–25.

Kelly DP, Scarpulla RC. 2004. Transcriptional regulatory circuits controlling mitochondrial biogenesis and function. *Genes Dev* **18:** 357–368.

Klar AJ, Fogel S, Macleod K. 1979. *MAR1*—A regulator of the *HMa* and *HMα* loci in *Saccharomyces cerevisiae*. *Genetics* **93:** 37–50.

Kleiner S, Nguyen-Tran V, Bare O, Huang X, Spiegelman B, Wu Z. 2009. PPARδ agonism activates fatty acid oxidation via PGC-1α but does not increase mitochondrial gene expression and function. *J Biol Chem* **284:** 18624–18633.

Libri V, Brown AP, Gambarota G, Haddad J, Shields GS, Dawes H, Pinato DJ, Hoffman E, Elliot PJ, Vlasuk GP, et al. 2012. A pilot randomized, placebo controlled, double blind phase I trial of the novel SIRT1 activator SRT2104 in elderly volunteers. *PloS ONE* **7:** e51395.

Lin J, Wu H, Tarr PT, Zhang CY, Wu Z, Boss O, Michael LF, Puigserver P, Isotani E, Olson EN, et al. 2002. Transcriptional co-activator PGC-1 α drives the formation of slow-twitch muscle fibres. *Nature* **418:** 797–801.

Liu Z, Butow RA. 2006. Mitochondrial retrograde signaling. *Annu Rev Genet* **40:** 159–185.

Lopez-Lluch G, Hunt N, Jones B, Zhu M, Jamieson H, Hilmer S, Cascajo MV, Allard J, Ingram DK, Navas P, et al. 2006. Calorie restriction induces mitochondrial biogenesis and bioenergetic efficiency. *Proc Natl Acad Sci* **103:** 1768–1773.

Lu B, Lee J, Nie X, Li M, Morozov YI, Venkatesh S, Bogenhagen DF, Temiakov D, Suzuki CK. 2013. Phosphorylation of human TFAM in mitochondria impairs DNA binding and promotes degradation by the AAA^+ Lon protease. *Mol Cell* **49:** 121–132.

Mayr B, Montminy M. 2001. Transcriptional regulation by the phosphorylation-dependent factor CREB. *Nat Rev Mol Cell Biol* **2:** 599–609.

Miller BF, Robinson MM, Bruss MD, Hellerstein M, Hamilton KL. 2012. A comprehensive assessment of mitochondrial protein synthesis and cellular proliferation with age and caloric restriction. *Aging Cell* **11:** 150–161.

Minor RK, Baur JA, Gomes AP, Ward TM, Csiszar A, Mercken EM, Abdelmohsen K, Shin YK, Canto C, Scheibye-Knudsen M, et al. 2011. SRT1720 improves survival and healthspan of obese mice. *Scientific Rep* **1:** 70.

Mootha VK, Lindgren CM, Eriksson KF, Subramanian A, Sihag S, Lehar J, Puigserver P, Carlsson E, Ridderstrale M, Laurila E, et al. 2003. PGC-1α-responsive genes involved in oxidative phosphorylation are coordinately downregulated in human diabetes. *Nat Genet* **34:** 267–273.

Mootha VK, Handschin C, Arlow D, Xie X, St Pierre J, Sihag S, Yang W, Altshuler D, Puigserver P, Patterson N, et al. 2004. Errα and Gabpa/b specify PGC-1α-dependent oxidative phosphorylation gene expression that is altered in diabetic muscle. *Proc Natl Acad Sci* **101:** 6570–6575.

Nakagawa T, Lomb DJ, Haigis MC, Guarente L. 2009. SIRT5 Deacetylates carbamoyl phosphate synthetase 1 and regulates the urea cycle. *Cell* **137:** 560–570.

* Narendra D, Walker JE, Youle R. 2012. Mitochondrial quality control mediated by PINK1 and Parkin: Links to parkinsonism. *Cold Spring Harb Perspect Biol* **4:** a011338.

Narkar VA, Downes M, Yu RT, Embler E, Wang YX, Banayo E, Mihaylova MM, Nelson MC, Zou Y, Juguilon H, et al. 2008. AMPK and PPARδ agonists are exercise mimetics. *Cell* **134:** 405–415.

Nasrin N, Wu X, Fortier E, Feng Y, Bare OC, Chen S, Ren X, Wu Z, Streeper RS, Bordone L. 2010. SIRT4 regulates fatty acid oxidation and mitochondrial gene expression in liver and muscle cells. *J Biol Chem* **285:** 31995–32002.

Owusu-Ansah E, Yavari A, Mandal S, Banerjee U. 2008. Distinct mitochondrial retrograde signals control the G1-S cell cycle checkpoint. *Nat Genet* **40:** 356–361.

Park SJ, Ahmad F, Philp A, Baar K, Williams T, Luo H, Ke H, Rehmann H, Taussig R, Brown AL, et al. 2012. Resveratrol ameliorates aging-related metabolic phenotypes by inhibiting cAMP phosphodiesterases. *Cell* **148:** 421–433.

Peralta S, Wang X, Morales CT. 2012. Mitochondrial transcription: Lessons from mouse models. *Biochim Biophys Acta* **1819:** 961–969.

Petersen KF, Befroy D, Dufour S, Dziura J, Ariyan C, Rothman DL, DiPietro L, Cline GW, Shulman GI. 2003. Mitochondrial dysfunction in the elderly: Possible role in insulin resistance. *Science* **300:** 1140–1142.

Price NL, Gomes AP, Ling AJ, Duarte FV, Martin-Montalvo A, North BJ, Agarwal B, Ye L, Ramadori G, Teodoro JS, et al. 2012. SIRT1 is required for AMPK activation and the beneficial effects of resveratrol on mitochondrial function. *Cell Metab* **15:** 675–690.

Puigserver P, Spiegelman BM. 2003. Peroxisome proliferator-activated receptor-γ coactivator 1α (PGC-1α): Transcriptional coactivator and metabolic regulator. *Endocr Rev* **24:** 78–90.

Puigserver P, Wu Z, Park CW, Graves R, Wright M, Spiegelman BM. 1998. A cold-inducible coactivator of nuclear receptors linked to adaptive thermogenesis. *Cell* **92:** 829–839.

Purushotham A, Schug TT, Xu Q, Surapureddi S, Guo X, Li X. 2009. Hepatocyte-specific deletion of SIRT1 alters fatty acid metabolism and results in hepatic steatosis and inflammation. *Cell Metab* **9:** 327–338.

Rasbach KA, Gupta RK, Ruas JL, Wu J, Naseri E, Estall JL, Spiegelman BM. 2010. PGC-1α regulates a HIF2α-dependent switch in skeletal muscle fiber types. *Proc Natl Acad Sci* **107:** 21866–21871.

Rebelo AP, Dillon LM, Moraes CT. 2011. Mitochondrial DNA transcription regulation and nucleoid organization. *J Inherit Metabol Dis* **34:** 941–951.

Reznick RM, Zong H, Li J, Morino K, Moore IK, Yu HJ, Liu ZX, Dong J, Mustard KJ, Hawley SA, et al. 2007. Aging-associated reductions in AMP-activated protein kinase activity and mitochondrial biogenesis. *Cell Metab* **5:** 151–156.

Rodgers JT, Lerin C, Haas W, Gygi SP, Spiegelman BM, Puigserver P. 2005. Nutrient control of glucose homeostasis through a complex of PGC-1α and SIRT1. *Nature* **434:** 113–118.

Salih DA, Brunet A. 2008. FoxO transcription factors in the maintenance of cellular homeostasis during aging. *Curr Opin Cell Biol* **20:** 126–136.

Scarpulla RC. 2008. Transcriptional paradigms in mammalian mitochondrial biogenesis and function. *Physiol Rev* **88:** 611–638.

Scarpulla RC, Vega RB, Kelly DP. 2012. Transcriptional integration of mitochondrial biogenesis. *Trends Endocrinol Metab:* **23:** 459–466.

Schieke SM, Phillips D, McCoy JP Jr, Aponte AM, Shen RF, Balaban RS, Finkel T. 2006. The mammalian target of rapamycin (mTOR) pathway regulates mitochondrial oxygen consumption and oxidative capacity. *J Biol Chem* **281:** 27643–27652.

Schmidt O, Pfanner N, Meisinger C. 2010. Mitochondrial protein import: From proteomics to functional mechanisms. *Nat Rev Mol Cell Biol* **11:** 655–667.

Schreiber SN, Emter R, Hock MB, Knutti D, Cardenas J, Podvinec M, Oakeley EJ, Kralli A. 2004. The estrogen-related receptor α (ERRα) functions in PPARγ coactivator 1α (PGC-1α)-induced mitochondrial biogenesis. *Proc Natl Acad Sci* **101:** 6472–6477.

Seale P, Kajimura S, Yang W, Chin S, Rohas LM, Uldry M, Tavernier G, Langin D, Spiegelman BM. 2007. Transcriptional control of brown fat determination by PRDM16. *Cell Metab* **6:** 38–54.

Sengupta S, Peterson TR, Laplante M, Oh S, Sabatini DM. 2010. mTORC1 controls fasting-induced ketogenesis and its modulation by ageing. *Nature* **468:** 1100–1104.

Shin JH, Ko HS, Kang H, Lee Y, Lee YI, Pletinkova O, Troconso JC, Dawson VL, Dawson TM. 2011. PARIS (ZNF746) repression of PGC-1α contributes to neurodegeneration in Parkinson's disease. *Cell* **144:** 689–702.

Silva JE. 1995. Thyroid hormone control of thermogenesis and energy balance. *Thyroid* **5:** 481–492.

Sohal RS, Weindruch R. 1996. Oxidative stress, caloric restriction, and aging. *Science* **273:** 59–63.

Soriano FX, Liesa M, Bach D, Chan DC, Palacin M, Zorzano A. 2006. Evidence for a mitochondrial regulatory pathway defined by peroxisome proliferator-activated receptor-gamma coactivator-1α, estrogen-related receptor-α, and mitofusin 2. *Diabetes* **55:** 1783–1791.

Spindler SR. 2001. Calorie restriction enhances the expression of key metabolic enzymes associated with protein renewal during aging. *Ann NY Acad Sci* **928:** 296–304.

Srivastava S, Diaz F, Iommarini L, Aure K, Lombes A, Moraes CT. 2009. PGC-1α/β induced expression partially compensates for respiratory chain defects in cells from patients with mitochondrial disorders. *Hum Mol Genet* **18:** 1805–1812.

Stowers RS, Megeath LJ, Gorska-Andrzejak J, Meinertzhagen IA, Schwarz TL. 2002. Axonal transport of mitochondria to synapses depends on milton, a novel *Drosophila* protein. *Neuron* **36:** 1063–1077.

Twig G, Hyde B, Shirihai OS. 2008. Mitochondrial fusion, fission and autophagy as a quality control axis: The bioenergetic view. *Biochim Biophys Acta* **1777:** 1092–1097.

Uldry M, Yang W, St-Pierre J, Lin J, Seale P, Spiegelman BM. 2006. Complementary action of the PGC-1 coactivators in mitochondrial biogenesis and brown fat differentiation. *Cell Metab* **3:** 333–341.

Venditti P, Bari A, Di Stefano L, Cardone A, Della Ragione F, D'Esposito M, Di Meo S. 2009. Involvement of PGC-1, NRF-1, and NRF-2 in metabolic response by rat liver to hormonal and environmental signals. *Mol Cell Endocrinol* **305:** 22–29.

Verdin E, Hirschey MD, Finley LW, Haigis MC. 2010. Sirtuin regulation of mitochondria: Energy production, apoptosis, and signaling. *Trends Biochem Sci* **35:** 669–675.

Viscomi C, Bottani E, Civiletto G, Cerutti R, Moggio M, Fagiolari G, Schon EA, Lamperti C, Zeviani M. 2011. In vivo correction of COX deficiency by activation of the AMPK/PGC-1α axis. *Cell Metab* **14:** 80–90.

Wallace DC, Fan W. 2009. The pathophysiology of mitochondrial disease as modeled in the mouse. *Genes Dev* **23:** 1714–1736.

Wang X, Winter D, Ashrafi G, Schlehe J, Wong YL, Selkoe D, Rice S, Steen J, LaVoie MJ, Schwarz TL. 2011. PINK1 and Parkin target Miro for phosphorylation and degradation to arrest mitochondrial motility. *Cell* **147:** 893–906.

Wenz T, Diaz F, Spiegelman BM, Moraes CT. 2008. Activation of the PPAR/PGC-1α pathway prevents a bioenergetic deficit and effectively improves a mitochondrial myopathy phenotype. *Cell Metab* **8:** 249–256.

Wenz T, Rossi SG, Rotundo RL, Spiegelman BM, Moraes CT. 2009. Increased muscle PGC-1α expression protects from sarcopenia and metabolic disease during aging. *Proc Natl Acad Sci* **106:** 20405–20410.

Westermann B. 2010. Mitochondrial fusion and fission in cell life and death. *Nat Rev Mol Cell Biol* **11:** 872–884.

Wilson BJ, Tremblay AM, Deblois G, Sylvain-Drolet G, Giguere V. 2010. An acetylation switch modulates the transcriptional activity of estrogen-related receptor α. *Mol Endocrinol* **24:** 1349–1358.

Wilson-Fritch L, Nicoloro S, Chouinard M, Lazar MA, Chui PC, Leszyk J, Straubhaar J, Czech MP, Corvera S. 2004. Mitochondrial remodeling in adipose tissue associated with obesity and treatment with rosiglitazone. *J Clin Invest* **114:** 1281–1289.

Wright DC, Geiger PC, Han DH, Jones TE, Holloszy JO. 2007. Calcium induces increases in peroxisome proliferator-activated receptor γ coactivator-1α and mitochondrial biogenesis by a pathway leading to p38 mitogen-activated protein kinase activation. *J Biol Chem* **282:** 18793–18799.

Wu Z, Puigserver P, Andersson U, Zhang C, Adelmant G, Mootha V, Troy A, Cinti S, Lowell B, Scarpulla RC, et al. 1999. Mechanisms controlling mitochondrial biogenesis and respiration through the thermogenic coactivator PGC-1. *Cell* **98:** 115–124.

Wu H, Kanatous SB, Thurmond FA, Gallardo T, Isotani E, Bassel-Duby R, Williams RS. 2002. Regulation of mitochondrial biogenesis in skeletal muscle by CaMK. *Science* **296:** 349–352.

Yan Z, Okutsu M, Akhtar YN, Lira VA. 2011. Regulation of exercise-induced fiber type transformation, mitochondrial biogenesis, and angiogenesis in skeletal muscle. *J Appl Physiol* **110:** 264–274.

Yatsuga S, Suomalainen A. 2012. Effect of bezafibrate treatment on late-onset mitochondrial myopathy in mice. *Hum Mol Genet* **21:** 526–535.

Youle RJ, van der Bliek AM. 2012. Mitochondrial fission, fusion, and stress. *Science* **337:** 1062–1065.

Yu W, Dittenhafer-Reed KE, Denu JM. 2012. SIRT3 protein deacetylates isocitrate dehydrogenase 2 (IDH2) and regulates mitochondrial redox status. *J Biol Chem* **287:** 14078–14086.

Zhao Q, Wang J, Levichkin IV, Stasinopoulos S, Ryan MT, Hoogenraad NJ. 2002. A mitochondrial specific stress response in mammalian cells. *EMBO J* **21:** 4411–4419.

Zheng B, Liao Z, Locascio JJ, Lesniak KA, Roderick SS, Watt ML, Eklund AC, Zhang-James Y, Kim PD, Hauser MA, et al. 2010. PGC-1α, a potential therapeutic target for early intervention in Parkinson's disease. *Sci Transl Med* **2:** 52ra73.

Zoncu R, Efeyan A, Sabatini DM. 2011. mTOR: From growth signal integration to cancer, diabetes and ageing. *Nat Rev Mol Cell Biol* **12:** 21–35.

Mechanisms of Mitochondrial Fission and Fusion

Alexander M. van der Bliek, Qinfang Shen, and Sumihiro Kawajiri

Department of Biological Chemistry, David Geffen School of Medicine, University of California, Los Angeles, Los Angeles, California 90095

Correspondence: avan@mednet.ucla.edu

Mitochondria continually change shape through the combined actions of fission, fusion, and movement along cytoskeletal tracks. The lengths of mitochondria and the degree to which they form closed networks are determined by the balance between fission and fusion rates. These rates are influenced by metabolic and pathogenic conditions inside mitochondria and by their cellular environment. Fission and fusion are important for growth, for mitochondrial redistribution, and for maintenance of a healthy mitochondrial network. In addition, mitochondrial fission and fusion play prominent roles in disease-related processes such as apoptosis and mitophagy. Three members of the Dynamin family are key components of the fission and fusion machineries. Their functions are controlled by different sets of adaptor proteins on the surface of mitochondria and by a range of regulatory processes. Here, we review what is known about these proteins and the processes that regulate their actions.

Mitochondrial movement and fission were first observed with light microscopy almost 100 years ago (Lewis and Lewis 1914). For a long time, these observations remained something of a curiosity and they were all but forgotten when electron microscopy popularized the idea that mitochondria exist as isolated sausage-shaped organelles floating in a sea of cytoplasm. Renewed appreciation for mitochondrial dynamics emerged some 20 or 30 years ago when technological advances made it much easier to track mitochondria in live cells. Careful observations, first with phase contrast microscopy, then with vital dyes and finally with targeted fluorescent proteins, showed that mitochondria continually divide and fuse, even in resting cells (Johnson et al. 1981; Bereiter-Hahn and Voth 1994; Rizzuto et al. 1996). Their lengths are determined by the balance between fission and fusion. Mitochondrial morphologies can change dramatically by shifting this balance. In some cells they fuse together, forming a single closed network, whereas in other cells or under different circumstances mitochondria convert into large numbers of small fragments. Because of these morphological changes mitochondria are now known to be very dynamic.

The importance of frequent mitochondrial fission and fusion events for cell survival was also not fully appreciated until fairly recently. Obvious reasons, such as accommodating cell growth, cell division, and the redistribution of mitochondria during differentiation, did not fully explain why mitochondria fuse nor did they explain the high frequencies of these occurrences. However, in more recent years, the biological relevance of these phenomena has become clear with the discovery of human diseases that

Cite this article as *Cold Spring Harb Perspect Biol* doi: 10.1101/cshperspect.a011072

are caused by mutations in fission and fusion proteins and the discovery of numerous connections with apoptosis and mitophagy (Westermann 2010; Chan 2012; Nunnari and Suomalainen 2012; Youle and van der Bliek 2012). Mitochondrial fission and fusion are now considered cornerstones for cell survival because of their contributions to health and disease.

DYNAMIN FAMILY MEMBERS THAT MEDIATE MITOCHONDRIAL FISSION AND FUSION

The main mitochondrial fission and fusion proteins are members of the Dynamin family (Fig. 1). The first mitochondrial Dynamin was named Mgm1, because mutations in the yeast gene cause a mitochondrial genome maintenance defect (Jones and Fangman 1992). This mitochondrial DNA (mtDNA) distribution defect was later shown to be a secondary consequence of defects in mitochondrial inner membrane fusion (Wong et al. 2000). Localization studies also showed that Mgm1 is anchored in the mitochondrial inner membrane with the bulk of the protein facing the mitochondrial inner membrane space (Wong et al. 2000). A human homolog of Mgm1 was discovered through sequence homologies and through genetic mapping of a late onset neurodegenerative eye disease, named Dominant Optic Atrophy (DOA). The human homolog was named Opa1 for the optic atrophy defects (Alexander et al. 2000; Delettre et al. 2000).

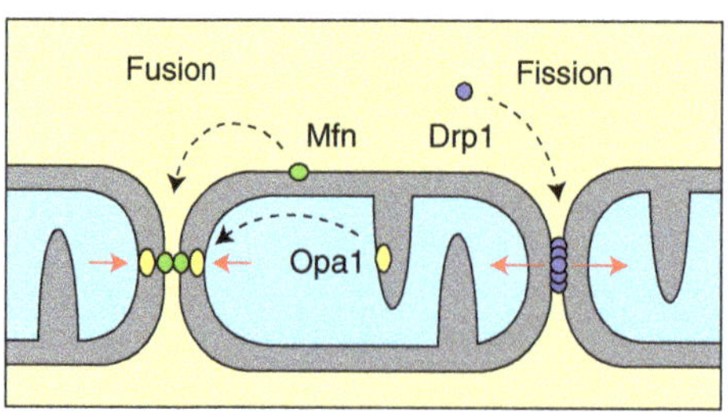

Figure 1. Functions of the mitochondrial Dynamin family members. Mitofusins mediate mitochondrial outer membrane fusion in mammals. Opa1 mediates mitochondrial inner membrane fusion. Drp1, which cycles between the cytosol and the mitochondrial outer membrane, mediates mitochondrial fission.

Fusion of mitochondrial outer membranes is mediated by a different set of Dynamin family members. The first of these was discovered in *Drosophila* sperm cells, where it was named fuzzy onions for the onion-like and fuzzy appearance of unfused mitochondria in electron micrographs of these mutants (Hales and Fuller 1997). A second mitochondrial outer membrane protein, named Marf, was later shown to mediate fusion in other cell types (Dorn et al. 2011). The sequences of Marf and fuzzy onions are similar, but their expression patterns are different so the mutants have different phenotypes: fuzzy onions mutants are sterile whereas Marf mutants are lethal (Dorn et al. 2011). Yeast and *Caenorhabditis elegans* each have only one fuzzy onions or Marf homolog (Fzo1p and FZO-1, respectively) (Hermann et al. 1998; Rapaport et al. 1998). Mammals do have two homologs (the Mitofusins Mfn1 and Mfn2), but those are often expressed in the same cells (Santel and Fuller 2001; Chen et al. 2003). Although some functional differences between Mfn1 and Mfn2 have been observed, both of these proteins are also able to support mitochondrial fusion by themselves, suggesting that they fulfill partially redundant functions in this process (Chen et al. 2003).

Mitochondrial outer membrane fusion is almost always coordinated with inner membrane fusion. There are, however, some instances in which outer membrane fusion can occur without concomitant inner membrane fusion as, for example, when mutations or loss of membrane potential block inner membrane fusion whereas outer membrane fusion can still occur (Olichon et al. 2003). Localized loss of membrane potential and its effects on inner membrane fusion may explain why the matrix compartments in larger mitochondrial networks sometimes have breaks, whereas their outer membranes appear to be fully fused (Twig et al. 2006). The physiological relevance of these observations has not yet been explored.

Unlike the fusion machinery, which has both outer and inner membrane components, it would appear that most higher eukaryotes and yeast use a single Dynamin family member for mitochondrial fission. This Dynamin-like protein is largely cytosolic, but can cycle on and off of the mitochondrial outer membrane. Homologs of this protein were discovered with genetic screens for mitochondrial morphology defects in yeast (yeast Dnm1) and through functional analyses of different Dynamin family members in *C. elegans* and mammalian cells (Bleazard et al. 1999; Labrousse et al. 1999; Smirnova et al. 2001). The mammalian protein was alternatively named Drp1, Dnm1L, DLP1, Dymple, and DVLP1 (Shin et al. 1997; Imoto et al. 1998; Kamimoto et al. 1998; Smirnova et al. 1998; Yoon et al. 1998). Here we refer to it as Drp1, because this is the most commonly used name.

The Drp1 homologs have a domain structure that is characteristic of the Dynamin family (Fig. 2). Each one has a GTPase domain followed by a middle domain, a variable domain, and a GTPase effector domain (GED) (Praefcke and McMahon 2004). The variable domain, also referred to as insert B, is generally required for binding to the target membrane. In the fission Dynamins, this domain binds to adaptor proteins on the mitochondrial surface. The lipid-binding domain in the fusion proteins, Mgm1 and Opa1, is specific for cardiolipin (Meglei and McQuibban 2009), whereas the outer membrane fusion proteins, Fzo1 and Mitofusins, each have two *trans*-membrane segments that serve as membrane anchors (Hoppins et al. 2007).

The conserved middle domain and the GED of Dynamin family members fold back on each other to form a stalk (van der Bliek and Payne 2010). As a result, the GTPase domain is directed away from the membrane. Biophysical, cryo-electron microscopy, and X-ray crystallography studies showed that Dynamin family members assemble into multimeric spirals through a series of binding interactions between the stalks of different subunits (Chappie et al. 2010; Gao et al. 2010). Once a spiral completes a turn, contacts between the GTPase domains of adjacent rungs trigger GTP hydrolysis (van der Bliek and Payne 2010). This enzymatic activity drives conformational changes within the multimers causing constriction of the spiral (Mears et al.

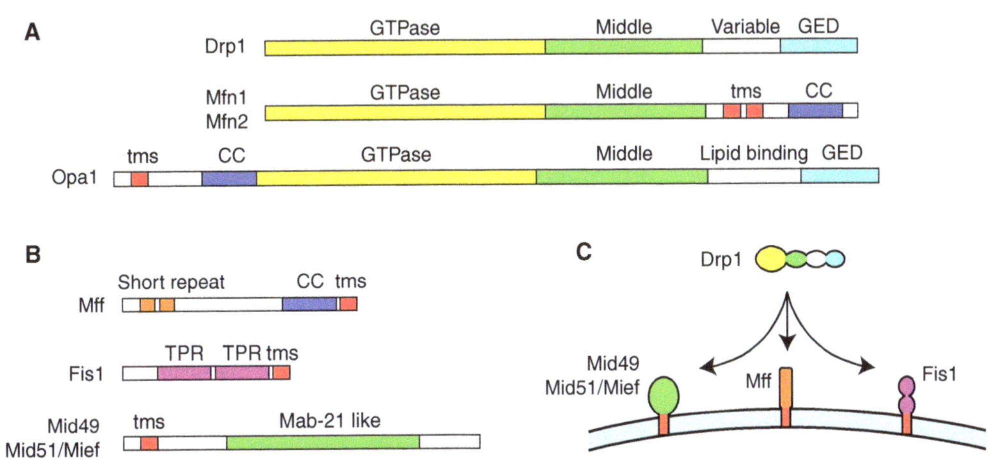

Figure 2. Schematic diagrams of fission and fusion proteins. (*A*) The main Dynamin family members in mammals, Drp1, Mfn1, Mfn2, and Opa1 are shown with their different protein domains. Each one has a GTPase domain, a middle domain, and a GED, which together form the canonical Dynamin structure. The GED in Mitofusins is not as well conserved, but it still has coiled-coil segments, suggesting that it can participate in multimeric assembly. The variable domain in Drp1, which most likely serves as a mitochondrial targeting sequence, is replaced by *trans*-membrane segments (tms) in the Mitofusins and by a cardiolipin binding domain in Opa1. (*B*) Adaptor proteins that can bind to drp1 on the surface of mitochondria. Each one has a membrane anchor (tms), but they have different protein interaction domains, suggesting that they are functionally distinct. (*C*) Illustration of the multiplicity of Drp1 receptors on the surface of mammalian mitochondria.

2011). Constriction driven by the mitochondrial fission Dynamins severs both mitochondrial inner and outer membranes.

Electron microscopy (EM) studies have provided insight into the mechanisms of mitochondrial fission. Like Dynamin, the yeast Drp1 homolog Dnm1 can tubulate membranes by assembling into large multimeric spirals that wrap around those tubules (Ingerman et al. 2005). There are, however, two striking differences with the Dynamin spirals. First, the diameter is much larger. Instead of the 50 nm for Dynamin, these spirals have an average diameter of 120 nm (Ingerman et al. 2005; Mears et al. 2011). This large diameter is enough to wrap around a double membrane structure, which would be formed by constrictions in mitochondria when both inner and outer membranes are still intact. How might the spiral generate the additional force needed to sever both membranes? Cryo-EM data show that Dnm1 forms "two-start" helices, perhaps accommodating the larger initial diameter, but it might also provide more force because the two intertwined spirals would have more contact points (Mears et al. 2011).

Although less is known about the mechanisms of fusion, genetic data show that the fusion Dynamins are needed on both of the opposing membranes. Current models suggest that the fusion Dynamins from opposing membranes dimerize and then pull the membranes together with a SNARE-like mechanism (Koshiba et al. 2004). Alternative mechanisms, however, have not yet been ruled out. It remains possible that the fusion Dynamins assemble into spirals on membranes like the other Dynamin family members. Protruding membrane tubules formed by this process might then be mated by contacts between opposing Dynamin spirals (Devay et al. 2009). Forced contacts between the convex lipid surfaces at the tips of the Dynamin spirals could be fusogenic, similar to the forced contacts between opposing convex membranes during SNARE-mediated fusion. Time will tell whether fusion Dynamins use either of these mechanisms or perhaps use a totally different mechanism.

Isolated mitochondria have been used to study biophysical requirements for mitochondrial inner and outer membrane fusion processes (Meeusen et al. 2006). As expected, both processes require GTP, but only inner membrane fusion requires membrane potential. The reasons for this last requirement are not yet clear. More importantly, it was shown that fusion Dynamins are required on both opposing membranes and that these Dynamins can form large protein complexes during the fusion process (Meeusen et al. 2006). There are, as of yet, no good in vitro assays to study mitochondrial fission, but the analogies with classical Dynamin, as well as EM studies of Drp1 homologs on membranes and studies with selective inhibitors have been remarkably informative. Several drugs, like Dynasore, inhibit both Dynamin and Drp1 (Macia et al. 2006), but there are also more selective drugs, like Mdivi-1, which only inhibit Drp1 homologs (Cassidy-Stone et al. 2008). Mdivi-1 has been used to study the role of Drp1 in cytochrome *c* release during apoptosis and it may represent a new class of drugs with therapeutic potential (Tanaka and Youle 2008).

ADAPTOR PROTEINS

Genetic screens for mutations that suppress fusion defects led to the discovery of additional fission proteins in yeast. One of the fission proteins that came out of these screens is Fis1, which has two tetratrico peptide repeat (TPR) domains and a carboxy-terminal hydrophobic segment that anchors this protein in the mitochondrial outer membrane (Mozdy et al. 2000). The second protein discovered with the genetic screens was Mdv1, which is a cytosolic protein with a poorly defined amino-terminal segment followed by a series of withdrawal (protein motif of approximately 40 amino acids ending with trypophan (W) and aspartic acid (D) [WD]) motifs (Tieu et al. 2002; Cerveny and Jensen 2003). The amino terminus of Mdv1 binds to Fis1 on the surface of mitochondria, whereas the WD repeats form a β propeller structure that binds to and recruits the yeast Drp1 homolog Dnm1 to the surface of mitochondria (Koirala et al. 2010). Yeast has a second Mdv1-like protein, Caf4, which is partially redundant with Mdv1 (Griffin et al. 2005). These proteins are the only known adaptors for Dnm1 on mitochondria in yeast.

The situation in higher eukaryotes is more confusing. Higher eukaryotes have no obvious Mdv1 homologs, but they do have Fis1 homologs. The functions of Fis1 in higher eukaryotes were, however, brought into question by genetic data showing that Fis1 is not required for fission (Otera et al. 2010). A number of alternative receptors for Drp1 have been proposed. Among these are Mff, Mid49, Mid51/Mief, and GDAP1 (Fig. 2). Mff was first isolated with a small interfering RNA (siRNA) screen for mitochondrial morphology defects (Gandre-Babbe and van der Bliek 2008). It is a relatively small protein with a short amino-terminal repeat, a coiled-coil segment, and a carboxy-terminal hydrophobic segment that anchors the protein in the mitochondrial outer membrane, similar to the tail anchor of Fis1. Subsequent work with a mistargeted protein in mammalian cells showed that Mff is sufficient to act as a Drp1 receptor (Otera et al. 2010). Knockdown and overexpression studies of Mid49, Mid51/Mief, and GDAP1 have independently shown effects on mitochondrial fission, suggesting that there are multiple ways to control fission (Niemann et al. 2005; Palmer et al. 2011; Zhao et al. 2011). This was confirmed in a recent paper showing that Fis1, Mff, Mid49, and Mid51/Mief act in a partially redundant way to promote Drp1-dependent fission (Loson et al. 2013). Interestingly, Fis1 and Mff are also present on peroxisomes, consistent with results showing that peroxisome fission is mediated by Drp1 in higher eukaryotes (Gandre-Babbe and van der Bliek 2008). It seems likely that this part of the fission machinery can act independently. It is not yet clear whether the other fission adaptors or receptors take part in redundant pathways or act at different steps in the same pathway.

No similar collection of adaptors or receptors has been found for mitochondrial fusion proteins with the possible exception of Ugo1 in yeast. Ugo1 is an integral mitochondrial outer membrane protein that coIPs with the inner and outer membrane fusion proteins (Sesaki and Jensen 2004). It may help coordinate the two membrane fusion events, but higher eukaryotes have no obvious Ugo1 homologs. The proapoptotic BH3 protein Bax is required to maintain Mitofusin tetramers in mammalian cells, but it is unclear whether Bax affects Mitofusin assembly or is directly involved in the fusion process (Karbowski et al. 2006). In addition, a cytosolic medium-chain dehydrogenase/reductase named MIB for Mitofusin binding protein was shown to selectively inhibit the functions of Mfn1 causing mitochondrial elongation with knockdown and fragmentation when it was overexpressed in mammalian cells (Eura et al. 2006). Last, it was found that the lipid metabolizing enzyme mitoPLD functions downstream from Mitofusins during mitochondrial fusion, potentially providing a link between fission and fusion (Choi et al. 2006). The role of this protein is discussed in more detail in the section on the regulation of fission. It remains possible that other, as yet undiscovered, proteins control fusion, but so far the data suggest that the control of fusion is much simpler than the control of fission.

REGULATION OF FISSION BY PHOSPHORYLATION AND UBIQUITINATION

Mitochondrial fission proteins are regulated by a range of protein modifications, including phosphorylation, ubiquitination, sumoylation, and nitrosylation. Different kinases control the activities of Drp1 by phosphorylating this protein at three different sites. These sites include Ser616, which is phosphorylated by protein kinase C (PKC) δ, Rock kinase, CDK1/Cyclin B or CAMK-Ia (Taguchi et al. 2007; Han et al. 2008; Qi et al. 2011; Wang et al. 2012), and Ser637, which is phosphorylated by protein kinase A (PKA) (Chang and Blackstone 2007; Cribbs and Strack 2007). Phosphorylation of Ser616 is likely to activate fission, because it promotes binding to other fission proteins, whereas phosphorylation of Ser637 could be an inactivating step (Chang and Blackstone 2010). The third site in Drp1 is Ser693, which is phosphorylated by GSK3β to inhibit mitochondrial fission during apoptosis (Chou et al. 2012). This residue is in the GED, where it most likely affects Drp1 oligomerization or GTP hydrolysis.

Ser616 and Ser637 are both toward the end of the variable domain. It is likely that the variable domain is in close contact with mitochondrial outer membrane proteins, so phosphorylation at these sites may regulate the association of Drp1 with mitochondria by changing the contacts with mitochondrial outer membrane proteins or with the membrane itself. This concept is supported by the enhanced binding of Drp1 to Fis1 when Ser616 is phosphorylated by CAMK-I (Han et al. 2008), followed by the putative inactivation step by PKA, which is localized to mitochondria through the A-kinase activating protein (AKAP) Rab32 in mammals (Alto et al. 2002; Merrill et al. 2011). The phosphate on Ser637 is removed by Calcineurin (PP2A), which is a calcium-activated cytosolic phosphatase (Cribbs and Strack 2007; Cereghetti et al. 2008). These observations suggest that phosphorylation regulates the cycling of Drp1 between cytosol and the mitochondrial outer membrane.

After Drp1 is primed for another round of fission by phosphatases, it needs a trigger to initiate the fission event, for example, by the phosphorylation of Ser616. Because PKC is one of the kinases that can phosphorylate Ser616 and different PKC isoforms are generally activated by calcium and diacylglycerol, this trigger may react to localized signals. Diacylglycerol is produced on mitochondria by Lipin1, which can be activated by phosphatidic acid on mitochondria (Huang et al. 2011). Phosphatidic acid in turn is produced by mitoPLD during mitochondrial fusion (Choi et al. 2006). These different steps suggest a cycle in which fusion is quickly followed by fission through the sequential actions of mitoPLD, Lipin1, PKC, and Drp1. This sequence fits remarkably well with live cell imaging studies in which it was shown that fission often follows fusion, more or less at the same place and within 20 min (Twig et al. 2008). It is tempting to speculate that these live images show results of the regulatory pathway outlined in this paragraph.

Alternative fission-inducing pathways may also exist because the conditions that activate some of the other kinases that phosphorylate Drp1 do not necessarily come from fusion. There is, for example, massive fragmentation when cells are treated with carbonyl cyanide *m*-chlorophenyl hydrazone (CCCP) or with apoptosis-inducing drugs (Frank et al. 2001; Ishihara et al. 2003). This fragmentation results from all but complete mobilization of Drp1 and a large number of fission events without corresponding fusion events. It has been suggested that calcium release from depolarized mitochondria may activate Calcineurin (Cribbs and Strack 2007; Cereghetti et al. 2008), whereas apoptosis-induced fission may rely on calcium release from the ER (Scorrano et al. 2003). Other pathways may exist, but they all seem to lead to Drp1 at the center of the mitochondrial fission apparatus.

Several other forms of protein modification may regulate mitochondrial fission. The membrane-bound E3 ligase March5 may regulate Drp1 through ubiquitination (Karbowski et al. 2007). Another membrane-bound E3 ligase, Mulan or Mapl, was proposed to regulate Drp1 through sumoylation (Braschi et al. 2009), and NO-dependent nitrosylation of a Drp1 has been proposed to activate fission in a way that promotes neurodegeneration in Alzheimer's patients (Cho et al. 2009). These alternative forms of regulation are still somewhat controversial. March5 and Mulan were both reported to have other targets, whereas the effects of nitrosylation were brought into question by independent investigators (Nakamura et al. 2006; Bossy et al. 2010). Further investigations may help determine which of these mechanisms are physiologically relevant regulators of Drp1.

COORDINATION OF MITOCHONDRIAL FISSION WITH THE CYTOSKELETON, ER, AND ITS ROLE IN APOPTOSIS

Frequent close contacts between mitochondria and ER in mitochondrion-associated membranes (MAMs) have been studied for years (Rowland and Voeltz 2012). They are known to contribute to lipid and calcium transfer between these organelles, but their relevance for mitochondrial fission was only recently discovered through high-resolution light microscopy of live cells (Friedman et al. 2011). These images showed that ER tubules often cross paths with mitochondria at points of impending fission or they wrap around mitochondria at those points (Friedman et al. 2011). In this report, it was also shown that the contacts between ER and mitochondria are made and that mitochondria are still constricted at those sights when the recruitment of Drp1 to mitochondria is prevented by siRNA for Drp1 or for Mff. These results suggested that ER might play an active role during the earliest stages of fission, well before Drp1-mediated severing of the membrane. This suggestion was nicely confirmed by the recent discovery that the ER protein INF-2 can promote mitochondrial fission by inducing constriction before Drp1 is recruited to mitochondria (Korobova et al. 2013). INF-2 is a Formin-like protein that triggers actin filament assembly. Actin was previously shown to be important for mitochondrial fission through studies with inhibitors (De Vos et al. 2005), but the connections with the fission apparatus were only now coming into focus.

Evidence for contributions from ER to mitochondrial fission also comes from earlier work on apoptosis. Cytochrome *c* is released through Bax-lined pores at sites of Drp1-mediated mitochondrial fission (Montessuit et al. 2010). Interestingly, Bax also affects calcium release from the ER, which might then affect Drp1-mediated fission, and it induces cleavage of the ER protein Bap31 (Iwasawa et al. 2011). Bap31 has been shown to be enriched in MAMs and it can co-immunoprecipitate with Fis1, suggesting that it is part of a larger fission complex (Simmen et al. 2005; Iwasawa et al. 2011). Although the molecular details have not yet been worked out, these studies suggest that the MAM is somehow involved in the regulation of cytochrome *c* release during apoptosis and that Drp1 plays a central role in this process. Because apoptosis is an end point that does not require functional organelles, this process may have co-opted certain parts of the MAM in abnormal ways.

There may be other connections between the MAM and mitochondrial fission or fusion. The yeast Miro homolog Gem1p was shown to regulate contacts between the ER and mitochondria (Kornmann et al. 2011); but in higher eukaryotes, Miro regulates mitochondrial transport along microtubules through interactions with dynein and the adaptor Milton (Glater et al. 2006). It is not yet clear to what extent these functions overlap. The small GTP-binding protein Rab32, which recruits PKA for inactivation of Drp1, was also shown to regulate the MAM, perhaps through its ability to recruit PKA, but the targets of these proteins on the ER are not yet known (Bui et al. 2010). Although this area is still in its infancy, it is already clear that the MAM plays a central role in mitochondrial fission.

REGULATION OF MITOCHONDRIAL FUSION BY PROTEOLYTIC CLEAVAGE AND UBIQUITINATION

Mitochondrial fusion proteins are regulated by proteolysis and ubiquitination. The predominant form of regulation for mitochondrial outer membrane fusion Dynamins is inactivation through ubiquitin-mediated degradation, whereas the functions of inner membrane fusion Dynamins are altered by proteolytic cleavage at a select few sites. The ubiquitin-mediated degradation of yeast Fzo1 is triggered by the actions of two mitochondrial outer membrane F-box proteins (Fritz et al. 2003; Neutzner and Youle 2005; Durr et al. 2006). Fzo1 accumulates when either one of these F-box proteins is mutated, suggesting that they promote Fzo1 turnover. Fzo1 is degraded by the proteasome as part of the yeast mating response, suggesting that regulation through turnover is physiologically relevant (Neutzner and Youle 2005). In higher eukaryotes, ubiquitin-mediated degradation of the mitochondrial fusion machinery is induced by stress. When mitochondria lose membrane potential, Mitofusins are degraded by PTEN-induced putative kinase 1 (Pink1)- and Parkin-mediated ubiquitination during mitophagy (Youle and Narendra 2011). After ubiquitination, these proteins are extracted by the AAA protein p97 and then degraded by the proteasome (Tanaka et al. 2010), as discussed in more detail elsewhere in this issue. The degradation of Mfn2 can also be induced by other kinds of stress, such as Doxorubicin, which triggers ubiquitination by the E3 ligase Huwe1 (Leboucher et al. 2012), or during muscle wasting with increased Mfn2 ubiquitination by the E3 ligase Mulan (Lokireddy et al. 2012). These results show tight temporal control of Mitofusin levels through different stress-induced pathways.

The mitochondrial inner membrane fusion dynamins, Mgm1 in yeast, EAT-3 in *C. elegans*, and Opa1 in mammals are also proteolytically cleaved but the effects are very different. These proteins each have a mitochondrial targeting sequence that is cleaved on import, whereas the bulk of the protein remains exposed to the intermembrane space (Wong et al. 2000; Olichon et al. 2002). In yeast, approximately half of the protein is further cleaved by the mitochondrial inner membrane rhomboid protease Pcp1 to generate a short and a long form (Esser et al. 2002; Herlan et al. 2003; Sesaki et al. 2003). The long form retains a membrane anchor at the amino terminus, whereas the short form loses this membrane anchor. The ratio of short to long forms is determined by the energy status of mitochondria (Herlan et al. 2004). Although the functional consequences of this form of regulation are not yet known, both short and long forms of Mgm1 are needed for fusion between mitochondria. *C. elegans* and *Drosophila* homologs of Mgm1 are also cleaved by a rhomboid protease (van der Bliek AM, unpubl.) (McQuibban et al. 2006). It was initially reported that mammalian Opa1 is also cleaved by a rhomboid protease (Cipolat et al. 2006), but subsequent experiments with knockout cells for this rhomboid did not show an effect on Opa1 (Griparic et al. 2007). Instead, Opa1 is cleaved by a different set of proteases in a complex pattern governed by alternative splicing of the Opa1 mRNA.

Mammals have eight different Opa1 isoforms generated by alternative splicing (Olichon et al. 2007). Differences between the isoforms result from the presence or absence of several short exons near the amino terminus. These exons replace the single rhomboid cleavage site in lower eukaryotes with three different proteolytic cleavage sites in mammals. The S1 cleavage site is present in all isoforms, whereas S2 and S3 cleavage sites are present in approximately half of the isoforms (Fig. 3). Those with S2 or S3 sites are constitutively cleaved by the large intermembrane space AAA-protease Yme1L (Griparic et al. 2007; Song et al. 2007). As a result, mammalian cells typically have a 50/50 ratio of short (Yme1L cleavage products) and long (uncleaved) forms of Opa1, similar to the distribution in lower eukaryotes. However, mammalian cells also have an S1 site, which is present in all isoforms and is only cleaved when mitochondria lose membrane potential, have low levels of ATP, or lose some of the other quality-control mechanisms (mutations in the AFG3L2 protease or the Prohibitin chaperone) (Baricault et al. 2007; Merkwirth et al. 2008; Ehses et al. 2009). This inducible cleavage is mediated by the small inner membrane Zn-protease Oma1 (Ehses et al. 2009; Head et al. 2009). Oma1 cleaves all remaining long forms of Opa1 within minutes after treatment with CCCP or other drugs that affect mitochondrial ATP production (Fig. 3). Oma1-mediated cleavage of Opa1 prevents inner membrane fusion and it may cause some changes in cristae architecture, well before other stress-management systems, such as the Pink1 Parkin pathway, are activated. The reasons for first shutting down the inner membrane fusion machinery are not yet clear.

ADDITIONAL ROLES OF FUSION PROTEINS

The phenotypic effects of mutations in mitochondrial inner and outer membrane fusion proteins are strikingly different. This was noted for yeast Fzo1 and Mgm1, where Mgm1 appears to have additional functions in helping to preserve mtDNA and cristae morphology (Guan et al. 1993; Meeusen et al. 2006). Differences in viability and tissue-specific effects have also been noted with mutations in the analogous *C. elegans* and *Drosophila* proteins (Kanazawa et al. 2008; Yarosh et al. 2008). Whereas homozygous mutations in mice are lethal, there are differences in the most severely affected stages and in heterozygous animals (Chen et al. 2003; White et al. 2009). Humans with disease-causing mutations in Mitofusins or Opa1 are gen-

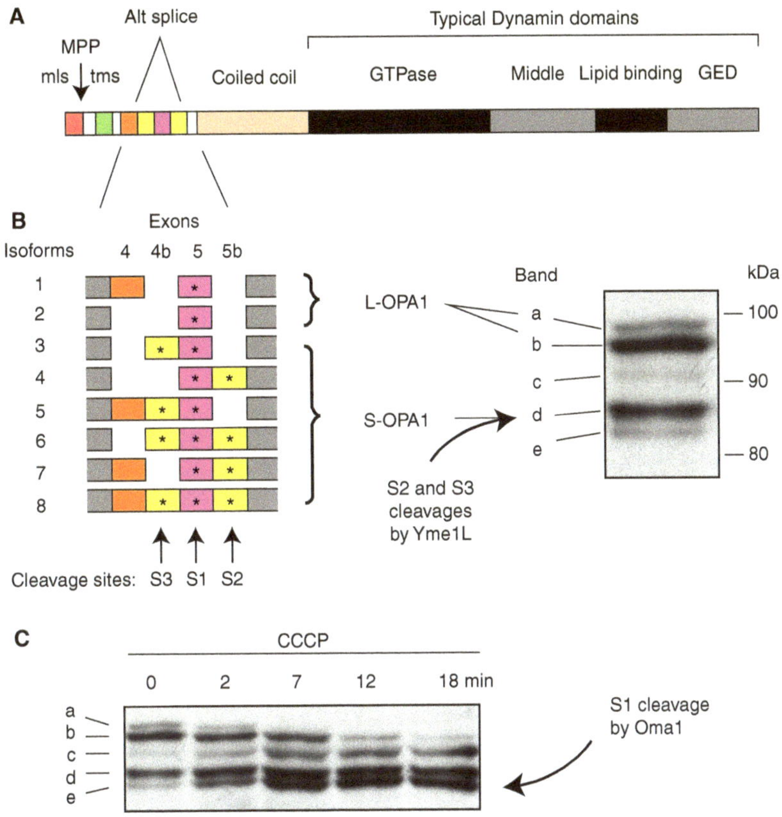

Figure 3. Proteolytic cleavage sites in mammalian Opa1. (*A*) Protein domains in Opa1. The amino terminus has a mitochondrial leader sequence, followed by a mitochondrial processing peptidase cleavage site (MPP), a *trans*-membrane segment (tms) and several alternatively spliced exons. (*B*) Alternatively spliced exons have three more cleavage sites. S2 and S3 sites are constitutively cleaved by Yme1L, giving rise to S-Opa1 (band *d*). Isoforms lacking exons 4b and 5b are not normally cleaved, giving rise to L-Opa1 (bands *a* and *b*). (*C*) Oma1 cleaves the S1 site in isoforms that were not already cleaved by Yme1L, but only when mitochondria lose membrane potential. Bands *a* and *b* disappear, giving rise to the intermediate form band *c* and the final product band *e*. (The blots were from Griparic et al. 2007; reprinted, with permission, from the authors.)

erally heterozygous, but the diseases caused by these mutations are different: mutations in Opa1 cause optic atrophy through progressive loss of retinal ganglion cells, whereas mutations in Mfn2 cause a form of peripheral neuropathy (Charcot-Marie-Tooth disease CMT2A) (Chan 2012).

Mammalian cell culture experiments and work with model organisms, such as *C. elegans*, *Drosophila*, and mice, shed light on the functional differences between mitochondrial inner and outer membrane fusion proteins. In mammals, loss of both inner and outer membrane fusion proteins promotes cytochrome *c* release during apoptosis, but the effects of losing the inner membrane fusion protein Opa1 are much more severe (Olichon et al. 2003; Arnoult et al. 2005). Mutations in *C. elegans* and *Drosophila* Opa1 homologs also severely impact growth and survival, much more so than mutations in the outer membrane fusion proteins, but not through apoptosis. These organisms do not release cytochrome *c* release as part of their apoptotic program. Instead, the *C. elegans* and *Drosophila* mutants become very sensitive to reactive oxygen species (ROS) producing agents such as Paraquat (Kanazawa et al. 2008; Yarosh et al. 2008). This effect is distinct from the fusion defect, because complete loss of the *C. elegans* Opa1 homolog is not suppressed by fission defects (Kanazawa et al. 2008). Similar dual functions were proposed for mammalian Opa1 and yeast Mgm1 (Cipolat et al. 2006; Meeusen et al. 2006).

Clues to this second function of Opa1 come from mammalian cell culture experiments. Immunogold-labeled EM shows that only 8% of Opa1 localizes to the rim, where fusion is likely to take place (Griparic et al. 2004). The remaining 92% of the Opa1 protein is inside cristae. Partial loss of Opa1 at early time points after transfection with siRNA causes some changes in mitochondrial morphology (alternate swelling and constriction of mitochondrial tubules and swollen cristae), consistent with defects in cristae maintenance (Griparic et al. 2004). It has been suggested that Opa1 may reside at the necks of cristae (cristae junctions), because those have the same diameter as Dynamin rings (Frezza et al. 2006), but this is not supported by the even distribution of Opa1 distributed throughout cristae as detected with Immuno-EM (Griparic et al. 2004). Instead, it seems likely that Opa1 somehow helps maintain cristae morphology, similar to the proposed functions of Mitofilin and ATP synthase (Rabl et al. 2009). Swelling of cristae morphologies could promote ROS production, for example, by dislodging different parts of the respiratory supercomplex (Schafer et al. 2006). Increased ROS production might also promote apoptosis in mammalian cells by oxidizing cardiolipin, which then promotes the release of cytochrome *c* from the inner membrane (Ott et al. 2007).

PATHOGENIC EFFECTS OF MUTATIONS IN FISSION AND FUSION PROTEINS

With live cell microscopy of mammalian cells, it was shown that mitochondria primarily exist as solitary units that periodically fuse with other mitochondria (Twig et al. 2008) (Fig. 4). These fusion events are often rapidly followed by fission, which can be asymmetric, because the daughter mitochondria often have different membrane potentials. This could to be a sorting event, giving rise to some mitochondria with a heavy load of defective components (Youle and van der Bliek 2012). Mitochondria that cannot recover membrane potential because of these defects become targets for Pink1- and Parkin-mediated degradation (Narendra et al. 2010). As discussed in more detail elsewhere in this collection and also touched on in this paper, a range

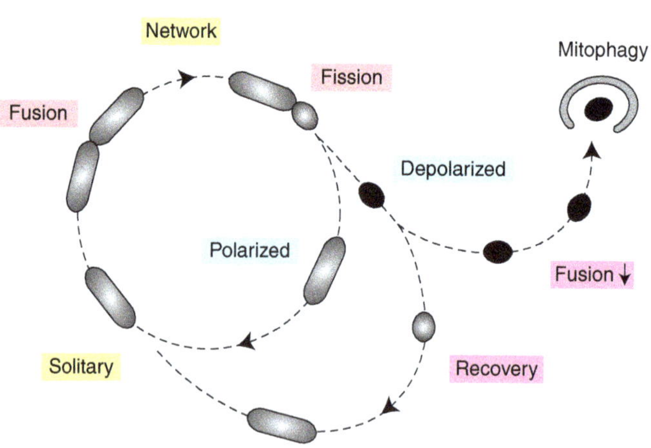

Figure 4. Mitochondrial life cycle. Live cell imaging showed that mitochondria often exist as solitary units. They do occasionally fuse to other mitochondria but fusion is often followed within 20 min by fission at the same place. Fission can give rise to daughter mitochondria with different membrane potentials, suggesting that fission is preceded by a sorting event. Daughter mitochondria with lower membrane potential often recover, allowing them to rejoin the mitochondrial network by fusion, but persistently low membrane potential will inhibit fusion and cause elimination through mitophagy. (The figure was created from data adapted from Twig et al. 2008.)

 Cite this article as *Cold Spring Harb Perspect Biol* doi: 10.1101/cshperspect.a011072

of human diseases has been connected with mutations in these pathways. Many of these diseases are neurodegenerative (Parkinson's, Charcot-Marie-Tooth, Dominant Optic Atrophy, and Alzheimer's), but effects on other tissues have also been observed (diabetes, ischemia reperfusion, and liver disease) (Liesa et al. 2009; Chan 2012; Nunnari and Suomalainen 2012).

Two opposing trends in fission and fusion rates are used to counter different levels of stress: increased fusion and/or decreased fission helps overcome low levels of stress, whereas decreased fusion and/or increased fission occurs with high levels of stress. Low levels of stress are seen with starvation, during which mitochondrial fission is inhibited to protect cells from excessive autophagosomal degradation of mitochondria (Gomes et al. 2011; Rambold et al. 2011). Similar effects are achieved by increasing the rates of fusion, which occurs when cells are treated with low levels of toxins. This phenomenon is called stress-induced mitochondrial hyperfusion (SIMH) (Tondera et al. 2009). SIMH is distinct from other forms of fusion because it relies on a specific subset of fusion proteins (Mfn1, L-Opa1, SLP2). Interestingly, knockdown of Yme1L, which increases the proportion of L-Opa1 in a cell, also increases the numbers of fused mitochondria, suggesting that this treatment leads to SIMH as well (Griparic et al. 2007). Increased fusion and decreased fission are likely beneficial because the functions of partially defective mitochondrial can be restored by mixing the contents as a form of complementation.

As discussed before, fission rates are drastically increased during apoptosis or when mitochondria lose membrane potential. In mammalian cells, fusion is then also blocked by Oma1-mediated cleavage of Opa1 (Ehses et al. 2009; Head et al. 2009). Similar steps may occur, but much less frequently, in normal cells to eliminate unsalvageable mitochondria through mitophagy (Frank et al. 2012). In these defective mitochondria, the Mitofusins are eventually eliminated by Pink1- and Parkin-mediated degradation (Youle and Narendra 2011). Parkinson's disease can be caused by mutations in Pink1 or Parkin, which most likely lead to the accumulation of damaged mitochondria. These damaged mitochondria may eventually kill the cell through ROS or other toxic agents in the dopaminergic neurons of Parkinson's patients (Chinta and Andersen 2008). Paradoxically, however, another neurodegenerative disease, the peripheral neuropathy Charcot-Marie-Tooth disease CMT2A, is caused by mutations in a Mitofusin (Bradbury 2004). So the Mitofusins that inappropriately remain active in Parkinson's disease are actually defective in patients with Charcot-Marie-Tooth disease. Perhaps the initial fusogenic stress response is more important for cell survival in peripheral neurons, whereas the elimination of malfunctioning parts through mitophagy is more important for the survival of dopaminergic neurons.

SOME EVOLUTIONARY CONSIDERATIONS

Lower eukaryotes, such as Dictyostelium and Trypanosomes, have Dynamin homologs that affect mitochondrial fission, but there is no obvious distinction in these organisms between Dynamins involved in organelle fission and those involved in membrane traffic, unlike the yeast Dynamins Vps1 and Dnm1 or Dynamin and Drp1 in higher eukaryotes (Wienke et al. 1999; Chanez et al. 2006). This lack of specialization may reflect an early stage in eukaryotic evolution. Bacteria also have Dynamin-like proteins. The closest Dynamin relatives appear to be a bacterial Dynamin-like protein (BDLP) in Nostoc punctiforme and dynA in *Bacillus subtilus*. Structural studies with BDLP suggest that it can form spirals around membrane tubules, similar to the Dynamin spirals in higher eukaryotes (Low et al. 2009), whereas dynA is able to fuse membranes in vitro (Burmann et al. 2011). Although bacteria are not known to have internal organelles or membrane traffic and these proteins are on the wrong side of the membrane for Drp1-like fission, membrane fusion may be needed at a late stage of cytokineses. Moreover, it was recently shown that dynA affects ftsZ-mediated bacterial fission, but this was only detected with additional mutations in bacterial Flotillin homologs (Dempwolff et al. 2012). These new data suggest that bacterial Dynamins play a nonessential role during cytokinesis. It will be interesting to see whether these proteins are directly involved in bacterial cell division and whether this might help us better understand mitochondrial fission and fusion.

The apparent lack of a mitochondrial inner membrane fission apparatus in higher eukaryotes raises some evolutionary questions as well. The endosymbiont that gave rise to mitochondria most likely had an elaborate inner membrane fission machinery, similar to the ftsZ-based fission apparatus of present-day α proteobacteria. Indeed, some lower eukaryotes like *Dictyostelium* and primitive algae still have mitochondrial ftsZ homologs, along with a Dynamin homolog for outer membrane fission (Takahara et al. 1999; Gilson et al. 2003; Kiefel et al. 2004). The presence of ftsZ in the mitochondria of certain algae, but not in the mitochondria of higher plants or animals, suggests that ftsZ was independently lost in plants and animals after these kingdoms diverged, but well before the emergence of the most modern phyla. It is tempting to speculate that the loss of inner membrane fission proteins was facilitated by a change in the environment, such as the dramatic increase in oxygen levels that more or less coincides with this time (Canfield et al. 2007), but other environmental factors have not been ruled out.

CONCLUDING REMARKS

The main components of the mitochondrial fission and fusion machinery are Dynamin family members. Keeping these machineries fully functional is important for cell survival. The plethora of regulatory mechanisms that control their actions are just now being pieced together. Disruptions in these processes lead to a range of neurodegenerative and other diseases.

REFERENCES

Alexander C, Votruba M, Pesch UE, Thiselton DL, Mayer S, Moore A, Rodriguez M, Kellner U, Leo-Kottler B, Auburger G, et al. 2000. OPA1, encoding a dynamin-related GTPase, is mutated in autosomal dominant optic atrophy linked to chromosome 3q28. *Nat Genet* **26:** 211–215.

Alto NM, Soderling J, Scott JD. 2002. Rab32 is an A-kinase anchoring protein and participates in mitochondrial dynamics. *J Cell Biol* **158:** 659–668.

Arnoult D, Grodet A, Lee YJ, Estaquier J, Blackstone C. 2005. Release of OPA1 during apoptosis participates in the rapid and complete release of cytochrome *c* and subsequent mitochondrial fragmentation. *J Biol Chem* **280:** 35742–35750.

Baricault L, Segui B, Guegand L, Olichon A, Valette A, Larminat F, Lenaers G. 2007. OPA1 cleavage depends on decreased mitochondrial ATP level and bivalent metals. *Exp Cell Res* **313:** 3800–3808.

Bereiter-Hahn J, Voth M. 1994. Dynamics of mitochondria in living cells: Shape changes, dislocations, fusion, and fission of mitochondria. *Microsc Res Tech* **27:** 198–219.

Bleazard W, McCaffery JM, King EJ, Bale S, Mozdy A, Tieu Q, Nunnari J, Shaw JM. 1999. The dynamin-related GTPase Dnm1 regulates mitochondrial fission in yeast. *Nat Cell Biol* **1:** 298–304.

Bossy B, Petrilli A, Klinglmayr E, Chen J, Lutz-Meindl U, Knott AB, Masliah E, Schwarzenbacher R, Bossy-Wetzel E. 2010. *S*-nitrosylation of DRP1 does not affect enzymatic activity and is not specific to Alzheimer's disease. *J Alzheimers Dis* **20:** S513–S526.

Bradbury J. 2004. Mitochondrial fusion protein mutated in CMT2A. *Lancet Neurol* **3:** 326.

Braschi E, Zunino R, McBride HM. 2009. MAPL is a new mitochondrial SUMO E3 ligase that regulates mitochondrial fission. *EMBO Rep* **10:** 748–754.

Bui M, Gilady SY, Fitzsimmons RE, Benson MD, Lynes EM, Gesson K, Alto NM, Strack S, Scott JD, Simmen T. 2010. Rab32 modulates apoptosis onset and mitochondria-associated membrane (MAM) properties. *J Biol Chem* **285:** 31590–31602.

Burmann F, Ebert N, van Baarle S, Bramkamp M. 2011. A bacterial dynamin-like protein mediating nucleotide-independent membrane fusion. *Mol Microbiol* **79:** 1294–1304.

Canfield DE, Poulton SW, Narbonne GM. 2007. Late-Neoproterozoic deep-ocean oxygenation and the rise of animal life. *Science* **315:** 92–95.

Cassidy-Stone A, Chipuk JE, Ingerman E, Song C, Yoo C, Kuwana T, Kurth MJ, Shaw JT, Hinshaw JE, Green DR, et al. 2008. Chemical inhibition of the mitochondrial division dynamin reveals its role in Bax/Bak-dependent mitochondrial outer membrane permeabilization. *Dev Cell* **14:** 193–204.

Cereghetti GM, Stangherlin A, Martins de Brito O, Chang CR, Blackstone C, Bernardi P, Scorrano L. 2008. Dephosphorylation by calcineurin regulates translocation of Drp1 to mitochondria. *Proc Natl Acad Sci* **105:** 15803–15808.

Cerveny KL, Jensen RE. 2003. The WD-repeats of Net2p interact with Dnm1p and Fis1p to regulate division of mitochondria. *Mol Biol Cell* **14:** 4126–4139.

Chan DC. 2012. Fusion and fission: Interlinked processes critical for mitochondrial health. *Annu Rev Genet* **46:** 265–287.

Chanez AL, Hehl AB, Engstler M, Schneider A. 2006. Ablation of the single dynamin of *T. brucei* blocks mitochondrial fission and endocytosis and leads to a precise cytokinesis arrest. *J Cell Sci* **119:** 2968–2974.

Chang CR, Blackstone C. 2007. Cyclic AMP-dependent protein kinase phosphorylation of Drp1 regulates its GTPase activity and mitochondrial morphology. *J Biol Chem* **282:** 21583–21587.

Chang CR, Blackstone C. 2010. Dynamic regulation of mitochondrial fission through modification of the dynamin-related protein Drp1. *Ann NY Acad Sci* **1201:** 34–39.

Chappie JS, Acharya S, Leonard M, Schmid SL, Dyda F. 2010. G domain dimerization controls dynamin's assembly-stimulated GTPase activity. *Nature* **465:** 435–440.

Chen H, Detmer SA, Ewald AJ, Griffin EE, Fraser SE, Chan DC. 2003. Mitofusins Mfn1 and Mfn2 coordinately regulate mitochondrial fusion and are essential for embryonic development. *J Cell Biol* **160:** 189–200.

Chinta SJ, Andersen JK. 2008. Redox imbalance in Parkinson's disease. *Biochim Biophys Acta* **1780:** 1362–1367.

Cho DH, Nakamura T, Fang J, Cieplak P, Godzik A, Gu Z, Lipton SA. 2009. S-nitrosylation of Drp1 mediates β-amyloid-related mitochondrial fission and neuronal injury. *Science* **324:** 102–105.

Choi SY, Huang P, Jenkins GM, Chan DC, Schiller J, Frohman MA. 2006. A common lipid links Mfn-mediated mitochondrial fusion and SNARE-regulated exocytosis. *Nat Cell Biol* **8:** 1255–1262.

Chou CH, Lin CC, Yang MC, Wei CC, Liao HD, Lin RC, Tu WY, Kao TC, Hsu CM, Cheng JT, et al. 2012. GSK3β-mediated Drp1 phosphorylation induced elongated mitochondrial morphology against oxidative stress. *PloS ONE* **7:** e49112.

Cipolat S, Rudka T, Hartmann D, Costa V, Serneels L, Craessaerts K, Metzger K, Frezza C, Annaert W, D'Adamio L, et al. 2006. Mitochondrial rhomboid PARL regulates cytochrome *c* release during apoptosis via OPA1-dependent cristae remodeling. *Cell* **126:** 163–175.

Cribbs JT, Strack S. 2007. Reversible phosphorylation of Drp1 by cyclic AMP-dependent protein kinase and calcineurin regulates mitochondrial fission and cell death. *EMBO Rep* **8:** 939–944.

Delettre C, Lenaers G, Griffoin JM, Gigarel N, Lorenzo C, Belenguer P, Pelloquin L, Grosgeorge J, Turc-Carel C, Perret E, et al. 2000. Nuclear gene OPA1, encoding a mitochondrial dynamin-related protein, is mutated in dominant optic atrophy. *Nat Genet* **26:** 207–210.

Dempwolff F, Wischhusen HM, Specht M, Graumann PL. 2012. The deletion of bacterial dynamin and flotillin genes results in pleiotrophic effects on cell division, cell growth and in cell shape maintenance. *BMC Microbiol* **12:** 298.

Devay RM, Dominguez-Ramirez L, Lackner LL, Hoppins S, Stahlberg H, Nunnari J. 2009. Coassembly of Mgm1 isoforms requires cardiolipin and mediates mitochondrial inner membrane fusion. *J Cell Biol* **186** 793–803.

De Vos KJ, Allan VJ, Grierson AJ, Sheetz MP. 2005. Mitochondrial function and actin regulate dynamin-related protein 1-dependent mitochondrial fission. *Curr Biol* **15:** 678–683.

Dorn GW 2nd, Clark CF, Eschenbacher WH, Kang MY, Engelhard JT, Warner SJ, Matkovich SJ, Jowdy CC. 2011. MARF and Opa1 control mitochondrial and cardiac function in *Drosophila*. *Circ Res* **108:** 12–17.

Durr M, Escobar-Henriques M, Merz S, Geimer S, Langer T, Westermann B. 2006. Nonredundant roles of mitochondria-associated F-box proteins Mfb1 and Mdm30 in maintenance of mitochondrial morphology in yeast. *Mol Biol Cell* **17:** 3745–3755.

Ehses S, Raschke I, Mancuso G, Bernacchia A, Geimer S, Tondera D, Martinou JC, Westermann B, Rugarli EI, Langer T. 2009. Regulation of OPA1 processing and mitochondrial fusion by *m*-AAA protease isoenzymes and OMA1. *J Cell Biol* **187:** 1023–1036.

Esser K, Tursun B, Ingenhoven M, Michaelis G, Pratje E. 2002. A novel two-step mechanism for removal of a mitochondrial signal sequence involves the mAAA complex and the putative rhomboid protease Pcp1. *J Mol Biol* **323:** 835–843.

Eura Y, Ishihara N, Oka T, Mihara K. 2006. Identification of a novel protein that regulates mitochondrial fusion by modulating mitofusin (Mfn) protein function. *J Cell Sci* **119:** 4913–4925.

Frank S, Gaume B, Bergmann-Leitner ES, Leitner WW, Robert EG, Catez F, Smith CL, Youle RJ. 2001. The role of dynamin-related protein 1, a mediator of mitochondrial fission, in apoptosis. *Dev Cell* **1:** 515–525.

Frank M, Duvezin-Caubet S, Koob S, Occhipinti A, Jagasia R, Petcherski A, Ruonala MO, Priault M, Salin B, Reichert AS. 2012. Mitophagy is triggered by mild oxidative stress in a mitochondrial fission dependent manner. *Biochim Biophys Acta* **1823:** 2297–2310.

Frezza C, Cipolat S, Martins de Brito O, Micaroni M, Beznoussenko GV, Rudka T, Bartoli D, Polishuck RS, Danial NN, De Strooper B, et al. 2006. OPA1 controls apoptotic Cristae remodeling independently from mitochondrial fusion. *Cell* **126:** 177–189.

Friedman JR, Lackner LL, West M, DiBenedetto JR, Nunnari J, Voeltz GK. 2011. ER tubules mark sites of mitochondrial division. *Science* **334:** 358–362.

Fritz S, Weinbach N, Westermann B. 2003. Mdm30 is an F-box protein required for maintenance of fusion-competent mitochondria in yeast. *Mol Biol Cell* **14:** 2303–2313.

Gandre-Babbe S, van der Bliek AM. 2008. The novel tail-anchored membrane protein Mff controls mitochondrial and peroxisomal fission in mammalian cells. *Mol Biol Cell* **19:** 2402–2412.

Gao S, von der Malsburg A, Paeschke S, Behlke J, Haller O, Kochs G, Daumke O. 2010. Structural basis of oligomerization in the stalk region of dynamin-like MxA. *Nature* **465:** 502–506.

Gilson PR, Yu XC, Hereld D, Barth C, Savage A, Kiefel BR, Lay S, Fisher PR, Margolin W, Beech PL. 2003. Two Dictyostelium orthologs of the prokaryotic cell division protein FtsZ localize to mitochondria and are required for the maintenance of normal mitochondrial morphology. *Eukaryot Cell* **2:** 1315–1326.

Glater EE, Megeath LJ, Stowers RS, Schwarz TL. 2006. Axonal transport of mitochondria requires milton to recruit kinesin heavy chain and is light chain independent. *J Cell Biol* **173:** 545–557.

Gomes LC, Di Benedetto G, Scorrano L. 2011. During autophagy mitochondria elongate, are spared from degradation and sustain cell viability. *Nat Cell Biol* **13:** 589–598.

Griffin EE, Graumann J, Chan DC. 2005. The WD40 protein Caf4p is a component of the mitochondrial fission machinery and recruits Dnm1p to mitochondria. *J Cell Biol* **170:** 237–248.

Griparic L, van der Wel NN, Orozco IJ, Peters PJ, van der Bliek AM. 2004. Loss of the intermembrane space protein Mgm1/OPA1 induces swelling and localized constrictions along the lengths of mitochondria. *J Biol Chem* **279:** 18792–18798.

Griparic L, Kanazawa T, van der Bliek AM. 2007. Regulation of the mitochondrial dynamin-like protein Opa1 by proteolytic cleavage. *J Cell Biol* **178:** 757–764.

Guan K, Farh L, Marshall TK, Deschenes RJ. 1993. Normal mitochondrial structure and genome maintenance in yeast requires the dynamin-like product of the MGM1 gene. *Curr Genet* **24:** 141–148.

Hales KG, Fuller MT. 1997. Developmentally regulated mitochondrial fusion mediated by a conserved, novel, predicted GTPase. *Cell* **90:** 121–129.

Han XJ, Lu YF, Li SA, Kaitsuka T, Sato Y, Tomizawa K, Nairn AC, Takei K, Matsui H, Matsushita M. 2008. CaM kinase I α-induced phosphorylation of Drp1 regulates mitochondrial morphology. *J Cell Biol* **182:** 573–585.

Head B, Griparic L, Amiri M, Gandre-Babbe S, van der Bliek AM. 2009. Inducible proteolytic inactivation of OPA1 mediated by the OMA1 protease in mammalian cells. *J Cell Biol* **187:** 959–966.

Herlan M, Vogel F, Bornhövd C, Neupert W, Reichert AS. 2003. Processing of Mgm1 by the rhomboid-type protease Pcp1 is required for maintenance of mitochondrial morphology and of mitochondrial DNA. *J Biol Chem* **278:** 27781–27788.

Herlan M, Bornhovd C, Hell K, Neupert W, Reichert AS. 2004. Alternative topogenesis of Mgm1 and mitochondrial morphology depend on ATP and a functional import motor. *J Cell Biol* **165:** 167–173.

Hermann GJ, Thatcher JW, Mills JP, Hales KG, Fuller MT, Nunnari J, Shaw JM. 1998. Mitochondrial fusion in yeast requires the *trans*-membrane GTPase Fzo1p. *J Cell Biol* **143:** 359–373.

Hoppins S, Lackner L, Nunnari J. 2007. The machines that divide and fuse mitochondria. *Annu Rev Biochem* **76:** 751–780.

Huang H, Gao Q, Peng X, Choi SY, Sarma K, Ren H, Morris AJ, Frohman MA. 2011. piRNA-associated germline nuage formation and spermatogenesis require MitoPLD profusogenic mitochondrial-surface lipid signaling. *Dev Cell* **20:** 376–387.

Imoto M, Tachibana I, Urrutia R. 1998. Identification and functional characterization of a novel human protein highly related to the yeast dynamin-like GTPase vps1p. *J Cell Sci* **111:** 1341–1349.

Ingerman E, Perkins EM, Marino M, Mears JA, McCaffery JM, Hinshaw JE, Nunnari J. 2005. Dnm1 forms spirals that are structurally tailored to fit mitochondria. *J Cell Biol* **170:** 1021–1027.

Ishihara N, Jofuku A, Eura Y, Mihara K. 2003. Regulation of mitochondrial morphology by membrane potential, and DRP1-dependent division and FZO1-dependent fusion reaction in mammalian cells. *Biochem Biophys Res Commun* **301:** 891–898.

Iwasawa R, Mahul-Mellier AL, Datler C, Pazarentzos E, Grimm S. 2011. Fis1 and Bap31 bridge the mitochondria–ER interface to establish a platform for apoptosis induction. *EMBO J* **30:** 556–568.

Johnson LV, Walsh ML, Bockus BJ, Chen LB. 1981. Monitoring of relative mitochondrial membrane potential in living cells by fluorescence microscopy. *J Cell Biol* **88:** 526–535.

Jones BA, Fangman WL. 1992. Mitochondrial DNA maintenance in yeast requires a protein containing a region related to the GTP-binding domain of dynamin. *Genes Dev* **6:** 380–389.

Kamimoto T, Nagai Y, Onogi H, Muro Y, Wakabayashi T, Hagiwara M. 1998. Dymple, a novel dynamin-like high molecular weight GTPase lacking a proline-rich carboxyl-terminal domain in mammalian cells. *J Biol Chem* **273:** 1044–1051.

Kanazawa T, Zappaterra MD, Hasegawa A, Wright AP, Newman-Smith ED, Buttle KF, McDonald K, Mannella CA, van der Bliek AM. 2008. The *C. elegans* Opa1 homologue EAT-3 is essential for resistance to free radicals. *PLoS Genet* **4:** e1000022.

Karbowski M, Norris KL, Cleland MM, Jeong SY, Youle RJ. 2006. Role of Bax and Bak in mitochondrial morphogenesis. *Nature* **443:** 658–662.

Karbowski M, Neutzner A, Youle RJ. 2007. The mitochondrial E3 ubiquitin ligase MARCH5 is required for Drp1 dependent mitochondrial division. *J Cell Biol* **178:** 71–84.

Kiefel BR, Gilson PR, Beech PL. 2004. Diverse eukaryotes have retained mitochondrial homologs of the bacterial division protein FtsZ. *Protist* **155:** 105–115.

Koirala S, Bui HT, Schubert HL, Eckert DM, Hill CP, Kay MS, Shaw JM. 2010. Molecular architecture of a dynamin adaptor: Implications for assembly of mitochondrial fission complexes. *J Cell Biol* **191:** 1127–1139.

Kornmann B, Osman C, Walter P. 2011. The conserved GTPase Gem1 regulates endoplasmic reticulum-mitochondria connections. *Proc Natl Acad Sci* **108:** 14151–14156.

Korobova F, Ramabhadran V, Higgs HN. 2013. An actin-dependent step in mitochondrial fission mediated by the ER-associated formin INF2. *Science* **339:** 464–467.

Koshiba T, Detmer SA, Kaiser JT, Chen H, McCaffery JM, Chan DC. 2004. Structural basis of mitochondrial tethering by mitofusin complexes. *Science* **305:** 858–862.

Labrousse AM, Zapaterra M, Rube DA, van der Bliek AM. 1999. *C. elegans* dynamin-related protein *drp-1* controls severing of the mitochondrial outer membrane. *Mol Cell* **4:** 815–826.

Leboucher GP, Tsai YC, Yang M, Shaw KC, Zhou M, Veenstra TD, Glickman MH, Weissman AM. 2012. Stress-induced phosphorylation and proteasomal degradation of mitofusin 2 facilitates mitochondrial fragmentation and apoptosis. *Mol Cell* **47:** 547–557.

Lewis MR, Lewis WR. 1914. Mitochondria (and other cytoplasmic structures) in tissue cultures. *Am J Anat* **17:** 339–401.

Liesa M, Palacin M, Zorzano A. 2009. Mitochondrial dynamics in mammalian health and disease. *Physiol Rev* **89:** 799–845.

Lokireddy S, Wijesoma IW, Teng S, Bonala S, Gluckman PD, McFarlane C, Sharma M, Kambadur R. 2012. The ubiquitin ligase Mul1 induces mitophagy in skeletal muscle in response to muscle-wasting stimuli. *Cell Metab* **16:** 613–624.

Loson OC, Song Z, Chen H, Chan DC. 2013. Fis1, Mff, MiD49 and MiD51 mediate Drp1 recruitment in mitochondrial fission. *Mol Biol Cell* **24:** 659–667.

Low HH, Sachse C, Amos LA, Lowe J. 2009. Structure of a bacterial dynamin-like protein lipid tube provides a mechanism for assembly and membrane curving. *Cell* **139:** 1342–1352.

Macia E, Ehrlich M, Massol R, Boucrot E, Brunner C, Kirchhausen T. 2006. Dynasore, a cell-permeable inhibitor of dynamin. *Dev Cell* **10:** 839–850.

McQuibban GA, Lee JR, Zheng L, Juusola M, Freeman M. 2006. Normal mitochondrial dynamics requires rhomboid-7 and affects *Drosophila* lifespan and neuronal function. *Curr Biol* **16:** 982–989.

Mears JA, Lackner LL, Fang S, Ingerman E, Nunnari J, Hinshaw JE. 2011. Conformational changes in Dnm1 support a contractile mechanism for mitochondrial fission. *Nat Struct Mol Biol* **18:** 20–26.

Meeusen S, DeVay R, Block J, Cassidy-Stone A, Wayson S, McCaffery JM, Nunnari J. 2006. Mitochondrial inner-membrane fusion and crista maintenance requires the dynamin-related GTPase Mgm1. *Cell* **127:** 383–395.

Meglei G, McQuibban GA. 2009. The dynamin-related protein Mgm1p assembles into oligomers and hydrolyzes GTP to function in mitochondrial membrane fusion. *Biochemistry* **48:** 1774–1784.

Merkwirth C, Dargazanli S, Tatsuta T, Geimer S, Lower B, Wunderlich FT, von Kleist-Retzow JC, Waisman A, Westermann B, Langer T. 2008. Prohibitins control cell proliferation and apoptosis by regulating OPA1-dependent cristae morphogenesis in mitochondria. *Genes Dev* **22:** 476–488.

Merrill RA, Dagda RK, Dickey AS, Cribbs JT, Green SH, Usachev YM, Strack S. 2011. Mechanism of neuroprotective mitochondrial remodeling by PKA/AKAP1. *PLoS Biol* **9:** e1000612.

Montessuit S, Somasekharan SP, Terrones O, Lucken-Ardjomande S, Herzig S, Schwarzenbacher R, Manstein DJ, Bossy-Wetzel E, Basanez G, Meda P, et al. 2010. Membrane remodeling induced by the dynamin-related protein Drp1 stimulates Bax oligomerization. *Cell* **142:** 889–901.

Mozdy AD, McCaffery JM, Shaw JM. 2000. Dnm1p GTPase-mediated mitochondrial fission is a multi-step process requiring the novel integral membrane component Fis1p. *J Cell Biol* **151:** 367–380.

Nakamura N, Kimura Y, Tokuda M, Honda S, Hirose S. 2006. MARCH-V is a novel mitofusin 2- and Drp1-binding protein able to change mitochondrial morphology. *EMBO Rep* **7:** 1019–1022.

Narendra DP, Jin SM, Tanaka A, Suen DF, Gautier CA, Shen J, Cookson MR, Youle RJ. 2010. PINK1 is selectively stabilized on impaired mitochondria to activate Parkin. *PLoS Biol* **8:** e1000298.

Neutzner A, Youle RJ. 2005. Instability of the mitofusin Fzo1 regulates mitochondrial morphology during the mating response of the yeast *Saccharomyces cerevisiae*. *J Biol Chem* **280:** 18598–18603.

Niemann A, Ruegg M, La Padula V, Schenone A, Suter U. 2005. Ganglioside-induced differentiation associated protein 1 is a regulator of the mitochondrial network: New implications for Charcot–Marie–Tooth disease. *J Cell Biol* **170:** 1067–1078.

Nunnari J, Suomalainen A. 2012. Mitochondria: In sickness and in health. *Cell* **148:** 1145–1159.

Olichon A, Emorine LJ, Descoins E, Pelloquin L, Brichese L, Gas N, Guillou E, Delettre C, Valette A, Hamel CP, et al. 2002. The human dynamin-related protein OPA1 is anchored to the mitochondrial inner membrane facing the inter-membrane space. *FEBS Lett* **523:** 171–176.

Olichon A, Baricault L, Gas N, Guillou E, Valette A, Belenguer P, Lenaers G. 2003. Loss of OPA1 perturbates the mitochondrial inner membrane structure and integrity, leading to cytochrome *c* release and apoptosis. *J Biol Chem* **278:** 7743–7746.

Olichon A, Elachouri G, Baricault L, Delettre C, Belenguer P, Lenaers G. 2007. OPA1 alternate splicing uncouples an evolutionary conserved function in mitochondrial fusion from a vertebrate restricted function in apoptosis. *Cell Death Differ* **14:** 682–692.

Otera H, Wang C, Cleland MM, Setoguchi K, Yokota S, Youle RJ, Mihara K. 2010. Mff is an essential factor for mitochondrial recruitment of Drp1 during mitochondrial fission in mammalian cells. *J Cell Biol* **191:** 1141–1158.

Ott M, Zhivotovsky B, Orrenius S. 2007. Role of cardiolipin in cytochrome *c* release from mitochondria. *Cell Death Differ* **14:** 1243–1247.

Palmer CS, Osellame LD, Laine D, Koutsopoulos OS, Frazier AE, Ryan MT. 2011. MiD49 and MiD51, new components of the mitochondrial fission machinery. *EMBO Rep* **12:** 565–573.

Praefcke GJ, McMahon HT. 2004. The dynamin superfamily: Universal membrane tubulation and fission molecules? *Nat Rev Mol Cell Biol* **5:** 133–147.

Qi X, Disatnik MH, Shen N, Sobel RA, Mochly-Rosen D. 2011. Aberrant mitochondrial fission in neurons induced by protein kinase Cδ under oxidative stress conditions in vivo. *Mol Biol Cell* **22:** 256–265.

Rabl R, Soubannier V, Scholz R, Vogel F, Mendl N, Vasiljev-Neumeyer A, Korner C, Jagasia R, Keil T, Baumeister W, et al. 2009. Formation of cristae and crista junctions in mitochondria depends on antagonism between Fcj1 and Su e/g. *J Cell Biol* **185:** 1047–1063.

Rambold AS, Kostelecky B, Elia N, Lippincott-Schwartz J. 2011. Tubular network formation protects mitochondria from autophagosomal degradation during nutrient starvation. *Proc Natl Acad Sci* **108:** 10190–10195.

Rapaport D, Brunner M, Neupert W, Westermann B. 1998. Fzo1p is a mitochondrial outer membrane protein essential for the biogenesis of functional mitochondria in Saccharomyces cerevisiae. *J Biol Chem* **273:** 20150–20155.

Rizzuto R, Brini M, De Giorgi F, Rossi R, Heim R, Tsien RY, Pozzan T. 1996. Double labelling of subcellular structures with organelle-targeted GFP mutants in vivo. *Curr Biol* **6:** 183–188.

Rowland AA, Voeltz GK. 2012. Endoplasmic reticulum-mitochondria contacts: Function of the junction. *Nat Rev Mol Cell Biol* **13:** 607–625.

Santel A, Fuller MT. 2001. Control of mitochondrial morphology by a human mitofusin. *J Cell Sci* **114:** 867–874.

Schafer E, Seelert H, Reifschneider NH, Krause F, Dencher NA, Vonck J. 2006. Architecture of active mammalian respiratory chain supercomplexes. *J Biol Chem* **281:** 15370–15375.

Scorrano L, Oakes SA, Opferman JT, Cheng EH, Sorcinelli MD, Pozzan T, Korsmeyer SJ. 2003. BAX and BAK regulation of endoplasmic reticulum Ca^{2+}: A control point for apoptosis. *Science* **300:** 135–139.

Sesaki H, Jensen RE. 2004. Ugo1p links the Fzo1p and Mgm1p GTPases for mitochondrial fusion. *J Biol Chem* **279:** 28298–28303.

Sesaki H, Southard SM, Hobbs AE, Jensen RE. 2003. Cells lacking Pcp1p/Ugo2p, a rhomboid-like protease required for Mgm1p processing, lose mtDNA and mitochondrial structure in a Dnm1p-dependent manner, but remain competent for mitochondrial fusion. *Biochem Biophys Res Commun* **308:** 276–283.

Shin HW, Shinotsuka C, Torii S, Murakami K, Nakayama K. 1997. Identification and subcellular localization of a novel mammalian dynamin-related protein homologous to yeast Vps1p and Dnm1p. *J Biochem* **122:** 525–530.

Simmen T, Aslan JE, Blagoveshchenskaya AD, Thomas L, Wan L, Xiang Y, Feliciangeli SF, Hung CH, Crump CM, Thomas G. 2005. PACS-2 controls endoplasmic reticulum-mitochondria communication and Bid-mediated apoptosis. *EMBO J* **24:** 717–729.

Smirnova E, Shurland DL, Ryazantsev SN, van der Bliek AM. 1998. A human dynamin-related protein controls the distribution of mitochondria. *J Cell Biol* **143:** 351–358.

Smirnova E, Griparic L, Shurland DL, van der Bliek AM. 2001. Dynamin-related protein Drp1 is required for mitochondrial division in mammalian cells. *Mol Biol Cell* **12:** 2245–2256.

Song Z, Chen H, Fiket M, Alexander C, Chan DC. 2007. OPA1 processing controls mitochondrial fusion and is regulated by mRNA splicing, membrane potential, and Yme1L. *J Cell Biol* **178:** 749–755.

Taguchi N, Ishihara N, Jofuku A, Oka T, Mihara K. 2007. Mitotic phosphorylation of dynamin-related GTPase Drp1 participates in mitochondrial fission. *J Biol Chem* **282:** 11521–11529.

Takahara M, Takahashi H, Matsunaga S, Sakai A, Kawano S, Kuroiwa T. 1999. Two types of ftsZ genes isolated from the unicellular primitive red alga *Galdieria sulphuraria*. *Plant Cell Physiol* **40:** 784–791.

Tanaka A, Youle RJ. 2008. A chemical inhibitor of DRP1 uncouples mitochondrial fission and apoptosis. *Mol Cell* **29:** 409–410.

Tanaka A, Cleland MM, Xu S, Narendra DP, Suen DF, Karbowski M, Youle RJ. 2010. Proteasome and p97 mediate mitophagy and degradation of mitofusins induced by Parkin. *J Cell Biol* **191:** 1367–1380.

Tieu Q, Okreglak V, Naylor K, Nunnari J. 2002. The WD repeat protein, Mdv1p, functions as a molecular adaptor by interacting with Dnm1p and Fis1p during mitochondrial fission. *J Cell Biol* **158:** 445–452.

Tondera D, Grandemange S, Jourdain A, Karbowski M, Mattenberger Y, Herzig S, Da Cruz S, Clerc P, Raschke I, Merkwirth C, et al. 2009. SLP-2 is required for stress-induced mitochondrial hyperfusion. *EMBO J* **28:** 1589–1600.

Twig G, Graf SA, Wikstrom JD, Mohamed H, Haigh SE, Elorza A, Deutsch M, Zurgil N, Reynolds N, Shirihai OS. 2006. Tagging and tracking individual networks within a complex mitochondrial web with photoactivatable GFP. *Am J Physiol Cell Physiol* **291:** C176–C184.

Twig G, Elorza A, Molina AJ, Mohamed H, Wikstrom JD, Walzer G, Stiles L, Haigh SE, Katz S, Las G, et al. 2008. Fission and selective fusion govern mitochondrial segregation and elimination by autophagy. *EMBO J* **27:** 433–446.

van der Bliek AM, Payne GS. 2010. Dynamin subunit interactions revealed. *Dev Cell* **18:** 687–688.

Wang W, Wang Y, Long J, Wang J, Haudek SB, Overbeek P, Chang BH, Schumacker PT, Danesh FR. 2012. Mitochondrial fission triggered by hyperglycemia is mediated by ROCK1 activation in podocytes and endothelial cells. *Cell Metab* **15:** 186–200.

Westermann B. 2010. Mitochondrial fusion and fission in cell life and death. *Nat Rev Mol Cell Biol* **11:** 872–884.

White KE, Davies VJ, Hogan VE, Piechota MJ, Nichols PP, Turnbull DM, Votruba M. 2009. OPA1 deficiency associated with increased autophagy in retinal ganglion cells in a murine model of dominant optic atrophy. *Invest Ophthalmol Vis Sci* **50:** 2567–2571.

Wienke DC, Knetsch MLW, Neuhaus EM, Reedy MC, Manstein DJ. 1999. Disruption of a dynamin homologue affects endocytosis, organelle morphology, and cytokinesis in dictyostelium discoideum. *Mol Biol Cell* **10:** 225–243.

Wong ED, Wagner JA, Gorsich SW, McCaffery JM, Shaw JM, Nunnari J. 2000. The dynamin-related GTPase, Mgm1p, is an intermembrane space protein required for maintenance of fusion competent mitochondria. *J Cell Biol* **151:** 341–352.

Yarosh W, Monserrate J, Tong JJ, Tse S, Le PK, Nguyen K, Brachmann CB, Wallace DC, Huang T. 2008. The molecular mechanisms of OPA1-mediated optic atrophy in Drosophila model and prospects for antioxidant treatment. *PLoS Genet* **4:** e6.

Yoon Y, Pitts KR, Dahan S, McNiven MA. 1998. A novel dynamin-like protein associates with cytoplasmic vesicles and tubules of the endoplasmic reticulum in mammalian cells. *J Cell Biol* **140:** 779–793.

Youle RJ, Narendra DP. 2011. Mechanisms of mitophagy. *Nat Rev Mol Cell Biol* **12:** 9–14.

Youle RJ, van der Bliek AM. 2012. Mitochondrial fission, fusion, and stress. *Science* **337:** 1062–1065.

Zhao J, Liu T, Jin S, Wang X, Qu M, Uhlen P, Tomilin N, Shupliakov O, Lendahl U, Nister M. 2011. Human MIEF1 recruits Drp1 to mitochondrial outer membranes and promotes mitochondrial fusion rather than fission. *EMBO J* **30:** 2762–2778.

Mitochondrial Quality Control Mediated by PINK1 and Parkin: Links to Parkinsonism

Derek Narendra[1,2], John E. Walker[2], and Richard Youle[1]

[1]Biochemistry Section, Surgical Neurology Branch, National Institute of Neurological Disorders and Stroke, National Institutes of Health, Bethesda, Maryland 20892

[2]The Medical Research Council Mitochondrial Biology Unit, Hills Road, Cambridge CB0 2XY, United Kingdom

Correspondence: youler@ninds.nih.gov

Mutations in Parkin or PINK1 are the most common cause of recessive familial parkinsonism. Recent studies suggest that PINK1 and Parkin form a mitochondria quality control pathway that identifies dysfunctional mitochondria, isolates them from the mitochondrial network, and promotes their degradation by autophagy. In this pathway the mitochondrial kinase PINK1 senses mitochondrial fidelity and recruits Parkin selectively to mitochondria that lose membrane potential. Parkin, an E3 ligase, subsequently ubiquitinates outer mitochondrial membrane proteins, notably the mitofusins and Miro, and induces autophagic elimination of the impaired organelles. Here we review the recent rapid progress in understanding the molecular mechanisms of PINK1- and Parkin-mediated mitophagy and the identification of Parkin substrates suggesting how mitochondrial fission and trafficking are involved. We also discuss how defects in mitophagy may be linked to Parkinson's disease.

Parkinson's disease (PD) is the second most common neurodegenerative disorder and is characterized by cardinal motor symptoms: slowness of movement, rigidity, rest tremor, and postural instability (Ropper et al. 2009). Although these symptoms initially respond to drugs that modulate dopamine metabolism or surgeries that alter basal ganglia circuitry, the disease eventually progresses. With a modest exception (Olanow et al. 2009), no therapy has been shown to alter the disease course.

The pathogenesis of sporadic Parkinson's disease is likely complex involving altered metabolism of the protein α-synuclein, lysosomal dysfunction, and a dysregulated inflammatory response (reviewed in Shulman et al. 2011). Several lines of evidence also point to mitochondrial dysfunction as a central player in the pathogenesis of PD. Complex I dysfunction is associated with sporadic PD and is sufficient to induce parkinsonism (reviewed in Schapira 2008). The inhibitors of complex I, MPTP (Langston et al. 1983) and rotenone (Betarbet et al. 2000), replicate the symptoms of PD, and rotenone recapitulates key pathognomonic features of PD, such as the α-synuclein-rich inclusion bodies (Betarbet et al. 2000). The cause of mitochondrial dysfunction in sporadic PD is not entirely clear, but laser capture microdissection of substantia nigra neurons from patients with PD reveal a higher burden of mitochondrial DNA deletions relative to age-matched

Cite this article as *Cold Spring Harb Perspect Biol* doi: 10.1101/cshperspect.a011338

controls (Bender et al. 2006). That such deletions are sufficient to cause parkinsonism is suggested by the occurrence of parkinsonism in patients with rare mutations in their mtDNA replication machinery (e.g., the catalytic subunit of the mtDNA polymerase POLG [Luoma et al. 2004] or the mtDNA helicase Twinkle [Baloh et al. 2007]). The defective mtDNA replicative machinery generates high levels of mtDNA deletions throughout the body that are qualitatively similar to those observed in the substantia nigra in patients with sporadic PD (Reeve et al. 2008). Thus, mitochondrial dysfunction is both associated with sporadic PD and sufficient to cause the parkinsonian syndrome.

As is discussed in this review, recent insights from certain genetic forms of PD—resulting from mutations in Parkin or PINK1—support the model that mitochondrial damage is a central driver of PD pathogenesis. Additionally, they provide a rationale for targeting mitochondrial quality control pathways in patients with PD.

GENETIC STUDIES SUPPORT A COMMON PINK1/PARKIN PATHWAY

Mutations in Parkin or PINK1 are the leading cause of recessive parkinsonism (Hardy 2010). The parkinsonian syndrome caused by loss of Parkin (Kitada et al. 1998) or PINK1 (Valente et al. 2004) function is clinically indistinguishable from idiopathic Parkinson's disease at its outset, although it tends to present before the age of 40 and to progress more slowly. Additionally, only some patients with Parkin mutations develop the α-synuclein-rich inclusions, known as Lewy bodies and Lewy neurites, which are pathognomonic for idiopathic Parkinson's disease (Pramstaller et al. 2005).

The clinical similarity between patients with Parkin or PINK1 mutations suggests that they might function in a common pathway. Supporting this notion, deletion of either dParkin or dPink1 in *Drosophila melanogaster* leads to a similar (and relatively unique) phenotype with the triad of muscle degeneration, disrupted spermatogenesis, and loss of dopaminergic neurons within a specific cluster (Greene et al. 2003; Whitworth et al. 2005; Clark et al. 2006; Park et al. 2006; Yang et al. 2006). Moreover, overexpression of dParkin in the dPink1 null *Drosophila* and not the converse partially compensates for this phenotype, indicating that dPink1 is upstream of dParkin in the same pathway (Clark et al. 2006; Park et al. 2006; Yang et al. 2006).

The function of the dPink1/dParkin pathway at least in *Drosophila* is suggested by ultrastructural examination of the tissues before their degeneration. Mitochondria in the flight muscles become swollen and fragmented early in the course of the degeneration, suggesting that a mitochondrial defect may be the primary event (Greene et al. 2003; Whitworth et al. 2005; Clark et al. 2006; Park et al. 2006; Yang et al. 2006). Overexpression of components of the glutathione-dependent detoxification pathway partially prevents the mitochondrial damage, implying that mitochondria in *Drosophila* lacking the dPink1/dParkin pathway are less able to cope with oxidative stress (Whitworth et al. 2005). The apparent damage to mitochondria in the flight muscles correlates with a severe bioenergetic deficit as well as increased oxidative damage of the surrounding tissue, suggesting a mechanism for the subsequent apoptosis of the muscle fibers (Greene et al. 2003; Whitworth et al. 2005; Clark et al. 2006; Park et al. 2006; Yang et al. 2006).

Studies in mice are largely consistent with the notion of a common Pink1/Parkin pathway important for maintaining mitochondrial integrity. Loss of Parkin or Pink1 causes one phenotype similar to, albeit milder than, that in man, in which nigrostriatal dysfunction correlates with mitochondrial dysfunction in the striatum (Palacino et al. 2004; Gautier et al. 2008; Kitada et al. 2009). Interestingly, PINK1 knock out (KO) mice also display pronounced cardiac hypertrophy (Billia et al. 2011) and Parkin KO mice display less subcutaneous fat, are resistant to high-fat-diet-induced weight gain (Kim et al. 2011a), and may be predisposed to hepatocellular carcinomas (Fujiwara et al. 2008). It remains to be determined if these non-neuronal phenotypes are shared between PINK1 and Parkin deficient mice.

Although Parkin and PINK1 may have independent functions in some cellular pathways,

 Cite this article as *Cold Spring Harb Perspect Biol* doi: 10.1101/cshperspect.a011338

parsimony would suggest that they function together in pathways relevant to PD pathogenesis. Additionally, as is discussed in more depth below, the difference in phenotype between insects and mammals suggests that, although features (including substrates) of PINK1/Parkin pathway are shared among insects and mammals, likely there are differences in the physiologic functions they serve.

PINK1 SENSES MITOCHONDRIAL STRESS

PINK1 is a serine/threonine protein kinase targeted to mitochondria (Valente et al. 2004; Silvestri et al. 2005; Zhou et al. 2008). Under basal conditions, mammalian Parkin is an E3 ubiquitin ligase residing (likely inert) in the cytosol. Upon mitochondrial damage, however, Parkin is activated and recruited selectively to bioenergetically impaired mitochondria (Narendra et al. 2008) by an elegant mechanism in which PINK1 functions as a sort of molecular scout reporting the internal state of individual mitochondria to Parkin in the cytosol (Geisler et al. 2010b; Matsuda et al. 2010; Narendra et al. 2010b; Vives-Bauza et al. 2010) (Fig. 1).

Like most other mitochondrial proteins, PINK1 is encoded in the nuclear genome and synthesized in the cytosol before its import via the TOM/TIM complexes (Silvestri et al. 2005). And like most other mitochondrial proteins expressing a mitochondrial targeting presequence, PINK1 requires the electrical component of the inner mitochondrial membrane potential ($\Delta\Psi$) for its import into mitochondria (Silvestri et al. 2005). Two features of PINK1 processing and targeting, however, make PINK1 uniquely suited for its role in signaling mitochondrial dysfunction. First, unlike most other mitochondrial proteins, imported PINK1 is cleaved into an unstable product by PARL (Whitworth et al. 2008; Jin et al. 2010; Deas et al. 2011; Meissner et al. 2011; Shi et al. 2011), an inner mitochondrial membrane protease (McQuibban et al. 2003), and other mitochondrial proteases (Greene et al. 2012) and consequently is present at very low levels in mitochondria that are healthy enough to maintain the $\Delta\Psi$ dependent import pathway (Fig. 1A). (As an indication of its low abundance, PINK1 is usually not detected in proteomics studies of mitochondrial proteins despite its relatively large size [see for instance the MitoCarta database (Pagliarini et al. 2008)].) Second, PINK1 contains a second weaker targeting signal that directs it to the outer mitochondrial membrane (Zhou et al. 2008), if the TOM/TIM import pathway is blocked by collapse of $\Delta\Psi$ (Narendra et al. 2010b). Thus, PINK1 acts as a kind of scout, dispatched centrally to all the mitochondria in the cell to assess their internal state. The healthy mitochondria with a strong inner membrane potential quickly dispose of PINK1, whereas failing mitochondria, unable to import and degrade the kinase, accumulate it on their cysotolic face, in effect displaying their inner dysfunction on their outer surface. The accumulation appears to be very rapid with a several fold increase in PINK1 protein expression within minutes of $\Delta\Psi$ collapse (Narendra et al. 2010b). However, an alternative model has been presented in which PARL-mediated PINK1 cleavage activates PINK1 rather than eliminates it (Whitworth et al. 2008; Deas et al. 2011). Once on the outer mitochondrial membrane, PINK1 constitutively associates with the TOM complex. This association has been suggested to mediate rapid reimport and degradation of PINK1 if mitochondrial membrane potential is restored thereby yielding a quick on/off signal of mitochondrial dysfunction (Lazarou et al. 2012).

PINK1 RECRUITS PARKIN TO MITOCHONDRIA AND ACTIVATES ITS LATENT E3 LIGASE ACTIVITY

PINK1, when exposed on the cytosolic surface, directs Parkin specifically to the ailing mitochondrion resulting in ubiquitination of several outer mitochondrial membrane proteins on that dysfunctional mitochondrion (Geisler et al. 2010b; Matsuda et al. 2010; Narendra et al. 2010b; Vives-Bauza et al. 2010) (Fig. 1B). Precisely how PINK1 activates and directs Parkin selectively to the mitochondria on which PINK1 accumulates is not resolved fully. Although there is evidence that Parkin and PINK1 directly bind each other (Parkin and PINK1 co-immunoprecipitate [Xiong et al. 2009; Sha et al. 2010]). The bulk

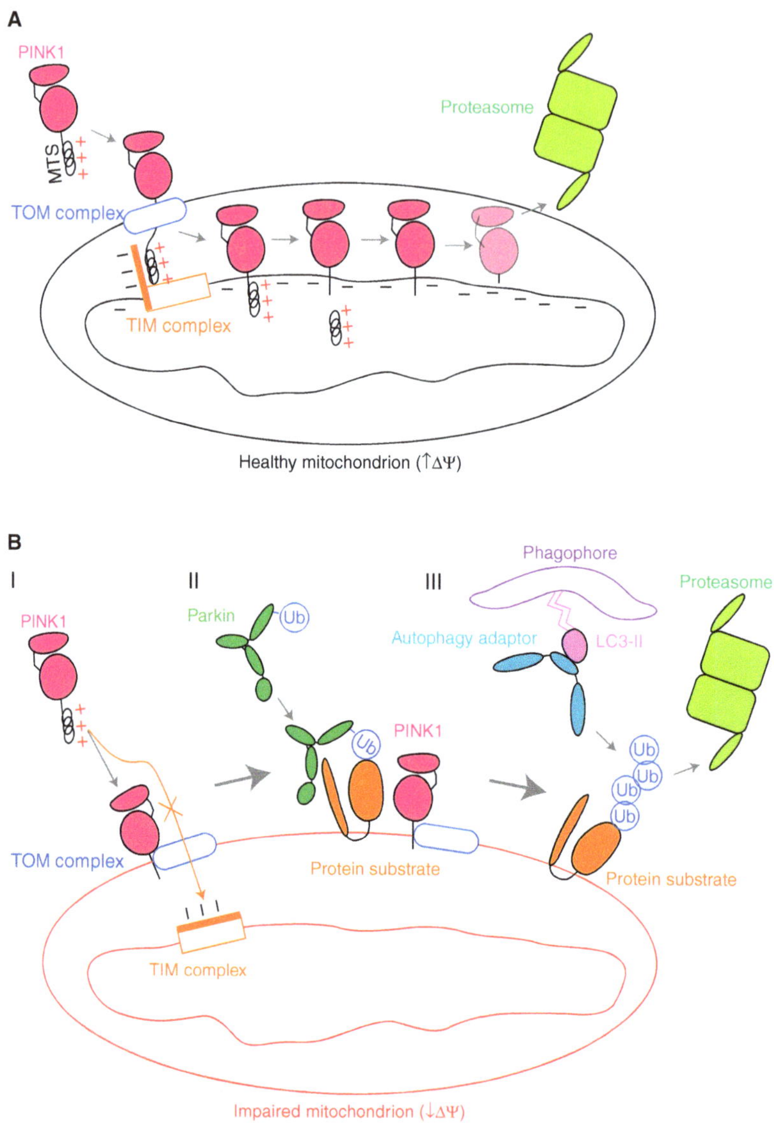

Figure 1. Regulation of the PINK1-Parkin mitochondrial quality control pathway by inner mitochondrial membrane potential. (*A*) In healthy mitochondria, PINK1 is imported to the inner mitochondrial membrane by the TOM (translocase of the outer membrane) and TIM23 (translocase of the inner membrane 23) complexes, along a path that depends on the voltage component of the mitochondrial inner membrane potential ($\Delta\Psi$). At the inner membrane, the mitochondrial-targeting signal (MTS) of PINK1 is cleaved by the matrix metalloprotease. PINK1 is cleaved again in its transmembrane domain by the rhomboid protease PARL. The resulting PINK1 cleavage product is unstable and is degraded in a manner that depends on proteasomal activity. (*B*, I) In impaired mitochondria with collapsed $\Delta\Psi$, PINK1 cannot be imported to the inner membrane along the $\Delta\Psi$-dependent TIM23 pathway. Instead, PINK1 is directed to the outer mitochondrial membrane by a cryptic signal amino terminal to its transmembrane domain, in which it associates with the TOM complex. (*B*, II) The accumulation of PINK1 on the outer mitochondrial membrane recruits Parkin from the cytosol and activates its ubiquitin ligase activity. Parkin ubiquitinates outer membrane proteins preferentially on the mitochondrion on which PINK1 has accumulated. (*B*, III) Ubiquitination of outer mitochondrial membrane proteins by Parkin leads either to their degradation by the proteasome or to the recruitment of ubiquitin-binding adaptor proteins to effect the removal of the damaged mitochondrion by autophagy.

 Cite this article as *Cold Spring Harb Perspect Biol* doi: 10.1101/cshperspect.a011338

of PINK1 and Parkin appear to reside in complexes of different sizes by size exclusion chromatography and native PAGE electrophoresis (Thomas et al. 2011; Lazarou et al. 2012). There is also mixed evidence that PINK1 directly phosphorylates Parkin (Kim et al. 2008; Narendra et al. 2010b; Sha et al. 2010; Vives-Bauza et al. 2010).

An alternative (and not mutually exclusive) hypothesis suggests that PINK1 may direct Parkin to particular substrates on the outer mitochondrial membrane by priming the substrates through phosphorylation. A recent study of one of the substrates of Parkin, Miro1, lends support for this shared substrate model. The investigators report that PINK1 phosphorylates Miro1 on a specific serine residue, S156 (Wang et al. 2011). In addition, the investigators found that if phosphorylation of S156 is blocked by mutating S156 to an alanine, Miro1 is no longer degraded in a Parkin-dependent manner. However, a recent study was unable to confirm that dMiro is phosphorylated by dPink1 in an in vitro assay or that the Miro1 S156A mutant is degraded by Parkin less efficiently than wild-type Miro1 (Liu et al. 2012). Additional studies will be required to validate whether Miro1 is a substrate of PINK1, although, as discussed below, it appears that Miro1 is a physiological substrate of Parkin.

If confirmed, the shared substrate model implies that there may be a motif that is recognized in the unphosphorylated form by PINK1 and the phosphorylated form by Parkin. It will be interesting to learn whether other Parkin substrates (in particular the mitofusins) are similarly phosphorylated by PINK1 and whether a common motif among PINK1/Parkin substrates is found. A preferred peptide substrate for a recombinant insect PINK1 has been identified in vitro, and it is notable for a proline in the 1+ position relative to the phosphorylated serine and positively charged residues in other positions: KKWIPYRRSPRRR (Woodroof et al. 2011). Of note, the context of the phosphorylated serine in this peptide is not obviously similar to that in the proposed Miro1 site, NLKNISELFYYAQK (Wang et al. 2011). This could reflect differences between the in vitro system and the cell-based system (e.g., the efficiency of phospho-transfer within the cell will likely be influenced by the local concentration of the kinase and substrate on the mitochondrial membrane as well as binding between kinase and substrate outside of the phosphorylated region.) Alternatively, the in vitro findings may cast doubt on the validity of the proposed Miro site.

Study of ectopic targeting of PINK1 to cellular membranes has provided additional insight into Parkin recruitment by PINK1 (Lazarou et al. 2012). Ectopic targeting of PINK1 to alternate organelles can recruit Parkin to those same alternate subcellular locations. Interestingly, when cytosolic PINK1 is recruited to mitochondria or other membranes using a heterodimerization system, Parkin recruitment to the same membranes lags by tens of minutes. This suggests that PINK1 assumes a form competent for binding Parkin after it associates with membranes or that Parkin recruitment to membranes reflects an interaction with membrane components other than PINK1 (e.g., binding to its protein substrates).

Failing mitochondria that are selectively tagged by activated Parkin are altered to promote their isolation within the cell and their degradation by selective macroautophagy (hereafter, "autophagy"). Several events in cultured mammalian cells overexpressing Parkin are observed following mitochondrial uncoupling: (1) The mitochondria fragment and fail to refuse when the membrane potential is restored (Tanaka et al. 2010); (2) the spherical mitochondria form clumps that resemble bunches of grapes (Narendra et al. 2010a; Okatsu et al. 2010); (3) the clumps of mitochondria are gathered in a perinuclear localization possibly because of increased minus end transport along microtubules (Lee et al. 2010; Vives-Bauza et al. 2010); and (4) mitochondria are degraded by a selective form of autophagy (mitophagy) (Narendra et al. 2008). Within 24 h, all of the mitochondria within mammalian cells overexpressing Parkin can be eliminated by autophagy. The specific targeting of Parkin to damaged mitochondria by PINK1 together with the autophagic elimination of such mitochondria suggests that this pathway mediates mitochondrial quality control.

Although this hypothesis has yet to be validated in animal models, a study of cultured cybrid cells containing a stable heteroplasmic mixture of wild-type mitochondrial DNA and mitochondrial DNA mutated in the *COX1* gene supports the idea. Stable overexpression of Parkin at levels 3–6-fold above endogenous Parkin levels found in some normal tissues mediated the selective removal of the mutated mitochondrial DNA and enrichment of cytochrome oxidase activity (Suen et al. 2010). In a complementary study, mitophagy of dysfunctional mitochondria in cells homoplasmic for mtDNA mutations was found to be dependent on both Parkin expression and mTORC1 inhibition (Gilkerson et al. 2012). Consistent with a housekeeping function of Parkin, it is widely distributed throughout mitochondria-rich tissues, including brain, muscle, and liver (Suen et al. 2010).

PARKIN UBIQUITINATES MITOCHONDRIAL OUTER MEMBRANE PROTEINS FOLLOWING PINK1 RECRUITMENT

Parkin mediates directly or indirectly the ubiquitination of mitochondrial outer membrane proteins with various ubiquitin linkages (Geisler et al. 2010b; Narendra et al. 2010a; Chan et al. 2011). Some of these proteins with predominantly K48-linked ubiquitin chains are eliminated by proteosomal degradation, whereas others with K63-linked ubiquitin chains may recruit adaptor proteins from the cytosol. Changes in mitochondrial dynamics and transport appear to be mediated by elimination of specific outer mitochondrial membrane proteins, whereas mitochondrial clumping and mitophagy appear to depend on recruitment of specific adaptor proteins to ubiquitinated substrates on the outer mitochondrial membrane (as discussed briefly below and in more depth elsewhere).

Proteins that accumulate on mitochondria concurrent with Parkin translocation and ubiquitination of mitochondrial substrates include p97 and proteasome subunits consistent with an ERAD type pathway of mitochondrial outer membrane protein degradation by the proteasome (Tanaka et al. 2010; Xu et al. 2011). Other proteins such as p62 accumulate on mitochondria by binding ubiquitin chains (with a preference for binding K63-linked chains [Kirkin et al. 2009]). p62 is an adaptor protein that recruits unfolded protein aggregates to autophagosomes, and its role in mitophagy will be discussed later. Which process, proteosomal elimination versus adaptor protein recruitment, mediates autophagosome engulfment of mitochondria is still not clear.

UNBIASED IDENTIFICATION OF SUBSTRATES DEGRADED BY PARKIN

Parkin likely mediates its effect on mitochondrial dynamics, distribution, and turnover through ubiquitination of specific substrates in the outer mitochondrial membrane. These effects could be mediated either through elimination or inhibition of the normal function of the outer membrane proteins or by recruitment of adaptor proteins to the post-translationally modified OMM proteins. Particularly when Parkin is present at supraphysiologic levels many proteins on the OMM appear to be ubiquitinated in a manner that depends on activated Parkin (Chan et al. 2011; Yoshii et al. 2011). Although it is clear that Parkin can promote the ubiquitination of many OMM proteins under special circumstances (e.g., in the presence of high levels of Parkin and PINK1), which substrate is ubiquitinated most efficiently in vivo and which is of greatest physiological importance is less clear.

At least for the subset of Parkin substrates that are degraded after ubiquitination their specificity as substrates in vivo may be reflected in the ease with which they are eliminated by activated Parkin in cultured cells. To help identify the subset of substrates targeted for degradation following Parkin ubiquitination as well as adaptor proteins recruited to OMM proteins ubiquitinated by Parkin, we measured changes in membrane protein abundance using a stable isotope labeling in cell culture (SILAC) proteomics approach (DP Narendra, unpubl.). The experiment was designed to assess changes in membrane protein abundance in response to two variables: the presence or absence of Parkin and the presence or absence of mitochondrial uncoupler CCCP (which activates Parkin by increasing endogenous PINK1 protein levels). Assessing changes along these two dimensions helps to separate changes caused by activated Parkin from secondary effects of the uncoupler CCCP (Fig. 2).

For each protein sequenced, two ratios, heavy labeled (H)/light labeled (L) and heavy labeled (H)/medium labeled (M), reflect changes in membrane protein abundance caused by CCCP treatment and Parkin expression, respectively (Tables 1 and 2). Proteins that increase in response to both Parkin expression and CCCP treatment may be adaptor proteins that are recruited to mitochondria from the cytosol following ubiquitination of outer mitochondrial membrane (OMM) proteins, whereas proteins that decrease in response to both Parkin expression and CCCP treatment may be substrates of Parkin degraded by the proteasome. Of note, this approach does not identify proteins ubiquitinated by Parkin that are not subsequently degraded in disproportion to other mitochondrial proteins (Fig. 3).

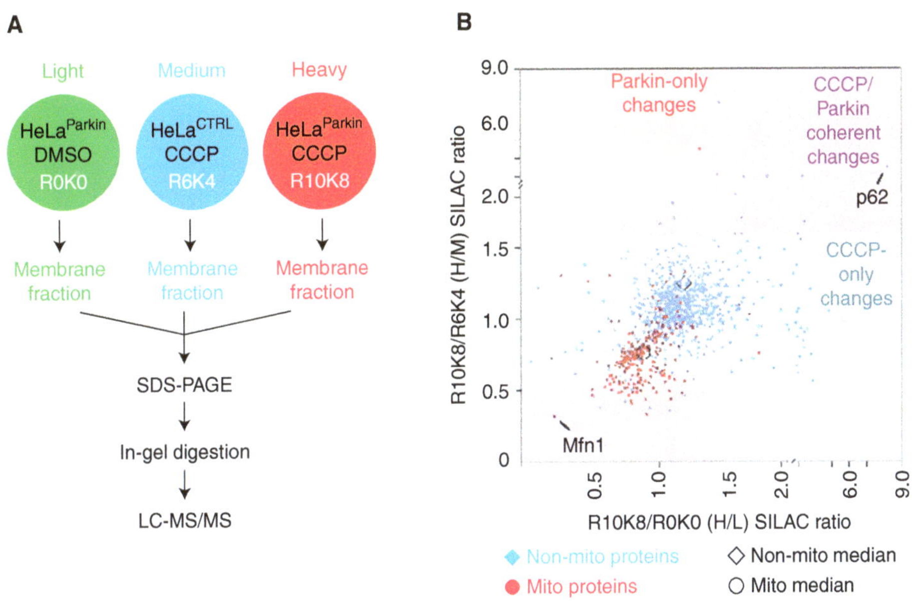

Figure 2. Quantitative proteomics-based screen for the identification of Parkin substrates and adaptors in the PINK1/Parkin pathway. (*A*) Schematic depicting the experimental setup. Three cell lines were labeled with stable isotopes of arginine and lysine to give a light (L) labeled HeLa cell line stably expressing YFP-Parkin treated with vehicle ($\text{HeLa}^{\text{Parkin}}$); a medium (M) labeled HeLa cell line without Parkin expression treated with the mitochondrial uncoupler CCCP for 2 h ($\text{HeLa}^{\text{CTRL}}$); and a heavy (H) labeled $\text{HeLa}^{\text{Parkin}}$ cell line treated with CCCP. A crude membrane fraction was extracted from each cell line by permeabilizing the membranes with a low dose of digitonin, isolating the membranes by centrifugation, and then extracting proteins from the remaining membranes with DDM (*n*-dodecyl-β-maltoside). The protein extracts from each cell line were mixed in a 1:1:1 ratio, before separation on a SDS-PAGE gel, in-gel digestion, and LC-MS/MS on an Orbitrap instrument with a top 5 duty cycle. The data was analyzed in MaxQuant. (*B*) Scatter plot depicting H/L (*x*-axis) and H/M (*y*-axis) ratios for 992 proteins identified in the experiment. Proteins with "mitochondrial" in the protein name appear as red circles, whereas the others appear as blue diamonds. The median for proteins annotated as mitochondrial is shifted down and to the left relative to the median for other proteins consistent with the *en masse* degradation of mitochondrial proteins following treatment with CCCP in the presence of Parkin expression. Proteins in the upper right quadrant represent potential adaptors recruited to mitochondria. Proteins in the lower left quadrant represent potential substrates of Parkin. The protein with the largest increase along the *x*- and *y*-axes, p62, and the largest decrease, Mfn1, are labeled on the plot.

Table 1. Proteins with coherent changes in abundance

Protein	Localization	Unique peptides	PEP	Ratio H/L	Ratio H/M	Ratio H/L (Chan et al. 2011)
Proteins up-regulated						
p62/SQSTM1	Cyto	4	3.37E-05	7.58	2.59	5.8
Heme oxygenase 1	ER	3	5.65E-07	2.42	2.11	NR
Heat shock protein β-1	Cyto/Nuc	9	0	2.54	1.68	NR
Ferritin light chain	Cyto	4	4.64E-32	1.79	2.07	NR
Thioredoxin		9	0	2.15	1.59	3.2
Proteins down-regulated						
Mitofusin1	OMM	3	4.54E-09	0.19[a]	0.28[a]	0.09[b]
GDAP1	OMM	14	2.49E-08	0.53	0.34	0.40
MitoNEET	OMM	18	0	0.48[a]	0.42	0.26[b]
MOSC1	OMM	3	0	0.47	0.44	0.39
TOM70	OMM	5	0	0.47[a]	0.48	0.13[b]
Additional proteins down-regulated in Chan et al. (2011)						
Mitofusin2	OMM			ND	ND	0.1[b]
MIRO1/RHOT1	OMM			ND	ND	0.13[b]
CPT1A	OMM	3	2.76E-24	0.73	0.60	0.23[b]
MOSC2	OMM			ND	ND	0.26[b]
GPAM	OMM			ND	ND	0.42[b]
MIRO2/RHOT2	OMM			ND	ND	0.44[b]
FIS1	OMM	2	9.93E-14	0.59	0.64	0.55[b]
Additional substrates identified by candidate approaches						
VDAC1	OMM	5	5.27E-29	0.50	0.68	NR
Tom20				ND	ND	NR
Bcl-2	OMM	3	1.61E-23	0.99	0.87	NR

Proteins with coherent changes in abundance (i.e., the H/L and H/M ratios are both high or both low) are listed. These represent proteins with a change in abundance that is dependent on both CCCP treatment and Parkin expression.

ND, not detected; NR, not reported.

[a]Ratios reaching statistical significance after correction for multiple comparisons in this study.

[b]Ratios reaching statistical significance after correction for multiple corrections in Chan et al. (2011).

After filtering out contaminants according to published guidelines (Cox et al. 2009), a total of 992 membrane proteins were identified and quantified, roughly a quarter of which were annotated as mitochondrial in their protein name. Proteins annotated as mitochondrial exhibited a 43.8% and 30.9% decrease in abundance along both the Parkin-dependent and CCCP-dependent axes relative to all other proteins (two-tailed Student's *t*-test; *p*-values $= 3.0 \times 10^{-47}$ and 2.5×10^{-29}, respectively), consistent with Parkin promoting the selective degradation of uncoupled mitochondria under these conditions.

p62/SQSTM1, a cytosolic ubiquitin-binding adaptor protein, exhibited the largest increase in abundance, consistent with previous studies showing that it binds proteins ubiquitinated by Parkin (Geisler et al. 2010a; Lee et al. 2010; Narendra et al. 2010a; Okatsu et al. 2010). Its possible role as an adaptor in the PINK1/Parkin pathway is discussed further below.

The five proteins exhibiting the largest decrease in abundance along both dimensions are resident to the OMM, consistent with their potential ubiquitination and degradation by Parkin. These are Mfn1, MitoNEET, GDAP1, MOSC1, and Tom70. Mfn1, MitoNEET, and Tom70 reached statistical significance along the CCCP-treated axis, following correction for multiple comparisons, whereas the others trended toward significance. Mfn1, which also reached statistical significance along the Parkin-dependent axis, exhibited an almost twofold greater decrease at 2 h relative to the other four OMM proteins

 Cite this article as *Cold Spring Harb Perspect Biol* doi: 10.1101/cshperspect.a011338

Table 2. Proteins increased or decreased along a single dimension

Protein	Function	Unique peptides	PEP	Ratio H/L	Ratio H/M	H/L (*p*-value)	H/M (*p*-value)
CCCP-only changes							
Subunits of ETC complexes							
Complex I-ESSS	Supernumary subunit of β subcomplex	2	0.00055987	0.23986	1.3004	4.92E-07	0.20850717
Complex I-PDSW	Supernumary subunit of β subcomplex	4	1.97E-21	0.44766	1.1732	3.74E-05	0.362282139
Complex IV - subunit 4 isoform 1	Regulatory subunit downregulated by hypoxia	9	0	0.43761	1.2583	2.33E-05	0.181333382
Subunits of vacuolar proton pump							
ATP6V1H	Component of vacuolar proton pump	3	4.63E-06	2.9342	0.72304	7.85E-06	0.347560475
ATP6B2	Component of vacuolar proton pump	14	0	2.6621	0.90669	7.59E-06	0.722666052
ATP6A1	Component of vacuolar proton pump	18	0	2.5312	0.88796	2.34E-05	0.671846651
ATP6G	Component of vacuolar proton pump	3	3.22E-31	2.4512	0.82108	0.000228262	0.56428798
ATP6E	Component of vacuolar proton pump	5	9.67E-22	2.3293	0.91277	0.000131918	0.739212408
ATP6M	Component of vacuolar proton pump	2	0.01103	2.2548	0.70783	0.000895118	0.317310509
Cholestrol metabolism							
2,3-epoxysqualene–lanosterolcyclase	Cholestrol synthesis	11	1.11E-39	2.4972	0.77465	3.13E-05	0.38264962
LDL receptor	Cell-surface receptor for LDL particles	11	1.46E-34	2.164	1.0538	0.0016753	0.816681405
Sterol-4-α-carboxylate 3-dehydrogenase	Cholestrol synthesis	13	0	2.0767	1.0301	0.001109734	0.904184966
Misc. increased							
α-SGT	Co-chaperone of HSP70	4	3.20E-28	4.0283	1.04	4.91E-09	0.866974414
Glyceraldehyde-3-phosphate dehydrogenase	Glycolysis	14	0	2.8911	1.1959	1.06E-06	0.304181529
Adult folate-binding protein	Folate transport	5	3.01E-39	2.1425	0.58908	0.000640711	0.075974088
Histone H4	Core component of nucelosome	6	0	2.0654	1.3682	0.001218908	0.064947946
Misc. decreased							
Nucleobindin-1	Multidomain calcium binding protein	18	0	0.27213	0.74534	8.74E-11	0.317310509
Cysteine-rich FGF receptor	FGF binding intracellular glycoprotein	10	0	0.34651	0.58858	9.31E-08	0.075513331
Mesothelin	GPI-linked cell surface protein	4	1.35E-13	0.35673	0.76751	0.000233983	0.441886649

Continued

Table 2. *Continued*

Protein	Function	Unique peptides	PEP	Ratio H/L	Ratio H/M	H/L (*p*-value)	H/M (*p*-value)
Retinol dehydrogenase 13	Oxidoreductase	4	3.30E-22	0.39441	0.75978	0.000845024	0.42493045
Cytochrome *c* type heme lyase	Covalently attaches heme to cytochrome *c*	13	0	0.39837	0.81099	2.91E-06	0.470230202
Proliferating cell nuclear antigen	DNA replication/repair	5	0	0.41676	1.3729	0.0016321	0.127915585
Parkin-only changes							
Increased							
Integrin α-1	Cell adhesion	19	0	1.4191	6.3657	0.162165287	4.54E-29
OPA1	Mitochondrial fusion	3	0	1.2757	3.7024	0.401069865	1.99E-10
GPRC5C	G-protein signaling	3	9.69E-09	1.4069	2.1807	0.205413008	0.000156889
Decreased							
Adenylate kinase isoenzyme 4, mitochondrial	ATP/AMP transphosphorylase	3	0.034025	0.87783	0.29612	0.537881606	0.00048901

Proteins with an increase or decrease along either the *x*- or *y*-axis, indicating an increase or decrease in response to either CCCP treatment or Parkin expression but not both in the same direction, are listed.

Cite this article as *Cold Spring Harb Perspect Biol* doi: 10.1101/cshperspect.a011338

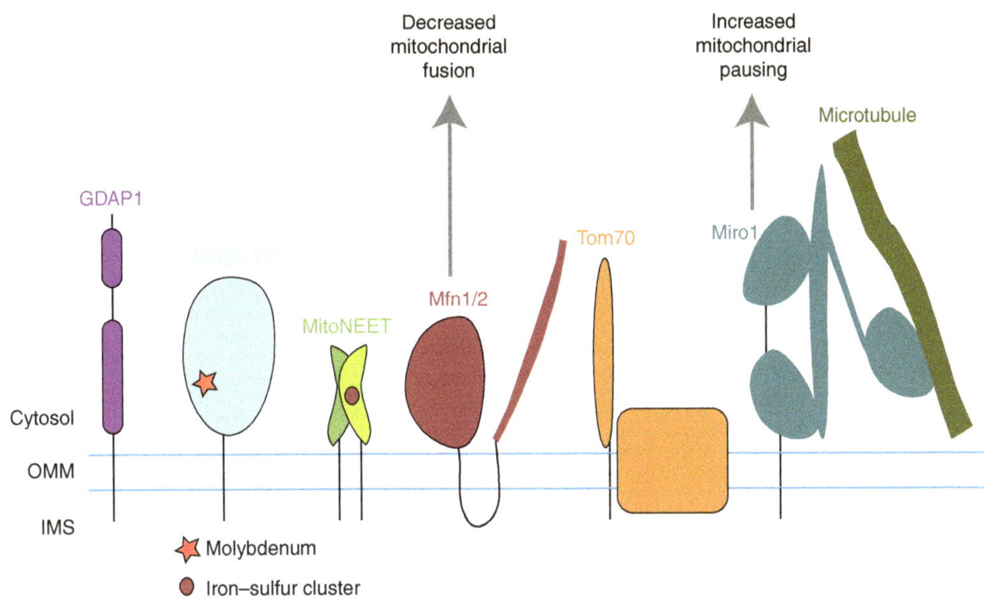

Figure 3. Figure depicting the five mitochondrial proteins with the largest relative decrease in abundance in response to Parkin activation in unbiased quantitative proteomics experiments. They differ in the number of transmembrane domains they possess, their size, and their association with other proteins, indicating a diversity of mitochondrial substrates toward which Parkin has high activity. Three of the substrates, Mfn1, Mfn2, and Miro1, have been extensively validated as physiologic substrates of Parkin in both insects and mammals by multiple groups.

that decreased in this experiment (DP Narendra, unpubl.).

These results are largely in concordance with those from a quantitative proteomics experiment recently published by an independent group (Chan et al. 2011). Similar to our results, Chan et al. (2011) report that p62, as well as its homolog NBR1 (not identified in our experiment), increased in the membrane fraction of HeLa cells expressing exogenous Parkin following depolarization with CCCP for 2 h. Chan et al. also observed an increase in autophagy related proteins, such as LC3, consistent with ongoing mitophagy. Of note although both groups (Chan et al. 2011; DP Narendra, unpubl.) observed an increase in subunits of the vacuolar proton pump in response to CCCP treatment, this increase did not depend on the presence on Parkin in our experiment, indicating that it is because of CCCP treatment alone and is not an effect mediated by Parkin.

Additionally, five of the top ten OMM proteins Chan et al identified as significantly decreased in the membrane fraction following treatment with CCCP (namely, Mfn1, Mfn2, MitoNEET, MOSC2, and Tom70) are the same or homologs of the five proteins we identified as significantly decreased under these conditions. As in our experiment, Mfn1 (as well as its homolog Mfn2) exhibited the largest decrease. Additionally, Chan et al found that Miro1 (not identified in our experiment) exhibited a decrease of similar magnitude to Mfn1/Mfn2. In both the Chan et al. and our experiments, other OMM proteins identified as potential substrates of Parkin by candidate approaches (such as VDAC1 [Geisler et al. 2010a] and Bcl-2 [Chen et al. 2010a]) were not significantly altered in abundance following depolarization with CCCP for 2 h. The concordance of these findings suggests that Parkin exhibits considerable specificity for Mfn1/Mfn2, Miro1, and Tom70 (and possibly a limited number of other mitochondrial proteins, such as MitoNEET). Mfn1/Mfn2, Miro1, and Tom70 emerge from these unbiased proteomics screens as the most likely mitochondrial candidates for physiologic substrates of Parkin.

Detailed studies of mitofusins and Miro1 support their relative physiological importance among proposed Parkin substrates. To our knowledge, Mfn1/Mfn2 and Miro1 are the only mitochondrial substrates shown to be ubiquitinated or degraded by endogenous levels of Parkin (Poole et al. 2010; Tanaka et al. 2010; Wang et al. 2011; Liu et al. 2012). Additionally, they are the only proteins reported to be targeted by the PINK1/Parkin pathway in both insects and mammals (Gegg et al. 2010; Poole et al. 2010; Tanaka et al. 2010; Ziviani et al. 2010; Chan et al. 2011; Rakovic et al. 2011; Wang et al. 2011). Finally, Mfn1/Mfn2 and Miro1 are quantitatively modified following PINK1/Parkin activation, whereas a relatively small portion of the protein pool appears to be modified in the case of other proposed substrates, such as VDAC1. Together these findings argue that Mfn1/2 and Miro1 are physiologically relevant substrates of the PINK1/Parkin mitochondrial quality control pathway.

PINK1 AND PARKIN REGULATE MITOFUSINS TO ISOLATE IMPAIRED MITOCHONDRIA

The mitofusin orthologs (Mfn1 and Mfn2 in mammals, Fzo1 and MARF in flies, and Fzo1 in yeast) are GTPases that mediate outer mitochondrial membrane fusion (Hales and Fuller 1997). Their disruption in mice and yeast leads to a highly fragmented mitochondrial network as well as loss of mitochondrial fidelity, as reflected in decreased mtDNA copy number, an increased number of mtDNA mutations, and an accompanying bioenergetic deficit (Hermann et al. 1998; Rapaport et al. 1998; Chen et al. 2010b).

The degradation of the mitofusins by Parkin appears to help isolate failing mitochondria from the mitochondrial network by preventing their outer membrane fusion with more bioenergetically competent mitochondria. If the isolation of the dysfunctional mitochondria is inhibited (e.g., by disruption of the mitochondrial fission machinery), then mitophagy of the dysfunctional mitochondria is delayed (Tanaka et al. 2010).

Whereas the PINK1/Parkin pathway appears to regulate the fusion machinery selectively on dysfunctional mitochondria as a quality control measure, it does not appear to play an important role in the basal regulation of the fusion machinery in mammalian cells. Mitofusin levels are not substantially altered following knockdown or knockout of endogenous Parkin or PINK1 in the absence of mitochondrial stress (Gegg et al. 2010; Tanaka et al. 2010; Chan et al. 2011; Rakovic et al. 2011). Likewise, several reports have shown a minimally changed or more fragmented mitochondrial network following knockdown or elimination of Parkin or PINK1 in cultured mammalian cells (Exner et al. 2007) or in tissues analyzed ex vivo (Gautier et al. 2008).

In *Drosophila*, in contrast, the dPink1/dParkin pathway does appear to play a critical role in the basal regulation of the fusion/fission machinery. Depletion of Parkin in *Drosophila* myocytes or cultured S2 cells leads to a substantial decrease in mitofusin ubiquitination and increase in mitofusin levels (Poole et al. 2010; Ziviani et al. 2010). Coinciding with the decrease in ubiquitinated mitofusin the mitochondrial network becomes hyperfused, indicating that the dPink1/dParkin pathway has a strong effect on the overall mitochondrial fission/fusion balance under basal conditions (Ziviani et al. 2010). Additionally, a loss of the basal regulation of mitochondrial dynamics by disruption of the dPink1/dParkin pathway appears to be at least partially responsible for the abnormal muscle and sperm phenotype observed, as evidenced by the partial amelioration of the dParkin or dPink1 null phenoytpe by genetic alterations that promote increased fission (Deng et al. 2008; Poole et al. 2008; Yang et al. 2008).

PINK1/PARKIN PATHWAY REGULATES MIRO LEVELS TO IMMOBILIZE MITOCHONDRIA

Miro1 and Miro2, homologous EF-hand-containing adaptor proteins anchored in the OMM, form a complex with Milton and kinesin to promote the transport of mitochondria toward the plus-end of microtubules (Stowers et al. 2002;

Guo et al. 2005). Miro1/2 are regulated by calcium via their EF-hands to promote pausing in the presence of high calcium concentrations (Saotome et al. 2008; Macaskill et al. 2009; Wang and Schwarz 2009). The dwelling of mitochondria in pools of higher calcium may help buffer the local calcium keeping the calcium rise spatially and temporally restricted. Miro1 is also a protein component of ER-mitochondrial contact sites (Kornmann et al. 2011). These sites also help to hem in calcium transients by forming a synapse between a major site of calcium release (the ER) and a major site of calcium buffering (the mitochondria) (Rizzuto et al. 1998). In addition, these contact sites allow phospholipid exchange between the organelles, which is necessary for the synthesis of certain phospholipids (reviewed in Osman et al. 2011).

The first indication that the Parkin/PINK1 pathway may regulate the Miro1/Milton complex came from the observation that a portion of the mitochondrial PINK1 pool exists in complex with Miro1 and Milton (Weihofen et al. 2009). Subsequently, it was shown that Miro1 is degraded under conditions that activate Parkin (e.g., treatment with a mitochondrial uncoupler) and that its degradation depends on the presence of endogenous levels of Parkin and PINK1 (Wang et al. 2011; Liu et al. 2012). The degradation of Miro1 leads to arrest of the dysfunctional mitochondria in neurons, which may prevent their transport to energy-hungry areas of the cell at the expense of bioenergetically viable mitochondria (Wang et al. 2011; Liu et al. 2012). The degradation of Miro1 also appears to promote the subsequent mitophagy of the impaired mitochondria (Liu et al. 2012).

In cultured mammalian cells, activation of the Parkin/PINK1 pathway promotes net retrograde movement of mitochondria resulting in the concentration of mitochondria near the microtubule-organizing center rather than mitochondrial pausing (Vives-Bauza et al. 2010). This transport is inhibited by nocodazole, which causes microtubule depolymerization, suggesting that it occurs along the microtubular tracks (Vives-Bauza et al. 2010). Similarly, loss of Parkin causes clustering of mitochondria around the plus-end of microtubules in *Drosophila* follicular cells, suggesting that Parkin also promotes net mitochondrial transport toward the minus end of microtubules in this tissue (Cox and Spradling 2009). Degradation of Miro1 may also be responsible for these Parkin-induced changes in mitochondrial transport. They may also depend on HDAC6, a microtubule binding protein that is recruited to mitochondria following the activation of Parkin by PINK1 (Lee et al. 2010).

Similar to the situation with mitochondrial fission, Parkin/PINK1 appears to be more important for basal mitochondrial transport in *Drosophila* relative to mice. Knockdown of dParkin or dPink1 in *Drosophila* leads to an increase in anterograde and retrograde mitochondrial transport within neurons, presumably caused by stabilization of the Miro1/Milton/Kinesin complex (Wang et al. 2011). However, knockout of Parkin or Pink1 in mice does not appear to significantly increase the likelihood of mitochondrial transport in the absence of mitochondrial stress, consistent with a quiescent PINK1/Parkin pathway when the mitochondria are healthy (Wang et al. 2011).

The relative importance of the Pink1/Parkin pathway for the basal regulation of mitochondrial transport and mitochondria fission/fusion in insects versus mammals may help explain the disparity in phenotype severity between the two classes of organisms. Pink1/Parkin appears to play an important housekeeping function in flies by controlling levels of the mitofusins and Miro, which are produced in excess. Consistent with a basal requirement for dPink1 and dParkin, their loss leads to severe disruption of multiple tissues. In mammals, in contrast, the PINK1/Parkin pathway appears to be kept in reserve until mitochondrial damage is suffered, thus triggering its activation. Consequently, in mice and humans dysfunction appears only in the most vulnerable tissues and at a relatively late stage of life.

MECHANISM OF PARKIN-INDUCED MITOPHAGY

Although the degradation of mitofusins and Miro by Parkin may explain the effect of the

PINK1/Parkin pathway on mitochondrial dynamics and transport and their degradation may promote the subsequent mitophagy (Tanaka et al. 2010; Liu et al. 2012), likely they do not fully explain how Parkin promotes mitophagy. Parkin, for instance, still induces the autophagy of depolarized mitochondria in mouse fibroblasts null for Mfn1 and Mfn2 (Tanaka et al. 2010; Chan et al. 2011).

The leading hypothesis for the mechanism of Parkin-induced mitophagy is that the stable ubiquitination of some OMM proteins by Parkin serves as a positive signal for autophagy through an interaction with adaptor proteins that mediate selective autophagy. Such as mechanism has been proposed for the selective autophagy of other organelles, such as peroxisomes, as well as some intracellular bacteria, and ribosomes (reviewed in Johansen and Lamark 2011).

To date, no Parkin substrate has been convincingly shown to be required for mitophagy. A single report proposed that VDAC1 is required for mitophagy (Geisler et al. 2010a). However, this report, which used siRNA knockdown of VDAC1, has not been reproduced. A subsequent report by our group found that Parkin-induced mitophagy occurs in mouse fibroblasts null for VDAC1 and VDAC3 (Narendra et al. 2010a). Additionally, only a small fraction of the VDAC1 pool appears to be ubiquitinated by overexpressed Parkin, and VDAC1 has not been shown to be ubiquitinated by endogenous levels of Parkin and PINK1, calling into question whether VDAC1 is a specific substrate of Parkin (Narendra et al. 2010a). This is in marked contrast to the situation with the mitofusins and Miro, in which endogenous levels of Parkin and PINK1 are sufficient to ubiquitinate most of the substrate pool (Tanaka et al. 2010; Wang et al. 2011; Liu et al. 2012).

The identification of specific Parkin substrates that are not subsequently degraded might benefit from additional quantitative proteomics studies. Using a recently described antibody, the tryptic digests of proteins can now be enriched for diglycine-modified peptides (the stigmata of ubiquitination following protein digestion with trypsin), facilitating the direct identification of ubiquitinated proteins (Xu et al. 2010; Kim et al. 2011b). These substrates may have a ubiquitin chain with a specific linkage that does not bind the proteasome subunits, explaining why they are not degraded following ubiquitination. Polyubiquitin chains with a K63 linkage, for instance, usually do not direct the substrate to the proteasome for degradation (reviewed in Winget and Mayor 2010). Parkin has been shown to induce the K63 polyubiquitination of unidentified OMM proteins, using both antibodies specific for particular diubiquitin linkages (Narendra et al. 2010a; Okatsu et al. 2010; Chan et al. 2011) and quantitative mass spectrometry based techniques (Chan et al. 2011).

Once ubiquitinated, mitochondrial surface proteins may interact with ubiquitin-binding adaptor proteins that connect the ubiquitinated proteins with the autophagy machinery. p62 and the related NBR1 appear to act as autophagy adaptors for peroxisomes, intracellular bacteria, and protein aggregates (reviewed in Johansen and Lamark 2011). p62 has been shown to be recruited to mitochondria ubiquitinated by Parkin (Geisler et al. 2010a; Lee et al. 2010; Narendra et al. 2010a; Okatsu et al. 2010), and the quantitative proteomics studies described above show that p62 and NBR1 exhibit a larger fold increase in the membrane fraction than any other protein detected following Parkin activation (Chan et al. 2011). Although p62 mediates mitochondrial clumping, it does not appear to be necessary or sufficient for mitophagy (Narendra et al. 2010a; Okatsu et al. 2010). Parkin-induced mitophagy occurs in fibroblasts from p62 null mice to a similar extent as in wild-type cells (Narendra et al. 2010a; Okatsu et al. 2010), and recruitment of p62 to mitochondria by an OMM targeted ubiquitin fusion protein does not induce mitophagy (Narendra et al. 2010a) (but see Geisler et al. [2010a] and Lee et al. [2010] for RNA interference data suggesting that p62 is limiting for Parkin-induced mitophagy). Nonetheless, it is possible that p62 and NBR1 together are necessary and/or sufficient for Parkin-induced mitophagy. *Drosophila* have only one p62 ortholog, called ref(2)P. A recent report suggested that Parkin overexpression rescues mitochondrial damage induced by a misfolded matrix protein by a mechanism that requires both the autophagy gene *atg1* and ref(2)P, consistent with the p62 ortholog functioning as an autophagy adaptor in this pathway (Pimenta de Castro et al. 2012). Future studies may reveal whether p62 and NBR1 together mediate autophagy in mammalian cells after OMM proteins are tagged with ubiquitin by Parkin.

A recent study showed that autophagosome formation is initiated adjacent to mitochondria following Parkin recruitment. The most upstream complex in the autophagy pathway, the ULK1–Atg13–FIP200–Atg101 complex, appears to be recruited near mitochondria (Itakura et al. 2012). Although p62/NBR1 has traditionally been thought to engage the autophagy machinery only after LC3-II has been incorporated into the mature phagophore, this may not preclude its role in the PINK1/Parkin pathway as p62/NBR1 also appears at the site of autophagosome formation early in the process: It colocalizes with the ULK1–Atg13–FIP200–Atg101 complex even in the absence of LC3 lipidation. Its subsequent interaction with LC3-II may promote efficient incorporation of the cargo (Itakura and Mizushima 2011).

CONCLUDING REMARKS

Studies published over the last six years have established a model for the PINK1/Parkin pathway of mitochondria quality control. This model is centered on three reproducible observations: (1) Parkin displays little activity toward mitochondrial proteins under basal conditions; (2) PINK1 accumulation on an impaired mitochondrion activates Parkin and constrains its activity to substrates on the impaired mitochondrion; and (3) Parkin ubiquitinates outer mitochondrial membrane proteins, most notably the mitofusins and Miro1, and induces the autophagic removal of impaired organelles. Whereas these findings are robust in cell culture in the presence of supraphysiological concentrations of Parkin, it has yet to be convincingly shown in vivo.

Given the strong genetic interaction between PINK1 and a Parkin in model systems and the similar disease caused by mutations in either PINK1 or Parkin in humans, the PINK1/Parkin mitochondrial quality control pathway is likely relevant to the pathogenesis of early onset parkinsonism.

We propose that this model may also provide a means of evaluating the relevance of novel PINK1 and Parkin substrates to the pathogenesis of early-onset parkinsonism. A zinc-finger protein renamed PARIS (parkin interacting substrate), for instance, was recently proposed to be a substrate of Parkin (Shin et al. 2011). As a putative repressor of PGC1-α, its degradation by Parkin was found to promote mitochondrial biogenesis. As with other proposed Parkin substrates, demonstrating that its degradation by Parkin depends on PINK1 would support its relevance to the pathogenesis of early-onset parkinsonism. Likewise, PINK1 has been proposed to physically or genetically interact with OMI/HTRA2 (Plun-Favreau et al. 2007), TRAP1 (Pridgeon et al. 2007), PGAM5 (Imai et al. 2010), and others. Although these interactions may reflect physiological functions of PINK1, parsimony would suggest that their relevance to the pathogenesis of early-onset parkinsonism may hinge on a genetic interaction with Parkin as well.

The unraveling of the PINK1/Parkin mitochondrial quality control pathway also holds great promise for the development of therapies aimed at sporadic Parkinson's disease as well as other disorders in which mitochondrial dysfunction drives pathogenesis. The ability of Parkin overexpression to clear mitochondria harboring deleterious mtDNA mutations shows the therapeutic potential of PINK1/Parkin upregulation for the treatment of mitochondrial disorders. The robust recruitment of fluorescent protein tagged Parkin to mitochondria on activation by PINK1 and the robust degradation of mitochondria following Parkin activation provide read-outs for high-throughput screens for activators and inhibitors of the PINK1/Parkin pathway. Whole genome siRNA and large library small compounds screens for PINK1/Parkin activators and inhibitors are currently underway by groups in government, academia, and industry. It is hoped that these efforts will uncover novel therapeutics for the treatment of Parkinson's disease as well as other mitochondrial disorders for which few disease-modifying therapies now exist.

ACKNOWLEDGMENTS

This review was supported by the intramural program at the NINDS (R.Y. and D.N.), the Medical Research Council (J.E.W. and D.N.), and NIH grant 1F30AG039185-01 (D.N.). We thank Drs. Ian Fearnley, Joe Carroll, and Shu-jing Ding for many helpful discussions and assistance in preparing proteomics data for Tables 1 and 2 and Figure 2 of this review.

REFERENCES

Baloh RH, Salavaggione E, Milbrandt J, Pestronk A. 2007. Familial parkinsonism and ophthalmoplegia from a mutation in the mitochondrial DNA helicase twinkle. *Arch Neurol* **64:** 998–1000.

Bender A, Krishnan KJ, Morris CM, Taylor GA, Reeve AK, Perry RH, Jaros E, Hersheson JS, Betts J, Klopstock T, et al. 2006. High levels of mitochondrial DNA deletions in substantia nigra neurons in aging and Parkinson disease. *Nat Genet* **38:** 515–517.

Betarbet R, Sherer TB, MacKenzie G, Garcia-Osuna M, Panov AV, Greenamyre JT. 2000. Chronic systemic pesticide exposure reproduces features of Parkinson's disease. *Nat Neurosci* **3:** 1301–1306.

Billia F, Hauck L, Konecny F, Rao V, Shen J, Mak TW. 2011. PTEN-inducible kinase 1 (PINK1)/Park6 is indispensable for normal heart function. *Proc Natl Acad Sci* **108:** 9572–9577.

Chan NC, Salazar AM, Pham AH, Sweredoski MJ, Kolawa NJ, Graham RL, Hess S, Chan DC. 2011. Broad activation of the ubiquitin-proteasome system by Parkin is critical for mitophagy. *Hum Mol Genet* **20:** 1726–1737.

Chen D, Gao F, Li B, Wang H, Xu Y, Zhu C, Wang G. 2010a. Parkin mono-ubiquitinates Bcl-2 and regulates autophagy. *J Biol Chem* **285:** 38214–38223.

Chen H, Vermulst M, Wang YE, Chomyn A, Prolla TA, McCaffery JM, Chan DC. 2010b. Mitochondrial fusion is required for mtDNA stability in skeletal muscle and tolerance of mtDNA mutations. *Cell* **141:** 280–289.

Clark IE, Dodson MW, Jiang C, Cao JH, Huh JR, Seol JH, Yoo SJ, Hay BA, Guo M. 2006. *Drosophila pink1* is required for mitochondrial function and interacts genetically with *parkin*. *Nature* **441:** 1162–1166.

Cox RT, Spradling AC. 2009. *clueless*, a conserved *Drosophila* gene required for mitochondrial subcellular localization, interacts genetically with *parkin*. *Dis Model Mech* **2:** 490–499.

Cox J, Matic I, Hilger M, Nagaraj N, Selbach M, Olsen JV, Mann M. 2009. A practical guide to the MaxQuant computational platform for SILAC-based quantitative proteomics. *Nat Protoc* **4:** 698–705.

Deas E, Plun-Favreau H, Gandhi S, Desmond H, Kjaer S, Loh SH, Renton AE, Harvey RJ, Whitworth AJ, Martins LM, et al. 2011. PINK1 cleavage at position A103 by the mitochondrial protease PARL. *Hum Mol Genet* **20:** 867–879.

Deng H, Dodson MW, Huang H, Guo M. 2008. The Parkinson's disease genes *pink1* and *parkin* promote mitochondrial fission and/or inhibit fusion in *Drosophila*. *Proc Natl Acad Sci* **105:** 14503–14508.

Exner N, Treske B, Paquet D, Holmstrom K, Schiesling C, Gispert S, Carballo-Carbajal I, Berg D, Hoepken HH, Gasser T, et al. 2007. Loss-of-function of human PINK1 results in mitochondrial pathology and can be rescued by parkin. *J Neurosci* **27:** 12413–12418.

Fujiwara M, Marusawa H, Wang HQ, Iwai A, Ikeuchi K, Imai Y, Kataoka A, Nukina N, Takahashi R, Chiba T. 2008. Parkin as a tumor suppressor gene for hepatocellular carcinoma. *Oncogene* **27:** 6002–6011.

Gautier CA, Kitada T, Shen J. 2008. Loss of PINK1 causes mitochondrial functional defects and increased sensitivity to oxidative stress. *Proc Natl Acad Sci* **105:** 11364–11369.

Gegg ME, Cooper JM, Chau KY, Rojo M, Schapira AH, Taanman JW. 2010. Mitofusin 1 and mitofusin 2 are ubiquitinated in a PINK1/parkin-dependent manner upon induction of mitophagy. *Hum Mol Genet* **19:** 4861–4870.

Geisler S, Holmstrom KM, Skujat D, Fiesel FC, Rothfuss OC, Kahle PJ, Springer W. 2010a. PINK1/Parkin-mediated mitophagy is dependent on VDAC1 and p62/SQSTM1. *Nat Cell Biol* **12:** 119–131.

Geisler S, Holmstrom KM, Treis A, Skujat D, Weber SS, Fiesel FC, Kahle PJ, Springer W. 2010b. The PINK1/Parkin-mediated mitophagy is compromised by PD-associated mutations. *Autophagy* **6:** 871–878.

Gilkerson RW, De Vries RL, Lebot P, Wikstrom JD, Torgyekes E, Shirihai OS, Przedborski S, Schon EA. 2012. Mitochondrial autophagy in cells with mtDNA mutations results from synergistic loss of transmembrane potential and mTORC1 inhibition. *Hum Mol Genet* **21:** 978–990.

Greene JC, Whitworth AJ, Kuo I, Andrews LA, Feany MB, Pallanck LJ. 2003. Mitochondrial pathology and apoptotic muscle degeneration in *Drosophila parkin* mutants. *Proc Natl Acad Sci* **100:** 4078–4083.

Greene AW, Grenier K, Aguileta MA, Muise S, Farazifard R, Haque ME, McBride HM, Park DS, Fon EA. 2012. Mitochondrial processing peptidase regulates PINK1 processing, import and Parkin recruitment. *EMBO Rep* **13:** 378–385.

Guo X, Macleod GT, Wellington A, Hu F, Panchumarthi S, Schoenfield M, Marin L, Charlton MP, Atwood HL, Zinsmaier KE. 2005. The GTPase dMiro is required for axonal transport of mitochondria to *Drosophila* synapses. *Neuron* **47:** 379–393.

Hales KG, Fuller MT. 1997. Developmentally regulated mitochondrial fusion mediated by a conserved, novel, predicted GTPase. *Cell* **90:** 121–129.

Hardy J. 2010. Genetic analysis of pathways to Parkinson disease. *Neuron* **68:** 201–206.

Hermann GJ, Thatcher JW, Mills JP, Hales KG, Fuller MT, Nunnari J, Shaw JM. 1998. Mitochondrial fusion in yeast requires the transmembrane GTPase Fzo1p. *J Cell Biol* **143:** 359–373.

Imai Y, Kanao T, Sawada T, Kobayashi Y, Moriwaki Y, Ishida Y, Takeda K, Ichijo H, Lu B, Takahashi R. 2010. The loss of PGAM5 suppresses the mitochondrial degeneration caused by inactivation of PINK1 in *Drosophila*. *PLoS Genet* **6:** e1001229.

Itakura E, Mizushima N. 2011. p62 Targeting to the autophagosome formation site requires self-oligomerization but not LC3 binding. *J Cell Biol* **192:** 17–27.

Itakura E, Kishi-Itakura C, Koyama-Honda I, Mizushima N. 2012. Structures containing Atg9A and the ULK1 complex independently target depolarized mitochondria at initial stages of Parkin-mediated mitophagy. *J Cell Sci* **125:** 1488–1499.

Jin SM, Lazarou M, Wang C, Kane LA, Narendra DP, Youle RJ. 2010. Mitochondrial membrane potential regulates PINK1 import and proteolytic destabilization by PARL. *J Cell Biol* **191:** 933–942.

Johansen T, Lamark T. 2011. Selective autophagy mediated by autophagic adapter proteins. *Autophagy* **7:** 279–296.

Kim Y, Park J, Kim S, Song S, Kwon SK, Lee SH, Kitada T, Kim JM, Chung J. 2008. PINK1 controls mitochondrial localization of Parkin through direct phosphorylation. *Biochem Biophys Res Commun* **377:** 975–980.

Kim KY, Stevens MV, Akter MH, Rusk SE, Huang RJ, Cohen A, Noguchi A, Springer D, Bocharov AV, Eggerman TL, et al. 2011a. Parkin is a lipid-responsive regulator of fat uptake in mice and mutant human cells. *J Clin Invest* **121:** 3701–3712.

Kim W, Bennett EJ, Huttlin EL, Guo A, Li J, Possemato A, Sowa ME, Rad R, Rush J, Comb MJ, et al. 2011b. Systematic and quantitative assessment of the ubiquitin-modified proteome. *Mol Cell* **44:** 325–340.

Kirkin V, Lamark T, Sou YS, Bjorkoy G, Nunn JL, Bruun JA, Shvets E, McEwan DG, Clausen TH, Wild P, et al. 2009. A role for NBR1 in autophagosomal degradation of ubiquitinated substrates. *Mol Cell* **33:** 505–516.

Kitada T, Asakawa S, Hattori N, Matsumine H, Yamamura Y, Minoshima S, Yokochi M, Mizuno Y, Shimizu N. 1998. Mutations in the *parkin* gene cause autosomal recessive juvenile parkinsonism. *Nature* **392:** 605–608.

Kitada T, Pisani A, Karouani M, Haburcak M, Martella G, Tscherter A, Platania P, Wu B, Pothos EN, Shen J. 2009. Impaired dopamine release and synaptic plasticity in the striatum of $parkin^{-/-}$ mice. *J Neurochem* **110:** 613–621.

Kornmann B, Osman C, Walter P. 2011. The conserved GTPase Gem1 regulates endoplasmic reticulum-mitochondria connections. *Proc Natl Acad Sci* **108:** 14151–14156.

Langston JW, Ballard P, Tetrud JW, Irwin I. 1983. Chronic Parkinsonism in humans due to a product of meperidine-analog synthesis. *Science* **219:** 979–980.

Lazarou M, Jin SM, Kane LA, Youle RJ. 2012. Role of PINK1 binding to the TOM complex and alternate intracellular membranes in recruitment and activation of the E3 ligase Parkin. *Dev Cell* **22:** 320–333.

Lee JY, Nagano Y, Taylor JP, Lim KL, Yao TP. 2010. Disease-causing mutations in parkin impair mitochondrial ubiquitination, aggregation, and HDAC6-dependent mitophagy. *J Cell Biol* **189:** 671–679.

Liu S, Sawada T, Lee S, Yu W, Silverio G, Alapatt P, Millan I, Shen A, Saxton W, Kanao T, et al. 2012. Parkinson's disease-associated kinase PINK1 regulates Miro protein level and axonal transport of mitochondria. *PLoS Genet* **8:** e1002537.

Luoma P, Melberg A, Rinne JO, Kaukonen JA, Nupponen NN, Chalmers RM, Oldfors A, Rautakorpi I, Peltonen L, Majamaa K, et al. 2004. Parkinsonism, premature menopause, and mitochondrial DNA polymerase γ mutations: Clinical and molecular genetic study. *Lancet* **364:** 875–882.

Macaskill AF, Rinholm JE, Twelvetrees AE, Arancibia-Carcamo IL, Muir J, Fransson A, Aspenstrom P, Attwell D, Kittler JT. 2009. Miro1 is a calcium sensor for glutamate receptor-dependent localization of mitochondria at synapses. *Neuron* **61:** 541–555.

Matsuda N, Sato S, Shiba K, Okatsu K, Saisho K, Gautier CA, Sou YS, Saiki S, Kawajiri S, Sato F, et al. 2010. PINK1 stabilized by mitochondrial depolarization recruits Parkin to damaged mitochondria and activates latent Parkin for mitophagy. *J Cell Biol* **189:** 211–221.

McQuibban GA, Saurya S, Freeman M. 2003. Mitochondrial membrane remodelling regulated by a conserved rhomboid protease. *Nature* **423:** 537–541.

Meissner C, Lorenz H, Weihofen A, Selkoe DJ, Lemberg MK. 2011. The mitochondrial intramembrane protease PARL cleaves human Pink1 to regulate Pink1 trafficking. *J Neurochem* **117:** 856–867.

Narendra D, Tanaka A, Suen DF, Youle RJ. 2008. Parkin is recruited selectively to impaired mitochondria and promotes their autophagy. *J Cell Biol* **183:** 795–803.

Narendra D, Kane LA, Hauser DN, Fearnley IM, Youle RJ. 2010a. p62/SQSTM1 is required for Parkin-induced mitochondrial clustering but not mitophagy; VDAC1 is dispensable for both. *Autophagy* **6:** 1090–1106.

Narendra DP, Jin SM, Tanaka A, Suen DF, Gautier CA, Shen J, Cookson MR, Youle RJ. 2010b. PINK1 is selectively stabilized on impaired mitochondria to activate Parkin. *PLoS Biol* **8:** e1000298.

Okatsu K, Saisho K, Shimanuki M, Nakada K, Shitara H, Sou YS, Kimura M, Sato S, Hattori N, Komatsu M, et al. 2010. p62/SQSTM1 cooperates with Parkin for perinuclear clustering of depolarized mitochondria. *Genes Cells* **15:** 887–900.

Olanow CW, Rascol O, Hauser R, Feigin PD, Jankovic J, Lang A, Langston W, Melamed E, Poewe W, Stocchi F, et al. 2009. A double-blind, delayed-start trial of rasagiline in Parkinson's disease. *N Engl J Med* **361:** 1268–1278.

Osman C, Voelker DR, Langer T. 2011. Making heads or tails of phospholipids in mitochondria. *J Cell Biol* **192:** 7–16.

Pagliarini DJ, Calvo SE, Chang B, Sheth SA, Vafai SB, Ong SE, Walford GA, Sugiana C, Boneh A, Chen WK, et al. 2008. A mitochondrial protein compendium elucidates complex I disease biology. *Cell* **134:** 112–123.

Palacino JJ, Sagi D, Goldberg MS, Krauss S, Motz C, Wacker M, Klose J, Shen J. 2004. Mitochondrial dysfunction and oxidative damage in *parkin*-deficient mice. *J Biol Chem* **279:** 18614–18622.

Park J, Lee SB, Lee S, Kim Y, Song S, Kim S, Bae E, Kim J, Shong M, Kim JM, et al. 2006. Mitochondrial dysfunction in *Drosophila* PINK1 mutants is complemented by *parkin*. *Nature* **441:** 1157–1161.

Pimenta de Castro I, Costa AC, Lam D, Tufi R, Fedele V, Moisoi N, Dinsdale D, Deas E, Loh SH, Martins LM. 2012. Genetic analysis of mitochondrial protein misfolding in *Drosophila melanogaster*. *Cell Death Differ* **19:** 1308–1316.

Plun-Favreau H, Klupsch K, Moisoi N, Gandhi S, Kjaer S, Frith D, Harvey K, Deas E, Harvey RJ, McDonald N, et al. 2007. The mitochondrial protease HtrA2 is regulated by Parkinson's disease-associated kinase PINK1. *Nat Cell Biol* **9:** 1243–1252.

Poole AC, Thomas RE, Andrews LA, McBride HM, Whitworth AJ, Pallanck LJ. 2008. The PINK1/Parkin pathway regulates mitochondrial morphology. *Proc Natl Acad Sci* **105:** 1638–1643.

Poole AC, Thomas RE, Yu S, Vincow ES, Pallanck L. 2010. The mitochondrial fusion-promoting factor mitofusin is a substrate of the PINK1/parkin pathway. *PLoS ONE* **5:** e10054.

Pramstaller PP, Schlossmacher MG, Jacques TS, Scaravilli F, Eskelson C, Pepivani I, Hedrich K, Adel S, Gonzales-McNeal M, Hilker R, et al. 2005. Lewy body Parkinson's disease in a large pedigree with 77 Parkin mutation carriers. *Ann Neurol* **58:** 411–422.

Pridgeon JW, Olzmann JA, Chin LS, Li L. 2007. PINK1 protects against oxidative stress by phosphorylating mitochondrial chaperone TRAP1. *PLoS Biol* **5:** e172.

Rakovic A, Grunewald A, Kottwitz J, Bruggemann N, Pramstaller PP, Lohmann K, Klein C. 2011. Mutations in PINK1 and Parkin impair ubiquitination of Mitofusins in human fibroblasts. *PLoS ONE* **6:** e16746.

Rapaport D, Brunner M, Neupert W, Westermann B. 1998. Fzo1p is a mitochondrial outer membrane protein essential for the biogenesis of functional mitochondria in *Saccharomyces cerevisiae*. *J Biol Chem* **273:** 20150–20155.

Reeve AK, Krishnan KJ, Elson JL, Morris CM, Bender A, Lightowlers RN, Turnbull DM. 2008. Nature of mitochondrial DNA deletions in substantia nigra neurons. *Am J Hum Genet* **82:** 228–235.

Rizzuto R, Pinton P, Carrington W, Fay FS, Fogarty KE, Lifshitz LM, Tuft RA, Pozzan T. 1998. Close contacts with the endoplasmic reticulum as determinants of mitochondrial Ca^{2+} responses. *Science* **280:** 1763–1766.

Ropper AH, Adams RD, Victor M, Samuels MA. 2009. *Adams and Victor's principles of neurology*. McGraw-Hill Medical, New York.

Saotome M, Safiulina D, Szabadkai G, Das S, Fransson A, Aspenstrom P, Rizzuto R, Hajnoczky G. 2008. Bidirectional Ca^{2+}-dependent control of mitochondrial dynamics by the Miro GTPase. *Proc Natl Acad Sci* **105:** 20728–20733.

Schapira AH. 2008. Mitochondria in the aetiology and pathogenesis of Parkinson's disease. *Lancet Neurol* **7:** 97–109.

Sha D, Chin LS, Li L. 2010. Phosphorylation of parkin by Parkinson disease-linked kinase PINK1 activates parkin E3 ligase function and NF-κB signaling. *Hum Mol Genet* **19:** 352–363.

Shi G, Lee JR, Grimes DA, Racacho L, Ye D, Yang H, Ross OA, Farrer M, McQuibban GA, Bulman DE. 2011. Functional alteration of PARL contributes to mitochondrial dysregulation in Parkinson's disease. *Hum Mol Genet* **20:** 1966–1974.

Shin JH, Ko HS, Kang H, Lee Y, Lee YI, Pletinkova O, Troconso JC, Dawson VL, Dawson TM. 2011. PARIS (ZNF746) repression of PGC-1α contributes to neurodegeneration in Parkinson's disease. *Cell* **144:** 689–702.

Shulman JM, De Jager PL, Feany MB. 2011. Parkinson's disease: Genetics and pathogenesis. *Annu Rev Pathol* **6:** 193–222.

Silvestri L, Caputo V, Bellacchio E, Atorino L, Dallapiccola B, Valente EM, Casari G. 2005. Mitochondrial import and enzymatic activity of PINK1 mutants associated to recessive parkinsonism. *Hum Mol Genet* **14:** 3477–3492.

Stowers RS, Megeath LJ, Gorska-Andrzejak J, Meinertzhagen IA, Schwarz TL. 2002. Axonal transport of mitochondria to synapses depends on milton, a novel *Drosophila* protein. *Neuron* **36:** 1063–1077.

Suen DF, Narendra DP, Tanaka A, Manfredi G, Youle RJ. 2010. Parkin overexpression selects against a deleterious mtDNA mutation in heteroplasmic cybrid cells. *Proc Natl Acad Sci* **107:** 11835–11840.

Tanaka A, Cleland MM, Xu S, Narendra DP, Suen DF, Karbowski M, Youle RJ. 2010. Proteasome and p97 mediate mitophagy and degradation of mitofusins induced by Parkin. *J Cell Biol* **191:** 1367–1380.

Thomas KJ, McCoy MK, Blackinton J, Beilina A, van der Brug M, Sandebring A, Miller D, Maric D, Cedazo-Minguez A, Cookson MR. 2011. DJ-1 acts in parallel to the PINK1/parkin pathway to control mitochondrial function and autophagy. *Hum Mol Genet* **20:** 40–50.

Valente EM, Abou-Sleiman PM, Caputo V, Muqit MM, Harvey K, Gispert S, Ali Z, Del Turco D, Bentivoglio AR, Healy DG, et al. 2004. Hereditary early-onset Parkinson's disease caused by mutations in PINK1. *Science* **304:** 1158–1160.

Vives-Bauza C, Zhou C, Huang Y, Cui M, de Vries RL, Kim J, May J, Tocilescu MA, Liu W, Ko HS, et al. 2010. PINK1-dependent recruitment of Parkin to mitochondria in mitophagy. *Proc Natl Acad Sci* **107:** 378–383.

Wang X, Schwarz TL. 2009. The mechanism of Ca^{2+}-dependent regulation of kinesin-mediated mitochondrial motility. *Cell* **136:** 163–174.

Wang X, Winter D, Ashrafi G, Schlehe J, Wong YL, Selkoe D, Rice S, Steen J, LaVoie MJ, Schwarz TL. 2011. PINK1 and Parkin target Miro for phosphorylation and degradation to arrest mitochondrial motility. *Cell* **147:** 893–906.

Weihofen A, Thomas KJ, Ostaszewski BL, Cookson MR, Selkoe DJ. 2009. Pink1 forms a multiprotein complex with Miro and Milton, linking Pink1 function to mitochondrial trafficking. *Biochemistry* **48:** 2045–2052.

Whitworth AJ, Theodore DA, Greene JC, Benes H, Wes PD, Pallanck LJ. 2005. Increased glutathione *S*-transferase activity rescues dopaminergic neuron loss in a *Drosophila* model of Parkinson's disease. *Proc Natl Acad Sci* **102:** 8024–8029.

Whitworth AJ, Lee JR, Ho VM, Flick R, Chowdhury R, McQuibban GA. 2008. Rhomboid-7 and HtrA2/Omi act in a common pathway with the Parkinson's disease factors Pink1 and Parkin. *Dis Model Mech* **1:** 168–174; discussion 173.

Winget JM, Mayor T. 2010. The diversity of ubiquitin recognition: Hot spots and varied specificity. *Mol Cell* **38:** 627–635.

Woodroof HI, Pogson JH, Begley M, Cantley LC, Deak M, Campbell DG, van Aalten DMF, Whitworth AJ, Alessi DR, Muqit MMK. 2011. Discovery of catalytically active orthologues of the Parkinson's disease kinase PINK1: Analysis of substrate specificity and impact of mutations. *Open Biol* **1:** 110012.

Xiong H, Wang D, Chen L, Choo YS, Ma H, Tang C, Xia K, Jiang W, Ronai Z, Zhuang X, et al. 2009. Parkin, PINK1, and DJ-1 form a ubiquitin E3 ligase complex promoting unfolded protein degradation. *J Clin Invest* **119:** 650–660.

Xu G, Paige JS, Jaffrey SR. 2010. Global analysis of lysine ubiquitination by ubiquitin remnant immunoaffinity profiling. *Nat Biotechnol* **28:** 868–873.

Xu S, Peng G, Wang Y, Fang S, Karbowski M. 2011. The AAA-ATPase p97 is essential for outer mitochondrial membrane protein turnover. *Mol Biol Cell* **22:** 291–300.

Yang Y, Gehrke S, Imai Y, Huang Z, Ouyang Y, Wang JW, Yang L, Beal MF, Vogel H, Lu B. 2006. Mitochondrial pathology and muscle and dopaminergic neuron degeneration caused by inactivation of *Drosophila* Pink1 is rescued by Parkin. *Proc Natl Acad Sci* **103:** 10793–10798.

Yang Y, Ouyang Y, Yang L, Beal MF, McQuibban A, Vogel H, Lu B. 2008. Pink1 regulates mitochondrial dynamics through interaction with the fission/fusion machinery. *Proc Natl Acad Sci* **105:** 7070–7075.

Yoshii SR, Kishi C, Ishihara N, Mizushima N. 2011. Parkin mediates proteasome-dependent protein degradation and rupture of the outer mitochondrial membrane. *J Biol Chem* **286:** 19630–19640.

Zhou C, Huang Y, Shao Y, May J, Prou D, Perier C, Dauer W, Schon EA, Przedborski S. 2008. The kinase domain of mitochondrial PINK1 faces the cytoplasm. *Proc Natl Acad Sci* **105:** 12022–12027.

Ziviani E, Tao RN, Whitworth AJ. 2010. *Drosophila parkin* requires PINK1 for mitochondrial translocation and ubiquitinates mitofusin. *Proc Natl Acad Sci* **107:** 5018–5023.

Mitochondrial Trafficking in Neurons

Thomas L. Schwarz

F.M. Kirby Neurobiology Center, Children's Hospital Boston, and Department of Neurobiology, Harvard Medical School, Boston, Massachusetts 02115

Correspondence: Thomas.schwarz@childrens.harvard.edu

Neurons, perhaps more than any other cell type, depend on mitochondrial trafficking for their survival. Recent studies have elucidated a motor/adaptor complex on the mitochondrial surface that is shared between neurons and other animal cells. In addition to kinesin and dynein, this complex contains the proteins Miro (also called RhoT1/2) and milton (also called TRAK1/2) and is responsible for much, although not necessarily all, mitochondrial movement. Elucidation of the complex has permitted inroads for understanding how this movement is regulated by a variety of intracellular signals, although many mysteries remain. Regulating mitochondrial movement can match energy demand to energy supply throughout the extraordinary architecture of these cells and can control the clearance and replenishing of mitochondria in the periphery. Because the extended axons of neurons contain uniformly polarized microtubules, they have been useful for studying mitochondrial motility in conjunction with biochemical assays in many cell types.

THE IMPORTANCE OF MITOCHONDRIAL MOVEMENT TO NEURONS

Live imaging of mitochondria has transformed our understanding of these organelles from static lumps afloat in a cytoplasmic soup to animated actors that slide and fuse and divide (Jakobs 2006). Their dance is captivating, even when we are uncertain of its purpose. In neurons, and especially in their axons and dendrites, the trafficking of mitochondria is essential and perhaps more orderly than in other cells. Long-range intracellular transport in animal cells is accomplished primarily by microtubule-based motors—kinesins and dynein—and this is true also for the movement of mitochondria (Ligon and Steward 2000b; Hollenbeck and Saxton 2005). Axons contain linear arrays of uniformly polarized microtubules, with the minus ends in the cell body and the plus ends in the distal tips. This uniform polarity has made neurons particularly useful for studying transport. Moreover, in neuronal cultures, the axons lie flat and are typically about a micrometer in diameter. The mitochondria therefore move within an easily visualized plane and, as long as the cell body or growth cone of the axon can be identified, it is easy to distinguish plus-end from minus-end directed movement and to follow mitochondria for 100 μm or more. Moreover, whereas mitochondria in many cell types form a complex reticulum, axonal mitochondria have separated from the reticulum and exist as discrete organelles of typically 1–3 μm in length; for unknown reasons, those in dendrites tend to be longer than those in axons (Chang et al. 2006). In some preparations, shorter mitochon-

Cite this article as *Cold Spring Harb Perspect Biol* doi: 10.1101/cshperspect.a011304

dria are more frequently in motion than longer ones (Misgeld et al. 2007), but the relationship of size to motility is not yet clear.

The urgency of proper mitochondrial trafficking in neurons arises from their exceptional cellular morphology. Whereas most cells are measured in micrometers or tens of micrometers, neurons extend their axons and dendrites for millimeters, centimeters, and, in the case of human peripheral nerves or corticospinal tracts, up to a meter. Thus the neuron poses an extreme case for mitochondrial distribution and the need to supply energy to far-flung cellular regions. This challenge is probably why mutations of motor proteins found in all cell types give rise to selectively neurological pathologies in humans (Baloh 2007; De Vos et al. 2008); the stress on the axonal transport system is such that perturbations that another cell could survive cause degeneration in neurons.

Indeed, the regions of the neuron that have the highest demand for mitochondrial ATP production, that is, the highest energy consumption, are the synapses, and these specializations are often located at the extremities of the cell. When a single synaptic vesicle releases a transmitter at a synapse, it opens tens to hundreds of ion channels on the postsynaptic membrane, which allows up to 70 million ions to enter the cell at that point. This influx will require comparable millions of ATP to be hydrolyzed to pump the ions back across the plasma membrane. With hundreds or thousands of synapses impinging on a cell, it is easy to see why a recent imaging study determined that a resting cortical neuron consumes 4.7 billion ATP molecules per second (Zhu et al. 2012). It is not only the synapses that consume energy; action potentials also require ATP to restore ion gradients, axonal transport is energy intensive, and, of course, there are all the "normal" requirements of a cell. Thus, achieving the correct distribution of mitochondria may be a particularly challenging biological problem for the cell. It is not a question of simply recognizing the polarity of the cell and distinguishing an axon from a dendrite, but rather of distributing mitochondria to all regions so as to match their density to the local demand for ATP, Ca^{2+} buffering, and other mitochondrial functions. The task is further complicated by changes in energy use that require changes in mitochondrial distribution. An interesting illustration of this phenomenon is the behavior of mitochondria in a myelinated nerve, a phenomenon recently analyzed in detail (Ohno et al. 2011). Most mitochondria in myelinated nerves reside in stationary pools in the internode and juxtaparanode. Electrical activity, however, will slow or arrest the movement of mitochondria as they traverse the node of Ranvier and a blockade of electrical activity increases the motile fraction. These modulations of movement probably represent a response to the change in energy supply necessitated by the opening of ion channels at the nodes when action potentials are firing.

Nor is the traffic of mitochondria a one-time event to populate the periphery. These postmitotic cells need to survive for the lifetime of the organism, although the lifetime of mitochondrial proteins is likely to be a matter of weeks. This means there must be constant turnover of mitochondria throughout the cell, including the clearance of older, damaged components and the delivery of new materials, most of which are encoded by nuclear genes. Three processes are likely to contribute to this maintenance. The first is the constant movement of a population of mitochondria in axons and dendrites. Determinations of the fraction moving at any given moment vary, but typically fall in the range of 10%–40% in motion, with 60%–90% stationary (for examples, see Morris and Hollenbeck 1993; Ligon and Steward 2000a; Misgeld et al. 2007; Russo et al. 2009; Wang and Schwarz 2009). Of those in motion, roughly half are moving anterograde (away from the cell body) and half retrograde (toward the cell body). This movement likely reflects a major mechanism for the delivery and exchange of newly synthesized mitochondrial components. One topic of ongoing debate is whether the anterograde-moving population comprises healthy, refreshed mitochondria and those moving retrograde represent older, damaged mitochondria with lower mitochondrial membrane potentials, as determined from voltage-sensing dyes. An early study found that the membrane potential was indeed diminished in retrogradely moving mitochondria (Miller and Sheetz 2004), but a subsequent study saw no difference (Verburg and Hollenbeck 2008). But whether or not these populations differ, the fact that there is such extensive movement means that there is opportunity for mitochondrial proteins synthesized in the soma to be delivered to the periphery.

The second important factor for maintaining peripheral mitochondria is mitochondrial fission and fusion. Because these processes are observed in axons and dendrites, as they are in all cells, they allow the exchange of materials between mitochondria (Amiri and Hollenbeck 2008). Even brief moments of contact can involve fusion and the extensive exchange of proteins in each compartment of the mitochondrion, as has been shown in non-neuronal cells (Liu et al. 2009). Therefore, the mitochondria that appear to be stationary, at least for the duration of what can be successfully imaged, may be refreshed in place by exchanging proteins with the motile fraction. Mitochondrial fission and fusion are important to the viability of neuronal mitochondria; mutations in mitofusin give rise to a peripheral neuropathy (Zuchner et al. 2004; Kijima et al. 2005) and cause progressive decreases in the integrity of mitochondrial DNA and membrane potential (Chen et al. 2007). Naturally, if there were no mitochondrial movement, fusions would be unlikely to occur and indeed, moving mitochondria have a higher rate of fusion than stationary mitochondria (Liu et al. 2009) and the mechanisms of movement and fusion may be mechanistically coupled (Misko et al. 2010).

The third factor is the possibility of local biogenesis of mitochondria in addition to what occurs in the cell body. Although the extent to which local synthesis contributes to the mitochondrial pool is unknown (much certainly occurs in the soma [Davis and Clayton 1996]), there are two strong lines of evidence that it occurs. In an elegant experiment, Amiri and Hollenbeck (2008) explanted sympathetic ganglia and, after axons had grown out from the ganglion, excised the cell bodies and observed BrdU incorporation into the mitochondrial DNA of the severed axons. In addition to DNA replication, local synthesis of mitochondrial proteins is also likely; several studies have found transcripts for local synthesis of mitochondrial proteins (Gioio et al. 2001; Aschrafi et al. 2010; Yoon et al. 2012). Thus local biogenesis of mitochondria may assist in maintaining healthy mitochondria in axons and dendrites. Moreover, local biogenesis would have the potential to respond rapidly to local changes brought about by either mitochondrial damage or increased demand. At present, it is unknown whether the nucleus is informed about the state of the mitochondrial supply in the neuronal periphery, what such a signal would be, or how quickly it could resupply a particular region in the neuronal arbor.

A KEY MOTOR/ADAPTOR COMPLEX FOR MITOCHONDRIAL MOTILITY

Recently, a motor/adaptor complex was described that mediates the transport of mitochondria in neurons and possibly most animal cells (Fig. 1) (Stowers et al. 2002; Fransson et al. 2003, 2006; Brickley et al. 2005; Guo et al. 2005; Glater et al. 2006; Brickley and Stephenson 2011). The core of this complex consists of three proteins: the heavy chain of the conventional kinesin-1, a protein anchored to the outer sur-

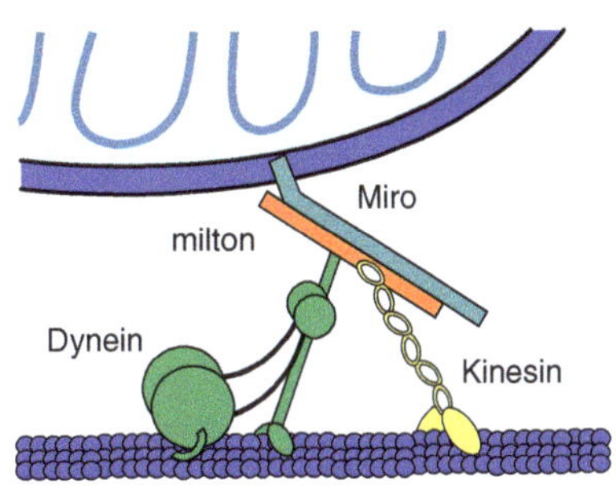

Figure 1. The mitochondrial motor/adaptor complex. The anterograde motor Kinesin-1 (also called kinesin heavy chain or Kif5) and the retrograde motor (the dynein/dynactin complex) are able to carry many cellular cargoes but are bound to mitochondria by their interactions with two mitochondrion-specific proteins: milton and Miro. Miro is carboxy-terminal anchored to the outer membrane of the mitochondrion. Milton serves as an adaptor between Miro and the motors. (Figure courtesy of Jarom Chung.)

face of the mitochondrion called Miro (also known as RhoT1 and RhoT2), and milton (also known as TRAK1 and TRAK2), a protein that links kinesin and Miro. Dynein, as will be discussed below, interacts with milton and Miro. The stoichiometry of the complex is not yet known, but the properties of the individual proteins and their relevance to mitochondrial transport have been elucidated through genetic, biochemical, and imaging studies.

The kinesin heavy chain (KHC) or kinesin-1 motor has three isoforms in mammals: Kif5A, B, and C. All have an amino-terminal motor domain that moves toward the plus end of the microtubule and a large coiled-coil tail that mediates its dimerization. Cargo binding occurs via the carboxy-terminal portion (reviewed in Vale 2003). The significance of KHC for mitochondrial movement is evident from the phenotypes of mutations in mouse (Tanaka et al. 1998) and *Drosophila* (Hurd and Saxton 1996) and from its association with the other proteins of the motor/adaptor complex (Stowers et al. 2002; Brickley et al. 2005; Glater et al. 2006).

The adaptor protein milton was named for the blind seventeenth century poet, John Milton, because it was identified in a *Drosophila* mutant screen for blind flies (Stowers et al. 2002). The significance of milton for mitochondrial transport was established by the phenotype of these mutants. In photoreceptors homozygous for the mutation, axons and terminals were completely devoid of mitochondria, although functional and morphologically normal mitochondria persisted in the somata of these cells (Stowers et al. 2002; Gorska-Andrzejak et al. 2003). Moreover, milton cofractionated with mitochondria, coprecipitated with KHC, and could cause the redistribution of mitochondria when overexpressed in COS cells (Stowers et al. 2002; Glater et al. 2006). The two mammalian isoforms of milton are called TRAK1 and TRAK2 (trafficking protein kinesin-binding) and sometimes GRIF1 and OIP106 for historical reasons, as discussed below (Beck et al. 2002; Iyer et al. 2003; Brickley et al. 2005), but for simplicity's sake will be discussed in aggregate as milton. Milton lacks an obvious domain structure other than an extensive coiled-coil region, which includes a significant region of homology with Huntingtin-associated protein 1 (HAP-1) (Stowers et al. 2002), a protein also implicated in organelle traffic (Gauthier et al. 2004). The ability of milton to associate with mitochondria when expressed in cells, to alter their distribution, and to bind KHC is shared by the fly and mammalian homologs (Stowers et al. 2002; Brickley et al. 2005, 2011; Glater et al. 2006; Smith et al. 2006). The interaction of TRAK1 with KHC is direct (Smith et al. 2006) and uses the cargo-binding region of KHC near the carboxyl terminus. Conventionally, the KHC binds cargo via the kinesin light chain (KLC), but milton is in competition with KLC for binding to this region of KHC (Glater et al. 2006). The role of KLC in mitochondrial trafficking is therefore unclear. Although it has been reported to be present in mitochondrial fractions (Khodjakov et al. 1998), it is not detected in the milton/Miro/KHC complex and is dispensable for mitochondrial transport in *Drosophila* photoreceptors (Glater et al. 2006) and cultured mammalian epithelial cells (Trejo et al. 2010). It seems likely that, for mitochondria, milton will replace KLC as the linker between cargo and motor.

Miro (Fig. 2) serves as the membrane anchor that links milton, and thereby kinesin, to the mitochondrial surface (Glater et al. 2006). Miro has two domains of homology with small GTPases and those domains are separated by a linker with two EF-hand motifs. At the carboxyl terminus there is a *trans*-membrane domain that tethers it to the outer surface of the outer

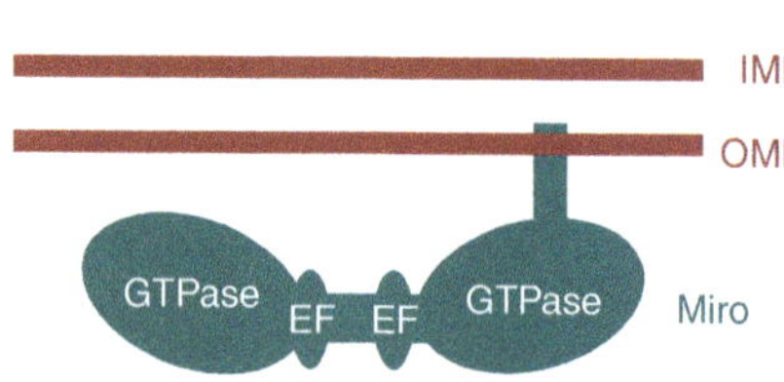

Figure 2. Domain structure of Miro (RhoT1/2). Miro has two GTPase domains that are distantly related to the small GTPases of the *ras* superfamily. These domains are separated by a pair of Ca^{2+}-binding EF hands. At the carboxyl terminus of the protein, a short *trans*-membrane domain anchors Miro in the outer mitochondrial membrane (OMM).

mitochondrial membrane (Fransson et al. 2003, 2006). There is a single fly Miro (Guo et al. 2005), but two mammalian isoforms that are sometimes called RhoT1 and RhoT2 (Aspenstrom et al. 2007). A more distant homolog in yeast, GEM1p, is necessary for proper mitochondrial morphology and inheritance (Frederick et al. 2004). Thus, the importance of the protein for regulating mitochondrial distribution appears to be very ancient, even though yeast does not use kinesin and microtubules for mitochondrial movement (Frederick and Shaw 2007). The importance of Miro for mitochondrial movement was hinted at by changes in mitochondrial distribution in overexpression studies in cells (Fransson et al. 2003), but established more directly by two parallel approaches. The first, a *Drosophila* mutant screen very similar to that which isolated *milton* alleles, isolated alleles of *miro* and found that they, too, lacked axonal and synaptic mitochondria (Guo et al. 2005). The untransported mitochondria formed unusual aggregates in the neuronal somata. In parallel, biochemical studies identified milton and Miro as direct binding partners (Fransson et al. 2006; Glater et al. 2006) and found that a truncated Miro lacking its membrane anchor could serve as a dominant negative construct that prevented milton from associating with mitochondria (Glater et al. 2006).

Biochemical data originally linked milton and Miro only to kinesin, but *Drosophila* phenotypes indicated that they may also be needed for dynein-mediated transport. The polarity of microtubules in *Drosophila* dendrites is the opposite of that in axons; dynein delivers cargo into dendrites and kinesins mediate movement back to the soma. In *milton* mutant flies, however, the neuropil regions of the central nervous system, although ~50% composed of dendrites, are lacking in mitochondria (Stowers et al. 2002), and their absence indicates that dynein-dependent movement of mitochondria was also disrupted. In addition, close examination of residual mitochondria in *miro* mutant axons indicated that both anterograde and retrograde movement was severely impaired (Guo et al. 2005). Recently, direct evidence for a biochemical interaction of the retrograde motor with the milton/Miro complex has been obtained (van Spronsen et al. 2013).

The abundant evidence linking milton and Miro to mitochondrial traffic does not preclude their involvement in the movement of other cargo and there are some indications that other cargos exist. Mammalian milton, for example, was also identified as a GABA receptor interacting factor (and consequently also referred to as GRIF-1) in a yeast two-hybrid screen (Beck et al. 2002); but in vivo evidence that it is required for GABA receptor transport is presently lacking. It has also been reported to interact with the endosomal protein Hrs (Kirk et al. 2006) and with a K^+ channel (Grishin et al. 2006).

OTHER MECHANISMS FOR MITOCHONDRIAL MOVEMENT

Is the milton/Miro/kinesin-heavy-chain complex the only mechanism by which mitochondria are transported? On the one hand, the phenotypes of *Drosophila* null alleles of *milton* and *miro* are extreme. In the cell types examined, mitochondria were absent from axons and dendrites. This finding, however, does not preclude the involvement of other motors and adaptors in other species, other cell types, or even within the cell bodies of neurons and there is some evidence for additional transport mechanisms. These include alternative kinesin-based mechanisms. Kif1Bα, for example, is a Kinesin-3 motor and was reported to associate with mitochondria and to be capable of transporting them in an in vitro assay (Nangaku et al. 1994), but in vivo evidence for their involvement remains scant and null alleles of the *Drosophila* homolog, *immaculate connections*, has normal transport of mitochondria (Pack-Chung et al. 2007). This family of motors is, however, critical for the transport of synaptic vesicle precursors (Hall and Hedgecock 1991; Yonekawa et al. 1998; Pack-Chung et al. 2007; Barkus et al. 2008). A distantly related member of the Kinesin-3 family, kinesin-like protein 6 (KLP6), was found to alter mitochondrial shape in *Caenorhabditis elegans* and mitochondrial motility in HeLa cells and a neural cell line (Tanaka et al. 2011) when mutated or knocked down with RNA interference (RNAi). Whereas loss of the conventional kinesin heavy chain prevented mitochondria from entering axons at all, mutant forms of KLP6 and Kif1Bα altered the velocity of axonal mitochondria (Tanaka et al. 2011). Both of these Kinesin-3 motors have been shown to bind to a candidate adaptor protein, the kinesin binding protein (KBP). Loss of KBP can alter the distribution of mitochondria in cultured cells (Wozniak et al. 2005) and in zebrafish (Lyons et al. 2008), although the phenotype in the latter indicated that the mitochondrial phenotype was likely to be secondary to disruption of the microtubule cytoskeleton. Three additional binding partners for KHC bear mentioning also: FEZ1, syntabulin, and Ran-binding protein 2 (Cai et al. 2005; Cho et al. 2007; Fujita et al. 2007; Ikuta et al. 2007). Interfering with the function of each of these can also alter mitochondrial distribution, although the mechanism of that effect is unclear. The function of these additional kinesins and binding partners, therefore, and their relationship to the milton/Miro/KHC complex remain an intriguing issue.

In addition to microtubule-based motility, actin-based mechanisms may have a role in distributing mitochondria (Morris and Hollenbeck 1993). In yeast, this is the primary mode of mitochondrial motility (Frederick and Shaw 2007), and in neurons it may mediate movement in regions where microtubules are interrupted or absent. A study of locust photoreceptors found that mitochondria moved toward and away from the phototransduction apparatus of the cell in a light-dependent manner. This movement was prevented by disruption of the actin cytoskeleton (Sturmer et al. 1995). More recently, Myo19 was shown to be a mitochondria-associated myosin and its overexpression in a neuronal cell line increased mitochondrial motility, whereas expression of a dominant-negative tail construct decreased their movement by 40% (Quintero et al. 2009). In contrast, Myosin V and Myosin VI may have the opposite role, resisting microtubule-based mechanisms for mitochondrial movement (Pathak et al. 2010). In cultured *Drosophila* neurons, RNAi knockdown of these proteins increased net movement and run lengths. The function of the myosins in these cells may be to remove mitochondria from microtubules and potentially tether them to the actin cytoskeleton to create a stationary pool (Pathak et al. 2010). How the myosins bind to mitochondria and the mechanisms by which they influence mitochondrial movement, however, remain unknown.

REGULATION OF MITOCHONDRIAL MOVEMENT: THE STATIONARY POOL

Movement alone cannot achieve the correct distribution of mitochondria within the neuronal arbor—mitochondrial movement must be closely regulated. Elucidating the machinery that moves mitochondria has begun to yield insights into how that movement is regulated, but many fundamental questions remain. One of the most striking features of mitochondria in axons is that they generally fall into three classes: those that move primarily anterograde (~15%), those that move primarily retrograde (~15%), and those that appear to be stationary for an extended period (~70%). What establishes these three categories or determines when a mitochondrion will switch its behavior remains largely mysterious. One possible mechanism would be that motors bind and unbind from the surface of the organelle and thereby determine whether and how it moves. At present, this mechanism seems unlikely. Neurons transfected with fluorescently tagged KHC, together with milton and Miro, revealed that KHC was present, at least under these circumstances, on the entire population of axonal mitochondria and at roughly equal levels, whether they were stationary, moving anterograde, or moving in the dynein-driven retrograde direction (Wang and Schwarz 2009). But some indications of how these populations are determined have been found. The above-mentioned studies of myosin V and VI in *Drosophila*, for example, suggest that tethering to the actin cytoskeleton may be one means of halting microtubule-based transport and creating a stationary pool (Pathak et al. 2010) (Fig. 3A).

Another likely anchoring mechanism may predominate in species that, unlike *Drosophila* (Goldstein and Gunawardena 2000), contain

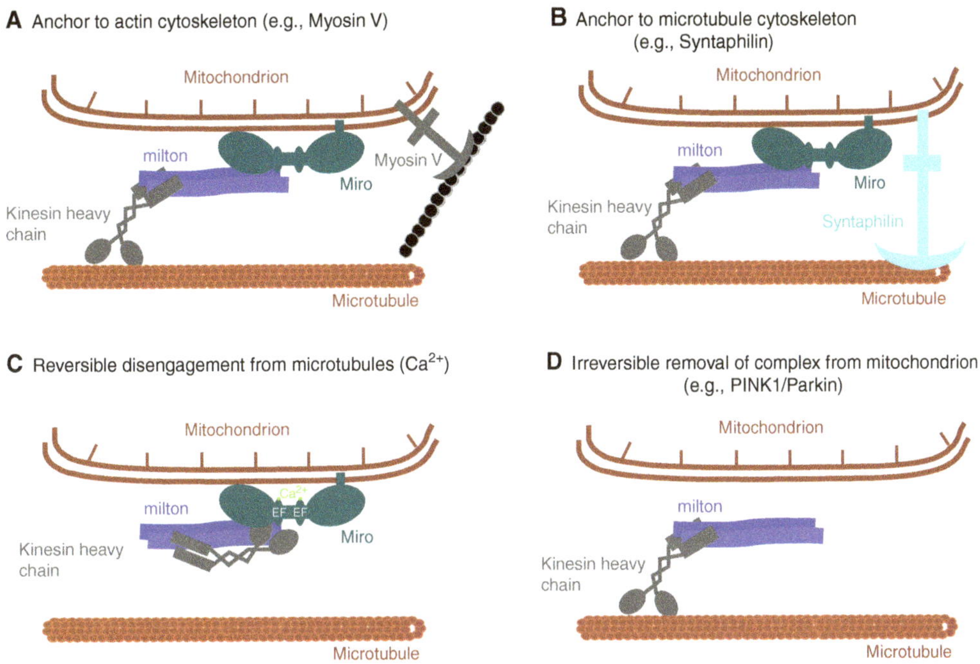

Figure 3. Four ways to stop a mitochondrion. (*A*) Myosin on mitochondria can inhibit their transport by tethering mitochondria to the axonal actin cytoskeleton (Pathak et al. 2010). (*B*) Syntaphilin is present on stationary axonal mitochondria and anchors them through interactions with microtubules (Kang et al. 2008). (*C*) Ca^{2+} binding to Miro causes a rearrangement of the complex such that the motor domain of kinesin directly binds to Miro and is blocked from binding to microtubules (Wang and Schwarz 2009). When cytosolic Ca^{2+} is lowered, this rearrangement can be reversed to permit continued kinesin-powered movement. (*D*) The PINK1/Parkin pathway causes an irreversible dissociation of the motors from the mitochondrial surface by causing Miro to be degraded. Miro degradation by the proteasome is triggered by PINK1 phosphorylation of Miro and Parkin ubiquitinylation of Miro (Wang et al. 2011).

neurofilaments. Mitochondria have been seen by electron microscopy to be associated with neurofilaments in frozen, deep-etched axonal tissue (Hirokawa 1982). This interaction may be mediated by the direct binding to mitochondria of the phosphorylated side arms of the neurofilament heavy chain (Wagner et al. 2003). Indeed, a substantial literature associates mitochondria with intermediate filament proteins in many vertebrate cell types (e.g., Milner et al. 2000; Tao et al. 2009; Nekrasova et al. 2011). An important factor in this association may be the intermediate-filament binding protein plectin (Reipert et al. 1999; Rezniczek et al. 2003; Winter et al. 2008).

One of the most intriguing anchoring mechanisms that have been proposed involves the association of mitochondria with microtubules via the protein syntaphilin (Kang et al. 2008) (Fig. 3B). Syntaphilin is anchored in the outer membrane of the mitochondrion and also contains a microtubule-binding domain. Syntaphilin-GFP was observed preferentially on stationary mitochondria, but not if the microtubule-binding region was deleted. The dynein light chain LC8 also appears to be involved with syntaphilin-based docking (Chen et al. 2009). Mice lacking syntaphilin were found to have substantially more mitochondria in the mobile fraction (Kang et al. 2008). However, although most mitochondria are stationary in both axons and dendrites, syntaphilin is exclusively axonal (Kang et al. 2008). Moreover, syntaphilin is absent from the genome of *Drosophila*, although flies have a stationary pool similar to mammals. Thus, although several potential mechanisms for creating a stationary pool have been reported, the relative importance of each may vary and their interplay is still unclear.

REGULATING MITOCHONDRIAL MOVEMENT: CYTOSOLIC Ca^{2+}

Cytosolic Ca^{2+} is one of the best-studied regulators of mitochondrial movement. Elevation of cytosolic Ca^{2+} stops both the anterograde and retrograde movement of mitochondria in neurons and in many cell lines (Rintoul et al. 2003; Yi et al. 2004; Chang et al. 2006; Szabadkai et al. 2006). From simultaneous measurements of free Ca^{2+} levels and mitochondrial dynamics, it has been determined that 50% reductions in movement will occur at concentrations of approximately 400 nM Ca^{2+} and complete arrest in the low micromolar range in H9c2 cells (Yi et al. 2004; Saotome et al. 2008). Elevations in Ca^{2+} can arise from a variety of physiological stimuli, including activation of glutamate receptors in dendrites, action potentials in axons, and neuromodulators that trigger Ca^{2+} release from intracellular stores (Rintoul et al. 2003; Yi et al. 2004; Saotome et al. 2008; Ohno et al. 2011). The purpose of this regulation is not firmly established, but is widely believed to be a homeostatic response to local Ca^{2+} influx: if passing mitochondria are arrested in areas where there is high cytosolic Ca^{2+}, they will be able to provide both Ca^{2+} buffering capacity and ATP for the active transport of Ca^{2+} from the cell. In addition, subcellular areas with inadequate cytosolic ATP are likely to have local elevations of Ca^{2+}, and the arrest of mitochondria in such areas will correct the lack of ATP (Yang and Steele 2000; Wang and Schwarz 2009). Tests of these possibilities have recently been made possible by manipulations of the motor/adaptor complex that selectively remove the ability of Ca^{2+} to regulate mitochondrial movement. Under these circumstances, mitochondrial will no longer stop at activated synapses (Macaskill et al. 2009) and neurons become more susceptible to excitotoxic cell death (Wang and Schwarz 2009), presumably because they can no longer cope with the Ca^{2+} that entered when glutamate receptors were activated.

The mechanism underlying this regulation has been examined in some detail. The presence of EF hands in Miro suggested a likely locus for the Ca^{2+}-binding site and these EF hands could be disrupted by the substitution of lysine residues for two key glutamates (Miro^{KK} [Fransson et al. 2006]). Overexpression in neurons of Miro^{KK} had little effect on the behavior of mitochondria in conditions of resting Ca^{2+}, but potently disrupted the ability of elevated Ca^{2+} to arrest their movement (Saotome et al. 2008; Macaskill et al. 2009; Wang and Schwarz 2009). A surprise from these studies was that both the kinesin-dependent anterograde movement and the dynein-dependent retrograde movement lost their regulation by Ca^{2+} when Miro was mutated. Biochemical evidence has now confirmed that the dynein motor indeed interacts with the milton/Miro complex (van Spronsen et al. 2013).

How then does Ca^{2+} binding to the EF hands of Miro stop the kinesin-mediated motility of the mitochondrion? The anterograde movement is driven by the interaction of the head domain of KHC with microtubules, but Miro binds to milton and milton binds to the opposite end of KHC. A possible mechanism would involve the Ca^{2+} triggering the dissociation of the complex by a conformational change (Macaskill et al. 2009) that would uncouple the kinesin from the mitochondrial surface. This could either be a disruption of the KHC/milton interaction or the milton/Miro interaction. Though this may occur under some circumstances, there is substantial evidence against such a model: the three proteins continue to coprecipitate as a complex even in the presence of very high Ca^{2+} (2 mM); isolated mitochondria washed in high Ca^{2+} do not lose KHC from their surface; and in living neurons transfected with fluorescently tagged KHC, equivalent amounts of KHC are present on mitochondria before and after their arrest by elevated Ca^{2+} (Wang and Schwarz 2009). The latter experiment contained another surprise: KHC was present on all mitochondria, including those that were in the stationary pool or moving retrograde, in addition to those that moved anterograde. Apparently, the choice of direction and whether or not to be stationary is not governed by the association and dissociation of the motor. An alternative mechanism for Ca^{2+}-dependent arrest has been proposed (Fig. 3C) (Wang and Schwarz 2009) in which the kinesin switches between an active state that interacts with the microtubules and an inhibited state where the motor "head" domain of kinesin binds to the Ca^{2+}-bound state of Miro and is thereby prevented from interacting with the microtubules. The isolated KHC head domain will indeed bind to wild-type Miro, but only when Ca^{2+} is present, and not to $Miro^{KK}$, and this interaction prevents the head domain from cosedimenting with microtubules (Wang and Schwarz 2009). Two aspects of this mechanism are noteworthy. The first is that it is readily reversible and will allow movement to begin again as soon as Ca^{2+} is unbound from the EF hands. The second is that it explains the selective arrest of mitochondrial motility because the regulation by Ca^{2+} is not intrinsic to the motor domain, but instead is based on the interaction of that domain with the mitochondrion-specific adaptor Miro.

REGULATING MITOCHONDRIAL MOVEMENT: PINK1 AND PARKIN

Recent studies have elucidated an additional means of arresting mitochondria—one that is very different from the readily reversible arrest by Ca^{2+}. This mechanism involves two genes, *PINK1* and *Parkin*, which are intensively studied because mutations in them are responsible for hereditary forms of Parkinsonism (Kitada et al. 1998; Valente et al. 2004). PINK1 (PTEN-induced putative kinase 1) is a protein kinase and Parkin is an E3 ubiquitin ligase, and together they form a pathway that initiates the autophagic clearance of mitochondria with poor membrane potentials (reviewed elsewhere and see Whitworth and Pallanck 2009; Youle and Narendra 2011; Narendra et al. 2012). In brief, PINK1 is normally inactivated by import into mitochondria and subsequent proteolytic cleavage. When mitochondria are damaged or poisoned and do not have a sufficient membrane potential, PINK1 instead accumulates on the mitochondrial surface and causes the recruitment of Parkin. In a manner still poorly understood, ubiquitination of proteins on the mitochondrial surface initiates mitophagy. Whether as a prelude to mitophagy or perhaps just as an independent action, the PINK1/Parkin pathway also regulates mitochondrial morphology (Greene et al. 2003; Clark et al. 2006; Park et al. 2006; Yang et al. 2006; Poole et al. 2008) and this is mediated, at least in part, by targeting mitofusin for degradation (Poole et al. 2010; Tanaka et al. 2010; Chan et al. 2011). PINK1 is only effective in bringing about these changes when Parkin is present, but in some cases Parkin overexpression can alter mitochondria even in the absence of PINK1 (Clark et al. 2006; Yang et al. 2006), as might be expected in a pathway where PINK1 is upstream of Parkin.

The potential involvement of mitochondrial movement in this pathway was first indicated by the observation of a biochemical association between PINK1 and the Miro/milton complex (Weihofen et al. 2009). It was subsequently found that Parkin can also bind to this complex and that, although this association may be scant in healthy cells, it can be greatly increased if mitochondria are depolarized with carbonyl cyanide *m*-chlorophenylhydrazone (CCCP) (Wang et al. 2011). PINK1 can phosphorylate Miro on several sites, at least one of which, S156, is important for the subsequent stimulation of the Parkin function and mitochondrial arrest (Wang et al. 2011). Miro is also a substrate of Parkin (Liu et al. 2012). The consequence of activating this PINK1/Parkin pathway is the proteasome-dependent degradation of Miro and the consequent release of kinesin and milton from the mitochondrial surface (Chan et al. 2011; Wang et al. 2011) (Fig. 3D). The loss of Miro explains why overexpression of PINK1 or Parkin stops mitochondrial motion in both mammalian and *Drosophila* neurons and, as in the case of mitophagy, PINK1 acts upstream of Parkin and in a manner that completely depends on Parkin activity (Wang et al. 2011). Although mitochondrial movement is normal in $Parkin^{-/-}$ mice, there is more movement in *Drosophila* axons expressing either PINK1 or Parkin RNAi relative to controls. Thus, whereas mammalian neurons may use this pathway only under conditions of mitochondrial damage, in *Drosophila* it may be part of the ongoing control of mitochondrial dynamics. It is also perhaps noteworthy that Miro and milton bind to mitofusin (Misko et al. 2010), and thus the degradation of these key regulators of mitochondrial dynamics may be coordinated.

As noted above, the association of PINK1 and Parkin with Miro is stimulated by depolarization of the mitochondrion. Mitochondrial depolarization also arrests mitochondrial transport in axons and this arrest is accompanied by the recruitment of cytosolic Parkin to the mitochondrion. Moreover, the degradation of Miro proceeds before the wholesale loss of mitochondrial proteins by mitophagy (Wang et al. 2011). It is therefore likely, although as yet undemonstrated, that halting mitochondria in some manner facilitates their clearance by mitophagy. One attractive model is that the degradation of Miro and mitofusin are both important as an early step to quarantine damaged mitochondria and prevent them from fusing with other healthy mitochondria, thereby mixing their damaged components in with the healthy population. Moving mitochondria are much more likely to fuse than stationary mitochondria (Liu et al. 2009) and therefore, together with the degradation of mitofusin, the loss of Miro will produce a population of stationary, fragmented mitochondria that are ready for engulfment by autophagosomes. Because it is a step toward clearance of the organelle, it is appropriate that the mechanism of mitochondrial arrest is by proteolysis and therefore not readily reversible.

REGULATING MITOCHONDRIAL MOVEMENT: OTHER FACTORS

The Ca^{2+} and PINK1/Parkin pathways are very far from the totality of regulatory influences, although for many others mechanistic information remains scant. One of the most intriguing is the influence of the nerve growh factor (NGF), which can cause a local accumulation of mitochondria where an NGF-coupled bead makes contact with a sensory axon (Chada and Hollenbeck 2003, 2004). Mitochondria that enter the portion of the axon closest to the bead tend to stop in their transit and thereby account for the accumulation. NGF activation of TrkA receptors is responsible for this effect and the pathway appears to involve PI3 kinase and the actin cytoskeleton. In light of the evidence mentioned above, that myosin interactions with actin may derail mitochondria from microtubule tracks and thereby anchor them (Pathak et al. 2010), it is possible that the NGF is increasing this set of cytoskeletal interactions.

A potentially related pathway that regulates mitochondrial motility is activated by the growth factor lysophosphatidic acid (LPA) and mediated by the small GTPase RhoA (Minin et al. 2006). Either LPA or a constitutively active RhoA will inhibit mitochondrial movement in cultured cells, including a neuron-like *Drosophila* cell line. RhoA can regulate interactions with the actin cytoskeleton through at least two pathways. RhoA, via Rho-kinase, can activate myosin motors (Kimura et al. 1996) and thereby potentially remove mitochondria from their microtubule tracks. In addition, RhoA can activate the actin-regulatory formin-like protein mDia (diaphanous in the fly) (Watanabe et al. 1997). The latter pathway is both necessary for and sufficient to explain the inhibitory effects of RhoA and LPA and entails the anchoring of mitochondria to the actin cytoskeleton (Minin et al. 2006). Despite the apparent importance of the actin cytoskeleton for these mechanisms of mitochondrial anchoring, the intermediate filament protein vimentin can also interact with mitochondria and influence their movement (Nekrasova et al. 2011).

Other proteins that associate with the milton/Miro complex may be significant regulators. These include HUMMR, a protein whose expression is induced by hypoxic conditions and that biases mitochondrial movement in the anterograde direction by an unknown means (Li et al. 2009), and O-GlcNAc transferase (OGT), an enzyme that catalyzes the transfer of single *N*-acetylglucosamine sugars to serine and threonine residues (Iyer et al. 2003). Although this enzyme typically has only low-affinity interactions with its many substrates, it binds strongly to milton, which is also a substrate (Iyer et al. 2003). The association of OGT and milton is conserved in *Drosophila* (Glater et al. 2006) and, by an unknown mechanism, OGT expression can alter the distribution of mitochondria in cultured cells (Brickley et al. 2011). Both HUMMR, as a sensor of hypoxic stress, and OGT, as a likely sensor of metabolic state (Lazarus et al. 2009), may help mitochondria redistribute themselves in response to changing environmental and energetic conditions. In this regard, they resemble the Ca^{2+}-sensing mechanism described above that can arrest mitochondria in regions of high-energy demand. A parallel mechanism for controlling the distribution of mitochondria entails the ability of ADP to decrease mitochondrial motility (Mironov 2007, 2009). Because ADP levels will be high where ATP consumption is high, the arrest of mitochondria by elevated ADP will cause ATP-producing mitochondria to concentrate where they are most needed.

The most recent addition to the list of motility regulators are the proteins of the Armcx family, including Alex3 (Lopez-Domenech et al. 2012). This family, characterized by multiple armadillo-like repeats, is unique to the placental mammals, localizes to mitochondria, and when overexpressed in cultured cells, including neurons, causes the mitochondria to aggregate near the nucleus. Alex3 binds to the Miro/milton complex and, when overexpressed, substantially decreases axonal transport of mitochondria (Lopez-Domenech et al. 2012). Also, as previously mentioned, mitofusins bind to the Miro/milton complex. This association could allow Miro to regulate mitochondrial fusion or allow mitofusin to regulate mitochondrial motility and, indeed, knockdown of the mitofusin2 expression decreased mitochondrial motility in axons (Misko et al. 2010). Mitochondrial movement and fusion are likely to be coordinated processes.

A very different form of regulation is likely to arise from the selection of the particular isoforms of milton and Miro that are found in a cell or on a subset of mitochondria. In mammals, there are two genes apiece for both milton and Miro (often called *TRAK1* and *TRAK2* and *RhoT1* and *RhoT2*). This diversity is augmented by alternative splicing. Both *Drosophila* and mammalian milton are found in multiple splice isoforms. Although there is no functional information at present on the significance of the mammalian splice variants, the splicing of *Drosophila* milton appears to be highly significant. In particular, of four known forms of the amino-terminal portion, one, milton-C, does not bind to kinesin heavy chain when expressed in cultured cells and does not recruit KHC to mitochondria, although it contains the KHC-binding sequence (Glater et al. 2006). This amino terminus, therefore, may be a regulatory domain that can influence the association and dissociation of the motor from the complex in vivo. Selective loss of splice isoforms in a *Drosophila* allele also suggested isoform-specific functions (Cox and Spradling 2006). The two mammalian genes, *TRAK1* and *TRAK2*, encode proteins that similarly differ in their interactions with the motors: *TRAK1* will bind both kinesin and dynein, but *TRAK2* preferentially interacts with dynein (van Spronsen et al. 2013).

Other regulatory signals may impinge on the motors themselves or on the microtubules and their associated proteins. These include microtubule-affinity regulating kinases (MARKs) (Mandelkow et al. 2004), JNK1 (Morfini et al. 2006, 2009), histone deacetylase 6 (Chen et al. 2010), and cyclin-dependent kinases (Ratner et al. 1998; Morfini et al. 2004; Holzbaur 2010). These influences, however, are unlikely to be selective for mitochondrial motility but rather will influence axonal transport broadly.

SUMMARY: DISTRIBUTING MITOCHONDRIA, COMPLEX SOLUTIONS TO A COMPLEX PROBLEM

Given the complexity of the task facing the neuron—distributing mitochondria over long distances and matching their distribution to an ever-changing demand for energy—perhaps it is not surprising that such a rich and complex regulatory machinery has evolved. This apparatus needs to contend not only with the delivery of healthy mitochondria to the appropriate regions, but also to their retention in areas with a high requirement for their Ca^{2+}-buffering capacity and energy production. What has emerged so far is a core apparatus of milton and Miro that will link the kinesin and dynein motors to the mitochondrial surface and an exceptional number of signaling pathways and regulatory proteins that will interact with that core and with the other cytoskeletal elements, particularly the actin network, which can stabilize mitochondria. When this machinery is compromised there are pathological consequences that may include the impaired ability of neurons to survive excitotoxic stimuli or to be cleared from cells when damaged. There is much still to decipher. The stoichiometry of the motor/adaptor complex is unknown and these proteins cry out for greater structural information. The significance of distinct isoforms and splice variants remains unclear and many modulatory pathways still lack a detailed mechanistic understanding. Most pressingly perhaps, the in vivo relevance of the regulatory pathways remains unclear. Nevertheless, studies of motility in animal cells have uncovered a remarkably complex and dynamic life for the mitochondrion.

REFERENCES

**Reference is also in this collection.*

Amiri M, Hollenbeck PJ. 2008. Mitochondrial biogenesis in the axons of vertebrate peripheral neurons. *Dev Neurobiol* **68:** 1348–1361.

Aschrafi A, Natera-Naranjo O, Gioio AE, Kaplan BB. 2010. Regulation of axonal trafficking of cytochrome *c* oxidase IV mRNA. *Mol Cell Neurosci* **43:** 422–430.

Aspenstrom P, Ruusala A, Pacholsky D. 2007. Taking Rho GTPases to the next level: The cellular functions of atypical Rho GTPases. *Exp Cell Res* **313:** 3673–3679.

Baloh RH. 2007. Mitochondrial dynamics and peripheral neuropathy. *Neuroscientist* **14:** 12–18.

Barkus RV, Klyachko O, Horiuchi D, Dickson BJ, Saxton WM. 2008. Identification of an axonal kinesin-3 motor for fast anterograde vesicle transport that facilitates retrograde transport of neuropeptides. *Mol Biol Cell* **19:** 274–283.

Beck M, Brickley K, Wilkinson HL, Sharma S, Smith M, Chazot PL, Pollard S, Stephenson FA. 2002. Identification, molecular cloning, and characterization of a novel GABAA receptor-associated protein, GRIF-1. *J Biol Chem* **277:** 30079–30090.

Brickley K, Stephenson FA. 2011. Trafficking kinesin protein (TRAK)-mediated transport of mitochondria in axons of hippocampal neurons. *J Biol Chem* **286:** 18079–18092.

Brickley K, Smith MJ, Beck M, Stephenson FA. 2005. GRIF-1 and OIP106, members of a novel gene family of coiled-coil domain proteins: Association in vivo and in vitro with kinesin. *J Biol Chem* **280:** 14723–14732.

Brickley K, Pozo K, Stephenson FA. 2011. *N*-acetylglucosamine transferase is an integral component of a kinesin-directed mitochondrial trafficking complex. *Biochim Biophys Acta* **1813:** 269–281.

Cai Q, Gerwin C, Sheng ZH. 2005. Syntabulin-mediated anterograde transport of mitochondria along neuronal processes. *J Cell Biol* **170:** 959–969.

Chada SR, Hollenbeck PJ. 2003. Mitochondrial movement and positioning in axons: The role of growth factor signaling. *J Exp Biol* **206:** 1985–1992.

Chada SR, Hollenbeck PJ. 2004. Nerve growth factor signaling regulates motility and docking of axonal mitochondria. *Curr Biol* **14:** 1272–1276.

Chan NC, Salazar AM, Pham AH, Sweredoski MJ, Kolawa NJ, Graham RL, Hess S, Chan DC. 2011. Broad activation of the ubiquitin-proteasome system by Parkin is critical for mitophagy. *Hum Mol Genet* **20:** 1726–1737.

Chang DT, Honick AS, Reynolds IJ. 2006. Mitochondrial trafficking to synapses in cultured primary cortical neurons. *J Neurosci* **26:** 7035–7045.

Chen H, McCaffery JM, Chan DC. 2007. Mitochondrial fusion protects against neurodegeneration in the cerebellum. *Cell* **130:** 548–562.

Chen YM, Gerwin C, Sheng ZH. 2009. Dynein light chain LC8 regulates syntaphilin-mediated mitochondrial docking in axons. *J Neurosci* **29:** 9429–9438.

Chen S, Owens GC, Makarenkova H, Edelman DB. 2010. HDAC6 regulates mitochondrial transport in hippocampal neurons. *PLoS ONE* **5:** e10848.

Cho KI, Cai Y, Yi H, Yeh A, Aslanukov A, Ferreira PA. 2007. Association of the kinesin-binding domain of RanBP2 to KIF5B and KIF5C determines mitochondria localization and function. *Traffic* **8:** 1722–1735.

Clark IE, Dodson MW, Jiang C, Cao JH, Huh JR, Seol JH, Yoo SJ, Hay BA, Guo M. 2006. *Drosophila pink1* is required for mitochondrial function and interacts genetically with *parkin*. *Nature* **441:** 1162–1166.

Cox RT, Spradling AC. 2006. Milton controls the early acquisition of mitochondria by *Drosophila* oocytes. *Development* **133:** 3371–3377.

Davis AF, Clayton DA. 1996. In situ localization of mitochondrial DNA replication in intact mammalian cells. *J Cell Biol* **135:** 883–893.

De Vos KJ, Grierson AJ, Ackerley S, Miller CC. 2008. Role of axonal transport in neurodegenerative diseases. *Annu Rev Neurosci* **31:** 151–173.

Fransson A, Ruusala A, Aspenstrom P. 2003. Atypical Rho GTPases have roles in mitochondrial homeostasis and apoptosis. *J Biol Chem* **278:** 6495–6502.

Fransson S, Ruusala A, Aspenstrom P. 2006. The atypical Rho GTPases Miro-1 and Miro-2 have essential roles in mitochondrial trafficking. *Biochem Biophys Res Commun* **344:** 500–510.

Frederick RL, Shaw JM. 2007. Moving mitochondria: Establishing distribution of an essential organelle. *Traffic* **8:** 1668–1675.

Frederick RL, McCaffery JM, Cunningham KW, Okamoto K, Shaw JM. 2004. Yeast Miro GTPase, Gem1p, regulates mitochondrial morphology via a novel pathway. *J Cell Biol* **167:** 87–98.

Fujita T, Maturana AD, Ikuta J, Hamada J, Walchli S, Suzuki T, Sawa H, Wooten MW, Okajima T, Tatematsu K, et al. 2007. Axonal guidance protein FEZ1 associates with tubulin and kinesin motor protein to transport mitochondria in neurites of NGF-stimulated PC12 cells. *Biochem Biophys Res Commun* **361:** 605–610.

Gauthier LR, Charrin BC, Borrell-Pages M, Dompierre JP, Rangone H, Cordelieres FP, De Mey J, MacDonald ME, Lessmann V, Humbert S, et al. 2004. Huntingtin controls neurotrophic support and survival of neurons by enhancing BDNF vesicular transport along microtubules. *Cell* **118:** 127–138.

Gioio AE, Eyman M, Zhang H, Lavina ZS, Giuditta A, Kaplan BB. 2001. Local synthesis of nuclear-encoded mitochondrial proteins in the presynaptic nerve terminal. *J Neurosci Res* **64:** 447–453.

Glater EE, Megeath LJ, Stowers RS, Schwarz TL. 2006. Axonal transport of mitochondria requires milton to recruit kinesin heavy chain and is light chain independent. *J Cell Biol* **173:** 545–557.

Goldstein LS, Gunawardena S. 2000. Flying through the *Drosophila* cytoskeletal genome. *J Cell Biol* **150:** F63–F68.

Gorska-Andrzejak J, Stowers RS, Borycz J, Kostyleva R, Schwarz TL, Meinertzhagen IA. 2003. Mitochondria are redistributed in *Drosophila* photoreceptors lacking milton, a kinesin-associated protein. *J Comp Neurol* **463:** 372–388.

Greene JC, Whitworth AJ, Kuo I, Andrews LA, Feany MB, Pallanck LJ. 2003. Mitochondrial pathology and apoptotic muscle degeneration in *Drosophila parkin* mutants. *Proc Natl Acad Sci* **100:** 4078–4083.

Grishin A, Li H, Levitan ES, Zaks-Makhina E. 2006. Identification of γ-aminobutyric acid receptor-interacting factor 1 (TRAK2) as a trafficking factor for the K^+ channel Kir2.1. *J Biol Chem* **281:** 30104–30111.

Guo X, Macleod GT, Wellington A, Hu F, Panchumarthi S, Schoenfield M, Marin L, Charlton MP, Atwood HL, Zinsmaier KE. 2005. The GTPase dMiro is required for axonal transport of mitochondria to *Drosophila* synapses. *Neuron* **47:** 379–393.

Hall DH, Hedgecock EM. 1991. Kinesin-related gene unc-104 is required for axonal transport of synaptic vesicles in *C. elegans*. *Cell* **65:** 837–847.

Hirokawa N. 1982. Cross-linker system between neurofilaments, microtubules, and membranous organelles in frog axons revealed by the quick-freeze, deep-etching method. *J Cell Biol* **94:** 129–142.

Hollenbeck PJ, Saxton WM. 2005. The axonal transport of mitochondria. *J Cell Sci* **118:** 5411–5419.

Holzbaur E. 2010. Axonal transport: CDKs as traffic signals for motor-ists along the axon? *Curr Biol* **20:** R641–R642.

Hurd DD, Saxton WM. 1996. Kinesin mutations cause motor neuron disease phenotypes by disrupting fast axonal transport in *Drosophila*. *Genetics* **144:** 1075–1085.

Ikuta J, Maturana A, Fujita T, Okajima T, Tatematsu K, Tanizawa K, Kuroda S. 2007. Fasciculation and elongation protein ζ-1 (FEZ1) participates in the polarization of hippocampal neuron by controlling the mitochondrial motility. *Biochem Biophys Res Commun* **353:** 127–132.

Iyer SP, Akimoto Y, Hart GW. 2003. Identification and cloning of a novel family of coiled-coil domain proteins that interact with O-GlcNAc transferase. *J Biol Chem* **278:** 5399–5409.

Jakobs S. 2006. High resolution imaging of live mitochondria. *Biochim Biophys* **1763:** 561–575.

Kang JS, Tian JH, Pan PY, Zald P, Li C, Deng C, Sheng ZH. 2008. Docking of axonal mitochondria by syntaphilin controls their mobility and affects short-term facilitation. *Cell* **132:** 137–148.

Khodjakov A, Lizunova EM, Minin AA, Koonce MP, Gyoeva FK. 1998. A specific light chain of kinesin associates with mitochondria in cultured cells. *Mol Biol Cell* **9:** 333–343.

Kijima K, Numakura C, Izumino H, Umetsu K, Nezu A, Shiiki T, Ogawa M, Ishizaki Y, Kitamura T, Shozawa Y, et al. 2005. Mitochondrial GTPase mitofusin 2 mutation in Charcot–Marie–Tooth neuropathy type 2A. *Human Genet* **116:** 23–27.

Kimura K, Ito M, Amano M, Chihara K, Fukata Y, Nakafuku M, Yamamori B, Feng J, Nakano T, Okawa K, et al. 1996. Regulation of myosin phosphatase by Rho and Rho-associated kinase (ρ-kinase). *Science* **273:** 245–248.

Kirk E, Chin LS, Li L. 2006. GRIF1 binds Hrs and is a new regulator of endosomal trafficking. *J Cell Sci* **119:** 4689–4701.

Kitada T, Asakawa S, Hattori N, Matsumine H, Yamamura Y, Minoshima S, Yokochi M, Mizuno Y, Shimizu N. 1998. Mutations in the *parkin* gene cause autosomal recessive juvenile parkinsonism. *Nature* **392:** 605–608.

Lazarus BD, Love DC, Hanover JA. 2009. O-GlcNAc cycling: Implications for neurodegenerative disorders. *Int J Biochem Cell Biol* **41:** 2134–2146.

Li Y, Lim S, Hoffman D, Aspenstrom P, Federoff HJ, Rempe DA. 2009. HUMMR, a hypoxia- and HIF-1α-inducible protein, alters mitochondrial distribution and transport. *J Cell Biol* **185:** 1065–1081.

Ligon LA, Steward O. 2000a. Movement of mitochondria in the axons and dendrites of cultured hippocampal neurons. *J Comp Neurol* **427:** 340–350.

Ligon LA, Steward O. 2000b. Role of microtubules and actin filaments in the movement of mitochondria in the axons and dendrites of cultured hippocampal neurons. *J Comp Neurol* **427:** 351–361.

Liu X, Weaver D, Shirihai O, Hajnoczky G. 2009. Mitochondrial "kiss-and-run": Interplay between mitochondrial motility and fusion–fission dynamics. *EMBO J* **28:** 3074–3089.

Liu S, Sawada T, Lee S, Yu W, Silverio G, Alapatt P, Millan I, Shen A, Saxton W, Kanao T, et al. 2012. Parkinson's disease-associated kinase PINK1 regulates Miro protein level and axonal transport of mitochondria. *PLoS Genet* **8:** e1002537.

Lopez-Domenech G, Serrat R, Mirra S, D'Aniello S, Somorjai I, Abad A, Vitureira N, Garcia-Arumi E, Alonso MT, Rodriguez-Prados M, et al. 2012. The Eutherian *Armcx* genes regulate mitochondrial trafficking in neurons and interact with Miro and Trak2. *Nat Commun* **3:** 814.

Lyons DA, Naylor SG, Mercurio S, Dominguez C, Talbot WS. 2008. KBP is essential for axonal structure, outgrowth and maintenance in zebrafish, providing insight into the cellular basis of Goldberg-Shprintzen syndrome. *Development* **135:** 599–608.

Macaskill AF, Rinholm JE, Twelvetrees AE, Arancibia-Carcamo IL, Muir J, Fransson A, Aspenstrom P, Attwell D, Kittler JT. 2009. Miro1 is a calcium sensor for glutamate receptor-dependent localization of mitochondria at synapses. *Neuron* **61:** 541–555.

Mandelkow EM, Thies E, Trinczek B, Biernat J, Mandelkow E. 2004. MARK/PAR1 kinase is a regulator of microtubule-dependent transport in axons. *J Cell Biol* **167:** 99–110.

Miller KE, Sheetz MP. 2004. Axonal mitochondrial transport and potential are correlated. *J Cell Sci* **117:** 2791–2804.

Milner DJ, Mavroidis M, Weisleder N, Capetanaki Y. 2000. Desmin cytoskeleton linked to muscle mitochondrial distribution and respiratory function. *J Cell Biol* **150:** 1283–1298.

Minin AA, Kulik AV, Gyoeva FK, Li Y, Goshima G, Gelfand VI. 2006. Regulation of mitochondria distribution by RhoA and formins. *J Cell Sci* **119:** 659–670.

Mironov SL. 2007. ADP regulates movements of mitochondria in neurons. *Biophys J* **92:** 2944–2952.

Mironov SL. 2009. Complexity of mitochondrial dynamics in neurons and its control by ADP produced during synaptic activity. *Int J Biochem Cell Biol* **41:** 2005–2014.

Misgeld T, Kerschensteiner M, Bareyre FM, Burgess RW, Lichtman JW. 2007. Imaging axonal transport of mitochondria in vivo. *Nature Methods* **4:** 559–561.

Misko A, Jiang S, Wegorzewska I, Milbrandt J, Baloh RH. 2010. Mitofusin 2 is necessary for transport of axonal mitochondria and interacts with the Miro/Milton complex. *J Neurosci* **30:** 4232–4240.

Morfini G, Szebenyi G, Brown H, Pant HC, Pigino G, DeBoer S, Beffert U, Brady ST. 2004. A novel CDK5-dependent pathway for regulating GSK3 activity and kinesin-driven motility in neurons. *EMBO J* **23:** 2235–2245.

Morfini G, Pigino G, Szebenyi G, You Y, Pollema S, Brady ST. 2006. JNK mediates pathogenic effects of polyglutamine-expanded androgen receptor on fast axonal transport. *Nat Neurosci* **9:** 907–916.

Morfini GA, You YM, Pollema SL, Kaminska A, Liu K, Yoshioka K, Bjorkblom B, Coffey ET, Bagnato C, Han D, et al. 2009. Pathogenic Huntingtin inhibits fast axonal transport by activating JNK3 and phosphorylating kinesin. *Nat Neurosci* **12:** 864–871.

Morris RL, Hollenbeck PJ. 1993. The regulation of bidirectional mitochondrial transport is coordinated with axonal outgrowth. *J Cell Sci* **104:** 917–927.

Nangaku M, Sato-Yoshitake R, Okada Y, Noda Y, Takemura R, Yamazaki H, Hirokawa N. 1994. KIF1B, a novel microtubule plus end-directed monomeric motor protein for transport of mitochondria. *Cell* **79:** 1209–1220.

* Narendra D, Walker JE, Youle R. 2012. Mitochondrial quality control mediated by PINK1 and Parkin: Links to Parkinsonism. *Cold Spring Harb Perspect Biol* **4:** a011338.

Nekrasova OE, Mendez MG, Chernoivanenko IS, Tyurin-Kuzmin PA, Kuczmarski ER, Gelfand VI, Goldman RD, Minin AA. 2011. Vimentin intermediate filaments modulate the motility of mitochondria. *Mol Biol Cell* **22:** 2282–2289.

Ohno N, Kidd GJ, Mahad D, Kiryu-Seo S, Avishai A, Komuro H, Trapp BD. 2011. Myelination and axonal electrical activity modulate the distribution and motility of mitochondria at CNS nodes of Ranvier. *J Neurosci* **31:** 7249–7258.

Pack-Chung E, Kurshan PT, Dickman DK, Schwarz TL. 2007. A *Drosophila* kinesin required for synaptic bouton formation and synaptic vesicle transport. *Nat Neurosci* **10:** 980–989.

Park J, Lee SB, Lee S, Kim Y, Song S, Kim S, Bae E, Kim J, Shong M, Kim JM, et al. 2006. Mitochondrial dysfunction in *Drosophila PINK1* mutants is complemented by *parkin*. *Nature* **441:** 1157–1161.

Pathak D, Sepp KJ, Hollenbeck PJ. 2010. Evidence that myosin activity opposes microtubule-based axonal transport of mitochondria. *J Neurosci* **30:** 8984–8992.

Poole AC, Thomas RE, Andrews LA, McBride HM, Whitworth AJ, Pallanck LJ. 2008. The PINK1/Parkin pathway regulates mitochondrial morphology. *Proc Natl Acad Sci* **105:** 1638–1643.

Poole AC, Thomas RE, Yu S, Vincow ES, Pallanck L. 2010. The mitochondrial fusion-promoting factor mitofusin is a substrate of the PINK1/parkin pathway. *PLoS ONE* **5:** e10054.

Quintero OA, DiVito MM, Adikes RC, Kortan MB, Case LB, Lier AJ, Panaretos NS, Slater SQ, Rengarajan M, Feliu M, et al. 2009. Human Myo19 is a novel myosin that associates with mitochondria. *Curr Biol* **19:** 2008–2013.

Ratner N, Bloom GS, Brady ST. 1998. A role for cyclin-dependent kinase(s) in the modulation of fast anterograde axonal transport: Effects defined by olomoucine and the APC tumor suppressor protein. *J Neurosci* **18:** 7717–7726.

Reipert S, Steinbock F, Fischer I, Bittner RE, Zeold A, Wiche G. 1999. Association of mitochondria with plectin and desmin intermediate filaments in striated muscle. *Exp Cell Res* **252:** 479–491.

Rezniczek GA, Abrahamsberg C, Fuchs P, Spazierer D, Wiche G. 2003. Plectin 5′-transcript diversity: Short alternative sequences determine stability of gene products, initiation of translation and subcellular localization of isoforms. *Human Mol Genet* **12:** 3181–3194.

Rintoul GL, Filiano AJ, Brocard JB, Kress GJ, Reynolds IJ. 2003. Glutamate decreases mitochondrial size and movement in primary forebrain neurons. *J Neurosci* **23:** 7881–7888.

Russo GJ, Louie K, Wellington A, Macleod GT, Hu F, Panchumarthi S, Zinsmaier KE. 2009. *Drosophila* Miro is required for both anterograde and retrograde axonal mitochondrial transport. *J Neurosci* **29:** 5443–5455.

Saotome M, Safiulina D, Szabadkai G, Das S, Fransson A, Aspenstrom P, Rizzuto R, Hajnoczky G. 2008. Bidirectional Ca^{2+}-dependent control of mitochondrial dynamics by the Miro GTPase. *Proc Natl Acad Sci* **105:** 20728–20733.

Smith MJ, Pozo K, Brickley K, Stephenson FA. 2006. Mapping the GRIF-1 binding domain of the kinesin, KIF5C, substantiates a role for GRIF-1 as an adaptor protein in the anterograde trafficking of cargoes. *J Biol Chem* **281:** 27216–27228.

Stowers RS, Megeath LJ, Gorska-Andrzejak J, Meinertzhagen IA, Schwarz TL. 2002. Axonal transport of mitochondria to synapses depends on milton, a novel *Drosophila* protein. *Neuron* **36:** 1063–1077.

Sturmer K, Baumann O, Walz B. 1995. Actin-dependent light-induced translocation of mitochondria and ER cisternae in the photoreceptor cells of the locust *Schistocerca gregaria*. *J Cell Sci* **108:** 2273–2283.

Szabadkai G, Simoni AM, Bianchi K, De Stefani D, Leo S, Wieckowski MR, Rizzuto R. 2006. Mitochondrial dynamics and Ca^{2+} signaling. *Biochim Biophys Acta* **1763:** 442–449.

Tanaka Y, Kanai Y, Okada Y, Nonaka S, Takeda S, Harada A, Hirokawa N. 1998. Targeted disruption of mouse conventional kinesin heavy chain, kif5B, results in abnormal perinuclear clustering of mitochondria. *Cell* **93:** 1147–1158.

Tanaka A, Cleland MM, Xu S, Narendra DP, Suen DF, Karbowski M, Youle RJ. 2010. Proteasome and p97 mediate mitophagy and degradation of mitofusins induced by Parkin. *J Cell Biol* **191:** 1367–1380.

Tanaka K, Sugiura Y, Ichishita R, Mihara K, Oka T. 2011. KLP6: A newly identified kinesin that regulates the morphology and transport of mitochondria in neuronal cells. *J Cell Sci* **124:** 2457–2465.

Tao GZ, Looi KS, Toivola DM, Strnad P, Zhou Q, Liao J, Wei Y, Habtezion A, Omary MB. 2009. Keratins modulate the shape and function of hepatocyte mitochondria: A mechanism for protection from apoptosis. *J Cell Sci* **122:** 3851–3855.

Trejo HE, Lecuona E, Grillo D, Szleifer I, Nekrasova OE, Gelfand VI, Sznajder JI. 2010. Role of kinesin light chain-2 of kinesin-1 in the traffic of Na,K-ATPase-containing vesicles in alveolar epithelial cells. *FASEB J* **24:** 374–382.

Vale RD. 2003. The molecular motor toolbox for intracellular transport. *Cell* **112:** 467–480.

Valente EM, Abou-Sleiman PM, Caputo V, Muqit MM, Harvey K, Gispert S, Ali Z, Del Turco D, Bentivoglio AR, Healy DG, et al. 2004. Hereditary early-onset Parkinson's disease caused by mutations in PINK1. *Science* **304:** 1158–1160.

van Spronsen M, Mikhaylova M, Lipka J, Schlager MA, van den Heuvel DJ, Kuijpers M, Wulf PS, Keijzer N, Demmers J, Kapitein LC, et al. 2013. TRAK/Milton motor-adaptor proteins steer mitochondrial trafficking to axons and dendrites. *Neuron* **77:** 485–502.

Verburg J, Hollenbeck PJ. 2008. Mitochondrial membrane potential in axons increases with local nerve growth factor or semaphorin signaling. *J Neurosci* **28:** 8306–8315.

Wagner OI, Lifshitz J, Janmey PA, Linden M, McIntosh TK, Leterrier JF. 2003. Mechanisms of mitochondria-neurofilament interactions. *J Neurosci* **23:** 9046–9058.

Wang X, Schwarz TL. 2009. The mechanism of Ca^{2+}-dependent regulation of kinesin-mediated mitochondrial motility. *Cell* **136:** 163–174.

Wang X, Winter D, Ashrafi G, Schlehe J, Wong YL, Selkoe D, Rice S, Steen J, LaVoie MJ, Schwarz TL. 2011. PINK1 and Parkin target Miro for phosphorylation and degradation to arrest mitochondrial motility. *Cell* **147:** 893–906.

Watanabe N, Madaule P, Reid T, Ishizaki T, Watanabe G, Kakizuka A, Saito Y, Nakao K, Jockusch BM, Narumiya S. 1997. p140mDia, a mammalian homolog of *Drosophila* diaphanous, is a target protein for Rho small GTPase and is a ligand for profilin. *EMBO J* **16:** 3044–3056.

Weihofen A, Thomas KJ, Ostaszewski BL, Cookson MR, Selkoe DJ. 2009. Pink1 forms a multiprotein complex with Miro and Milton, linking Pink1 function to mitochondrial trafficking. *Biochemistry* **48:** 2045–2052.

Whitworth AJ, Pallanck LJ. 2009. The PINK1/Parkin pathway: A mitochondrial quality control system? *J Bioenerg Biomembr* **41:** 499–503.

Winter L, Abrahamsberg C, Wiche G. 2008. Plectin isoform 1b mediates mitochondrion-intermediate filament network linkage and controls organelle shape. *J Cell Biol* **181:** 903–911.

Wozniak MJ, Melzer M, Dorner C, Haring HU, Lammers R. 2005. The novel protein KBP regulates mitochondria localization by interaction with a kinesin-like protein. *BMC Cell Biol* **6:** 35.

Yang Z, Steele DS. 2000. Effects of cytosolic ATP on spontaneous and triggered Ca^{2+}-induced Ca^{2+} release in permeabilised rat ventricular myocytes. *J Physiol* **523:** 29–44.

Yang Y, Gehrke S, Imai Y, Huang Z, Ouyang Y, Wang JW, Yang L, Beal MF, Vogel H, Lu B. 2006. Mitochondrial pathology and muscle and dopaminergic neuron degeneration caused by inactivation of *Drosophila* Pink1 is rescued by Parkin. *Proc Natl Acad Sci* **103:** 10793–10798.

Yi M, Weaver D, Hajnoczky G. 2004. Control of mitochondrial motility and distribution by the calcium signal: A homeostatic circuit. *J Cell Biol* **167:** 661–672.

Yonekawa Y, Harada A, Okada Y, Funakoshi T, Kanai Y, Takei Y, Terada S, Noda T, Hirokawa N. 1998. Defect in synaptic vesicle precursor transport and neuronal cell death in KIF1A motor protein-deficient mice. *J Cell Biol* **141:** 431–441.

Yoon BC, Jung H, Dwivedy A, O'Hare CM, Zivraj KH, Holt CE. 2012. Local translation of extranuclear lamin B promotes axon maintenance. *Cell* **148:** 752–764.

Youle RJ, Narendra DP. 2011. Mechanisms of mitophagy. *Nat Rev Mol Cell Biol* **12:** 9–14.

Zhu XH, Qiao H, Du F, Xiong Q, Liu X, Zhang X, Ugurbil K, Chen W. 2012. Quantitative imaging of energy expenditure in human brain. *Neuroimage* **60:** 2107–2117.

Zuchner S, Mersiyanova IV, Muglia M, Bissar-Tadmouri N, Rochelle J, Dadali EL, Zappia M, Nelis E, Patitucci A, Senderek J, et al. 2004. Mutations in the mitochondrial GTPase mitofusin 2 cause Charcot–Marie–Tooth neuropathy type 2A. *Nat Genet* **36:** 449–451.

Where Killers Meet—Permeabilization of the Outer Mitochondrial Membrane during Apoptosis

Tom Bender and Jean-Claude Martinou

Department of Cell Biology, University of Geneva, 1211 Geneva 4, Switzerland

Correspondence: jean-claude.martinou@unige.ch

Although mitochondria are usually considered as supporters of life, they are also involved in cellular death. Mitochondrial outer membrane permeabilization (MOMP) is a crucial event during apoptosis because it causes the release of proapoptotic factors from the mitochondrial intermembrane space to the cytosol. MOMP is mainly controlled by the Bcl-2 family of proteins, which consists of both proapoptotic and antiapoptotic members. We discuss the current understanding of how activating and inhibitory interactions within this family lead to the activation and oligomerization of MOMP effectors Bax and Bak, which result in membrane permeabilization. The order of events leading to MOMP is then highlighted step by step, emphasizing recent discoveries regarding the formation of Bax/Bak pores on the outer mitochondrial membrane. Besides the Bcl-2 proteins, the mitochondrial organelle contributes to and possibly regulates MOMP, because mitochondrial resident proteins and membrane lipids are prominently involved in the process.

Mitochondria are essential for the life of the cell. They produce most of the ATP via oxidative phosphorylation thanks to the respiratory chain that is embedded in the inner mitochondrial membrane. Consequently, mitochondrial dysfunction is implicated in the development of many human diseases, in particular, neurodegenerative disorders (Lin and Beal 2006). Mitochondria are also prominently involved in cell death, because they play a crucial role in many apoptotic responses. Apoptosis is a self-destruction program that is essential during the development of multicellular organisms. Its dysregulation has also been recognized as a main feature of many pathological conditions, especially cancer (Llambi and Green 2011).

The executioners of apoptosis are a family of cysteine proteases termed caspases that cleave a variety of cellular targets, resulting in morphological changes, degradation of genomic DNA, and, ultimately, phagocytic removal of the apoptotic cell (Taylor et al. 2008). Caspases are synthesized as inactive zymogens that become activated after regulated limited proteolysis. Two different pathways of apoptotic signaling that result in the activation of executioner caspases 3 and 7 can be distinguished. In the extrinsic pathway, binding of ligands such as FasL or

Cite this article as *Cold Spring Harb Perspect Biol* doi: 10.1101/cshperspect.a011106

TNFα to a death receptor on the plasma membrane leads to the activation of initiator caspase 8. Active caspase 8 propagates the signal by directly cleaving and thereby activating caspases 3 and 7, which continue a proteolytic cascade ultimately leading to the removal of the cell.

The intrinsic pathway, on the other hand, is initiated upon exposure to a number of stress situations, including DNA damage. A subclass of the Bcl-2 protein family termed BH3-only proteins (see below) becomes activated after an internal stress stimulus and translocates to the outer mitochondrial membrane (OMM), where they orchestrate a process called mitochondrial outer membrane permeabilization (MOMP). As an outcome of this process, pores are formed in the OMM, membrane integrity is lost, and contents of the intermembrane space gain access to the cytosol. One of the molecules that is rapidly released to the cytosol is cytochrome *c*, which is normally a soluble electron carrier between respiratory complexes III and IV. Together with the proapoptotic cytosolic factor APAF1, cytochrome *c* assembles into a caspase-activating complex termed the "apoptosome." This complex subsequently activates caspase 9, which is able to cleave caspases 3 and 7, proceeding with the same downstream cascade as in the extrinsic pathway. Other intermembrane space proteins also contribute to cell death after being released into the cytosol (e.g., SMAC/Diablo, which blocks the caspase inhibitor protein XIAP).

Remarkably, the two pathways are not completely independent. Cross talk between the extrinsic and intrinsic pathways exists because of caspase 8-dependent cleavage of the BH3-only protein Bid. Upon cleavage, Bid becomes activated, and the truncated version, tBid, translocates to the surface of mitochondria to induce MOMP. In so-called type II cells, this mitochondrial feedback loop is needed to induce apoptosis through the extrinsic pathway, because of the requirement of XIAP antagonism by SMAC.

The loss of OMM integrity caused by MOMP is usually considered the point of no return in the whole process, because cells are committed to die once MOMP is initiated. Therefore, this process represents a major checkpoint of apoptosis and must be tightly controlled to ensure that it is initiated at the right time and place. The main molecular players of MOMP belong to the Bcl-2 protein family. Integration of proapoptotic and antiapoptotic signals by the network of Bcl-2 proteins determines whether or not the OMM is permeabilized. In the following sections, we describe in detail the stimulatory and inhibitory protein–protein interactions within this family, discussing various models of how the MOMP effectors, Bax and Bak, become activated. Furthermore, we focus on the actual event of membrane permeabilization, summarizing the current understanding of how pores are formed in the OMM by Bax and Bak oligomers.

MITOCHONDRIA AS A PLATFORM FOR INTEGRATION OF APOPTOTIC SIGNALS

Models for Interaction between Bcl-2 Proteins

The family of Bcl-2 proteins can be subdivided both on the basis of the number of Bcl-2 homology (BH) domains they possess and on their proapoptotic or antiapoptotic function (Youle and Strasser 2008; Chipuk et al. 2010). Both the antiapoptotic Bcl-2-like proteins (Bcl-2, Bcl-XL, MCL-1, Bcl-w, A1) and proapoptotic members Bax and Bak share four short homologous regions of high conservation, termed BH1 to 4. A variety of other Bcl-2 proteins like Bid, Bim, Bad, Puma, Noxa, and others, belong to a third subgroup called the BH3-only proteins. As the name indicates, they contain the conserved BH3 region only and are, apart from that, very heterogeneous. BH3-only proteins, generally speaking, serve as sensors of intrinsic stimuli because in response to DNA damage or ER stress, specific members become activated by transcriptional induction (Puma and Noxa), posttranslational modification (Bad), or limited proteolysis (Bid). BH3-only proteins are thus proapoptotic and transmit the signal by translocating to the OMM, where they either activate Bax and Bak or inhibit the antiapoptotic family members. The subsequent step is the homo-oligomerization of activated Bax and Bak at the OMM. Because this oligomerization is essential for membrane permeabilization, Bax and Bak are often termed the "effectors" of MOMP. The essential role of Bax and Bak at the downstream end of the signaling cascade was shown in cells lacking both proteins, because those cells are resistant to MOMP following diverse apoptotic stimuli (Wei et al. 2001).

Proapoptotic effectors Bax and Bak and the antiapoptotic Bcl-2-like proteins share a similar structure, being small globular proteins that display a characteristic arrangement of seven amphipathic α-helices surrounding the central hydrophobic helix 5 (Muchmore et al. 1996; Suzuki et al. 2000; Moldoveanu et al. 2006). Interactions between individual proteins commonly involve the BH3 region located in helix 2. This region is also present in the BH3-only proteins, which, except for Bid, are believed to be intrinsically unstructured in solution and do not share any common features with other Bcl-2 family proteins except for the BH3 region. In nonapoptotic cells, Bax is primarily a cytosolic protein, whereas Bak and most of its antiapoptotic relatives are already attached to the OMM via a carboxy-terminal tail anchor (Lindsay et al. 2011). The carboxy-terminal α-helix 9 of Bax also serves as a tail anchor that is able to direct the protein to the OMM. However, the helix is sequestered in a hydrophobic groove on the surface of the same molecule, composed of the BH1 and 3 regions. The tail anchor is displaced from the groove upon activation, leading to a cytosol-to-mitochondria redistribution of the protein (Hsu et al. 1997; Eskes et al. 2000). Further structural rearrangements then cause oligomerization of Bax or Bak so that MOMP can proceed.

Several models have been discussed for the activation of MOMP effectors that involve protein–protein interaction between the subgroups of Bcl-2 proteins (Fig. 1). In the derepressor model, Bax and Bak are considered active by default but are kept in check by Bcl-XL or another antiapoptotic protein (Chen et al. 2005; Willis et al. 2005, 2007). Activation of the effectors, according to this model, is achieved if a BH3-only protein replaces Bax or Bak at the binding site on Bcl-XL so that the active effector

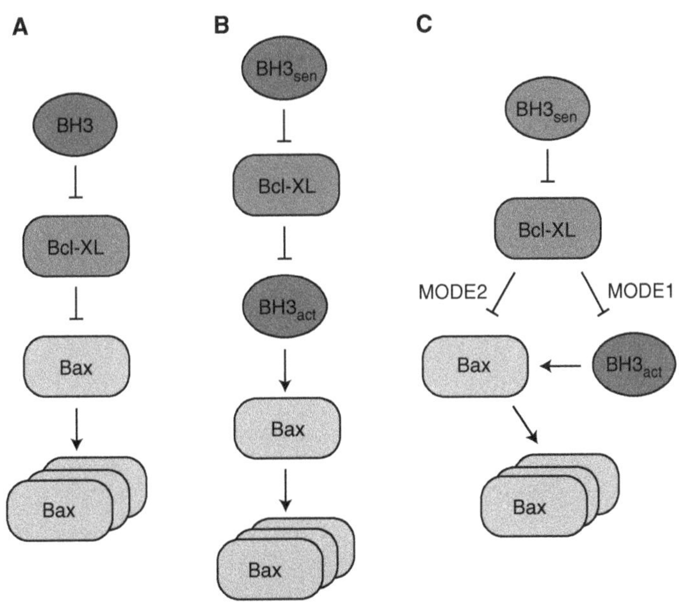

Figure 1. Models for activation of MOMP effectors. (*A*) Derepressor model. (*B*) Direct activator model. (*C*) Current model, which is a refinement of the two former models. Antiapoptotic Bcl-2-like Bcl-XL proteins inhibit both direct activator BH3-proteins (MODE1) and effectors Bax and Bak (MODE2). Inhibition through Bcl-XL is relieved by sensitizer BH3-only proteins.

becomes liberated. The direct activator model, however, proposes that Bax and Bak must be activated by direct contact with a BH3-only protein (Desagher et al. 1999; Kuwana et al. 2002; Letai et al. 2002; Terradillos et al. 2002; Kim et al. 2006). This model postulates two classes of BH3-only proteins. "Direct activators" such as tBid and Bim are able to activate MOMP effectors unless they are sequestered by antiapoptotic Bcl-2 proteins. To relieve this inhibition, "sensitizer" BH3-only proteins such as Bad or Noxa replace the direct activators by binding to Bcl-XL, so that Bax or Bak activation can occur. Indeed, by reconstituting membrane permeabilization in liposomes, direct activation of Bax by the BH3-only protein tBid could be shown (Kuwana et al. 2002; Lucken-Ardjomande et al. 2008b). Besides tBid, Bim and Puma have also been described as direct activators of Bax (Kuwana et al. 2005; Kim et al. 2006; Gallenne et al. 2009). Although the mode of action of Puma is more controversial and some investigators classify it as a sensitizer (Chipuk et al. 2008), the fact that a Bid/Bim/Puma triple knockout displays the same phenotype as a Bax/Bak double knockout strongly supports the notion of Puma being a direct activator (Ren et al. 2010).

Another model that has been proposed is termed "embedded together" because it emphasizes the role of the lipid bilayer in the interactions between proapoptotic and antiapoptotic Bcl-2 proteins (Leber et al. 2007, 2010). The model can be regarded as a refinement of the conflicting two former models because it integrates concepts from both (Fig. 1). In particular, antiapoptotic proteins are not seen as either inhibitors of activator BH3-only proteins (as in the "direct activator" model) or inhibitors of Bax and Bak (as in the "derepressor" model), but engage both. Like the "direct activator" model, the "embedded together" model differentiates between activator and sensitizer BH3 proteins. An elegant study using fluorescently labeled proteins in a reconstituted system in liposomes was able to establish a clear order of events during Bax activation (Lovell et al. 2008). Using FRET measurements between a variety of protein-attached fluorophores, different protein–protein interactions as well as membrane integration of proteins could be followed simultaneously and in real time. In this experimental setting, the first step is the targeting of tBid, the caspase 8-cleaved active form of Bid, to the membrane. Bax is subsequently recruited by tBid and integrates into the membrane, which can be inhibited by the addition of Bcl-XL. When Bax becomes activated at the membrane, the rate-limiting step is its oligomerization, which involves the recruitment and activation of more Bax molecules (Tan et al. 2006; Lovell et al. 2008). While Bax oligomerizes, liposome contents are released because of membrane permeabilization at the same time. By examining several Bax mutants, it could be confirmed that activation of Bax proceeds in a stepwise manner, and membrane targeting and oligomerization are separate events (Lucken-Ardjomande et al. 2008a; Kim et al. 2009). Interestingly, Bad could displace tBid from Bcl-XL, again confirming its role as a sensitizer (Lovell et al. 2008). Thus, both the initial recruitment of tBid and the activation of Bax require the presence of a membrane. We discuss the role of different membrane lipids in Bax activation below, including the controversial issue of whether or not cardiolipin is involved in this process.

The "embedded together" model not only proposes an important role for the mitochondrial outer membrane in the activation of Bax by direct activators Bid and Bim, but also for the recruitment of antiapoptotic Bcl-2 proteins. Just like Bax, its antiapoptotic opponent Bcl-2 undergoes extensive structural rearrangements when it becomes active at the OMM. These structural changes, which are required for inhibiting Bax oligomerization, appear to be initialized upon binding to activator BH3 proteins (Dlugosz et al. 2006). Furthermore, both Bax and Bcl-XL were shown to be recruited to the membrane by tBid in a comparable way in another study (Billen et al. 2008). Based on these observations, it was proposed that Bcl-XL acts as a dominant-negative version of Bax, providing an explanation for why two structurally similar proteins have opposing activities (Billen et al. 2008). Whereas Bax becomes activated by tBid, Bcl-XL remains associated with this activator, thus inhibiting its function. Likewise,

 Cite this article as *Cold Spring Harb Perspect Biol* doi: 10.1101/cshperspect.a011106

Bax could interact with Bcl-XL as it would with another Bax molecule during oligomerization. However, in this case, Bax oligomerization would be prevented because no membrane-perforating higher-order structure can be formed by Bax/Bcl-XL heterodimers.

Further refinement to the "embedded together" model was made by a recent study that used an elegant domain-shuffling approach between homologous proteins to dissect the antiapoptotic proteins' binding to activator versus effector proteins (Llambi et al. 2011). Because tBid is inhibited by Bcl-XL, Bcl-2, and MCL-1 equally well, the investigators constructed chimeric proteins in which the BH3 domain of tBid was replaced by the BH3 domain of Bax or Bak. This yielded direct activator protein versions that could still activate both Bax and Bak, but which were inhibited only by a subset of the three antiapoptotic proteins. This allowed the investigators to distinguish inhibition of direct activators (termed MODE1) from inhibition of effectors (termed MODE2), showing that the latter mode of inhibition is more efficient. A scenario was proposed in which direct activators are under MODE1 inhibition at low stress levels, whereas MODE2 inhibition prevents the initiation of MOMP when the amount of direct activator rises. Under MODE2 inhibition, Bax was shown to be in an active conformation and tightly bound to mitochondria. This must be distinguished from the weaker association with mitochondria of unactivated Bax, as described in another recent study (Edlich et al. 2011). Cytosolic Bax is in a constant equilibrium with the loosely mitochondria-associated form in the presence of Bcl-XL because Bcl-XL promotes its retrotranslocation back to the cytosol by a yet-undiscovered mechanism. However, even if a mutation in Bcl-XL prevents its interaction with Bax and Bax localizes to the OMM because of blocked retrotranslocation, MOMP is not readily caused (Edlich et al. 2011), presumably because of "MODE1" sequestration of direct activators (Llambi et al. 2011). Interestingly, in cancer cells, levels of BH3-only proteins are high, but inhibition by antiapoptotic Bcl-2 proteins still causes cell survival (Certo et al. 2006). This condition has been termed "primed to death," because BH3-only levels in these cancer cells are constantly close to the threshold that would initiate apoptosis. Therefore, BH3-only mimetics like the drug ABT-737 provide an attractive tool to push "primed" cells over the limit in favor of apoptosis (Certo et al. 2006). Indeed, a recent study showed that the responsiveness of cancer cells to chemotherapeutic treatment correlates with "priming" by BH3-only levels (Ni Chonghaile et al. 2011).

Structural Reorganization of Bax/Bak during Activation

The discussed models of MOMP initiation agree that BH3-only proteins provide the death signal for tipping the balance between proapoptotic and antiapoptotic Bcl-2 family members. Reconstitution of the whole process in liposomes, using fluorescently labeled recombinant proteins, has established the order of events leading to the activation of effectors and, ultimately, membrane permeabilization, as described above. But what is the structural basis for the interactions taking place in vivo on the mitochondrial surface? Which structural rearrangements lead to the transition of inactive cytosolic Bax to the active killer in the OMM, and what triggers them?

As observed for complexes of antiapoptotic Bcl-2 proteins with BH3 peptides derived from either activators or effectors, interactions involve the binding of a BH3 domain in the hydrophobic groove built from BH1 and BH3 (Sattler et al. 1997; Petros et al. 2000). The sequestration of a BH3 domain of a proapoptotic protein in the hydrophobic groove of an antiapoptotic inhibitor thus seems to be a common feature. However, it might also be assumed that all interactions between Bcl-2 proteins involve the BH3/groove contact, because the proapoptotic multidomain proteins also contain the groove structure on their surface. Interestingly, the hydrophobic groove of Bax in its cytosolic state is occupied by its own transmembrane helix, thereby preventing translocation to mitochondria. Therefore, in the case of Bax, the BH3 domain of an activator BH3-only protein would have to replace the transmembrane helix in the groove. However, it was shown that the stepwise activation of Bax by tBid leads to structural rearrangements at the amino terminus first (Kim et al. 2009). In the same study, it was proposed, based on the analysis of Bax mutants, that the displacement of helix 1 is required to release the tail-anchor helix 9. Hence, it is very likely that the first contact between an activating BH3 domain and Bax occurs at its amino-terminal region (Kim et al. 2009). This view is supported by the discovery of a binding region at helices 1 and 6, using an α-helical Bim-BH3-derived peptide stabilized by hydrocarbon stapling (Walensky et al. 2006; Gavathiotis et al. 2008). This novel interaction site is often termed the "rear pocket," because it is located on the opposite surface of Bax compared with the canonical BH1/BH3 groove. Binding of the stapled Bim BH3 helix leads to the displacement of the loop between helices 1 and 2 (Gavathiotis et al. 2008). This rearrangement at the amino terminus of Bax has been recognized for a long time as a structural feature accompanying Bax activation (Hsu and Youle 1997). Subsequently, the amino-terminal rearrangements allow the tail-anchor helix 9 to pop out of the groove and to integrate into the OMM (Fig. 2) (Gavathiotis et al. 2008; Kim et al. 2009).

The BH3 domain of Bax is exposed as a consequence of the amino- and carboxy-terminal rearrangements triggered by BH3 binding to the rear pocket (Gavathiotis et al. 2010). On the basis of similarity in primary structures of the Bim and Bax BH3 domains, it was proposed that the exposure of the Bax BH3 domain could provide yet another device that is able to recruit and activate further Bax molecules. In that case, the exposed Bax BH3 domain would act just like the BH3 domain of a direct activator BH3-only protein. This would be consistent with the observation that a small number of BH3-only proteins can recruit a molar excess of Bax to the membrane through a Bax-autoactivating mechanism (Tan et al. 2006). Indeed, experimental evidence obtained with a stapled Bax BH3 peptide similar to the Bim peptide described above suggests self-propagating autoactivation of Bax (Gavathiotis et al. 2010). These results provide a very attractive model of Bax oligomerization. After initial activation of Bax through interaction with a BH3 domain at the rear pocket, not only is the membrane-anchor helix 9 released, but also the BH3 domain becomes exposed. By engaging another Bax molecule at its rear pocket, the activation could be propagated from molecule to molecule, leading to a higher-order structure of Bax molecules in a homo-oligomer

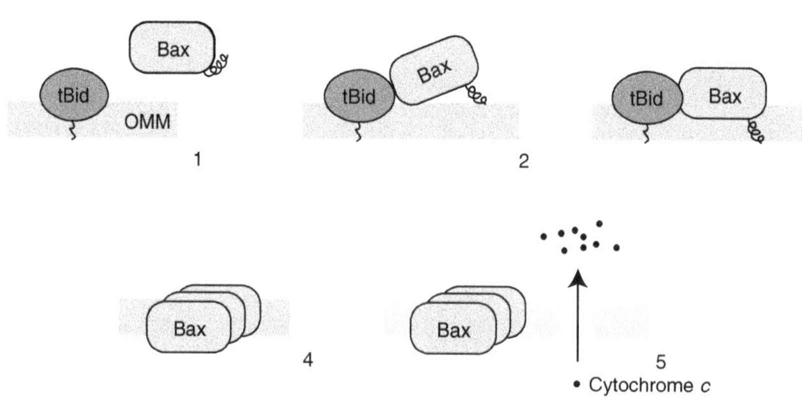

Figure 2. The order of events leading to MOMP. At first, an activator BH3 protein like tBid is targeted to the membrane (Step 1), which binds and recruits Bax to the membrane (Step 2). Upon activation, Bax exposes its carboxy-terminal membrane-anchor helix and becomes membrane integrated (Step 3). Activated Bax homo-oligomerizes (Step 4), which causes the formation of pores so that proapoptotic contents of the mitochondrial intermembrane space like cytochrome *c* can be released (Step 5).

 Cite this article as *Cold Spring Harb Perspect Biol* doi: 10.1101/cshperspect.a011106

at the OMM (Fig. 3A). Interestingly, if the release of helix 9 is blocked by mutation, the subsequent BH3 exposure does not lead to oligomerization (Gavathiotis et al. 2010), emphasizing that Bax must be targeted to the membrane in order to oligomerize.

In contrast to this asymmetric Bax oligomeric assembly, the formation of symmetric dimers has been suggested for Bak (Fig. 3B) (Dewson et al. 2008). The search for mutations in Bak that inhibit oligomer formation revealed that the BH1 and BH3 regions are prominently involved in this process. Moreover, cysteine residues were engineered in Bak, which were designed to match in a proposed reciprocal interaction of the BH3 domain of one monomer with the groove region of another. Because disulfide cross-linking stabilized Bak dimers, this is a strong indication of symmetric dimer formation. Finally, reciprocal mutations that allow noncovalent interactions between Bak monomers by the symmetric BH3/groove contact could complement the loss-of-function mutations in oligomerization, pointing to a functional significance of this interaction (Dewson et al. 2008). This evidence is supported by another study aimed at determining the minimal requirements for Bax oligomerization, which

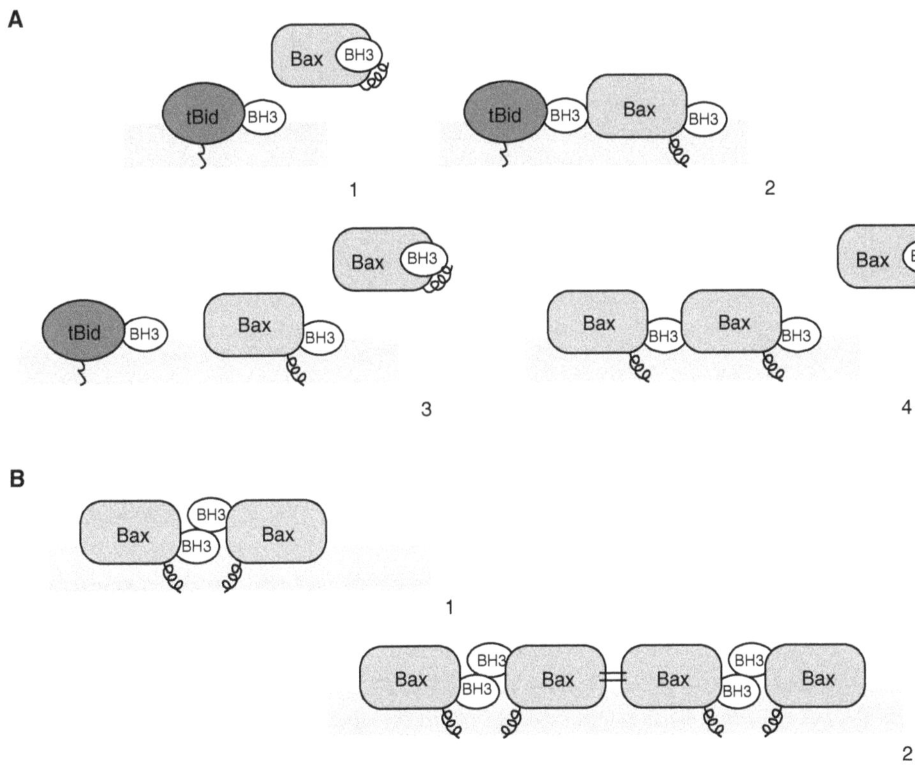

Figure 3. Bax activation and oligomerization. (*A*) Asymmetric autoactivation of Bax. Membrane-embedded tBid interacts through its BH3 domain with Bax at its rear pocket (Step 1). This results not only in the exposure of the Bax carboxy-terminal helix, but also in the exposure of the BH3 domain of Bax (Step 2). According to the asymmetric autoactivation model, the exposed BH3 domain of Bax can recruit further Bax molecules to the OMM (Step 3) and activate them (Step 4). (*B*) Symmetric dimer formation causes Bax oligomerization. Activated Bax, which exposes its BH3 domain, forms symmetric dimers by a reciprocal interaction of one molecule's BH3 domain with the hydrophobic groove of another, and vice versa (Step 1). Bax dimers can then form higher-order oligomers through another symmetric interaction involving the α-helices 6 at the surface opposite from the BH3–groove interaction (Step 2).

found helices 2, 4, and 5 to be sufficient for this process (George et al. 2007). This region overlaps with the BH1 and BH3 domains found to be essential in symmetric dimer formation. Furthermore, several biochemical studies confirmed symmetric dimers involving the BH3/groove interaction for both Bax and Bak (Bleicken et al. 2010; Oh et al. 2010; Zhang et al. 2010). This potential interaction, however, would be reciprocal, resulting in both the BH3 domains and hydrophobic grooves of Bax being occupied. Therefore, to form higher-order oligomers, as observed during MOMP, dimers would need to be connected by yet another intermolecular contact. This additional contact is provided by another symmetric contact, because helix 6 of one Bax molecule was shown to interact with helix 6 of another (Dewson et al. 2009). Therefore, dimers formed by the BH3/groove contact could oligomerize by helix 6/helix 6 interactions (Fig. 3B). A previously reported oligomerization mechanism via the coordination of zinc ions (Moldoveanu et al. 2006), on the other hand, could be ruled out (Dewson et al. 2009).

Thus, two conflicting views exist regarding Bax or Bak oligomer formation, based on the interaction of monomers in an either symmetric or asymmetric fashion. However, it is likely that neither model describes the mode of oligomerization exclusively. The symmetric model raises the question whether the BH3 exposure of one monomer, provoked by a direct activator, is sufficient to also activate the other monomer, or if this second monomer also has to be activated by a BH3-only direct activator binding to its rear pocket. The observation that BH3 ligands leave the rear pocket in a hit-and-run fashion and then seem to engage the effector at another site, possibly the BH1/BH3 region (Kim et al. 2009), suggests that the initial interactions are highly dynamic. Thus, it is possible to envision a scenario in which the exposed BH3 domain of activated Bax can either activate other Bax molecules via the asymmetric rear-pocket activation site or interact with an already activated Bax molecule via the symmetric BH3/groove interface to initialize oligomerization.

Pore Formation by Bax/Bak Oligomers

The activation and oligomerization of proapoptotic Bcl-2 family effector proteins Bax and Bak are the prerequisites for MOMP. Although over the last few years models and concepts have emerged regarding the interactions between different Bcl-2 family members, as well as the structural rearrangements of Bax and Bak causing their oligomerization, the molecular connection between Bax or Bak oligomers and the actual event of membrane permeabilization is less evident (Fig. 1). Experiments with fluorescent dextrans of different sizes encapsulated by liposomes have shown that pores formed by activated Bax allow the release of molecules with a broad range of molecular mass from 10 kDa to 2 MDa (Kuwana et al. 2002). Although another study suggests that the maximum size of intermembrane space contents released in vivo is lower (Rehm et al. 2003), it is generally agreed that Bax pores accommodate the release of all proapoptotic intermembrane space (IMS) molecules like cytochrome *c* or SMAC/Diablo.

So what is the mode of pore formation by oligomerized Bax? Bax alone is able to form pores in artificial planar lipid membranes (Antonsson et al. 1997; Schlesinger et al. 1997), although the experimental setup used in these experiments is differing substantially from the in vivo situation. The structural similarities of multidomain Bcl-2 proteins like Bax or Bcl-XL to bacterial pore-forming toxins like colicins, in particular, the arrangement of amphipathic helices surrounding a central hydrophobic helix, suggest that Bax itself is able to porate membranes (Muchmore et al. 1996). Indeed, following activation, helices 5 and 6 of Bax were shown to insert into the membrane (Annis et al. 2005), emphasizing that these central helices might play an important role in membrane permeabilization. Moreover, replacement of helix 5 of Bcl-XL by the respective Bax helix enabled pore formation by the chimeric protein, thereby turning Bcl-XL into a killer molecule (George et al. 2007). This supports the notion that Bax oligomers are forming a proteinaceous channel in the OMM. Patch-clamp measurements of pores formed by Bak have led to the description of a channel structure termed the mitochondrial apoptosis-induced channel (MAC) by Martinez-Caballero et al. (2009). This channel shows an increase in conductance over time, being consistent with the stepwise incorporation of activated Bax homodimers into an oligomeric Bax pore. Alternatively, Bax oligomers could provoke the opening of a preexisting mitochondrial pore structure like the permeability-transition pore. However, this was effectively ruled out by experiments in cells lacking the central components of this pore structure (Baines et al. 2005; Nakagawa et al. 2005). Likewise, an involvement of the voltage-dependent anion channel (VDAC) was ruled out (Baines et al. 2007), although Bak was shown to interact with VDAC at the OMM, VDAC having an inhibitory function on Bak (Cheng et al. 2003).

Apart from observations in favor of a proteinaceous channel formed directly by a complex of activated Bax or Bak subunits, evidence also exists that a lipidic pore is formed. In a scenario like this, Bax or Bak could act like another bacterial toxin called melittin, which induces the formation of inverted micelle structures in membranes, resulting in lipidic pores. This view is supported by observations that pore formation is facilitated by the incorporation of lipids into the membrane that induce positive membrane curvature (Basanez et al. 2002). Remarkably, the pores formed by isolated Bax helix 5 were shown to be lipidic in nature (Garcia-Saez et al. 2007; Qian et al. 2008), although, of course, a single helix might have different effects from full-length Bax. The analysis of pores formed by activated Bax in liposomes by cryo-electron microscopy, on the other hand, revealed that the pore edges are devoid of proteinaceous particles, thus strongly supporting the lipidic pore model (Schafer et al. 2009). Furthermore, a recent structural study using small-angle neutron scattering revealed that activated Bax in liposomes adopts a shape reminiscent of bacterial toxins known to form lipidic pores (Satsoura et al. 2012).

Yet another intriguing possibility for the mode of action of Bax oligomers was recently proposed by Montessuit et al. (2010). The researchers observed that the mitochondrial fission factor Drp-1 facilitates pore formation by Bax in liposomes. Interestingly, a direct interaction of Bax and Drp-1 was not responsible for this effect, but the shape of the membrane played a key role. In particular, the existence of hemifusion intermediates promotes Bax oligomerization in this study (Montessuit et al. 2010). This type of intermediate is formed transiently during both the mitochondrial fusion or fission process and is characterized by an interconnected stalk between the outer leaflets of the two lipid bilayers, whereas the inner leaflets remain intact (Chernomordik and Kozlov 2008). Complete separation of the membranes or the formation of a fusion pore then completes membrane fission or fusion, respectively. Although Drp1 has so far only been described as a fission factor, it seems that it can also support vesicle fusion to the intermediate stage under the experimental conditions used (Montessuit et al. 2010). Interestingly, intermediate structures common to membrane fusion or fission have been described to be transiently leaky. Moreover, any perturbations in the fusion process in general can lead to permanent membrane leakage and thus the loss of membrane integrity (Engel and Walter 2008). It is thus conceivable that transient leakage also occurs during mitochondrial fission. Because mitochondria commonly fragment during apoptosis, Bax oligomers might take advantage of membrane vulnerability, insert into fragile sections at or near hemifission intermediates, and cause further perturbation during the fission process, leading to a loss of mitochondrial integrity (Martinou and Youle 2011). Consistent with this model, Bax has formerly been shown to insert at constriction sites, where it colocalizes with components of both the fusion and fission machineries (Karbowski et al. 2002).

The important role of membrane properties, especially membrane shape, not only in the recruitment of Bcl-2 proteins to the OMM but also in pore formation itself, raises the question as to whether specific membrane lipids influence the course or the outcome of MOMP. The role of cardiolipin, a lipid restricted to mitochondrial membranes, has been controversially discussed in the literature. Several studies show that cardiolipin is required for tBid/Bax-induced permeabilization of liposomes (Kuwana et al. 2002; Terrones et al. 2004; Lucken-Ardjomande et al. 2008b). The targeting of tBid, in particular, seems to require an interaction with this membrane lipid (Lutter et al. 2000). It remains questionable, however, if cardiolipin is required for MOMP in vivo, because it is primarily found in the inner mitochondrial membrane and its concentration in the outer membrane is low. One possible explanation for this obvious discrepancy is that cardiolipin is enriched in contact sites between the outer and inner membrane (Lutter et al. 2001), in such a way that local concentrations could be high enough to promote Bax activation. Other investigators have suggested that in vivo outer membrane proteins can take over the role that cardiolipin has in liposomes (Schafer et al. 2009). Candidate proteins include components of the TOM receptor complex, components of the fusion or fission machineries, or the mitochondrial carrier homolog MCTH2/MIMP (Ott et al. 2009; Zaltsman et al. 2010).

An additional study introduced yet another possibility for the requirement of membrane lipids in MOMP. Chipuk et al. (2012) discovered that contaminants originating from the endoplasmic reticulum, which are usually present in the mitochondrial preparations used for in vitro experiments, contain an MOMP-promoting factor. By stepwise purification of this factor from the contaminating microsomal fraction, a neutral sphingomyelinase was identified that promoted membrane permeabilization by tBid and Bak or Bax. Further investigation of the sphingolipid metabolism then revealed that Bak and Bax interact in the membrane with the sphingolipid metabolites sphigosin-1-phosphate and hexadecenal, respectively (Chipuk et al. 2012). Interestingly, this interaction resulted in the exposure of the 6A7 epitope in Bax, indicating activating structural rearrangements (see above). One might thus speculate that interaction of MOMP effectors with specific lipid components cooperates with direct activator BH3-only proteins by lowering the activation energy for the required structural remodeling. Furthermore, the researchers showed that cardiolipin is not needed for Bax activation in liposomes if the respective sphingolipid metabolite, hexadecenal, is present (Chipuk et al. 2012). All of these observations give rise to a potentially important role of the contact between mitochondria and the endoplasmic reticulum during apoptosis.

CONCLUDING REMARKS

Besides hosting numerous essential metabolic pathways, mitochondria are also prominently involved in cell death. In vertebrate cells, death signals arising from internal stress-sensing pathways converge on the outer mitochondrial membrane, where the effector Bcl-2 proteins Bax and Bak become activated, integrate into the membrane, homo-oligomerize, and thereby cause the loss of membrane integrity. The participation in the execution of apoptosis is just one example of how mitochondria are part of an interconnected cellular network that not only maintains life but also determines a cell's fate in a variety of situations. In this respect, it is not surprising that, besides dysregulation of cellular proliferation, the inhibition of cell death through apoptosis is also a common feature in the development of cancer (for review, see Llambi and Green 2011). Still, because regulation involves the interplay of proapoptotic and antiapoptotic Bcl-2 family proteins, apoptosis is not completely abolished in cancer cells. On the contrary, cancer cells have been described as "primed" for death, because a high level of BH3-only proteins is counteracted by up-regulated antiapoptotic Bcl-2-like proteins. BH3-only mimetics like ABT-737 represent tools for tipping the balance in favor of death, thereby providing a promising possibility for therapeutic intervention (Vogler et al. 2009).

However, although our knowledge of the early steps of MOMP (i.e., Bax targeting, activation, and oligomerization) has expanded substantially over the past few years, the actual event of membrane permeabilization by Bax or Bak oligomers remains, in part, mysterious. It has become evident that the outer membrane of mitochondria itself is not only providing the grounds for MOMP, but actively participates in this process. Membrane lipids and also resident mitochondrial proteins seem to be involved in all of the steps leading to MOMP, from the recruitment of tBid or other direct activators to the pore formation itself. Like cytochrome *c*, which normally serves as an electron carrier between respiratory complexes, other mitochondrial proteins are likely to have dual functions both in mitochondrial biology and apoptosis. On the other hand, there is growing evidence that Bcl-2 proteins also participate in other cellular processes besides apoptosis, like the regulation of mitochondrial fusion and fission (Martinou and Youle 2011). Therefore, it is most likely that we are awaiting many more exciting discoveries regarding the role of mitochondria in cell death in the near future.

ACKNOWLEDGMENTS

We thank Dr. C. Rodley for critical reading of the manuscript and all members of the laboratory for their support. Work in our laboratory is supported by subsidy 31003A-141068/1 of the Swiss National Foundation. T.B. receives a fellowship from the Deutsche Forschungsgemeinschaft (BE 5103/1-1).

REFERENCES

Annis MG, Soucie EL, Dlugosz PJ, Cruz-Aguado JA, Penn LZ, Leber B, Andrews DW. 2005. Bax forms multispanning monomers that oligomerize to permeabilize membranes during apoptosis. *EMBO J* **24:** 2096–2103.

Antonsson B, Conti F, Ciavatta A, Montessuit S, Lewis S, Martinou I, Bernasconi L, Bernard A, Mermod JJ, Mazzei G, et al. 1997. Inhibition of Bax channel-forming activity by Bcl-2. *Science* **277:** 370–372.

Baines CP, Kaiser RA, Purcell NH, Blair NS, Osinska H, Hambleton MA, Brunskill EW, Sayen MR, Gottlieb RA, Dorn GW, et al. 2005. Loss of cyclophilin D reveals a critical role for mitochondrial permeability transition in cell death. *Nature* **434:** 658–662.

Baines CP, Kaiser RA, Sheiko T, Craigen WJ, Molkentin JD. 2007. Voltage-dependent anion channels are dispensable for mitochondrial-dependent cell death. *Nat Cell Biol* **9:** 550–555.

Basanez G, Sharpe JC, Galanis J, Brandt TB, Hardwick JM, Zimmerberg J. 2002. Bax-type apoptotic proteins porate pure lipid bilayers through a mechanism sensitive to intrinsic monolayer curvature. *J Biol Chem* **277:** 49360–49365.

Billen LP, Kokoski CL, Lovell JF, Leber B, Andrews DW. 2008. Bcl-XL inhibits membrane permeabilization by competing with Bax. *PLoS Biol* **6:** e147.

Bleicken S, Classen M, Padmavathi PV, Ishikawa T, Zeth K, Steinhoff HJ, Bordignon E. 2010. Molecular details of Bax activation, oligomerization, and membrane insertion. *J Biol Chem* **285:** 6636–6647.

Certo M, Del Gaizo Moore V, Nishino M, Wei G, Korsmeyer S, Armstrong SA, Letai A. 2006. Mitochondria primed by death signals determine cellular addiction to antiapoptotic BCL-2 family members. *Cancer Cell* **9:** 351–365.

Chen L, Willis SN, Wei A, Smith BJ, Fletcher JI, Hinds MG, Colman PM, Day CL, Adams JM, Huang DC. 2005. Differential targeting of prosurvival Bcl-2 proteins by their BH3-only ligands allows complementary apoptotic function. *Mol Cell* **17:** 393–403.

Cheng EH, Sheiko TV, Fisher JK, Craigen WJ, Korsmeyer SJ. 2003. VDAC2 inhibits BAK activation and mitochondrial apoptosis. *Science* **301:** 513–517.

Chernomordik LV, Kozlov MM. 2008. Mechanics of membrane fusion. *Nat Struct Mol Biol* **15:** 675–683.

Chipuk JE, Fisher JC, Dillon CP, Kriwacki RW, Kuwana T, Green DR. 2008. Mechanism of apoptosis induction by inhibition of the anti-apoptotic BCL-2 proteins. *Proc Natl Acad Sci* **105:** 20327–20332.

Chipuk JE, Moldoveanu T, Llambi F, Parsons MJ, Green DR. 2010. The BCL-2 family reunion. *Mol Cell* **37:** 299–310.

Chipuk JE, McStay GP, Bharti A, Kuwana T, Clarke CJ, Siskind LJ, Obeid LM, Green DR. 2012. Sphingolipid metabolism cooperates with BAK and BAX to promote the mitochondrial pathway of apoptosis. *Cell* **148:** 988–1000.

Desagher S, Osen-Sand A, Nichols A, Eskes R, Montessuit S, Lauper S, Maundrell K, Antonsson B, Martinou JC. 1999. Bid-induced conformational change of Bax is responsible for mitochondrial cytochrome *c* release during apoptosis. *J Cell Biol* **144:** 891–901.

Dewson G, Kratina T, Sim HW, Puthalakath H, Adams JM, Colman PM, Kluck RM. 2008. To trigger apoptosis, Bak exposes its BH3 domain and homodimerizes via BH3: groove interactions. *Mol Cell* **30:** 369–380.

Dewson G, Kratina T, Czabotar P, Day CL, Adams JM, Kluck RM. 2009. Bak activation for apoptosis involves oligomerization of dimers via their α6 helices. *Mol Cell* **36:** 696–703.

Dlugosz PJ, Billen LP, Annis MG, Zhu W, Zhang Z, Lin J, Leber B, Andrews DW. 2006. Bcl-2 changes conformation to inhibit Bax oligomerization. *EMBO J* **25:** 2287–2296.

Edlich F, Banerjee S, Suzuki M, Cleland MM, Arnoult D, Wang C, Neutzner A, Tjandra N, Youle RJ. 2011. Bcl-x(L) retrotranslocates Bax from the mitochondria into the cytosol. *Cell* **145:** 104–116.

Engel A, Walter P. 2008. Membrane lysis during biological membrane fusion: Collateral damage by misregulated fusion machines. *J Cell Biol* **183:** 181–186.

Eskes R, Desagher S, Antonsson B, Martinou JC. 2000. Bid induces the oligomerization and insertion of Bax into the outer mitochondrial membrane. *Mol Cell Biol* **20:** 929–935.

Gallenne T, Gautier F, Oliver L, Hervouet E, Noel B, Hickman JA, Geneste O, Cartron PF, Vallette FM, Manon S, et al. 2009. Bax activation by the BH3-only protein Puma promotes cell dependence on antiapoptotic Bcl-2 family members. *J Cell Biol* **185:** 279–290.

Garcia-Saez AJ, Chiantia S, Salgado J, Schwille P. 2007. Pore formation by a Bax-derived peptide: Effect on the line tension of the membrane probed by AFM. *Biophys J* **93:** 103–112.

Gavathiotis E, Suzuki M, Davis ML, Pitter K, Bird GH, Katz SG, Tu HC, Kim H, Cheng EH, Tjandra N, et al. 2008. BAX activation is initiated at a novel interaction site. *Nature* **455:** 1076–1081.

Gavathiotis E, Reyna DE, Davis ML, Bird GH, Walensky LD. 2010. BH3-triggered structural reorganization drives the activation of proapoptotic BAX. *Mol Cell* **40:** 481–492.

George NM, Evans JJ, Luo X. 2007. A three-helix homo-oligomerization domain containing BH3 and BH1 is responsible for the apoptotic activity of Bax. *Genes Dev* **21:** 1937–1948.

Hsu YT, Youle RJ. 1997. Nonionic detergents induce dimerization among members of the Bcl-2 family. *J Biol Chem* **272:** 13829–13834.

Hsu YT, Wolter KG, Youle RJ. 1997. Cytosol-to-membrane redistribution of Bax and Bcl-X(L) during apoptosis. *Proc Natl Acad Sci* **94:** 3668–3672.

Karbowski M, Lee YJ, Gaume B, Jeong SY, Frank S, Nechushtan A, Santel A, Fuller M, Smith CL, Youle RJ. 2002. Spatial and temporal association of Bax with mitochondrial fission sites, Drp1, and Mfn2 during apoptosis. *J Cell Biol* **159:** 931–938.

Kim H, Rafiuddin-Shah M, Tu HC, Jeffers JR, Zambetti GP, Hsieh JJ, Cheng EH. 2006. Hierarchical regulation of mitochondrion-dependent apoptosis by BCL-2 subfamilies. *Nat Cell Biol* **8:** 1348–1358.

Kim H, Tu HC, Ren D, Takeuchi O, Jeffers JR, Zambetti GP, Hsieh JJ, Cheng EH. 2009. Stepwise activation of BAX and BAK by tBID, BIM, and PUMA initiates mitochondrial apoptosis. *Mol Cell* **36:** 487–499.

Kuwana T, Mackey MR, Perkins G, Ellisman MH, Latterich M, Schneiter R, Green DR, Newmeyer DD. 2002. Bid, Bax, and lipids cooperate to form supramolecular openings in the outer mitochondrial membrane. *Cell* **111:** 331–342.

Kuwana T, Bouchier-Hayes L, Chipuk JE, Bonzon C, Sullivan BA, Green DR, Newmeyer DD. 2005. BH3 domains of BH3-only proteins differentially regulate Bax-mediated mitochondrial membrane permeabilization both directly and indirectly. *Mol Cell* **17:** 525–535.

Leber B, Lin J, Andrews DW. 2007. Embedded together: The life and death consequences of interaction of the Bcl-2 family with membranes. *Apoptosis* **12:** 897–911.

Leber B, Lin J, Andrews DW. 2010. Still embedded together binding to membranes regulates Bcl-2 protein interactions. *Oncogene* **29:** 5221–5230.

Letai A, Bassik MC, Walensky LD, Sorcinelli MD, Weiler S, Korsmeyer SJ. 2002. Distinct BH3 domains either sensitize or activate mitochondrial apoptosis, serving as prototype cancer therapeutics. *Cancer Cell* **2:** 183–192.

Lin MT, Beal MF. 2006. Mitochondrial dysfunction and oxidative stress in neurodegenerative diseases. *Nature* **443:** 787–795.

Lindsay J, Esposti MD, Gilmore AP. 2011. Bcl-2 proteins and mitochondria—Specificity in membrane targeting for death. *Biochim Biophys Acta* **1813:** 532–529.

Llambi F, Green DR. 2011. Apoptosis and oncogenesis: Give and take in the BCL-2 family. *Curr Opin Genet Dev* **21:** 12–20.

Llambi F, Moldoveanu T, Tait SW, Bouchier-Hayes L, Temirov J, McCormick LL, Dillon CP, Green DR. 2011. A unified model of mammalian BCL-2 protein family interactions at the mitochondria. *Mol Cell* **44:** 517–531.

Lovell JF, Billen LP, Bindner S, Shamas-Din A, Fradin C, Leber B, Andrews DW. 2008. Membrane binding by tBid initiates an ordered series of events culminating in membrane permeabilization by Bax. *Cell* **135:** 1074–1084.

Lucken-Ardjomande S, Montessuit S, Martinou JC. 2008a. Bax activation and stress-induced apoptosis delayed by the accumulation of cholesterol in mitochondrial membranes. *Cell Death Differ* **15:** 484–493.

Lucken-Ardjomande S, Montessuit S, Martinou JC. 2008b. Contributions to Bax insertion and oligomerization of lipids of the mitochondrial outer membrane. *Cell Death Differ* **15:** 929–937.

Lutter M, Fang M, Luo X, Nishijima M, Xie X, Wang X. 2000. Cardiolipin provides specificity for targeting of tBid to mitochondria. *Nat Cell Biol* **2:** 754–761.

Lutter M, Perkins GA, Wang X. 2001. The pro-apoptotic Bcl-2 family member tBid localizes to mitochondrial contact sites. *BMC Cell Biol* **2:** 22.

Martinez-Caballero S, Dejean LM, Kinnally MS, Oh KJ, Mannella CA, Kinnally KW. 2009. Assembly of the mitochondrial apoptosis-induced channel, MAC. *J Biol Chem* **284:** 12235–12245.

Martinou JC, Youle RJ. 2011. Mitochondria in apoptosis: Bcl-2 family members and mitochondrial dynamics. *Dev Cell* **21:** 92–101.

Moldoveanu T, Liu Q, Tocilj A, Watson M, Shore G, Gehring K. 2006. The X-ray structure of a BAK homodimer reveals an inhibitory zinc binding site. *Mol Cell* **24:** 677–688.

Montessuit S, Somasekharan SP, Terrones O, Lucken-Ardjomande S, Herzig S, Schwarzenbacher R, Manstein DJ, Bossy-Wetzel E, Basanez G, Meda P, et al. 2010. Membrane remodeling induced by the dynamin-related protein Drp1 stimulates Bax oligomerization. *Cell* **142:** 889–901.

Muchmore SW, Sattler M, Liang H, Meadows RP, Harlan JE, Yoon HS, Nettesheim D, Chang BS, Thompson CB, Wong SL, et al. 1996. X-ray and NMR structure of human Bcl-xL, an inhibitor of programmed cell death. *Nature* **381:** 335–341.

Nakagawa T, Shimizu S, Watanabe T, Yamaguchi O, Otsu K, Yamagata H, Inohara H, Kubo T, Tsujimoto Y. 2005. Cyclophilin D–dependent mitochondrial permeability transition regulates some necrotic but not apoptotic cell death. *Nature* **434:** 652–658.

Ni Chonghaile T, Sarosiek KA, Vo TT, Ryan JA, Tammareddi A, Moore Vdel G, Deng J, Anderson KC, Richardson P, Tai YT, et al. 2011. Pretreatment mitochondrial priming correlates with clinical response to cytotoxic chemotherapy. *Science* **334:** 1129–1133.

Oh KJ, Singh P, Lee K, Foss K, Lee S, Park M, Aluvila S, Kim RS, Symersky J, Walters DE. 2010. Conformational changes in BAK, a pore-forming proapoptotic Bcl-2 family member, upon membrane insertion and direct evidence for the existence of BH3–BH3 contact interface in BAK homo-oligomers. *J Biol Chem* **285:** 28924–28937.

Ott M, Norberg E, Zhivotovsky B, Orrenius S. 2009. Mitochondrial targeting of tBid/Bax: A role for the TOM complex? *Cell Death Differ* **16:** 1075–1082.

Petros AM, Nettesheim DG, Wang Y, Olejniczak ET, Meadows RP, Mack J, Swift K, Matayoshi ED, Zhang H, Thompson CB, et al. 2000. Rationale for Bcl-xL/Bad peptide complex formation from structure, mutagenesis, and biophysical studies. *Protein Sci* **9:** 2528–2534.

Qian S, Wang W, Yang L, Huang HW. 2008. Structure of transmembrane pore induced by Bax-derived peptide: Evidence for lipidic pores. *Proc Natl Acad Sci* **105:** 17379–17383.

Rehm M, Dussmann H, Prehn JH. 2003. Real-time single cell analysis of Smac/DIABLO release during apoptosis. *J Cell Biol* **162:** 1031–1043.

Ren D, Tu HC, Kim H, Wang GX, Bean GR, Takeuchi O, Jeffers JR, Zambetti GP, Hsieh JJ, Cheng EH. 2010. BID, BIM, and PUMA are essential for activation of the BAX- and BAK-dependent cell death program. *Science* **330:** 1390–1393.

Satsoura D, Kucerka N, Shivakumar S, Pencer J, Griffiths C, Leber B, Andrews DW, Katsaras J, Fradin C. 2012. Interaction of the full-length Bax protein with biomimetic mitochondrial liposomes: A small-angle neutron scattering and fluorescence study. *Biochim Biophys Acta* **1818:** 384–401.

Sattler M, Liang H, Nettesheim D, Meadows RP, Harlan JE, Eberstadt M, Yoon HS, Shuker SB, Chang BS, Minn AJ, et al. 1997. Structure of Bcl-xL–Bak peptide complex: Recognition between regulators of apoptosis. *Science* **275:** 983–986.

Schafer B, Quispe J, Choudhary V, Chipuk JE, Ajero TG, Du H, Schneiter R, Kuwana T. 2009. Mitochondrial outer membrane proteins assist Bid in Bax-mediated lipidic pore formation. *Mol Biol Cell* **20:** 2276–2285.

Schlesinger PH, Gross A, Yin XM, Yamamoto K, Saito M, Waksman G, Korsmeyer SJ. 1997. Comparison of the ion channel characteristics of proapoptotic BAX and anti-apoptotic BCL-2. *Proc Natl Acad Sci* **94:** 11357–11362.

Suzuki M, Youle RJ, Tjandra N. 2000. Structure of Bax: Coregulation of dimer formation and intracellular localization. *Cell* **103:** 645–654.

Tan C, Dlugosz PJ, Peng J, Zhang Z, Lapolla SM, Plafker SM, Andrews DW, Lin J. 2006. Auto-activation of the apoptosis protein Bax increases mitochondrial membrane permeability and is inhibited by Bcl-2. *J Biol Chem* **281:** 14764–14775.

Taylor RC, Cullen SP, Martin SJ. 2008. Apoptosis: Controlled demolition at the cellular level. *Nat Rev Mol Cell Biol* **9:** 231–241.

Terradillos O, Montessuit S, Huang DC, Martinou JC. 2002. Direct addition of BimL to mitochondria does not lead to cytochrome *c* release. *FEBS Lett* **522:** 29–34.

Terrones O, Antonsson B, Yamaguchi H, Wang HG, Liu J, Lee RM, Herrmann A, Basanez G. 2004. Lipidic pore formation by the concerted action of proapoptotic BAX and tBID. *J Biol Chem* **279:** 30081–30091.

Vogler M, Weber K, Dinsdale D, Schmitz I, Schulze-Osthoff K, Dyer MJS, Cohen GM. 2009. Different forms of cell death induced by putative BCL-2 inhibitors. *Cell Death Differ* **16:** 1030–1039.

Walensky LD, Pitter K, Morash J, Oh KJ, Barbuto S, Fisher J, Smith E, Verdine GL, Korsmeyer SJ. 2006. A stapled BID BH3 helix directly binds and activates BAX. *Mol Cell* **24:** 199–210.

Wei MC, Zong WX, Cheng EH, Lindsten T, Panoutsakopoulou V, Ross AJ, Roth KA, MacGregor GR, Thompson CB, Korsmeyer SJ. 2001. Proapoptotic BAX and BAK: A requisite gateway to mitochondrial dysfunction and death. *Science* **292:** 727–730.

Willis SN, Chen L, Dewson G, Wei A, Naik E, Fletcher JI, Adams JM, Huang DC. 2005. Proapoptotic Bak is sequestered by Mcl-1 and Bcl-xL, but not Bcl-2, until displaced by BH3-only proteins. *Genes Dev* **19:** 1294–1305.

Willis SN, Fletcher JI, Kaufmann T, van Delft MF, Chen L, Czabotar PE, Ierino H, Lee EF, Fairlie WD, Bouillet P, et al. 2007. Apoptosis initiated when BH3 ligands engage multiple Bcl-2 homologs, not Bax or Bak. *Science* **315:** 856–859.

Youle RJ, Strasser A. 2008. The BCL-2 protein family: Opposing activities that mediate cell death. *Nat Rev Mol Cell Biol* **9:** 47–59.

Zaltsman Y, Shachnai L, Yivgi-Ohana N, Schwarz M, Maryanovich M, Houtkooper RH, Vaz FM, De Leonardis F, Fiermonte G, Palmieri F, et al. 2010. MTCH2/MIMP is a major facilitator of tBID recruitment to mitochondria. *Nat Cell Biol* **12:** 553–562.

Zhang Z, Zhu W, Lapolla SM, Miao Y, Shao Y, Falcone M, Boreham D, McFarlane N, Ding J, Johnson AE, et al. 2010. Bax forms an oligomer via separate, yet interdependent, surfaces. *J Biol Chem* **285:** 17614–17627.

The Role of Mitochondria in Cellular Iron–Sulfur Protein Biogenesis: Mechanisms, Connected Processes, and Diseases

Oliver Stehling[1] **and Roland Lill**[1,2,3]

[1]Institut für Zytobiologie, Philipps-Universität Marburg, 35032 Marburg, Germany
[2]Max-Planck-Institut für Terrestrische Mikrobiologie, 35043 Marburg, Germany
[3]LOEWE Zentrum für Synthetische Mikrobiologie SynMikro, 35043 Marburg, Germany
Correspondence: lill@staff.uni-marburg.de

Iron–sulfur (Fe/S) clusters belong to the most ancient protein cofactors in life, and fulfill functions in electron transport, enzyme catalysis, homeostatic regulation, and sulfur activation. The synthesis of Fe/S clusters and their insertion into apoproteins requires almost 30 proteins in the mitochondria and cytosol of eukaryotic cells. This review summarizes our current biochemical knowledge of mitochondrial Fe/S protein maturation. Because this pathway is essential for various extramitochondrial processes, we then explain how mitochondria contribute to the mechanism of cytosolic and nuclear Fe/S protein biogenesis, and to other connected processes including nuclear DNA replication and repair, telomere maintenance, and transcription. We next describe how the efficiency of mitochondria to assemble Fe/S proteins is used to regulate cellular iron homeostasis. Finally, we briefly summarize a number of mitochondrial "Fe/S diseases" in which various biogenesis components are functionally impaired owing to genetic mutations. The thorough understanding of the diverse biochemical disease phenotypes helps with testing the current working model for the molecular mechanism of Fe/S protein biogenesis and its connected processes.

Mitochondria perform crucial roles in many biochemical processes. They generate ATP by oxidative phosphorylation and participate in numerous metabolic pathways such as citric acid cycle, fatty acid degradation, urea cycle, and the biosynthesis of lipids and amino acids. Moreover, the organelles are involved in the biosynthesis of various protein cofactors such as heme, Moco, biotin, lipoic acid, and, last but not least, iron–sulfur (Fe/S) clusters. Fe/S clusters are ancient protein cofactors, and they are involved in electron transfer reactions, participate in catalytic and regulatory processes, and serve as sulfur donors during the synthesis of lipoic acid and biotin (Beinert et al. 1997). The most common and simplest forms of Fe/S clusters are of the [2Fe-2S] and [4Fe-4S] type, but also [3Fe-4S] forms or more complex clusters containing additional heavy metal ions are known (Hu and Ribbe 2012; Peters and Broderick 2012). The Fe ion of the cluster is typically coordinated by the sulfur of protein-

Cite this article as *Cold Spring Harb Perspect Biol* doi: 10.1101/cshperspect.a011312

bound cysteine residues or the nitrogen of histidine residues, but in rare cases other amino acid residues or cofactors such as *S*-adenosylmethionine (SAM) are used as coordinating ligands (Lanz and Booker 2012).

Despite the chemical simplicity of Fe/S clusters their biosynthesis is rather complicated and requires more than two dozen components in eukaryotes (for detailed previous reviews see Lill and Mühlenhoff 2008; Lill 2009; Stemmler et al. 2010; Alfonzo and Lukes 2011; Balk and Pilon 2011; Lill et al. 2012; Rouault 2012). The pathway is initiated by the mitochondrial iron–sulfur cluster (ISC) assembly machinery, which comprises more than 15 components (Fig. 1). The core of this machinery is required not only for the biosynthesis of Fe/S proteins inside but also outside mitochondria. Maturation of these latter Fe/S proteins is assisted by the cytosolic iron–sulfur protein assembly (CIA) machinery consisting of eight known proteins (Lill 2009; Sharma et al. 2010). The function of the CIA machinery and some of its components strictly depends on the core mitochondrial ISC assembly system, which serves as a sulfur donor for extramitochondrial Fe/S protein maturation (Fig. 1). A still unknown sulfur-containing compound is generated by the ISC assembly system and exported by the mitochondrial ABC transporter Atm1 (human ABCB7) to the cytosol for use in the CIA system.

In eukaryotes, known Fe/S proteins are located in mitochondria, cytosol, and nucleus where they perform rather diverse functions (Fig. 1). In mitochondria they are involved in the TCA cycle (aconitase), the electron transfer chain (respiratory complexes I–III), fatty acid oxidation (ETF-ubiquinone oxidoreductase), and in lipoate and biotin biosynthesis (lipoate and biotin synthases). In the cytosol, Fe/S proteins function in amino acid biosynthesis (isopropylmalate isomerase), tRNA modification

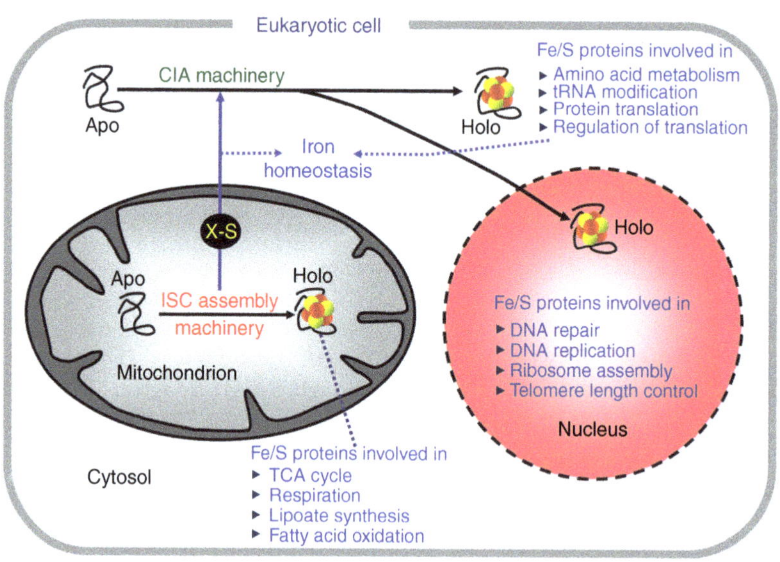

Figure 1. The biogenesis of cellular Fe/S proteins in eukaryotes and the links to cellular iron homeostasis, protein translation, and nuclear genome integrity. Eukaryotic Fe/S proteins are located in mitochondria, cytosol, and nucleus where they perform diverse functions in cellular metabolism and regulation. The mitochondrial Fe/S cluster (ISC) assembly machinery matures all organellar Fe/S proteins, and additionally contributes to the biogenesis of cytosolic and nuclear Fe/S proteins by producing an unknown sulfur-containing compound (X-S) that is exported to the cytosol and used by the cytosolic Fe/S protein assembly (CIA) machinery. Hence, mitochondria are directly responsible for the essential functions (e.g., of nuclear Fe/S proteins involved in DNA metabolism and genome maintenance). Additionally, the ISC assembly machinery exerts a regulatory role on cellular iron homeostasis (see text for details). Red and yellow circles indicate iron and sulfur ions, respectively.

(e.g., Tyw1 generating the wybutosine base), posttranscriptional regulation of iron metabolism (iron regulatory protein 1), or ribosomal protein translation (ABC protein Rli1, aka ABCE1). Important nuclear Fe/S proteins participate in DNA replication (DNA polymerases and primases) and DNA repair or telomere length regulation (ATP-dependent DNA helicases) (Lill and Mühlenhoff 2008; Lill et al. 2012). It can readily be seen from the scheme in Fig. 1 that mitochondria play a crucial role in the biogenesis of a number of extramitochondrial Fe/S proteins that are essential for life such as those involved in DNA maintenance and protein translation. Because without these proteins any eukaryotic cell is unviable, this provides a satisfactory explanation for the essential character of mitochondria within the cell, and shows that mitochondrial function in Fe/S protein biogenesis directly affects nuclear function in gene expression.

Failure to assemble Fe/S proteins is associated with severe and frequently fatal neurodegenerative, metabolic, or hematological diseases (Camaschella et al. 2007; Calvo et al. 2010; Sheftel et al. 2010a; Cameron et al. 2011; Navarro-Sastre et al. 2011; Rouault 2012). The founding example of an "Fe/S disease" is Friedreich's ataxia, which is characterized by a functional defect in the mitochondrial protein frataxin (yeast Yfh1). It is generally agreed that frataxin is a key component of Fe/S protein biogenesis (Stemmler et al. 2010). Cells with defective frataxin not only show a low assembly efficiency of Fe/S proteins but also a characteristic accumulation of iron in mitochondria. Although initially this was attributed to a specific role of frataxin in mitochondrial iron storage and/or iron exit from mitochondria (Radisky et al. 1999; Cavadini et al. 2002), it turned out that this mitochondrial iron-overload phenotype is generally shared by deficiencies or diseases in other core members of the ISC assembly system (Kispal et al. 1999; Schilke et al. 1999; Lange et al. 2000; Chen et al. 2002; Rodriguez-Manzaneque et al. 2002; Biederbick et al. 2006; Wiedemann et al. 2006; Sheftel et al. 2010a; Rouault 2012) pointing to the primary role of frataxin in Fe/S protein biogenesis. It should be noted at this point that other more recently discovered Fe/S diseases do not show any conspicuous alterations of the cellular iron metabolism. This raises the interesting question of how these differences may be biochemically explained.

In this review, we will first summarize the mechanisms of mitochondrial Fe/S protein biogenesis by describing in some detail the current functional knowledge of the mitochondrial ISC assembly components. Because mitochondria are essential for life owing to their role in the maturation of indispensable cytosolic and nuclear Fe/S proteins (Fig. 1), we next will explain the mechanism of how mitochondria, in concert with the components of the CIA machinery, facilitate the assembly of these proteins. We then will lay out how the assembly process affects important pathways of life such as cytosolic protein translation, nuclear genome stability, and DNA maintenance. Next, we will briefly address how an ISC assembly deficiency affects the cellular iron metabolism in both yeast and humans. Finally, we describe several Fe/S diseases in which ISC components are genetically altered. The biochemical analyses of the diseased cells both contribute to and serve as a rigorous test of our current mechanistic understanding of Fe/S protein biogenesis.

BIOGENESIS OF MITOCHONDRIAL Fe/S PROTEINS BY THE ISC ASSEMBLY MACHINERY

The majority of functional studies of mitochondrial Fe/S protein biogenesis have been performed in the yeast *Saccharomyces cerevisiae*, which has served as a simple model organism to dissect this complicated biosynthesis pathway. The fact that the basic mechanism of mitochondrial ISC assembly has been inherited in evolution from the bacterial ancestor of the organelles has been useful to transfer functional insights from studying the related bacterial ISC assembly system to the eukaryotic situation (Johnson et al. 2005; Bandyopadhyay et al. 2008a; Py and Barras 2010). More recent studies in human cells (using both cultured cell lines and patient material) have underlined the striking conservation of both the ISC components and the biosynthetic mechanisms from yeast to man (Sheftel et al. 2010a; Rouault 2012).

The ISC assembly machinery consists of 17 known components. Functionally, the biosynthetic process has been dissected experimentally into three major steps (Fig. 2) (Mühlenhoff et al. 2003). First, a [2Fe-2S] cluster is synthesized on the highly conserved scaffold protein Isu1, which provides both an assembly platform and a transient binding site for the nascent Fe/S cluster. Next, the association of the Fe/S cluster on Isu1 is labilized by the interaction of holo-Isu1 with a dedicated Hsp70 chaperone system, followed by the transient binding of the cluster to so-called Fe/S cluster transfer proteins. The biogenesis proteins participating up to this step comprise the core ISC assembly components and are also crucially involved in the biogenesis of extramitochondrial Fe/S proteins (Fig. 2). In the third major step, the Fe/S cluster is specifically targeted to mitochondrial apoproteins and inserted into the polypeptide chains by coordination with dedicated amino acid residues. The latter reactions are supported by the specific function of ISC-targeting factors. In the following sections, we will describe the molecular

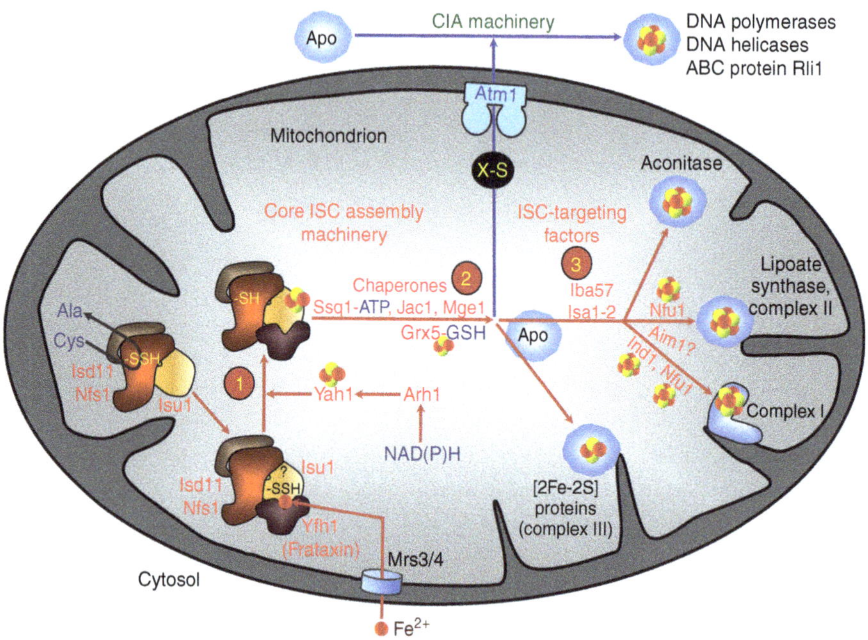

Figure 2. The three key steps of mitochondrial Fe/S protein assembly. In the first step (1), a [2Fe-2S] cluster is assembled on the Isu1 scaffold protein, which tightly interacts with the cysteine desulfurase complex Nfs1-Isd11 serving as the sulfur donor. The initially formed persulfide (-SSH) on Nfs1 by conversion of cysteine to alanine is possibly next transferred to Isu1. Fe/S cluster assembly on Isu1 then requires Yfh1 (frataxin) and the electron transfer chain consisting of the [2Fe-2S] ferredoxin Yah1 and ferredoxin reductase Arh1, which receives electrons from NAD(P)H. The mitochondrial carrier proteins Mrs3-Mrs4 supply ferrous iron. In the second step (2), the Fe/S cluster is released from Isu1, which involves the specialized Hsp70 chaperone system comprised of Ssq1, Jac1, and Mge1. This may lead to Fe/S cluster transfer to the monothiol glutaredoxin Grx5, which binds the metallo-cofactor in a glutathione (GSH)-dependent fashion. All of these ISC components form the core ISC assembly machinery, and are required for biogenesis of all cellular Fe/S proteins including those assembled by the CIA machinery (see Fig. 3) in the cytosol and nucleus (blue arrows). Finally, the third step (3) involves dedicated ISC-targeting factors, which transfer the cluster to specific apoproteins and assemble it into the polypeptide chains. Biogenesis of [4Fe-4S] clusters is facilitated by Isa1-Isa2 and Iba57, which may bind folate. Nfu1 and Aim1 are essential for the specific maturation of respiratory complexes I and II, and for lipoate synthase. Complex I assembly further needs the P-loop ATPase Ind1, which, similar to Nfu1, transiently coordinates a [4Fe-4S] cluster.

functions of the individual ISC assembly components participating in the three biogenesis steps in more detail.

Fe/S cluster assembly on Isu1 (in yeast also on the functionally redundant Isu2) requires the cysteine desulfurase complex Nfs1-Isd11, which releases sulfur from free cysteine to form both alanine and an Nfs1 cysteine residue-bound persulfide as an intermediate (Fig. 2) (Kispal et al. 1999; Adam et al. 2006; Biederbick et al. 2006; Wiedemann et al. 2006; Shi et al. 2009). The Nfs1 protein is related to the founding member of cysteine desulfurases, NifS, the sulfur donor for the Fe/S clusters of the complex metalloprotein nitrogenase of nitrogen-fixing bacteria. In vivo, the Nfs1-Isd11 complex rather than Nfs1 alone is the functional entity for sulfur activation and for Fe/S cluster synthesis on the Isu1 scaffold. To achieve the latter reaction, Nfs1-Isd11 and Isu1 directly interact and it is thought, but not yet proven, that the persulfide is transferred from Nfs1 to one of the three conserved cysteine residues of Isu1. After ferrous iron import via the mitochondrial carriers Mrs3-Mrs4 (human mitoferrin 1/2; Fig. 2) and iron binding to Isu1 the Fe/S cluster is formed by a still unknown chemical mechanism. It is generally agreed on that frataxin participates in this partial reaction of Fe/S protein maturation. One suggestion for the molecular function of frataxin is that it functions as an iron-binding protein and delivers the metal ions to Isu1. Consistent with this idea, frataxin tightly interacts with the Nfs1-Isd11-Isu1 complex (Gerber et al. 2003; Tsai and Barondeau 2010; Bridwell-Rabb et al. 2011; Schmucker et al. 2011; Colin et al. 2013). Whether this complex formation is stimulated by the presence of iron or not is controversial. An alternative, not necessarily mutually exclusive suggestion is that frataxin functions as a regulator of the Nfs1 desulfurase activity. Frataxin was found to be required for sulfide production by human Nfs1-Isd11, and stimulated the desulfurase activity maximally when all components of the reaction including iron were present. Frataxin therefore was proposed to act as an allosteric regulator of this initial step of Fe/S cluster formation. Further details of frataxin and the relation to Friedreich's ataxia will be discussed below. Formally, for the reduction of the persulfidic sulfur (S^0) on Nfs1 (or Isu1) to sulfide (S^{2-}) present in Fe/S clusters, a reduction step is needed but it is unclear how this reaction may take place. Because the [2Fe-2S] ferredoxin Yah1 (Fdx2 in mammals) is necessary in vivo for Fe/S cluster generation on Isu1 it seems likely that Yah1, together with ferredoxin (adrenodoxin) reductase and NAD(P)H, provides the needed electrons (Fig. 2) (Lange et al. 2000; Sheftel et al. 2010b). Direct functional evidence for this idea is still lacking.

The second major step of mitochondrial Fe/S protein biogenesis involves the release of the Isu1-bound Fe/S cluster and its transient binding by Fe/S transfer proteins (Fig. 2). The dissociation of the Fe/S cluster from Isu1 is facilitated by a dedicated Hsp70 chaperone system comprised of the Hsp70 ATPase Ssq1 (human mortalin), the DnaJ-like cochaperone Jac1 (human Hsc20), and the nucleotide exchange factor Mge1 (human GRPE-L1/2) (Schilke et al. 2006; Vickery and Cupp-Vickery 2007; Uhrigshardt et al. 2010). The molecular mechanism of the chaperones in this reaction is strikingly similar to Hsp70 chaperone function in protein folding, and has been worked out in both bacteria and yeast (Vickery and Cupp-Vickery 2007; Kampinga and Craig 2010). First, Jac1 specifically binds to Isu1 (Ciesielski et al. 2012). The complex of holo-Isu1 and Jac1 recruits the ATP-bound form of the Hsp70 (Pukszta et al. 2010). This leads to a highly specific interaction of the peptide-binding domain of Ssq1 with the conserved LPPVK loop of Isu1 followed by ATP hydrolysis on Ssq1. The conformational change of Isu1 during complex formation is thought to labilize Fe/S cluster binding to Isu1 thus facilitating its dissociation. The constituents of the Ssq1-ADP–apo-Isu1 complex are then recycled by ADP-ATP exchange on Ssq1, a step facilitated by the nucleotide exchange factor Mge1. Dissociation enables Isu1 and Ssq1 to undergo the next round of Fe/S cluster synthesis and transfer.

The released Fe/S cluster may then be transferred to the mitochondrial monothiol glutaredoxin Grx5 or directly to apoproteins (Chandramouli and Johnson 2006; Bonomi et al. 2008). Grx5 possibly serves as an Fe/S cluster transfer protein by transiently coordinating the Isu1-derived [2Fe-2S] cluster between two protein subunits and two protein-bound glutathione molecules (Bandyopadhyay et al. 2008b; Shakamuri et al. 2012). Recently, Grx5 was shown to specifically associate with Ssq1 at a site different from the chaperone peptide-binding pocket (Uzarska et al. 2013). The simultaneous binding of holo-Isu1 and apo-Grx5 to Ssq1 potentially increases the efficiency of Fe/S cluster transfer. However, at least to some extent, a direct delivery of the Fe/S cluster to target apoproteins seems possible, because Grx5, unlike most of the ISC proteins discussed so far, is not essential for viability of yeast cells. Nevertheless, deletion of the yeast *GRX5* gene leads to a severe Fe/S protein assembly defect both inside and outside mitochondria and is associated with a pronounced sensitivity to oxidative stress, possibly as a result of the iron accumulation in mitochondria (see below). In both humans and zebrafish, GLRX5 is essential for life (Wingert et al. 2005; Camaschella et al. 2007; Ye et al. 2010). In humans, a mutation leading to decreased amounts of GLRX5 causes a severe iron-storage disease with a characteristic cellular Fe/S protein and heme synthesis defect as well as with an iron accumulation in mitochondria as indicated by the occurrence of ringed sideroblasts (Cazzola and Invernizzi 2011).

The third major step of mitochondrial Fe/S protein biogenesis involves the delivery of the Fe/S cluster to specific target apoproteins and the dedicated integration of the cluster into the polypeptide chain by coordination of its iron ions with specific amino acid ligands. For the formation of [2Fe-2S] proteins, no other factors have been identified in addition to the mentioned members of the core ISC assembly machinery (Fig. 2). For all mitochondrial [4Fe-4S] proteins, on the other hand, cofactor insertion must be preceded or accompanied by conversion of the [2Fe-2S] cluster that has been synthesized on Isu1 and transferred by Grx5. This reaction is accomplished by the A-type ISC proteins Isa1 and Isa2 (human ISCA1 and ISCA2) and the tetrahydrofolate-binding protein Iba57 (Mühlenhoff et al. 2007, 2011; Gelling et al. 2008; Song et al. 2009; Long et al. 2011; Sheftel et al. 2012). The three proteins functionally interact with each other, and deletion of the individual genes elicits highly similar phenotypes indicating that they cooperate in the same reaction (Gelling et al. 2008; Waller et al. 2010; Mühlenhoff et al. 2011; Sheftel et al. 2012). How the three proteins mechanistically help in generating the [4Fe-4S] cluster is currently unresolved. This is mainly owing to the fact that it is still unclear what the physiological meaning of different forms of iron bound to different members of the Isa protein family may be. The yeast Isa1 and Isa2 proteins were shown to bind iron in vitro and in vivo, even under conditions when Fe/S cluster synthesis is blocked (Lu et al. 2010; Mühlenhoff et al. 2011). However, by which mechanism the iron-binding Isa proteins may assist in the conversion of the Isu1-generated [2Fe-2S] cluster into a [4Fe-4S] moiety remains unclear. In bacteria, the related A-type ISC proteins IscA, ErpA, and SufA were shown to bind either iron or a [2Fe-2S] cluster (see, e.g., Gupta et al. 2009; Py and Barras 2010; Wang et al. 2010; Mapolelo et al. 2013). However, the physiological relevance of the two different bound iron cofactors remains to be resolved.

The final step of specific insertion of the [4Fe-4S] cluster into the polypeptide chains of various apoproteins depends on a number of ISC assembly factors that show target protein specificity (Fig. 2). Unlike the core ISC assembly components, these factors are required for maturation of subsets of Fe/S proteins, yet are dispensable for others. Some of the factors appear to show overlapping substrate specificity explaining why deletion of their genes is associated with a comparatively mild phenotype. For instance, complex I maturation involves the P-loop NTPase Ind1, and to date, no other targets of Ind1 are known. Ind1 is a close homolog of the CIA proteins Cfd1 and Nbp35 discussed below, and similarly to these proteins, Ind1 transiently binds [4Fe-4S] clusters and transfers them to the Fe/S subunits of complex I (Bych et al. 2008; Sheftel et al. 2009). The ISC assembly protein Nfu1 shows a broader target specificity including complexes I and II of the respiratory chain and lipoate synthase, whereas aconitase does not depend on this maturation factor (Cameron et al. 2011; Navarro-Sastre et al. 2011). Like Ind1, Nfu1 also binds a [4Fe-4S] cluster (Tong et al. 2003), and may transfer it to its targets (Navarro-Sastre et al. 2011). Interestingly, the target protein specificity of Nfu1 was initially revealed from the biochemical phenotype of patients with genetic mutations in this gene (Cameron et al. 2011; Navarro-Sastre et al. 2011). In addition to defects in complexes I and II, affected patients show severe deficits in lipoate-containing proteins such as pyruvate dehydrogenase, 2-ketoglutarate dehydrogenase, and the H protein of glycine cleavage system. This suggested that human NFU1 acts as a specific targeting factor for these mitochondrial [4Fe-4S] proteins. Based on these results, the Nfu1 substrate specificity was confirmed in yeast indicating that its function is conserved in eukaryotes (Navarro-Sastre et al. 2011). The role of Ind1 and Nfu1 as late-acting ISC proteins also becomes obvious from the dependence of the assembly of their own transiently bound Fe/S cluster on earlier acting ISC assembly proteins such as Nfs1, Isu1, and Ssq1.

A similar role as for NFU1 was suggested for human BOLA3 (yeast relative Aim1) (Cameron et al. 2011). This protein belongs to a family of typically three members, two of which are located in mitochondria (BOLA1 and BOLA3) and another one representing a cytosolic protein (yeast Fra2 and human BOLA2) (Kumanovics et al. 2008; Li and Outten 2012). Patients with mutations in BOLA3 show a similar biochemical phenotype as NFU1-deficient individuals. The molecular function of the protein and its potential cooperation with the other late-acting ISC factors remains to be resolved.

ASSEMBLY OF CYTOSOLIC AND NUCLEAR Fe/S PROTEINS INVOLVES MITOCHONDRIA AND THE CIA MACHINERY

As mentioned above, the cytosol and nucleus of eukaryotic cells contain many Fe/S proteins with utmost importance for cell survival. It has been recognized as early as in 1999 that their maturation strictly depends on the function of members of the mitochondrial ISC assembly machinery (Figs. 1 and 2) (Kispal et al. 1999). More recent studies have indicated that only the members of the core ISC assembly machinery are assisting biogenesis steps 1 and 2, but not the ISC-targeting factors that facilitate step 3 only (cf. Fig. 2). In particular, the cysteine desulfurase Nfs1 and the scaffold Isu1 have been shown to be required inside mitochondria for Fe/S protein assembly in the cytosol and nucleus (Gerber et al. 2004; Mühlenhoff et al. 2004; Biederbick et al. 2006). Hence, mitochondria seem to produce the sulfur moiety for cytosolic Fe/S clusters. Why all other members of the ISC core machinery are required as well, and what the chemical nature of the activated sulfur species may be, has remained unclear over many years of research. Numerous studies in yeast, man, mice and zebrafish have shown that deletion of the ABC transporter Atm1 (human ABCB7) of the mitochondrial inner membrane leads to a functional deficiency in cytosolic and nuclear Fe/S proteins, whereas mitochondrial Fe/S proteins are largely unaffected (Kispal et al. 1999; Pondarre et al. 2006; Miao et al. 2009). Hence, it is proposed that Atm1 facilitates the export of the sulfur-containing compound (X-S in Fig. 2) from the mitochondrial matrix to the cytosol for integration into extramitochondrial Fe/S proteins. Additionally, depletion of the sulfhydryl oxidase Erv1 of the intermembrane space and of the tripeptide glutathione (GSH) leads to a similar phenotype as deletion of Atm1, namely, cytosolic-nuclear Fe/S protein defects, yet no severe effects on mitochondrial Fe/S proteins. Therefore, these three components have been coined as "ISC export machinery" (Lill and Mühlenhoff 2005).

Research of the past 10 years has identified eight cytosolic proteins with essential functions in the biogenesis of cytosolic and nuclear Fe/S proteins (Fig. 3). Together, they comprise the CIA machinery (Sharma et al. 2010). The proteins can be attributed to various partial reactions that are formally similar to those performed by the mitochondrial ISC machinery (Netz et al. 2007, 2010, 2012a). First, a [4Fe-4S] cluster is transiently assembled on the P-loop NTPases Cfd1 and Nbp35, which serve as a scaffold. This step requires the core mitochon-

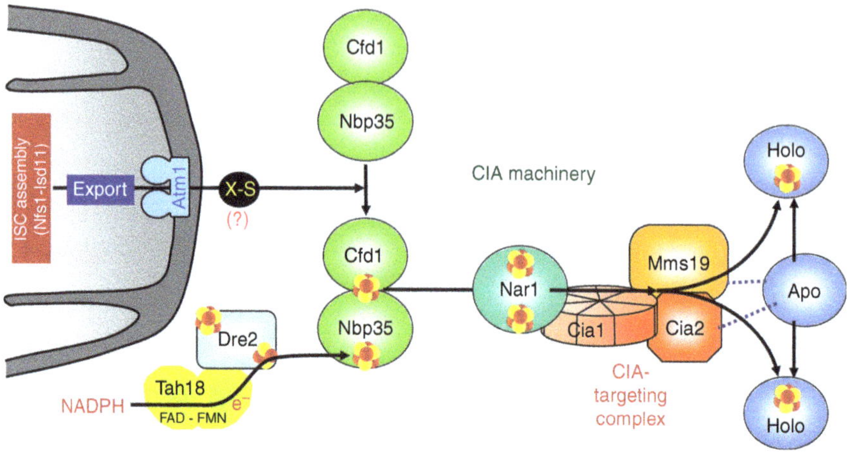

Figure 3. A model for assembly of cytosolic and nuclear Fe/S proteins. The CIA machinery comprises eight known proteins. In a first step, a bridging [4Fe-4S] cluster is assembled on the Cfd1-Nbp35 scaffold complex. This reaction requires a sulfur source (X-S) generated by the mitochondrial ISC assembly machinery and exported by the mitochondrial ABC transporter Atm1 (human ABCB7). Generation of the functionally essential amino-terminal Fe/S cluster of Nbp35 (bottom) depends on the flavoprotein Tah18 and the Fe/S protein Dre2, which serve as an NADPH-dependent electron transfer chain. In the second step, the bridging Fe/S cluster is released from Cfd1-Nbp35, a reaction mediated by Nar1 and the CIA-targeting complex Cia1-Cia2-Mms19. The latter two proteins interact with target (apo)proteins and assure specific Fe/S cluster insertion. Biogenesis further requires, at an unknown step, the cytosolic multidomain monothiol glutaredoxins Grx3-Grx4 (human PICOT), which bind a glutathione-coordinated, bridging [2Fe-2S] cluster (not shown). These proteins also play a role in intracellular iron trafficking.

drial ISC assembly machinery, including Nfs1, as a sulfur donor (see Fig. 2) (Nakai et al. 2001; Mühlenhoff et al. 2004; Biederbick et al. 2006). Cfd1-Nbp35 form a heterotetramer and bind the [4Fe-4S] cluster in a bridging manner on two conserved cysteine residues of two monomers. Mutation of the Walker motif of Cfd1 and Nbp35 showed that it is essential for function of these proteins (Sharma et al. 2010; Netz et al. 2012a). It is therefore generally assumed that NTP hydrolysis is needed for Fe/S cluster assembly, but hitherto neither nucleotide binding nor hydrolysis has been proven experimentally. Therefore, the precise role of the nucleotide-binding motif for Cfd1-Nbp35 function remains to be determined. In addition to the transient Fe/S cluster, Nbp35 contains another [4Fe-4S] cluster at its amino terminus, which is stably bound and essential for function. Its assembly depends on the electron transfer chain from NADPH as an electron source, to the flavin-containing oxidoreductase Tah18, and, finally, to the Fe/S protein Dre2 (Zhang et al. 2008; Netz et al. 2010; Banci et al. 2013). What the precise role of reduction in this early step of CIA function may be is unclear. One possibility is the reduction of the sulfur moiety exported from mitochondria to sulfide, but other options are equally likely.

The transiently bound, bridging [4Fe-4S] cluster assembled on Cfd1-Nbp35 is then transferred to apoproteins. This reaction is mechanistically poorly understood. However, it is known that the CIA protein Nar1 interacts with Nbp35 and therefore may be involved in Fe/S cluster mobilization (Balk et al. 2004; Song and Lee 2008). Nar1 shows similarity to iron-only hydrogenases and binds two [4Fe-4S] clusters, which are assembled with the help of Cfd1-Nbp35 (Urzica et al. 2009). Hence, Nar1 is both a target and a component of the cellular Fe/S protein biogenesis machinery, and possibly is required for its own maturation (Balk et al. 2004). Because later-acting CIA factors (see below) are dispensable for the assembly of the two Nar1 Fe/S clusters, the protein may act as a mediator between early and late steps of the CIA biogenesis process (Fig. 3). The recently identified CIA components Cia1, Cia2, and Mms19 form the so-called CIA-targeting complex, as these proteins appear to mediate both Fe/S cluster transfer and target-specific cluster insertion into the various polypeptide chains (Fig. 3) (Srinivasan et al. 2007; Weerapana et al. 2010; Gari et al. 2012; Stehling et al. 2012). These reactions involve the direct physical interaction of the CIA-targeting complex proteins with the target Fe/S proteins, presumably their apoforms. This became most evident from affinity pull-down experiments in human cells where the various CIA-targeting factors interacted with a large number of cytosolic and nuclear Fe/S proteins (Stehling et al. 2012). This protein list includes DNA polymerases and primases, ATP-dependent DNA helicases, DNA glycosylases, and the ABC protein ABCE1. In yeast, the interactions between CIA components and Fe/S target proteins seem to be weaker or less stable, and only a few such interactions have been identified including the binding of Fe/S helicase Rad3 to Mms19 and of the helicase-nuclease Dna2 to Cia2 (Stehling et al. 2012). The precise molecular function of the late-acting CIA components remains to be determined.

In addition to the mentioned proteins, the cytosolic monothiol glutaredoxins Grx3-Grx4 (human PICOT) were shown to be crucial for cytosolic and nuclear Fe/S protein biogenesis (Mühlenhoff et al. 2010). Because these proteins are involved in intracellular iron trafficking and iron uptake regulation, they seem to play a more general function and hence are no typical CIA proteins. For instance, these glutaredoxins are also involved in the maturation of di-iron proteins such as ribonucleotide reductase or in heme biosynthesis in erythroid cells (Mühlenhoff et al. 2010; Zhang et al. 2011; Haunhorst et al. 2013). Failure of the glutaredoxins to participate in these processes or in Fe/S protein biogenesis explains the drastic effect on the viability of cells lacking both proteins. How the glutaredoxins mechanistically fulfill their iron-related function is currently unknown.

Overall, the process of cellular Fe/S protein biogenesis is surprisingly complex with some 30 components being directly involved. However, it seems likely that more components with a function in this essential process will be discovered in the future.

THE MITOCHONDRIAL CONTRIBUTION TO EXTRAMITOCHONDRIAL Fe/S PROTEIN MATURATION EXPLAINS THE INDISPENSABLE CHARACTER OF THE ORGANELLES

More than half of the known ISC and almost all CIA components are essential for the viability of yeast cells (Lill and Mühlenhoff 2005, 2008). In humans, even ISC genes that can be knocked out in yeast without major consequences are essential for life (e.g., *NFU1*; for details see below). Hence, their contribution to cellular Fe/S protein biogenesis renders mitochondria essential for life. This essentiality sharply contrasts with the importance of other mitochondrial functions such as ATP synthesis, citric acid cycle, and fatty acid oxidation that can be deleted without the loss of cell viability, at least in yeast. This organism can survive without respiratory competent mitochondria when cultivated on media containing fermentable carbon sources such as glucose. Strikingly, a nonessential function of mitochondrial respiration and ATP synthesis is evident also for human cells, at least under certain conditions. For instance, cultured human cells lacking their mitochondrial DNA (ρ^0 cells) are viable despite the absence of oxidative phosphorylation, as long as they are supplied with high glucose, which is used for ATP production by glycolysis. These points clearly indicate that respiration and other classical mitochondrial functions are not primarily responsible for the essential character of mitochondria in eukaryotes. Rather, the involvement in the maturation of cellular Fe/S proteins satisfactorily explains why mitochondria are needed for life.

This conclusion has received impressive biological support from the thorough genetic analysis of a number of diverse species that originally were thought to lack mitochondria altogether, and hence were called amitochondriates (for review see Molik and Lill 2012, and citations therein). These species contain either hydrogenosomes or mitosomes, both of which are double membrane-bounded organelles that are evolutionarily related to classical mitochondria. They have evolved from mitochondria-containing species by successive loss of mitochondrial genes and mitochondria-typical functions such as respiration, citric acid cycle, and heme synthesis (Embley and Martin 2006; van der Giezen 2009; Muller et al. 2012). As a consequence, mitosomes present in, for example, microsporidia, diplomonads, and amoebozoa, have lost virtually all classical functions of mitochondria, with the notable exception of Fe/S protein biogenesis (Tovar et al. 2003; Goldberg et al. 2008). Strikingly, mitosomes do not contain any Fe/S proteins, which would explain the maintenance of the ISC assembly machinery. The only known mitosomal Fe/S protein is the [2Fe-2S] ferredoxin Yah1, which itself is an ISC component involved in Fe/S protein maturation. Hence, it can be speculated that the remnant function of mitosomes and its "leftover" ISC system lies in the maturation of extramitochondrial Fe/S proteins.

Which extramitochondrial Fe/S proteins may then explain the essentiality of mitochondrial Fe/S protein biogenesis? The first known example of an essential extramitochondrial Fe/S protein was the ABC protein Rli1, a component involved in ribosome function and assembly (Kispal et al. 2005; Yarunin et al. 2005; Becker et al. 2012). Other recently identified essential Fe/S proteins include components involved in nucleotide excision repair (Rad3, human XPD), RNA primer synthesis during DNA replication (Pri2) (Rudolf et al. 2006; Klinge et al. 2007), DNA replication (DNA polymerases) (Netz et al. 2012b), and telomere length variation (RTEL1) (Gari et al. 2012). Recently, it was experimentally proven for some of these proteins that their maturation requires both the mitochondrial ISC and the cytosolic CIA components, and that assembly defects lead to increased sensitivity to DNA damage agents such as UV light and mutagenic chemicals (Gari et al. 2012; Stehling et al. 2012). These findings imply that mitochondria, via their function in Fe/S protein maturation, play a direct role in fundamental processes of life such as nuclear DNA maintenance, gene expression, and genome stability.

MITOCHONDRIAL Fe/S PROTEIN BIOGENESIS IS A KEY REGULATOR OF CELLULAR IRON HOMEOSTASIS

It has long been recognized that Fe/S protein biogenesis performs a regulatory role in cellular iron metabolism. This makes physiological sense, as the supply of the metal is directly controlled by an iron-consuming process. Nevertheless, this regulatory mechanism is remarkable as heme synthesis does not directly impact on the adjustment of iron supply, at least not in nonerythroid cells. In particular, it is the mitochondrial ISC system that plays a key role in cellular iron homeostasis. Functional depletion of the core components of the mitochondrial ISC assembly and export systems is associated with increased cellular iron uptake and usually with an iron overload of mitochondria. This phenomenon is similar in yeast and human cells, even though the underlying regulatory mechanisms differ substantially. In yeast, the iron regulatory function is mostly exerted by transcriptional activation of the Aft1-Aft2-transcription factor-dependent iron regulon genes, whereas other fungi use repressor systems (for comprehensive recent reviews, see Schrettl and Haas 2011; Lill et al. 2012; Philpott et al. 2012). In human cells, the major impact on cellular iron supply and distribution within the cell is mediated by the two iron regulatory proteins (IRP) 1 and 2 via complex posttranscriptional mechanisms (for comprehensive recent reviews, see Anderson et al. 2012; Thompson and Bruick 2012). Here, we will only briefly summarize which events may lead to cellular iron uptake and mitochondrial iron accumulation when (mitochondrial) Fe/S protein biogenesis is functionally impaired. Many details of this regulatory pathway are still poorly defined at the molecular level.

In yeast, depletion of any member of the core mitochondrial ISC assembly machinery or of the ISC export system including the ABC transporter Atm1 leads to activation of the iron-sensing transcription factors Aft1-Aft2 and an induction of genes of the iron regulon (Figs. 1 and 2). This includes some 30 genes that are involved in cellular iron acquisition and proper intracellular distribution of the metal to various compartments including mitochondria and vacuoles. What is the molecular basis for the role of the mitochondrial ISC system in iron regulation? Likely, the core ISC machinery produces and Atm1 exports a sensor molecule to the cytosol, which is then used by Aft1-Aft2 to interpret the mitochondrial iron status. It is reasonable to believe that the molecule X-S used for Fe/S protein maturation by the CIA machinery and the iron sensor are identical or at least chemically related. The Aft transcription factors interact with the monothiol glutaredoxins Grx3-Grx4 mentioned above and the BolA protein Fra2 (Mühlenhoff et al. 2010; Li and Outten 2012). Deletion of Grx3-Grx4 leads to strong activation of the Aft iron regulon and cellular iron uptake. The iron accumulates in the cytosol and not in mitochondria, thus differing from the situation with defects in ISC proteins where iron levels strongly increase in mitochondria (see above). The glutaredoxins bind a [2Fe-2S] cluster and it may be hypothesized that this cluster is the molecule sensed by Aft1-Aft2, but in vivo evidence for this hypothesis is lacking.

In human cells, intracellular iron regulation is accomplished by IRP1 and IRP2. IRP1 is a cytosolic Fe/S protein with aconitase activity, yet in its apoform it can bind to iron-responsive elements (IREs) of messenger RNAs (mRNAs), which encode proteins involved in iron trafficking (transferrin receptor and ferroportin), storage (ferritin), and utilization (aconitase and eALAS), thereby regulating the efficiency of translation or the mRNA stability (Anderson et al. 2012; Thompson and Bruick 2012). The equilibrium between the apo- and holoforms of IRP1 is shifted by Fe/S cluster assembly, which depends on both the mitochondrial ISC systems and the CIA machinery, which uses the mitochondria-exported molecule X-S for Fe/S cluster synthesis. Thus, unlike in yeast, depletion of both ISC and CIA components in human cells affects cellular iron regulation via IRP1. Strikingly and in similarity to the situation in yeast, iron accumulates in mitochondria on defects in the ISC systems. It is clear from these observations that both yeast and mammals regulate cellular iron uptake and supply to the mitochondrial matrix by the efficiency of mitochondria to generate Fe/S clusters.

DIFFERENT BIOCHEMICAL PHENOTYPES OF MITOCHONDRIAL Fe/S DISEASES PROVIDE INSIGHTS INTO THE MECHANISM OF Fe/S PROTEIN ASSEMBLY

Over the past decade, a number of disorders with diverse clinical phenotypes have been reported as a consequence of genetic defects in mitochondrial ISC components (Fig. 4) (Sheftel et al. 2010a; Rouault 2012). Many of these diseases are fatal, sometimes in early childhood, which is not surprising based on the essential character of Fe/S protein biogenesis. Although it is still difficult to explain the overall clinical phenotypes, it seems possible to reconcile the biochemical observations made in patient material with the mechanistic studies performed in tissue culture using RNAi-depletion technology. Hence, the disease phenotypes are a critical test of the molecular mechanism worked out for ISC assembly and can be used to extend the model for this process.

The founding member of Fe/S diseases is the neurodegenerative disorder Friedreich's ataxia, in which the early-acting ISC protein frataxin is functionally deficient, either by a more than 70% decrease in protein levels as a result of decreased transcription or by point mutations (Schmucker and Puccio 2010; Stemmler et al. 2010). This functional impairment is associated with decreased activities of respiratory complexes I–III and of mitochondrial aconitase. Moreover, frataxin-deficient cells show a defect in cytosolic IRP1 maturation and a mitochondrial iron accumulation. These biochemical phenotypes are well explained by the crucial function of frataxin as a core member of the ISC machinery (Figs. 2 and 4). It has been noted for both yeast and human cells that frataxin-deficient cells show increased sensitivity to nuclear DNA damage (Karthikeyan et al. 2002; Thierbach et al. 2010). This effect has now been explained as a general consequence of ISC (and CIA) protein defects, which hamper the assembly of critical Fe/S components of DNA metabolism such as

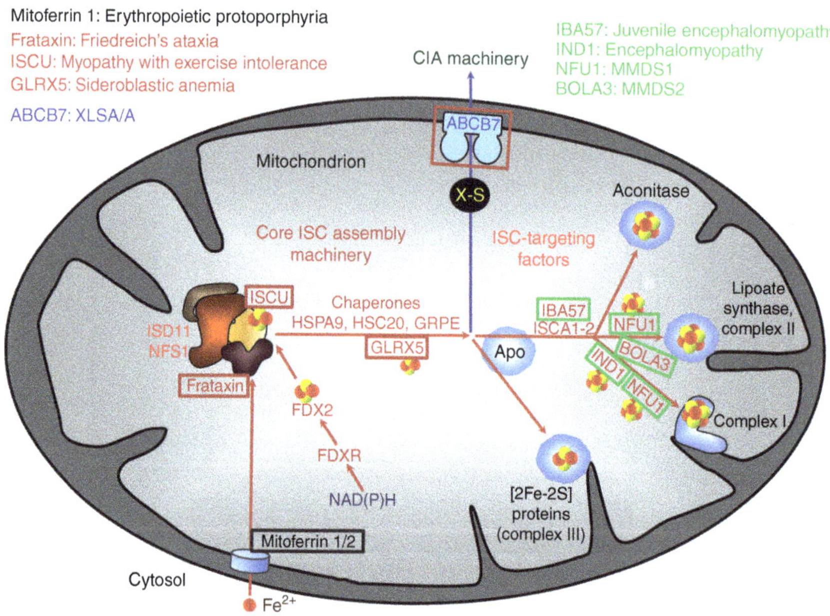

Figure 4. Diseases associated with the assembly of mitochondrial and cytosolic Fe/S proteins. The model is a simplified version of Fig. 2 showing the human ISC protein names. The boxes highlight ISC proteins, which are mutated in two different kinds of Fe/S diseases. Red boxes indicate Fe/S diseases that are associated with a mitochondrial iron accumulation, whereas mutations in ISC proteins indicated by green boxes do not affect the iron metabolism. In addition, the model depicts a mitochondrial iron import disorder associated with a mutation in the inner membrane carrier mitoferrin 1 (gray box). XLSA/A, X-linked sideroblastic anemia and cerebellar ataxia; MMDS, multiple mitochondrial dysfunction syndrome.

replicative DNA polymerases and numerous DNA helicases involved in DNA repair (Gari et al. 2012; Stehling et al. 2012).

A splice mutation in the mitochondrial glutaredoxin GLRX5 drastically decreases its protein level, and causes a clinical phenotype that is rather different from Friedreich's ataxia. The GLRX5 mutation most severely affects erythroblasts causing a sideroblastic (microcytic) anemia characterized by mitochondrial iron overload (Camaschella et al. 2007). The erythroblast cell-type specificity was explained by the unique combination of the IRP targets ALAS2 (IRP-repressible) and ferroportin 1b (non-IRP-repressible) (Ye et al. 2010). A third example with a mutation in a member of the core ISC assembly machinery is the muscle-specific splicing defect in the scaffold protein ISCU that causes a myopathy with exercise tolerance and lactic acidosis (Mochel et al. 2008; Olsson et al. 2008; Crooks et al. 2012; Nordin et al. 2012). Biochemically, this defect is associated with cellular Fe/S protein deficiencies and cellular iron accumulation as a result of an altered iron homeostasis. Despite striking differences in the clinical appearance, all three diseases show common biochemical alterations, which are characterized by a complex defect in mitochondrial and cytosolic Fe/S proteins (including those carrying a [2Fe-2S] cluster) and an iron accumulation. These phenotypes are typical for early-acting ISC components and hence support the current mechanistic model of Fe/S protein biogenesis.

Recently, several other pathogenic mutations affecting four mitochondrial late-acting ISC proteins were reported. Mutations in the P-loop NTPase IND1 specifically affect complex I activity and cause a disease phenotype that is typical for complex I defects (Sheftel et al. 2009; Calvo et al. 2010). Because IND1 can bind a [4Fe-4S] cluster, it has been suggested that IND1 may transiently bind the cluster synthesized on Isu1 before inserting it into the complex I Fe/S subunits (Bych et al. 2008). Mutations in the [4Fe-4S] protein NFU1 are associated with defects in respiratory complexes I and II, and, most strikingly, lipoate synthase (Cameron et al. 2011; Navarro-Sastre et al. 2011). The latter Fe/S protein produces lipoic acid for the four mitochondrial enzymes pyruvate dehydrogenase, α-ketoglutarate dehydrogenase, branched-chain ketoacid dehydrogenase (BCKDH), and the H protein of the glycine cleavage system (H-GCS). Patients with NFU1 mutations present with respiratory chain defects, hyperglycinemia, and low pyruvate dehydrogenase complex activities (Cameron et al. 2011; Navarro-Sastre et al. 2011). Strikingly, the [4Fe-4S] protein aconitase was not affected in NFU1 patients suggesting a target-specific function of NFU1. A similar pattern of Fe/S protein deficiencies is seen for mutations in BOLA3 (Cameron et al. 2011) indicating a characteristic target specificity for these two ISC assembly proteins. A fourth Fe/S disease case involving a point mutation in the late-acting ISC protein IBA57 affects both the activity and stability of the protein, and causes defects in mitochondrial [4Fe-4S] proteins including respiratory complexes I and II, aconitase, and lipoic acid synthase, whereas respiratory complex III was normal (Gelling et al. 2008; Ajit Bolar et al. 2013). Thus, the biochemical profiles of IBA57, NFU1, and BOLA3 patients substantially overlap, yet characteristically differ in the mitochondrial aconitase activity, which was normal in NFU1- and BOLA3-deficient individuals. These studies, together with mechanistic studies in yeast, provide the basis for the working model (Figs. 2 and 4) proposing that IBA57, together with ISCA1-ISCA2, assists the generation of all mitochondrial [4Fe-4S] proteins, whereas IND1, NFU1, and BOLA3 serve as more specific ISC-targeting factors.

Notably, for all four disease cases of late-acting ISC proteins, no major effects on the cellular or mitochondrial iron metabolism have been reported, which supports their late mechanistic action in the ISC assembly process (Fig. 4). Likewise, the diseased state did not show any impairment of cytosolic Fe/S protein maturation. This feature distinguishes these cases from mutations in frataxin, GLRX5, and ISCU, which show, in addition to general mitochondrial [2Fe-2S] and [4Fe-4S] protein defects, a strong impairment in cytosolic Fe/S protein maturation and a severe cellular iron accumulation as a consequence of diminished IRP1 maturation and hence activation of its IRE binding. All of these phenotypes, with the notable exception of mitochondrial Fe/S protein defects, are observed for mutations in the ABC transporter ABCB7 (Fig. 4), which cause X-linked sideroblastic anemia and cerebellar ataxia (XLSA/A) (Bekri et al. 2000; Cazzola and Invernizzi 2011). Collectively, the presence or absence of effects on cytosolic Fe/S proteins and on cellular iron metabolism may readily allow clinicians to distinguish the two groups of mitochondrial Fe/S diseases in which either core ISC assembly and export members or ISC-targeting factors, respectively, are functionally impaired.

CONCLUSIONS AND OUTLOOK

The assembly of mitochondrial, cytosolic, and nuclear Fe/S proteins has turned out to be a surprisingly complex biosynthetic process, despite the chemically simple nature of the Fe/S cofactor. The central contribution of mitochondria to cellular Fe/S protein maturation makes these organelles indispensable for life, even under situations when other more classical processes of mitochondria can be bypassed. Because biogenesis targets Fe/S proteins participating in many diverse cellular pathways, numerous downstream biochemical processes are affected when Fe/S protein assembly is impaired. Examples include, in addition to various steps of amino acid and nucleotide anabolism and catabolism, cellular iron homeostasis (which is regulated by mitochondrial Fe/S protein assembly), cytosolic protein translation, DNA synthesis and repair, as well as the complex issue of genome integrity. Undoubtedly, Fe/S protein biogenesis is one of the most basic biosynthetic processes of life.

Future studies on this important cellular process will be dedicated to the better biochemical definition of the function of individual biogenesis components. This will be supported by insights into the 3D structures of ISC and CIA components, knowledge about the dynamics of the protein interactions underlying the various biogenesis steps, and the elucidation of the regulation of the process. Even though it is safe to state that most biogenesis components have been identified to date, there are likely a few more of these components to be discovered and functionally characterized in the near future.

ACKNOWLEDGMENTS

We thank our group for fruitful discussions. Research is generously supported by grants from Deutsche Forschungsgemeinschaft (SFB 593, SFB-TR1, SFB 987, and GRK 1216), von Behring-Röntgen Stiftung, LOEWE program of state Hessen, and Max-Planck Gesellschaft.

REFERENCES

Adam AC, Bornhövd C, Prokisch H, Neupert W, Hell K. 2006. The Nfs1 interacting protein Isd11 has an essential role in Fe/S cluster biogenesis in mitochondria. *EMBO J* **25:** 174–183.

Ajit Bolar N, Vanlander AV, Wilbrecht C, Van der Aa N, Smet J, De Paepe B, Vandeweyer G, Kooy F, Eyskens F, De Latter E, et al. 2013. Mutation of the iron–sulfur cluster assembly gene IBA57 causes severe myopathy and encephalopathy. *Hum Mol Genet* **22:** 2590–2602.

Alfonzo JD, Lukes J. 2011. Assembling Fe/S-clusters and modifying tRNAs: Ancient co-factors meet ancient adaptors. *Trends Parasitol* **27:** 235–238.

Anderson CP, Shen M, Eisenstein RS, Leibold EA. 2012. Mammalian iron metabolism and its control by iron regulatory proteins. *Biochim Biophys Acta* **1823:** 1468–1483.

Balk J, Pilon M. 2011. Ancient and essential: The assembly of iron–sulfur clusters in plants. *Trends Plant Sci* **16:** 218–226.

Balk J, Pierik AJ, Aguilar Netz D, Mühlenhoff U, Lill R. 2004. The hydrogenase-like Nar1p is essential for maturation of cytosolic and nuclear iron–sulphur proteins. *EMBO J* **23:** 2105–2115.

Banci L, Bertini I, Calderone V, Ciofi-Baffoni S, Giachetti A, Jaiswal D, Mikolajczyk M, Piccioli M, Winkelmann J. 2013. Molecular view of an electron transfer process essential for iron–sulfur protein biogenesis. *Proc Natl Acad Sci* **110:** 7136–7141.

Bandyopadhyay S, Chandramouli K, Johnson MK. 2008a. Iron–sulfur cluster biosynthesis. *Biochem Soc Trans* **36:** 1112–1119.

Bandyopadhyay S, Gama F, Molina-Navarro MM, Gualberto JM, Claxton R, Naik SG, Huynh BH, Herrero E, Jacquot JP, Johnson MK, et al. 2008b. Chloroplast monothiol glutaredoxins as scaffold proteins for the assembly and delivery of [2Fe-2S] clusters. *EMBO J* **27:** 1122–1133.

Becker T, Franckenberg S, Wickles S, Shoemaker CJ, Anger AM, Armache JP, Sieber H, Ungewickell C, Berninghausen O, Daberkow I, et al. 2012. Structural basis of highly conserved ribosome recycling in eukaryotes and archaea. *Nature* **482:** 501–506.

Beinert H, Holm RH, Münck E. 1997. Iron–sulfur clusters: Nature's modular, multipurpose structures. *Science* **277:** 653–659.

Bekri S, Kispal G, Lange H, Fitzsimons E, Tolmie J, Lill R, Bishop DF. 2000. Human ABC7 transporter: Gene structure and mutation causing X-linked sideroblastic anemia with ataxia (XLSA/A) with disruption of cytosolic iron–sulfur protein maturation. *Blood* **96:** 3256–3264.

Biederbick A, Stehling O, Rösser R, Niggemeyer B, Nakai Y, Elsässer HP, Lill R. 2006. Role of human mitochondrial Nfs1 in cytosolic iron–sulfur protein biogenesis and iron regulation. *Mol Cell Biol* **26:** 5675–5687.

Bonomi F, Iametti S, Morleo A, Ta D, Vickery LE. 2008. Studies on the mechanism of catalysis of iron–sulfur cluster transfer from IscU[2Fe2S] by HscA/HscB chaperones. *Biochemistry* **47:** 12795–12801.

Bridwell-Rabb J, Winn AM, Barondeau DP. 2011. Structure-function analysis of Friedreich's ataxia mutants reveals determinants of frataxin binding and activation of the Fe-S assembly complex. *Biochemistry* **50:** 7265–7274.

Bych K, Kerscher S, Netz DJ, Pierik AJ, Zwicker K, Huynen MA, Lill R, Brandt U, Balk J. 2008. The iron–sulphur protein Ind1 is required for effective complex I assembly. *EMBO J* **27:** 1736–1746.

Calvo SE, Tucker EJ, Compton AG, Kirby DM, Crawford G, Burtt NP, Rivas M, Guiducci C, Bruno DL, Goldberger OA, et al. 2010. High-throughput, pooled sequencing identifies mutations in NUBPL and FOXRED1 in human complex I deficiency. *Nat Genet* **42:** 851–858.

Camaschella C, Campanella A, De Falco L, Boschetto L, Merlini R, Silvestri L, Levi S, Iolascon A. 2007. The human counterpart of zebrafish shiraz shows sideroblastic-like microcytic anemia and iron overload. *Blood* **110:** 1353–1358.

Cameron JM, Janer A, Levandovskiy V, Mackay N, Rouault TA, Tong WH, Ogilvie I, Shoubridge EA, Robinson BH. 2011. Mutations in iron–sulfur cluster scaffold genes NFU1 and BOLA3 cause a fatal deficiency of multiple respiratory chain and 2-oxoacid dehydrogenase enzymes. *Am J Hum Genet* **89:** 486–495.

Cavadini P, O'Neill HA, Benada O, Isaya G. 2002. Assembly and iron-binding properties of human frataxin, the protein deficient in Friedreich ataxia. *Hum Mol Genet* **11:** 217–227.

Cazzola M, Invernizzi R. 2011. Ring sideroblasts and sideroblastic anemias. *Haematologica* **96:** 789–792.

Chandramouli K, Johnson MK. 2006. HscA and HscB stimulate [2Fe-2S] cluster transfer from IscU to apoferredoxin in an ATP-dependent reaction. *Biochemistry* **45:** 11087–11095.

Chen OS, Hemenway S, Kaplan J. 2002. Inhibition of Fe-S cluster biosynthesis decreases mitochondrial iron export: Evidence that Yfh1p affects Fe-S cluster synthesis. *Proc Natl Acad Sci* **99:** 12321–12326.

Ciesielski SJ, Schilke BA, Osipiuk J, Bigelow L, Mulligan R, Majewska J, Joachimiak A, Marszalek J, Craig EA, Dutkiewicz R. 2012. Interaction of J-protein co-chaperone

Jac1 with Fe-S scaffold Isu is indispensable in vivo and conserved in evolution. *J Mol Biol* **417:** 1–12.

Colin F, Martelli A, Clemancey M, Latour JM, Gambarelli S, Zeppieri L, Birck C, Page A, Puccio H, Ollagnier de Choudens S. 2013. Mammalian frataxin controls sulfur production and iron entry during de novo Fe4S4 cluster assembly. *J Am Chem Soc* **135:** 733–740.

Crooks DR, Jeong SY, Tong WH, Ghosh MC, Olivierre H, Haller RG, Rouault TA. 2012. Tissue specificity of a human mitochondrial disease: Differentiation-enhanced mis-splicing of the Fe-S scaffold gene ISCU renders patient cells more sensitive to oxidative stress in ISCU myopathy. *J Biol Chem* **287:** 40119–40130.

Embley TM, Martin W. 2006. Eukaryotic evolution, changes and challenges. *Nature* **440:** 623–630.

Gari K, Leon Ortiz AM, Borel V, Flynn H, Skehel JM, Boulton SJ. 2012. MMS19 links cytoplasmic iron–sulfur cluster assembly to DNA metabolism. *Science* **337:** 243–245.

Gelling C, Dawes IW, Richhardt N, Lill R, Mühlenhoff U. 2008. Mitochondrial Iba57p is required for Fe/S cluster formation on aconitase and activation of radical SAM enzymes. *Mol Cell Biol* **28:** 1851–1861.

Gerber J, Mühlenhoff U, Lill R. 2003. An interaction between frataxin and Isu1/Nfs1 that is crucial for Fe/S cluster synthesis on Isu1. *EMBO Rep* **4:** 906–911.

Gerber J, Neumann K, Prohl C, Mühlenhoff U, Lill R. 2004. The yeast scaffold proteins Isu1p and Isu2p are required inside mitochondria for maturation of cytosolic Fe/S proteins. *Mol Cell Biol* **24:** 4848–4857.

Goldberg AV, Molik S, Tsaousis AD, Neumann K, Kuhnke G, Delbac F, Vivares CP, Hirt RP, Lill R, Embley TM. 2008. Localization and functionality of microsporidian iron–sulphur cluster assembly proteins. *Nature* **452:** 624–628.

Gupta V, Sendra M, Naik SG, Chahal HK, Huynh BH, Outten FW, Fontecave M, Ollagnier de Choudens S. 2009. Native *Escherichia coli* SufA, coexpressed with SufBCDSE, purifies as a [2Fe-2S] protein and acts as an Fe-S transporter to Fe-S target enzymes. *J Am Chem Soc* **131:** 6149–6153.

Haunhorst P, Hanschmann EM, Bräutigam L, Stehling O, Hoffmann B, Mühlenhoff U, Lill R, Berndt C, Lillig CH. 2013. Crucial function of vertebrate glutaredoxin 3 (PICOT) in iron homeostasis and hemoglobin maturation. *Mol Biol Cell* **24:** 1895–1903.

Hu Y, Ribbe MW. 2012. Nitrogenase assembly. *Biochim Biophys Acta* doi: 10.1016/j.bbabio.2012.12.001.

Johnson DC, Dean DR, Smith AD, Johnson MK. 2005. Structure, function and formation of biological iron–sulfur clusters. *Ann Rev Biochem* **74:** 247–281.

Kampinga HH, Craig EA. 2010. The HSP70 chaperone machinery: J proteins as drivers of functional specificity. *Nat Rev Mol Cell Biol* **11:** 579–592.

Karthikeyan G, Lewis LK, Resnick MA. 2002. The mitochondrial protein frataxin prevents nuclear damage. *Hum Mol Genet* **11:** 1351–1362.

Kispal G, Csere P, Prohl C, Lill R. 1999. The mitochondrial proteins Atm1p and Nfs1p are required for biogenesis of cytosolic Fe/S proteins. *EMBO J* **18:** 3981–3989.

Kispal G, Sipos K, Lange H, Fekete Z, Bedekovics T, Janaky T, Bassler J, Aguilar Netz DJ, Balk J, Rotte C, et al. 2005. Biogenesis of cytosolic ribosomes requires the essential iron–sulphur protein Rli1p and mitochondria. *EMBO J* **24:** 589–598.

Klinge S, Hirst J, Maman JD, Krude T, Pellegrini L. 2007. An iron–sulfur domain of the eukaryotic primase is essential for RNA primer synthesis. *Nat Struct Mol Biol* **14:** 875–877.

Kumanovics A, Chen O, Li L, Bagley D, Adkins E, Lin H, Dingra NN, Outten CE, Keller G, Winge D, et al. 2008. Identification of FRA1 and FRA2 as genes involved in regulating the yeast iron regulon in response to decreased mitochondrial iron–sulfur cluster synthesis. *J Biol Chem* **283:** 10276–10286.

Lange H, Kaut A, Kispal G, Lill R. 2000. A mitochondrial ferredoxin is essential for biogenesis of cellular iron–sulfur proteins. *Proc Natl Acad Sci* **97:** 1050–1055.

Lanz ND, Booker SJ. 2012. Identification and function of auxiliary iron–sulfur clusters in radical SAM enzymes. *Biochim Biophys Acta* **1824:** 1196–1212.

Li H, Outten CE. 2012. Monothiol CGFS glutaredoxins and BolA-like proteins: [2Fe-2S] binding partners in iron homeostasis. *Biochemistry* **51:** 4377–4389.

Lill R. 2009. Function and biogenesis iron–sulphur proteins. *Nature* **460:** 831–838.

Lill R, Mühlenhoff U. 2005. Iron–sulfur protein biogenesis in eukaryotes. *Trends Biochem Sci* **30:** 133–141.

Lill R, Mühlenhoff U. 2008. Maturation of iron–sulfur proteins in eukaryotes: Mechanisms, connected processes, and diseases. *Annu Rev Biochem* **77:** 669–700.

Lill R, Hoffmann B, Molik S, Pierik AJ, Rietzschel N, Stehling O, Uzarska MA, Webert H, Wilbrecht C, Mühlenhoff U. 2012. The role of mitochondria in cellular iron–sulfur protein biogenesis and iron metabolism. *Biochim Biophys Acta* **1823:** 1491–1508.

Long S, Changmai P, Tsaousis AD, Skalicky T, Verner Z, Wen YZ, Roger AJ, Lukes J. 2011. Stage-specific requirement for Isa1 and Isa2 proteins in the mitochondrion of *Trypanosoma brucei* and heterologous rescue by human and Blastocystis orthologues. *Mol Microbiol* **81:** 1403–1418.

Lu J, Bitoun JP, Tan G, Wang W, Min W, Ding H. 2010. Iron-binding activity of human iron–sulfur cluster assembly protein hIscA1. *Biochem J* **428:** 125–131.

Mapolelo DT, Zhang B, Randeniya S, Albetel AN, Li H, Couturier J, Outten CE, Rouhier N, Johnson MK. 2013. Monothiol glutaredoxins and A-type proteins: Partners in Fe-S cluster trafficking. *Dalton Trans* **42:** 3107–3115.

Miao R, Kim H, Koppolu UM, Ellis EA, Scott RA, Lindahl PA. 2009. Biophysical characterization of the iron in mitochondria from Atm1p-depleted *Saccharomyces cerevisiae. Biochemistry* **48:** 9556–9568.

Mochel F, Knight MA, Tong WH, Hernandez D, Ayyad K, Taivassalo T, Andersen PM, Singleton A, Rouault TA, Fischbeck KH, et al. 2008. Splice mutation in the iron–sulfur cluster scaffold protein ISCU causes myopathy with exercise intolerance. *Am J Hum Genet* **82:** 652–660.

Molik S, Lill R. 2012. Role of mitosomes in cellular iron–sulfur protein biogenesis. *J Endocyt Cell Res* **23:** 77–85.

Mühlenhoff U, Gerber J, Richhardt N, Lill R. 2003. Components involved in assembly and dislocation of iron–sulfur clusters on the scaffold protein Isu1p. *EMBO J* **22:** 4815–4825.

Mühlenhoff U, Balk J, Richhardt N, Kaiser JT, Sipos K, Kispal G, Lill R. 2004. Functional characterization of the eukaryotic cysteine desulfurase Nfs1p from *Saccharomyces cerevisiae*. *J Biol Chem* **279:** 36906–36915.

Mühlenhoff U, Gerl MJ, Flauger B, Pirner HM, Balser S, Richhardt N, Lill R, Stolz J. 2007. The ISC proteins Isa1 and Isa2 are required for the function but not for the *de novo* synthesis of the Fe/S clusters of biotin synthase in *Saccharomyces cerevisiae*. *Eukaryot Cell* **6:** 495–504.

Mühlenhoff U, Molik S, Godoy JR, Uzarska MA, Richter N, Seubert A, Zhang Y, Stubbe J, Pierrel F, Herrero E, et al. 2010. Cytosolic monothiol glutaredoxins function in intracellular iron sensing and trafficking via their bound iron–sulfur cluster. *Cell Metab* **12:** 373–385.

Mühlenhoff U, Richter N, Pines O, Pierik AJ, Lill R. 2011. Specialized function of yeast Isa1 and Isa2 proteins in the maturation of mitochondrial [4Fe-4S] proteins. *J Biol Chem* **286:** 41205–41216.

Müller M, Mentel M, van Hellemond JJ, Henze K, Woehle C, Gould SB, Yu RY, van der Giezen M, Tielens AG, Martin WF. 2012. Biochemistry and evolution of anaerobic energy metabolism in eukaryotes. *Microbiol Mol Biol Rev* **76:** 444–495.

Nakai Y, Nakai M, Hayashi H, Kagamiyama H. 2001. Nuclear localization of yeast Nfs1p is required for cell survival. *J Biol Chem* **276:** 8314–8320.

Navarro-Sastre A, Tort F, Stehling O, Uzarska MA, Arranz JA, Del Toro M, Labayru MT, Landa J, Font A, Garcia-Villoria J, et al. 2011. A fatal mitochondrial disease is associated with defective NFU1 function in the maturation of a subset of mitochondrial Fe-S proteins. *Am J Hum Genet* **89:** 656–667.

Netz DJ, Pierik AJ, Stümpfig M, Mühlenhoff U, Lill R. 2007. The Cfd1-Nbp35 complex acts as a scaffold for iron–sulfur protein assembly in the yeast cytosol. *Nat Chem Biol* **3:** 278–286.

Netz DJ, Stümpfig M, Dore C, Mühlenhoff U, Pierik AJ, Lill R. 2010. Tah18 transfers electrons to Dre2 in cytosolic iron–sulfur protein biogenesis. *Nat Chem Biol* **6:** 758–765.

Netz DJ, Pierik AJ, Stümpfig M, Bill E, Sharma AK, Pallesen LJ, Walden WE, Lill R. 2012a. A bridging [4Fe-4S] cluster and nucleotide binding are essential for the function of the Cfd1-Nbp35 complex as a scaffold in iron–sulfur protein maturation. *J Biol Chem* **287:** 12365–12378.

Netz DJ, Stith CM, Stümpfig M, Kopf G, Vogel D, Genau HM, Stodola JL, Lill R, Burgers PM, Pierik AJ. 2012b. Eukaryotic DNA polymerases require an iron–sulfur cluster for the formation of active complexes. *Nat Chem Biol* **8:** 125–132.

Nordin A, Larsson E, Holmberg M. 2012. The defective splicing caused by the ISCU intron mutation in patients with myopathy with lactic acidosis is repressed by PTBP1 but can be derepressed by IGF2BP1. *Hum Mutat* **33:** 467–470.

Olsson A, Lind L, Thornell LE, Holmberg M. 2008. Myopathy with lactic acidosis is linked to chromosome 12q23.3–24.11 and caused by an intron mutation in the ISCU gene resulting in a splicing defect. *Hum Mol Genet* **17:** 1666–1667.

Peters JW, Broderick JB. 2012. Emerging paradigms for complex iron–sulfur cofactor assembly and insertion. *Annu Rev Biochem* **81:** 429–450.

Philpott CC, Leidgens S, Frey AG. 2012. Metabolic remodeling in iron-deficient fungi. *Biochim Biophys Acta* **1823:** 1509–1520.

Pondarre C, Antiochos BB, Campagna DR, Clarke SL, Greer EL, Deck KM, McDonald A, Han AP, Medlock A, Kutok JL, et al. 2006. The mitochondrial ATP-binding cassette transporter Abcb7 is essential in mice and participates in cytosolic iron–sulphur cluster biogenesis. *Hum Mol Genet* **15:** 953–964.

Pukszta S, Schilke B, Dutkiewicz R, Kominek J, Moczulska K, Stepien B, Reitenga KG, Bujnicki JM, Williams B, Craig EA, et al. 2010. Co-evolution-driven switch of J-protein specificity towards an Hsp70 partner. *EMBO Rep* **11:** 360–365.

Py B, Barras F. 2010. Building Fe-S proteins: Bacterial strategies. *Nat Rev Microbiol* **8:** 436–446.

Radisky DC, Babcock MC, Kaplan J. 1999. The yeast frataxin homologue mediates mitochondrial iron efflux. Evidence for a mitochondrial iron cycle. *J Biol Chem* **274:** 4497–4499.

Rodriguez-Manzaneque MT, Tamarit J, Belli G, Ros J, Herrero E. 2002. Grx5 is a mitochondrial glutaredoxin required for the activity of iron/sulfur enzymes. *Mol Biol Cell* **13:** 1109–1121.

Rouault TA. 2012. Biogenesis of iron–sulfur clusters in mammalian cells: New insights and relevance to human disease. *Dis Model Mech* **5:** 155–164.

Rudolf J, Makrantoni V, Ingledew WJ, Stark MJ, White MF. 2006. The DNA repair helicases XPD and FancJ have essential iron–sulfur domains. *Mol Cell* **23:** 801–808.

Schilke B, Voisine C, Beinert H, Craig E. 1999. Evidence for a conserved system for iron metabolism in the mitochondria of *Saccharomyces cerevisiae*. *Proc Natl Acad Sci* **6:** 10206–10211.

Schilke B, Williams B, Knieszner H, Pukszta S, D'Silva P, Craig EA, Marszalek J. 2006. Evolution of mitochondrial chaperones utilized in Fe-S cluster biogenesis. *Curr Biol* **16:** 1660–1665.

Schmucker S, Puccio H. 2010. Understanding the molecular mechanisms of Friedreich's ataxia to develop therapeutic approaches. *Hum Mol Genet* **19:** R103–R110.

Schmucker S, Martelli A, Colin F, Page A, Wattenhofer-Donze M, Reutenauer L, Puccio H. 2011. Mammalian frataxin: An essential function for cellular viability through an interaction with a preformed ISCU/NFS1/ISD11 iron–sulfur assembly complex. *PLoS ONE* **6:** e16199.

Schrettl M, Haas H. 2011. Iron homeostasis—Achilles' heel of Aspergillus fumigatus? *Curr Opin Microbiol* **14:** 400–405.

Shakamuri P, Zhang B, Johnson MK. 2012. Monothiol glutaredoxins function in storing and transporting [Fe2S2] clusters assembled on IscU scaffold proteins. *J Am Chem Soc* **134:** 15213–15216.

Sharma AK, Pallesen LJ, Spang RJ, Walden WE. 2010. Cytosolic iron–sulfur cluster assembly (CIA) system: Factors, mechanism, and relevance to cellular iron regulation. *J Biol Chem* **285:** 26745–26751.

Sheftel AD, Stehling O, Pierik AJ, Netz DJ, Kerscher S, Elsässer HP, Wittig I, Balk J, Brandt U, Lill R. 2009. Human Ind1, an iron–sulfur cluster assembly factor for respiratory complex I. *Mol Cell Biol* **29:** 6059–6073.

Sheftel A, Stehling O, Lill R. 2010a. Iron–sulfur proteins in health and disease. *Trends Endocrinol Metab* **21:** 302–314.

Sheftel AD, Stehling O, Pierik AJ, Elsässer HP, Mühlenhoff U, Webert H, Hobler A, Hannemann F, Bernhardt R, Lill R. 2010b. Humans possess two mitochondrial ferredoxins, Fdx1 and Fdx2, with distinct roles in steroidogenesis, heme, and Fe/S cluster biosynthesis. *Proc Natl Acad Sci* **107:** 11775–11780.

Sheftel AD, Wilbrecht C, Stehling O, Niggemeyer B, Elsässer HP, Mühlenhoff U, Lill R. 2012. The human mitochondrial ISCA1, ISCA2, and IBA57 proteins are required for [4Fe-4S] protein maturation. *Mol Biol Cell* **23:** 1157–1166.

Shi Y, Ghosh MC, Tong WH, Rouault TA. 2009. Human ISD11 is essential for both iron–sulfur cluster assembly and maintenance of normal cellular iron homeostasis. *Hum Mol Genet* **18:** 3014–3025.

Song D, Lee FS. 2008. A role for IOP1 in mammalian cytosolic iron–sulfur protein biogenesis. *J Biol Chem* **283:** 9231–9238.

Song D, Tu Z, Lee FS. 2009. Human ISCA1 interacts with IOP1/NARFL and functions in both cytosolic and mitochondrial iron–sulfur protein biogenesis. *J Biol Chem* **284:** 35297–35307.

Srinivasan V, Netz DJA, Webert H, Mascarenhas J, Pierik AJ, Michel H, Lill R. 2007. Structure of the yeast WD40 domain protein Cia1, a component acting late in iron–sulfur protein biogenesis. *Structure* **15:** 1246–1257.

Stehling O, Vashisht AA, Mascarenhas J, Jonsson ZO, Sharma T, Netz DJ, Pierik AJ, Wohlschlegel JA, Lill R. 2012. MMS19 assembles iron–sulfur proteins required for DNA metabolism and genomic integrity. *Science* **337:** 195–199.

Stemmler TL, Lesuisse E, Pain D, Dancis A. 2010. Frataxin and mitochondrial FeS cluster biogenesis. *J Biol Chem* **285:** 26737–26743.

Thierbach R, Drewes G, Fusser M, Voigt A, Kuhlow D, Blume U, Schulz TJ, Reiche C, Glatt H, Epe B, et al. 2010. The Friedreich's ataxia protein frataxin modulates DNA base excision repair in prokaryotes and mammals. *Biochem J* **432:** 165–172.

Thompson JW, Bruick RK. 2012. Protein degradation and iron homeostasis. *Biochim Biophys Acta* **1823:** 1484–1490.

Tong WH, Jameson GN, Huynh BH, Rouault TA. 2003. Subcellular compartmentalization of human Nfu, an iron–sulfur cluster scaffold protein, and its ability to assemble a [4Fe-4S] cluster. *Proc Natl Acad Sci* **100:** 9762–9767.

Tovar J, Leon-Avila G, Sanchez LB, Sutak R, Tachezy J, Van Der Giezen M, Hernandez M, Muller M, Lucocq JM. 2003. Mitochondrial remnant organelles of *Giardia* function in iron–sulphur protein maturation. *Nature* **426:** 172–176.

Tsai CL, Barondeau DP. 2010. Human frataxin is an allosteric switch that activates the Fe-S cluster biosynthetic complex. *Biochemistry* **49:** 9132–9139.

Uhrigshardt H, Singh A, Kovtunovych G, Ghosh M, Rouault TA. 2010. Characterization of the human HSC20, an unusual DnaJ type III protein, involved in iron–sulfur cluster biogenesis. *Hum Mol Genet* **19:** 3816–3834.

Urzica E, Pierik AJ, Muhlenhoff U, Lill R. 2009. Crucial role of conserved cysteine residues in the assembly of two iron–sulfur clusters on the CIA protein Nar1. *Biochemistry* **48:** 4946–4958.

Uzarska MA, Dutkiewicz R, Freibert SA, Lill R, Muhlenhoff U. 2013. The mitochondrial Hsp70 chaperone Ssq1 facilitates Fe/S cluster transfer from Isu1 to Grx5 by complex formation. *Mol Biol Cell* **24:** 1830–1841.

van der Giezen M. 2009. Hydrogenosomes and mitosomes: Conservation and evolution of functions. *J Eukaryot Microbiol* **56:** 221–231.

Vickery LE, Cupp-Vickery JR. 2007. Molecular chaperones HscA/Ssq1 and HscB/Jac1 and their roles in iron–sulfur protein maturation. *Crit Rev Biochem Mol Biol* **42:** 95–111.

Waller JC, Alvarez S, Naponelli V, Lara-Nunez A, Blaby IK, Da Silva V, Ziemak MJ, Vickers TJ, Beverley SM, Edison AS, et al. 2010. A role for tetrahydrofolates in the metabolism of iron–sulfur clusters in all domains of life. *Proc Natl Acad Sci* **107:** 10412–10417.

Wang W, Huang H, Tan G, Si F, Liu M, Landry AP, Lu J, Ding H. 2010. In vivo evidence for the iron-binding activity of an iron–sulfur cluster assembly protein IscA in *Escherichia coli*. *Biochem J* **432:** 429–436.

Weerapana E, Wang C, Simon GM, Richter F, Khare S, Dillon MB, Bachovchin DA, Mowen K, Baker D, Cravatt BF. 2010. Quantitative reactivity profiling predicts functional cysteines in proteomes. *Nature* **468:** 790–795.

Wiedemann N, Urzica E, Guiard B, Müller H, Lohaus C, Meyer HE, Ryan MT, Meisinger C, Mühlenhoff U, Lill R, et al. 2006. Essential role of Isd11 in iron–sulfur cluster synthesis on Isu scaffold proteins. *EMBO J* **25:** 184–195.

Wingert RA, Galloway JL, Barut B, Foott H, Fraenkel P, Axe JL, Weber GJ, Dooley K, Davidson AJ, Schmid B, et al. 2005. Deficiency of glutaredoxin 5 reveals Fe-S clusters are required for vertebrate haem synthesis. *Nature* **436:** 1035–1039.

Yarunin A, Panse V, Petfalski E, Tollervey D, Hurt E. 2005. Functional link between ribosome formation and biogenesis of iron–sulfur proteins. *EMBO J* **24:** 580–588.

Ye H, Jeong SY, Ghosh MC, Kovtunovych G, Silvestri L, Ortillo D, Uchida N, Tisdale J, Camaschella C, Rouault TA. 2010. Glutaredoxin 5 deficiency causes sideroblastic anemia by specifically impairing heme biosynthesis and depleting cytosolic iron in human erythroblasts. *J Clin Invest* **120:** 1749–1761.

Zhang Y, Lyver ER, Nakamaru-Ogiso E, Yoon H, Amutha B, Lee D-W, Bi E, Ohnishi T, Daldal F, Pain D, et al. 2008. Dre2, a conserved eukaryotic Fe/S cluster protein, functions in cytosolic Fe/S protein biogenesis. *Mol Cell Biol* **28:** 5569–5582.

Zhang Y, Liu L, Wu X, An X, Stubbe J, Huang M. 2011. Investigation of in vivo diferric tyrosyl radical formation in *Saccharomyces cerevisiae* Rnr2 protein: Requirement of Rnr4 and contribution of Grx3/4 AND Dre2 proteins. *J Biol Chem* **286:** 41499–41509.

Altered Sulfide (H_2S) Metabolism in Ethylmalonic Encephalopathy

Valeria Tiranti and Massimo Zeviani

Pierfranco and Luisa Mariani Center for Research on Children's Mitochondrial Disorders, Unit of Molecular Neurogenetics, Institute of Neurology "Carlo Besta," Istituto di Ricovero e Cura a Carattere Scientifico (IRCCS) Foundation, Milan, Italy

Correspondence: zeviani@istituto-besta.it; tiranti@istituto-besta.it

Hydrogen sulfide (sulfide, H_2S) is a colorless, water-soluble gas with a typical smell of rotten eggs. In the past, it has been investigated for its role as a potent toxic gas emanating from sewers and swamps or as a by-product of industrial processes. At high concentrations, H_2S is a powerful inhibitor of cytochrome *c* oxidase; in trace amounts, it is an important signaling molecule, like nitric oxide (NO) and carbon monoxide (CO), together termed "gasotransmitters." This review will cover the physiological role and the pathogenic effects of H_2S, focusing on ethylmalonic encephalopathy, a human mitochondrial disorder caused by genetic abnormalities of sulfide metabolism. We will also discuss the options that are now conceivable for preventing genetically driven chronic H_2S toxicity, taking into account that a complete understanding of the physiopathology of H_2S has still to be achieved.

Because of its fame as a deadly gas, the physiological role of H_2S had been long overlooked, until Abe and Kimura in 1996 (Abe and Kimura 1996) demonstrated that H_2S is a neuromodulator that facilitates the induction of long-term potentiation (LTP) by enhancing the activity of *N*-methyl-D-aspartate (NMDA) receptors in hippocampal pyramidal neurons. H_2S was later shown to induce relaxation of smooth muscle, to regulate the release of insulin, and to act as a mediator of inflammation.

The physiological importance of H_2S has further been supported by demonstrating that H_2S is endogenously produced in mammals by two pyridoxal-5′-phosphate-dependent enzymes, cystathionine β-synthase (CBS, EC 4.2.1.22) and cystathionine γ-lyase (CSE, EC 4.4.1.1), which both utilize L-cysteine as a substrate (Stipanuk and Beck 1982; Hosoki et al. 1997; Zhao et al. 2001). L-cysteine is taken up with the diet, extracted from endogenous proteins, or synthesized endogenously via *trans*-sulfuration of serine by L-methionine. In humans, the end products of cysteine sulfur catabolism are sulfate (which accounts for 77%–92% of the total sulfur excreted in the urine), sulfate esters (7%–9%), and taurine (2%–6%) (Stipanuk 2004) (Fig. 1). In some tissues, both CBS and CSE can generate H_2S, whereas in others only one enzyme is present (Diwakar and Ravindranath 2007). For instance, CBS is predominantly expressed in the nervous system but

Cite this article as *Cold Spring Harb Perspect Biol* doi: 10.1101/cshperspect.a011437

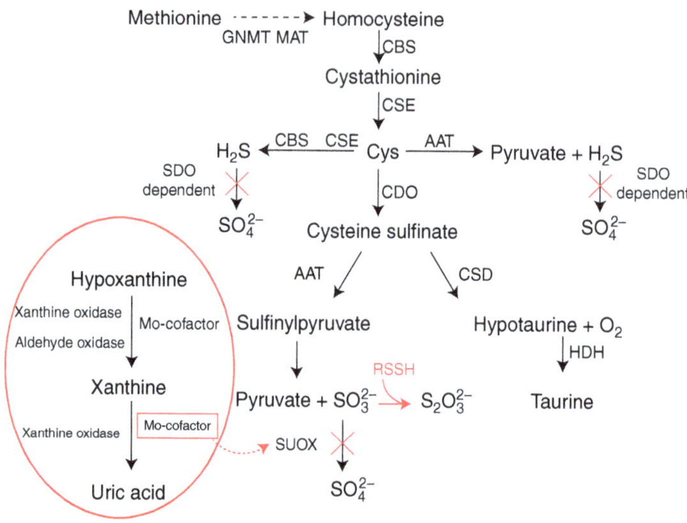

Figure 1. Sulfide production. Methionine, derived from alimentary sources, is converted into *S*-adenosylmethionine by methionine adenosyltransferase (MAT). *S*-adenosylmethionine is hydrolyzed to homocysteine by glycine *N*-methyltransferase (GNMT). Cystathionine β-synthase (CBS) transfers serine to homocysteine and produces cystathionine. Cystathionine γ-lyase (CSE) converts cystathionine to cysteine (Cys). One pathway of cysteine metabolism involves its oxidation to cysteine sulfinate by cysteine deoxygenase (CDO), which is then converted to hypotaurine by cysteine sulfinate decarboxylase (CSD); hypotaurine is eventually oxidated to taurine by hypotaurine dehydrogenase (HDH). In mitochondria, cysteine can be converted to 3-mercaptopyruvate by aspartate aminotransferase (AAT), which can then be converted to H_2S by 3-mercaptopyruvate sulfur transferase. The conversion of cysteine sulfinate to sulfinyl pyruvate by AAT, followed by a nonenzymatic reaction, can also yield sulfite, which is then oxidized to sulfate by sulfite oxidase (SUOX) in a glutathione (GSH)-dependent process. The H_2S produced by either AAT or CBS/CSE can be further oxidized by a sulfur dioxygenase (SDO/ETHE1)-dependent pathway (see Fig. 2). The metabolic pathway carried out by Moco-dependent enzymes is encircled in red (see text for details).

is also active in the liver and kidney (Enokido et al. 2005). CSE is mainly expressed in liver, as well as vascular and nonvascular smooth muscle (e.g., the small intestine and stomach [Awata et al. 1995; Ishii et al. 2004]).

Both CBS and CSE activities increase in rat brain after birth, reaching adult levels at week four (Awata et al. 1995).

An important additional source of H_2S is the enterobacterial anaerobic flora. The intestinal epithelium expresses specialized enzyme systems that efficiently convert H_2S into thiosulfate and sulfate, to both prevent the local increase of H_2S to toxic levels and its entry through the portal vein system in the liver and other organs. Excessive production and absorption of H_2S, as well as reduced detoxification by colonic epithelial cells (colonocytes), is thought to play an important role in the mucosal damage of ulcerative colitis (Rowan et al. 2009).

A third, however marginal, source of H_2S is the nonenzymatic reduction of elemental sulfur, which is present in traces in human (Westley and Westley 1991) and mouse (Buzaleh et al. 1991) blood, into H_2S by electrons provided through glycolysis (Searcy and Lee 1998).

The main catabolic pathway of H_2S takes place in mitochondria and consists of a series of oxidative reactions ultimately yielding sulfate (SO_4^{2-}), with thiosulfate ($S_2O_3^{2-}$) and sulfite (SO_3^{2-}) as intermediate compounds (Beauchamp et al. 1984; Szabò 2007). The existence of such a pathway has been known for more than 20 years (Powell and Somero 1986), although the enzymatic actors and detailed steps have been elucidated (Hildebrandt and Grieshaber

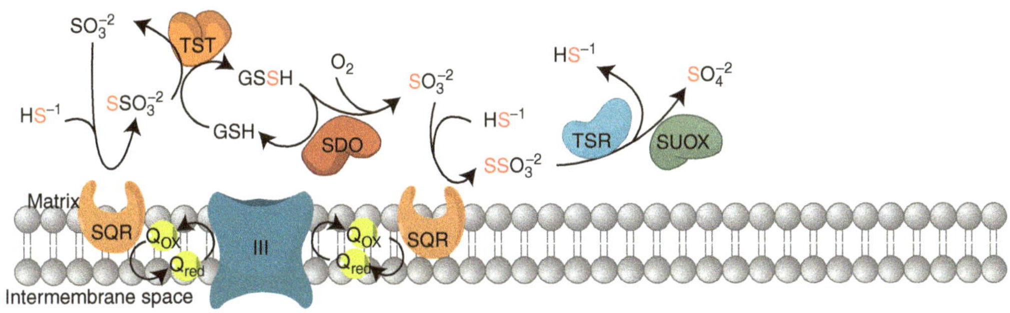

Figure 2. Proposed pathway for sulfide oxidation. H_2S is produced from cysteine (see Fig. 1) and is initially fixed by a membrane-bound sulfide quinone oxidoreductase (SQR). The electrons extracted from H_2S enter the respiratory chain at the level of the quinone pool (UQH/UQ), are then transferred to complex III (cIII), and are finally fixed by cytochrome *c* oxidase (complex IV, cIV) to molecular oxygen with the formation of water. A sulfur dioxygenase (SDO/ETHE1), present as an Fe-binding dimer in the mitochondrial matrix, oxidizes the persulfide to sulfite (H_2SO_3) in a reaction that includes molecular oxygen and water. See text for details. SUOX, sulfite oxidase; TST, thiosulfate sulfur transferase; TSR, thiosulfate reductase.

2008) and further refined (Jackson et al. 2012) only recently. The pathway proposed by Hildebrandt (2008), later modified by Jackson, is shown in Figure 2. The pathway consists of (1) a mitochondrial inner-membrane-bound sulfide quinone reductase (SQR) (Theissen and Martin 2008), which fixes H_2S to sulfite to produce thiosulfate; (2) mitochondrial thiosulfate sulfur transferase, also known as rhodanese, which reconstitutes sulphite from thiosulfate by fixating the sulfane sulfur of the latter onto an –SH-containing substrate, for instance glutathione, GSH, to form a persulfide (R-S-SH) species; (3) a mitochondrial matrix sulfur dioxygenase (SDO), which oxidizes the sulfur atom extracted from persulfide, converting it again into sulfite (SO_3^{2-}); and (4) mitochondrial sulfite oxidase, which further oxidizes sulfite into sulfate (SO_4^{2-}). Thiosulfate can also be converted into two molecules of sulfate by the combined action of a thiosulfate reductase (Müller et al. 2004) and sulfite oxidase (Feng et al. 2007). Additional mechanisms of detoxification rely on the fixation of H_2S to methemoglobin, which is formed from hemoglobin by oxidation of the Fe center, for instance by sodium nitrate, to produce sulfo-methemoglobin, or on the administration of oxidized glutathione (GSSG) (Beauchamp et al. 1984), which is able to fixate H_2S to produce glutathione persufide (GSSH), thus preventing the interaction of H_2S with critical enzymatic activities. Additional sources of sulfite are known, for instance by the cysteine dioxygenase/aspartate aminotransferase (CDO/AAT) system (Cooper et al. 2002; Stipanuk and Ueki 2010), converting cysteine into β-sulfinylpyruvate, which then decomposes into pyruvate and sulfite.

H_2S: PHYSICAL PROPERTIES

Sulfur, an essential element for life, occurs in different oxidation states, from −2 (as in sulfide, S^{2-}) to +6 (as in sulfate, SO_4^{2-}). Sulfur is present in living organisms as a constituent of proteins (mainly in sulfurated amino acids such as cysteine and methionine), in coenzymes (for instance CoA, biotine, and thiamine), and as a component of iron-sulfur clusters in metalloproteins. From a physical–chemical standpoint, H_2S is the sulfur analog of water, but because its intermolecular bonds are weaker than H_2O, at room temperature and atmospheric pressure, it is a gas. In mammals, at a physiological pH of 7.4, approximately one-third of H_2S exists as the undissociated form and two-thirds as the hydrosulfide anion (HS−) (Reiffenstein et al. 1992). In the undissociated form, H_2S can easily penetrate the plasma membranes of cells because of its high solubility in lipids.

PHYSIOLOGICAL ACTIONS OF H_2S

A preliminary question, however crucial for understanding the pathophysiology of sulfide metabolism, is how much free H_2S is present in tissues and body fluids of mammalian organisms. Free H_2S is difficult to measure, mainly because of its high volatility and the interference of cysteine residues contained in proteins, which hamper the results of the most common analytical methods based on spectrophotometric assays from tissue extracts. Levels as high as 30–150 μM have been reported in different tissues in normal conditions (Goodwin et al. 1989; Mitchell et al. 1993; Ogasawara et al. 1994; Wang 2002; Fiorucci et al. 2006;), a remarkable overestimation since these values are well above the olfactory threshold for H_2S and are clearly associated with severe toxicity in vivo, as shown in both cells and mammalian organisms. Using a gas-chromatographic assay that avoids interference by cysteines, Furne et al. (2008) have recently shown that the concentration of free sulfide in mouse liver and brain is 15 nM, that is $10^{-3} - 10^{-4}$ less than the values obtained by spectrophotometric assays.

Recent studies have demonstrated for H_2S a multiplicity of physiological functions, including up-regulation of antioxidant systems and amplification of the effects produced by known antioxidants such as *N*-acetylcysteine (NAC) and glutathione (Whiteman 2004; Yan 2006); up-regulation of anti-inflammatory and cytoprotective genes including heme oxygenase (HO1, also known as HMOX1) in pulmonary smooth-muscle cells (Qingyou et al. 2004) and macrophages (Oh et al. 2006). In the CNS, low concentrations of H_2S are sufficient to increase the production of cAMP, enhance NMDA-receptor-mediated responses, and facilitate the induction of long-term potentiation in hippocampal neurons (Kimura and Kimura 2004).

H_2S displays a number of different effects on myocardial cells and vascular smooth muscle. By activating K_{ATP} channels, H_2S increases the membrane potential of vascular smooth-muscle cells that become hyperpolarized, therefore causing vasorelaxation (Zhao et al. 2001; Elsey et al. 2010). Accordingly, hypertension is a consistent feature of CSE knockout mice, confirming that H_2S plays a role as a smooth-muscle relaxant and suggesting that it may physiologically regulate blood pressure (Yang et al. 2008).

This effect could, however, be mediated by a quenching action of H_2S on nitric oxide (NO), possibly via the formation of a nitrothiol intermediate (Whiteman et al. 2006). A counteracting action by H_2S on vasodilation induced by NO could explain why low concentrations of H_2S (10–100 μM) induce vasoconstriction, whereas concentrations over 100 μM induce vasorelaxation (Ali et al. 2006). A second mechanism of interaction between H_2S and NO may involve transcriptional regulation of CSE, which is induced by NO (Zhao et al. 2001). It is thus possible that the vasorelaxant effect of H_2S may be mediated, at least in part, by a local regulatory effect on the NO concentration.

H_2S induces the proliferation of smooth-muscle cells by activating the PI3 Kinase/Akt pathway, thus stimulating angiogenesis (Cai et al. 2007) and apoptosis by activating the mitogen-activated protein kinase (MAPK) pathway (Yang et al. 2004).

Finally, H_2S can directly modulate the function of the heart, with negative inotropic and chronotropic effects (Ferdinandy et al. 2007), which are mediated by activation of K_{ATP} channels. This mechanism results in the hyperpolarization of the membrane potential, closure of L-type calcium channels, and consequent reduction of contraction strength dependent on calcium-activated calcium release. In addition, the activation of K_{ATP} channels depresses the activity of the sinoatrial pacemaker, thus reducing the heart rate (Xu et al. 2008).

The several effects that H_2S exerts on the cardiovascular system suggest that it may have a protective role during ischemia/reperfusion injury. During ischemia, cells are in hypoxic conditions, and subsequent reperfusion and reoxygenation lead to inflammation and cell death by increasing reactive oxygen species (ROS) production, increasing mitochondrial Ca^{2+} overload, and opening the mitochondrial permeability transition pore (mPTP) (Ferdinandy et al. 2007). In experimental ischemia/

reperfusion tests, the treatment with H_2S is protective, either before the induction of ischemia (preconditioning) or before the reperfusion phase (postconditioning). These effects are mediated by activation of the cardioprotective PI3K/Akt pathway, activation of K_{ATP} channels, and increase of the free thiol pool, which acts as ROS scavenger, thus inhibiting the opening of mPTP. The protective effects of H_2S in ischemia/reperfusion experiments have been further confirmed in a mouse model characterized by a twofold increase of H_2S in the myocardium due to transgenic overexpression of CSE (Elrod et al. 2007).

TOXIC EFFECTS OF H_2S

Whereas at physiological (nanomolar) concentrations, Although H_2S acts as a cytoprotective agent, at micromolar concentrations, it can interfere with a variety of cellular functions, including mitochondrial respiration, via inhibition of cytochrome *c* oxidase (COX, EC 1.9.3.1) (Hill et al. 1984). COX activity is inhibited by the formation of a covalent bond between H_2S and the Fe atom coordinated by heme a, which is exquisitely sensitive to the toxic action of H_2S (and cyanide as well). The chronic exposure and binding of H_2S to COX causes accelerated degradation of its protein subunits (Di Meo et al. 2011), thus reducing the amount of fully assembled and functionally active enzyme.

H_2S is also an inhibitor of carbonic anhydrase (EC 4.2.1.1) (Coleman 1967), an effect that could explain the alterations of ventilatory dynamics in response to inhaled H_2S due to changes in the reactivity and distribution of intrapulmonary CO_2 receptors (Klentz and Fedde 1978).

A further effect of H_2S is inhibition of enzymatic activity of short-chain acylCoA dehydrogenase (SCAD), which can cause dicarboxylic aciduria (Pedersen et al. 2003).

Acute exposure to gaseous H_2S causes anoxic brain injury, pulmonary edema, and death (Yalamanchili and Smith 2008). Neurological impairment is consistent with prolonged brain hypoxia, whereas respiratory insufficiency and pulmonary edema are consistent with acute respiratory syndrome. These two injuries are the most likely cause of death after acute exposure to H_2S, whereas chronic exposure to low levels determines irritant effects including upper airway inflammation, conjunctivitis, and wheezing. A direct effect of H_2S on the respiratory center in the brainstem, resulting in apnea, is likely to be caused by inhibition of COX activity (Yalamanchili and Smith 2008).

H_2S IN ETHYLMALONIC ENCEPHALOPATHY (EE)

Ethylmalonic encephalopathy (EE) (Online Mendelian Inheritance in Man [OMIM] #602473) (Burlina et al. 1991) is a severe mitochondrial disease of early infancy, clinically characterized by a combination of progressive encephalopathy, vascular lesions determining bouts of petechial purpura and orthostatic acrocyanosis (Fig. 3A), and chronic hemorrhagic diarrhea. Neurological symptoms include psychomotor regression, spastic tetraparesis, axial hypotonia, dystonia, and seizures, reflecting the presence of patchy, bilateral lesions in the basal nuclei and brainstem gray matter (Fig. 3B). The onset is in the first months after birth; children usually die within the first decade.

Biochemically, EE is characterized by the unusual combination of severe deficiency of COX, the terminal component of the mitochondrial respiratory chain, in muscle (Fig. 3C) and brain, leading to high levels of lactate in blood, and accumulation of ethylmalonate (ethylmalonic acid, EMA) in urines (Fig. 3D). EMA, a dicarboxylic organic acid produced by the carboxylation of butyrate, is a hallmark of enzymatic defects of β-oxidation of fatty acids and branched-chain amino acids, for instance defects of the short-chain and branched-chain acylCoA dehydrogenase (SCAD, BCAD) (Burlina et al. 1991; Gregersen et al. 2000; Korman 2006). Like in SCAD and BCAD deficiency, accumulation of C4- and C5-acylcarnitines has been documented in blood of EE patients. However, SCAD and BCAD activities are normal in EE fibroblasts (Nowaczyk et al. 1998) and muscle, therefore excluding them from being the primary abnormality in this disease. For still

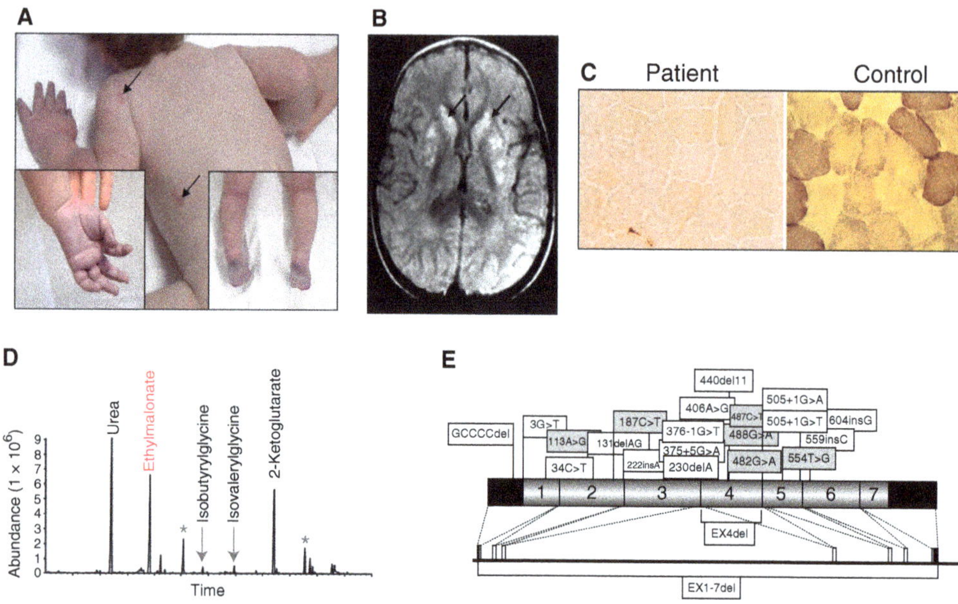

Figure 3. Ethylmalonic encephalopathy: clinical and molecular features. (*A*) Skin areas with petechial hemorrhages are indicated by arrows. The *insets* show the acrocyanosis. (*B*) T2-weighted MRI images of a transverse section of the brain: symmetrical, patchy, high-intensity signals are present in the head of nucleus caudatus and putamen (arrows). (*C*) COX activity is virtually absent in skeletal muscle of an EE patient as compared to normal reactivity present in control. (*D*) Gas-chromatographic profile of urinary organic acids in EE. The abnormal peak of ethylmalonic acid (EMA) is indicated in red. Asterisks indicate internal standards. (*E*) ETHE1 mutations identified in our patients. Missense (gray) and nonsense (white) mutations are indicated along the schematic representation of the ETHE1 cDNA subdivided in the corresponding exons (indicated by numbers). The black areas represent the 5′ and 3′ UTRs. The genomic organization is shown below the cDNA.

unexplained reasons, in virtually all cases, the origin of these EE families is from or could be traced to different regions of the Mediterranean basin or the Arabic peninsula.

In 2004, we demonstrated the presence of several pathogenic mutations in a gene dubbed *ETHE1* for EE 1 (Tiranti et al. 2004). We have reported a series of 29 patients, presenting a fairly homogeneous clinical and biochemical presentation, in spite of a wide spectrum of *ETHE1* mutations including missense, nonsense, frame-shift and deletion of single exons or of the entire gene (Tiranti et al. 2004, 2006) (Fig. 3E, shows only part of the mutations identified). Mutations in the *ETHE1* gene have been found in 32 patients affected by EE worldwide (Di Rocco et al. 2006; Merinero et al. 2006; Zafeiriou et al. 2007; Mineri et al. 2008). A total of 18 nonsense (Tiranti et al. 2004, 2006; Merinero et al. 2006) and nine missense (Tiranti et al. 2004, 2006; Di Rocco et al. 2006; Zafeiriou et al. 2007) mutations has so far been described. To date, a total of 70 mutant patients have been identified in our laboratory (V Tiranti and M Zeviani, unpubl.). Western-blot analysis of the Ethe1 protein (Ethe1p) indicated that some of the missense mutations are associated with the presence of the protein, suggesting that the corresponding wild-type amino acid residues have a catalytic function (Tiranti et al. 2006). In fact, a 3D model of the Ethe1p makes it possible to analyze the spatial location of the amino acid changes predicted by missense mutations and to classify the mutation as catalytic versus structural (Tiranti et al. 2006). More recently, the availability of the crystal structure of the Ethe1p from *Arabidopsis thaliana* (Holdorf et al. 2008) allowed us to demonstrate that the predicted 3D model of human Ethe1p was a faithful in silico reconstruction of the presumed real structure of the native protein (Mineri et al. 2008).

The Ethe1p, a 30 kDa polypeptide located in the mitochondrial matrix, functions in vivo as a homodimeric, Fe-containing sulfur dioxygenase (SDO) activity (Tiranti et al. 2009), which we have already mentioned as one of the reactions involved in the catabolic oxidation of sulfide to sulfate.

Impaired activity of ETHE1-SDO leads to the accumulation of H_2S in critical tissues, including colonic mucosa, liver, muscle, and brain, up to concentrations that inhibit SCAD and COX activities, therefore accounting for EMA aciduria and high levels of C4 and C5 acylcarnitines, EMA, and lactate in plasma (Tiranti et al. 2009). Chronic COX inhibition by accumulated H_2S leads to accelerated degradation of the COX protein backbone (Di Meo et al. 2011). Although ETHE1 activity is measurable in all organs, muscle and brain of *Ethe1*-less (*Ethe1*$^{-/-}$) mice show both defective COX activity and reduced COX amount, whereas liver does not. This may reflect the presence of alternative and highly effective metabolic pathways protecting hepatocytes from H_2S toxicity.

In addition to H_2S, which is a highly unstable, difficult-to-measure gas, its stable derivative thiosulfate is also present at very high concentrations in tissues and body fluids of *Ethe1*$^{-/-}$ mice and EE patients, and can well be considered a specific biomarker of the disease. The pathway outlined in Figure 1 explains these findings: H_2S is converted into thiosulfate by the SQR activity. In normal conditions, rhodanese-TST catalyzes the transfer of the sulfane sulfur from thiosulfate to GSH (or other R-SH substrate) to form persulfide, which is the substrate of the Ethe1-SDO activity, but since the latter is missing, the disposal of the sulfane sulfur is also blocked and consequently both thiosulfate and sulfide accumulate.

We have demonstrated that the ablation of *Ethe1* restricted to muscle or brain is clearly associated with an isolated COX deficiency in the targeted tissue but not in other, *Ethe1*-competent, tissues (Di Meo et al. 2011). This data unequivocally demonstrates that failure to neutralize the endogenous production of H_2S is sufficient for COX activity to decrease, but not for the animals to become sick or for thiosulfate to increase. This observation suggests that multiorgan accumulation of H_2S and diffusion of exogenously released H_2S from the bacterial flora are both needed to determine the severe metabolic impairment and the fatal clinical course of *Ethe1*-less mice and humans.

Besides COX and SCAD deficiency, other symptoms of EE are explained by accumulation of H_2S, including damage of endothelial cells and vasodilation, which account for the petechiae and the acrocyanosis. Since a substantial amount of H_2S derives from the anaerobic bacterial flora residing in the intestinal lumen, COX activity is markedly reduced in the luminal colonocytes of *Ethe1*$^{-/-}$ mice, whereas it is normal in cryptal colonocytes that are relatively secluded from the luminal surface. This difference is likely to reflect different exposure of the two cell populations to the inhibitory action of exogenous H_2S. Excessive production and absorption of H_2S, as well as reduced detoxification by colonocytes, is regarded to play an important role in the mucosal damage of ulcerative colitis. A similar mechanism can well account for the severe chronic diarrhea afflicting EE patients.

THERAPEUTIC APPROACHES IN EE

In principle, H_2S production from the anaerobes of the large intestine can be attenuated by reducing the bacterial contingent, and the excess of free H_2S can be buffered by suitable compounds in critical tissues. For instance, metronidazole is a drug widely used to combat anaerobic infections by reducing the bacterial load in the large intestine (Kang et al. 2008). Metronidazole is a nitroimidazole anti-infective compound that is in fact a pro-drug, being activated in anaerobic organisms by the redox enzyme pyruvate-ferredoxin oxidoreductase. Metronidazole is selectively taken up by anaerobic bacteria and sensitive protozoal organisms (Samuelson 1999) because of the unique ability of these organisms to reduce uptaken metronidazole to its active form within their cell bodies. The nitro group of metronidazole is chemically reduced by ferredoxin (or a ferredoxin-linked metabolic process), and the products are responsible for disrupting the DNA helical structure, thus inhibiting nucleic-acid synthesis (Eisenstein and Schaechter 2007). NAC is a cell-permeable precursor of glutathione (GSH), an abundant compound in mitochondria (Marí et al. 2009), where it acts as one of the physiological acceptors of the sulfur atom of H_2S operated by SQR. This reaction leads to the formation of GSSH persulfide, which acts in turn as a substrate for the Ethe1-SDO. If Ethe1-SDO activity is missing, GSH can buffer, at least in part, accumulating H_2S by forming GSSH, which unlike free H_2S, is a nontoxic compound (Atkuri et al. 2007).

Combined exposure to metronidazole and NAC has been effective in prolonging the survival of *Ethe1*$^{-/-}$ mice and in improving the main symptoms in a pilot study on EE patients (Viscomi et al. 2010), including marked attenuation or disappearance of the vascular lesions and diarrhea, as well as amelioration of some neurological abnormalities. These encouraging results suggest that the invariably fatal clinical course of EE could be modified by a pharmacological protocol based on the off-label use of low-cost, relatively safe drugs.

More recently, Hildebrandt (2011) demonstrated that sulfide toxicity can be modified by the addition of dehydroascorbic acid (DHA) in rat mitochondria. DHA significantly reduced the inhibitory effect of sulfide on COX, resulting in higher rates of respiration and sulfide oxidation. For this purpose, DHA is more suitable than ascorbic acid, the main functionally active form of vitamin C, because its concentration is lower and more flexible. The fraction of oxidized vitamin C reflects the redox state of the cell or organelle and can therefore induce a specific modification of sulfide metabolism. This observation has important clinical implications because DHA could be used during the treatment of sulfide poisoning either after accidental environmental exposure or in patients with EE.

Additional useful information, potentially relevant in the treatment of EE, was recently obtained by Palmfeldt and colleagues (2011), who performed a proteomics study on mitochondria derived from cultured patient fibroblast cells and compared it with healthy controls. Sulfide:quinone oxidoreductase (SQRDL), which takes part in the same sulfide pathway as ETHE1, was also underrepresented in EE patients. Two proteins, the mitochondrial superoxide dismutase (SOD2) and aldehyde dehydrogenase X (ALDH1B) involved in pathways of detoxification and oxidative/reductive stress, were down-regulated in EE. These authors propose that redox perturbation could be an additional factor in the molecular mechanism of EE and an additional therapeutic target.

Another approach to scavenge circulating sulfide coming from the gut could be restoration of the ETHE1-SDO activity in liver of constitutive *Ethe1*$^{-/-}$ individuals by using adeno-associated virus (AAV)-based gene replacement. Because liver is the main checkpoint organ against toxic compounds from the gut, AAV-mediated liver-specific ETHE1 expression may be beneficial to *Ethe1*$^{-/-}$ organisms. This very strategy has recently been successful in markedly prolonging the survival of constitutive *Ethe1*$^{-/-}$ mice (Di Meo et al. 2012).

A complementary strategy, based on the same rationale, is bone marrow transplantation (BMT). BMT has emerged as a major therapeutic option in the treatment of malignant diseases and has also become the treatment of choice for a number of nonmalignant disorders. As for liver-targeted, AAV-based gene replacement, the rationale of BMT-based therapy in the context of EE is to provide a substantial contingent of Ethe1-SDO proficient cells able to clear the excess of H_2S from plasma and other body fluids. A similar strategy has recently been proposed for mitochondrial neurogastrointestinal encephalomyopathy (MNGIE), due to loss of activity of thymidine phosphorylase (TP), resulting in high plasma and tissue levels of thymidine (Thd) and deoxyuridine (dUrd), which become mutagenic on mitochondrial DNA (Halter et al. 2011).

CONCLUDING REMARKS

Because of the multiple functions of mitochondria, not only does OXPHOS impairment determine various degrees of energy failure, but so also do a host of additional effects that account for the intricacy and complexity of the pathophysiology of mitochondrial disorders.

The recent elucidation of the pathogenetic basis of EE has provided evidence for the existence of yet another mechanism of mitochondrial energy impairment, namely the genetically determined failure to detoxify OXPHOS poisons, such as H_2S. In principle, accumulation of a harmful compound can be contrasted more effectively than intrinsic failure of OXPHOS energy metabolism. Therefore, we believe that the specific toxic mechanism underpinning EE makes effective therapy a realistic goal. The results of our effort will constitute a fundamental step toward the development of a cure for this fatal mitochondrial disease in humans.

REFERENCES

Abe K, Kimura H. 1996. The possible role of hydrogen sulfide as an endogenous neuromodulator. *J Neurosci* **16:** 1066–71.

Ali MY, Ping CY, Mok YY, Ling L, Whiteman M, Bhatia M, Moore PK. 2006. Regulation of vascular nitric oxide in vitro and in vivo; a new role for endogenous hydrogen sulphide? *Br J Pharmacol* **149:** 625–34.

Atkuri KR, Mantovani JJ, Herzenberg LA, Herzenberg LA. 2007. *N*-Acetylcysteine—a safe antidote for cysteine/glutathione deficiency. *Curr Opin Pharmacol* **7:** 355–359.

Awata S, Nakayama K, Suzuki I, Sugahara K, Kodama H. 1995. Changes in cystathionine γ-lyase in various regions of rat brain during development. *Biochem Mol Biol Int* **35:** 1331–1338.

Beauchamp RO Jr, Bus JS, Popp JA, Boreiko CJ, Andjelkovich DA. 1984. A critical review of the literature on hydrogen sulfide toxicity. *Crit Rev Toxicol* **13:** 25–97.

Burlina A, Zacchello F, Dionisi-Vici C, Bertini E, Sabetta G, Bennet MJ, Hale DE, Schmidt-Sommerfeld E, Rinaldo P. 1991. New clinical phenotype of branched-chain acyl-CoA oxidation defect. *Lancet* **338:** 1522–1523.

Buzaleh AM, Vazquez ES, del Carmen Battle AM. 1990. Cyanide intoxication—III: On the analogous and different effects provoked by non-lethal and lethal challenged doses. *Gen Pharmacol* **21:** 27–32.

Cai WJ, Wang MJ, Moore PK, Jin HM, Yao T, Zhu YC. 2007. The novel proangiogenic effect of hydrogen sulfide is dependent on Akt phosphorylation. *Cardiovasc Res* **76:** 29–40.

Coleman JE. 1967. Mechanism of action of carbonic anhydrase: Subtrate, sulfonamide, and anion binding. *J Biol Chem* **242:** 5212–5219.

Cooper AJ, Bruschi SA, Iriarte A, Martinez-Carrion M. 2002. Mitochondrial aspartate aminotransferase catalyses cysteine S-conjugate β-lyase reactions. *Biochem J* **368:** 253–261.

Di Meo I, Fagiolari G, Prelle A, Viscomi C, Zeviani M, Tiranti V. 2011. Chronic exposure to sulfide causes accelerated degradation of cytochrome *c* oxidase in ethylmalonic encephalopathy. *Antioxid Redox Signal* **15:** 353–362.

Di Meo I, Auricchio A, Lamperti C, Burlina A, Viscomi C, Zeviani M. 2012. Effective AAV-mediated gene therapy in a mouse model of ethylmalonic encephalopathy. *EMBO Mol Med* **4:** 1008–1014.

Di Rocco M, Caruso U, Briem E, Rossi A, Allegri AE, Buzzi D, Tiranti V. 2006. A case of ethylmalonic encephalopathy with atypical clinical and biochemical presentation. *Mol Genet Metab* **89:** 395–397.

Diwakar L, Ravindranath V. 2007. Inhibition of cystathionine γ-lyase leads to loss of glutathione and aggravation of mitochondrial dysfunction mediated by excitatory amino acid in the CNS. *Neurochem Int* **50:** 418–426.

Eisenstein BI, Schaechter M. 2007. DNA and chromosome mechanics. In *Schaechter's mechanisms of microbial disease* (ed. Engleberg NC, et al.), p. 28. Lippincott Williams & Wilkins: Hagerstown, MD.

Elrod JW, Calvert JW, Morrison J, Doeller JE, Kraus DW, Tao L, Jiao X, Scalia R, Kiss L, Szabo C, et al. 2007. Hydrogen sulfide attenuates myocardial ischemia-reperfusion injury by preservation of mitochondrial function. *Proc Natl Acad Sci* **104:** 15560–15565.

Elsey DJ, Fowkes RC, Baxter GF. 2010. Regulation of cardiovascular cell function by hydrogen sulfide (H_2S). *Cell Biochem Funct* **28:** 95–106.

Enokido Y, Suzuki E, Iwasawa K, Namekata K, Okazawa H, Kimura H. 2005. Cystathionine β-synthase, a key enzyme for homocysteine metabolism, is preferentially expressed in the radial glia/astrocyte lineage of developing mouse CNS. *FASEB J* **19:** 1854–1856.

Feng C, Tollin G, Enemark JH. 2007. Sulfite oxidizing enzymes. *Biochim Biophys Acta* **1774:** 527–539.

Ferdinandy P, Schulz R, Baxter GF. 2007. Interaction of cardiovascular risk factors with myocardial ischemia/reperfusion injury, preconditioning, and postconditioning. *Pharmacol Rev* **59:** 418–458.

Fiorucci S, Distrutti E, Cirino G, Wallace JL. 2006. The emerging roles of hydrogen sulfide in the gastrointestinal tract and liver. *Gastroenterology* **131:** 259–271.

Furne J, Saeed A, Levitt MD. 2008. Whole tissue hydrogen sulfide concentrations are orders of magnitude lower than presently accepted values. *Am J Physiol Regul Integr Comp Physiol* **295:** R1479–R1485.

Goodwin LR, Francom D, Dieken FP, Taylor JD, Warenycia MW. 1989. Determination of sulfide in brain tissue by gas dialysis/ion chromatography: Postmortem studies and two case reports. *J Anal Toxicol* **13:** 105–109.

Gregersen N, Andresen BS, Bross P. 2000. Prevalent mutations in fatty acid oxidation disorders: diagnostic considerations. *Eur J Pediatr* **159:** S213–S218.

Halter J, Schüpbach WM, Casali C, Elhasid R, Fay K, Hammans S, Illa I, Kappeler L, Krähenbühl S, Lehmann T, et al. 2011. Allogeneic hematopoietic SCT as treatment option for patients with mitochondrial neurogastrointestinal encephalomyopathy (MNGIE): A consensus conference proposal for a standardized approach. *Bone Marrow Transplant* **46:** 330–337.

Hildebrandt TM. 2011. Modulation of sulfide oxidation and toxicity in rat mitochondria by dehydroascorbic acid. *Biochim Biophys Acta* **1807:** 1206–1213.

Hildebrandt TM, Grieshaber MK. 2008. Three enzymatic activities catalyze the oxidation of sulfide to thiosulfate in mammalian and invertebrate mitochondria. *FEBS J* **275:** 3352–3361.

Hill BC, Woon TC, Nicholls P, Peterson J, Greenwood C, Thomson AJ. 1984. Interactions of sulphide and other ligands with cytochrome *c* oxidase. An electron-paramagnetic-resonance study. *Biochem J* **224:** 591–600.

Holdorf MM, Bennett B, Crowder MW, Makaroff CA. 2008. Spectroscopic studies on Arabidopsis ETHE1, a glyoxalase II-like protein. *J Inorg Biochem* **102:** 1825–1830.

Hosoki R, Matsuki N, Kimura H. 1997. The possible role of hydrogen sulfide as an endogenous smooth muscle relaxant in synergy with nitric oxide. *Biochem Biophys Res Commun* **237:** 527–531.

Ishii I, Akahoshi N, Yu XN, Kobayashi Y, Namekata K, Komaki G, Kimura H. 2004. Murine cystathionine γ-lyase: Complete cDNA and genomic sequences, promoter activity, tissue distribution and developmental expression. *Biochem J* **381:** 113–123.

Jackson MR, Melideo SL, Jorns MS. 2012. Human sulfide: Quinone oxidoreductase catalyzes the first step in hydrogen sulfide metabolism and produces a sulfane sulfur metabolite. *Biochemistry* **51:** 6804–6815.

Kang SS, Bloom SM, Norian LA, Geske MJ, Flavell RA, Stappenbeck TS, Allen PM. 2008. An antibiotic-responsive mouse model of fulminant ulcerative colitis. *PLoS Med* **4:** 5.

Kimura Y, Kimura H. 2004. Hydrogen sulfide protects neurons from oxidative stress. *FASEB J* **18:** 1165–1167.

Klentz RD, Fedde MR. 1978. Hydrogen sulfide: Effects on avian respiratory control and intrapulmonary CO_2 receptors. *Respir Physiol* **32:** 355–367.

Korman SH, Inborn errors of isoleucine degradation: A review. 2006. *Mol Genet Metab* **89:** 289–99.

Marí M, Morales A, Colell A, García-Ruiz C, Fernández-Checa JC. 2009. Mitochondrial glutathione, a key survival antioxidant. *Antioxid Redox Signal* **11:** 2685–2700.

Merinero B, Pérez-Cerdá C, Ruiz Sala P, Ferrer I, García MJ, Martínez Pardo M, Belanger-Quintana A, de la Mota JL, Martin-Hernández E, Vianey-Saban C, et al. 2006. Persistent increase of plasma butyryl/isobutyrylcarnitine concentrations as marker of SCAD defect and ethylmalonic encephalopathy. *J Inherit Metab Di* **29:** 685.

Mineri R, Rimoldi M, Burlina AB, Koskull S, Perletti C, Heese B, von Döbeln U, Mereghetti P, Di Meo I, Invernizzi F, et al. 2008. Identification of new mutations in the ETHE1 gene in a cohort of 14 patients presenting with ethylmalonic encephalopathy. *J Med Genet* **45:** 473–478.

Mitchell TW, Savage JC, Gould DH. 1993. High-performance liquid chromatography detection of sulfide in tissues from sulfide-treated mice. *J Appl Toxicol* **13:** 389–394.

Müller FH, Bandeiras TM, Urich T, Teixeira M, Gomes CM, Kletzin A. 2004. Coupling of the pathway of sulphur oxidation to dioxygen reduction: Characterization of a novel membrane-bound thiosulphate:quinone oxidoreductase. *Mol Microbiol* **53:** 1147–1160.

Nowaczyk MJ, Lehotay DC, Platt BA, Fisher L, Tan R, Phillips H, Clarke JT. 1998. Ethylmalonic and methylsuccinic aciduria in ethylmalonic encephalopathy arise from abnormal isoleucine metabolism. *Metabolism* **47:** 836–839.

Ogasawara Y, Isoda S, Tanabe S. 1994. Tissue and subcellular distribution of bound and acid-labile sulfur, and the enzymic capacity for sulfide production in the rat. *Biol Pharm Bull* **17:** 1535–1542.

Oh GS, Pae HO, Lee BS, Kim BN, Kim JM, Kim HR, Jeon SB, Jeon WK, Chae HJ, Chung HT. 2006. Hydrogen sulfide inhibits nitric oxide production and nuclear factor-κB via heme oxygenase-1 expression in RAW2647 macrophages stimulated with lipopolysaccharide. *Free Radic Biol Med* **41:** 106–119.

Palmfeldt J, Vang S, Stenbroen V, Pavlou E, Baycheva M, Buchal G, Monavari AA, Augoustides-Savvopoulou P, Mandel H, Gregersen N. 2011. Proteomics reveals that redox regulation is disrupted in patients with ethylmalonic encephalopathy. *J Proteome Res* **10:** 2389–2396.

Pedersen CB, Bross P, Winter VS, Corydon TJ, Bolund L, Bartlett K, Vockley J, Gregersen N. 2003. Misfolding, degradation, and aggregation of variant proteins. The molecular pathogenesis of short chain acyl-CoA dehydrogenase (SCAD) deficiency. *J Biol Chem* **278:** 47449–47458.

Powell MA, Somero GN. 1986. Hydrogen sulfide oxidation is coupled to oxidative phosphorylation in mitochondria of Solemya reidi. *Science* **233:** 563–566.

Qingyou Z, Junbao D, Weijin Z, Hui Y, Chaoshu T, Chunyu Z. 2004. Impact of hydrogen sulfide on carbon monoxide/heme oxygenase pathway in the pathogenesis of hypoxic pulmonary hypertension. *Biochem Biophys Res Commun* **317:** 30–37.

Reiffenstein RJ, Hulbert WC, Roth SH. 1992. Toxicology of hydrogen sulfide. *Annu Rev Pharmacol Toxicol* **32:** 109–134.

Rowan FE, Docherty NG, Coffey JC, O'Connell PR. 2009. Sulphate-reducing bacteria and hydrogen sulphide in the aetiology of ulcerative colitis. *Br J Surg* **96:** 151–158.

Searcy DG, Lee SH. 1998. Sulfur reduction by human erythrocytes. *J Exp Zool* **282:** 310–322.

Samuelson J. 1999. Why metronidazole is active against both bacteria and parasites. *Antimicrob Agents Chemother* **43:** 1533–1541.

Stipanuk MH. 2004. Role of the liver in regulation of body cysteine taurine levels: A brief review. *Neurochem Res* **29:** 105–110.

Stipanuk MH, Beck PW. 1982. Characterization of the enzymic capacity for cysteine desulphhydration in liver and kidney of the rat. *Biochem J* **206:** 267–277.

Stipanuk MH, Ueki I. 2010. Dealing with methionine/homocysteine sulfur: Cysteine metabolism to taurine and inorganic sulfur. *J Inherit Metab Dis* **34:** 17–32.

Szabó C. 2007. Hydrogen sulphide and its therapeutic potential. *Nat Rev Drug Discov* **6:** 917–935.

Theissen U, Martin W. 2008. Sulfide : quinone oxidoreductase (SQR) from the lugworm *Arenicola marina* shows cyanide- and thioredoxin-dependent activity. *FEBS J* **275:** 1131–1139.

Tiranti V, D'Adamo P, Briem E, Ferrari G, Mineri R, Lamantea E, Mandel H, Balestri P, Garcia-Silva MT, Vollmer B, et al. 2004. Ethylmalonic encephalopathy is caused by mutations in ETHE1, a gene encoding a mitochondrial matrix protein. *Am J Hum Genet* **74:** 239–252.

Tiranti V, Briem E, Lamantea E, Mineri R, Papaleo E, De Gioia L, Forlani F, Rinaldo P, Dickson P, Abu-Libdeh B, et al. 2006. ETHE1 mutations are specific to ethylmalonic encephalopathy. *J Med Genet* **43:** 340–346.

Tiranti V, Viscomi C, Hildebrandt T, Di Meo I, Mineri R, Tiveron C, Levitt MD, Prelle A, Fagiolari G, Rimoldi M, et al. 2009. Loss of ETHE1, a mitochondrial dioxygenase, causes fatal sulfide toxicity in ethylmalonic encephalopathy. *Nat Med* **15:** 200–205.

Viscomi C, Burlina AB, Dweikat I, Savoiardo M, Lamperti C, Hildebrandt T, Tiranti V, Zeviani M. 2010. Combined treatment with oral metronidazole and *N*-acetylcysteine is effective in ethylmalonic encephalopathy. *Nat Med* **16:** 869–871.

Wang R. 2002. Two's company, three's a crowd: Can H_2S be the third endogenous gaseous transmitter? *FASEB J* **16:** 1792–1798.

Westley AM, Westley J. 1991. Biological sulfane sulfur. *Biochemistry* **195:** 63–67.

Whiteman M, Armstrong JS, Chu SH, Jia-Ling S, Wong BS, Cheung NS, Halliwell B, Moore PK. 2004. The novel neuromodulator hydrogen sulfide: An endogenous peroxynitrite "scavenger?" *J Neurochem* **90:** 765–768.

Whiteman M, Li L, Kostetski I, Chu SH, Siau JL, Bhatia M, Moore PK. 2006. Evidence for the formation of a novel nitrosothiol from the gaseous mediators nitric oxide and hydrogen sulphide. *Biochem Biophys Res Commun* **343:** 303–310.

Xu M, Wu YM, Li Q, Wang X, He RR. 2008. Electrophysiological effects of hydrogen sulfide on pacemaker cells in sinoatrial nodes of rabbits. *Sheng Li Xue Bao* **60:** 175–180.

Yalamanchili C, Smith MD. 2008. Acute hydrogen sulfide toxicity due to sewer gas exposure. *Am J Emerg Med* **26:** 518.e5–7.

Yan SK, Chang T, Wang H, Wu L, Wang R, Meng QH. 2006. Effects of hydrogen sulfide on homocysteine-induced oxidative stress in vascular smooth muscle cells. *Biochem Biophys Res Commun* **361:** 485–491.

Yang G, Sun X, Wang R. 2004. Hydrogen sulfide-induced apoptosis of human aorta smooth muscle cells via the activation of mitogen-activated protein kinases and caspase-3. *FASEB J* **18:** 1782–1784.

Yang G, Wu L, Jiang B, Yang W, Qi J, Cao K, Meng Q, Mustafa AK, Mu W, Zhang S, et al. 2008. H_2S as a physiologic vasorelaxant: hypertension in mice with deletion of cystathionine γ-lyase. *Science* **322:** 587–590.

Zafeiriou DI, Augoustides-Savvopoulou P, Haas D, Smet J, Triantafyllou P, Vargiami E, Tamiolaki M, Gombakis N, van Coster R, Sewell AC, et al. 2007. Ethylmalonic encephalopathy: Clinical and biochemical observations. *Neuropediatrics* **38:** 78–82.

Zhao W, Zhang J, Lu Y, Wang R. 2001. The vasorelaxant effect of H_2S as a novel endogenous gaseous K_{ATP} channel opener. *EMBO J* **20:** 6008–6016.

Mitochondrial Dysfunction and Defective Autophagy in the Pathogenesis of Collagen VI Muscular Dystrophies

Paolo Bernardi and Paolo Bonaldo

Department of Biomedical Sciences, University of Padova, I-35121 Padova, Italy

Correspondence: bernardi@bio.unipd.it; bonaldo@bio.unipd.it

Ullrich Congenital Muscular Dystrophy (UCMD), Bethlem Myopathy (BM), and Congenital Myosclerosis are diseases caused by mutations in the genes encoding the extracellular matrix protein collagen VI. A dystrophic mouse model, where collagen VI synthesis was prevented by targeted inactivation of the *Col6a1* gene, allowed the investigation of pathogenesis, which revealed the existence of a Ca^{2+}-mediated dysfunction of mitochondria and sarcoplasmic reticulum, and of defective autophagy. Key events are dysregulation of the mitochondrial permeability transition pore, an inner membrane high-conductance channel that for prolonged open times causes mitochondrial dysfunction, and inadequate removal of defective mitochondria, which amplifies the damage. Consistently, the *Col6a1*$^{-/-}$ myopathic mice could be cured through inhibition of cyclophilin D, a matrix protein that sensitizes the pore to opening, and through stimulation of autophagy. Similar defects contribute to disease pathogenesis in patients irrespective of the genetic lesion causing the collagen VI defect. These studies indicate that permeability transition pore opening and defective autophagy represent key elements for skeletal muscle fiber death, and provide a rationale for the use of cyclosporin A and its nonimmunosuppressive derivatives in patients affected by collagen VI myopathies, a strategy that holds great promise for treatment.

Anchoring and adhesion complexes at the surface of cells link the cytoskeleton to the surrounding extracellular matrix (ECM), thus maintaining cell integrity and allowing signal transduction. These anchoring structures have a critical role in tissues undergoing extensive mechanical stress, like skeletal muscle; it comes as no coincidence that genetic defects of components of the anchoring complexes cause human muscular dystrophies. The ECM is directly involved in the molecular pathogenesis of various forms of muscular dystrophy, and there is accumulating evidence that ECM components, such as laminin-2 and collagen VI (ColVI), play a critical role in maintaining muscle integrity and function (Schessl et al. 2006).

ColVI is a main ECM protein forming a distinct microfilamentous network with a broad distribution in several organs including skeletal muscle, skin, cornea, lung, blood vessels, intervertebral disks, and joints (Keene et al. 1988). It consists of three major chains, $\alpha 1$(VI),

Cite this article as *Cold Spring Harb Perspect Biol* doi: 10.1101/cshperspect.a011387

α2(VI), and α3(VI), encoded by separate genes (*COL6A1*, *COL6A2*, and *COL6A3*, respectively). The 140-kD α1(VI) and α2(VI) chains are about 1000 amino acids long, while the α3(VI) chain is three times larger (250–300 kD) with several alternatively spliced variants ranging between ~2500 and 3100 amino acids (Bonaldo et al. 1989, 1990; Chu et al. 1990; Doliana et al. 1990). ColVI is a multimodular protein made of several domains; each chain contains a short triple helical domain of 335–336 amino acids and two large amino- and carboxy-terminal globular ends composed of repeated domains of 200 amino acids each, sharing similarity with the von Willebrand factor type A (vWF-A) module (Colombatti and Bonaldo 1991). Three additional ColVI chains, α4(VI), α5(VI), and α6(VI), were recently identified. These chains, coded by distinct genes (*COL6A4-COL6A6*), show a high degree of similarity with α3(VI), and their tissue distribution is more restricted, suggesting a specific role in the assembly and/or function of ColVI (Gara et al. 2008, 2011; Sabatelli et al. 2011).

ColVI has a complex pathway of intracellular assembly (Colombatti and Bonaldo 1987; Colombatti et al. 1987, 1995). Association of the α-chains into a triple helical monomer (three chains) is followed by formation of dimers (six chains) and tetramers (12 chains), stabilized by disulfide bonds. After secretion into the extracellular space, the tetramers interact in an end-to-end fashion and form "beaded" microfilaments with a typical 100-nm periodicity, which are deposited into the ECM (Furthmayr et al. 1983). ColVI microfilaments form an extended network particularly abundant in the close periphery of several cell types, and they have been suggested to bridge the surface of cells with the interstitial connective tissue (Keene et al. 1988; Bonaldo et al. 1990; Kuo et al. 1997). In skeletal muscle, ColVI is a major component of endomysium, where it is localized in the outer layer of the basement membrane (Kuo et al. 1997).

The different domains of ColVI are involved in a broad range of interactions with cell surface receptors and with other ECM components. The triple helical region of ColVI is a major cell-binding domain, and it promotes adhesion through interaction with several integrins and with NG2 proteoglycan (Pfaff et al. 1993; Burg et al. 1996). Both the triple helical region and the vWF-A domains of ColVI bind various ECM proteins including fibrillar collagens, basal lamina (type IV) collagen, decorin, biglycan, and fibronectin (Bonaldo et al. 1990; Bidanset et al. 1992; Kuo et al. 1997; Sabatelli et al. 2001; Wiberg et al. 2001). These interactions and the close association of beaded microfilaments with muscle basal lamina strongly suggest that ColVI provides an important mechanical link of muscle cells to the surrounding ECM.

ColVI is synthesized and secreted by cells organizing an ECM, such as fibroblasts and muscle cells. Studies in transfected cells and in transgenic mice demonstrated that transcriptional regulation is the major determinant for the highly regulated and dynamic pattern expression of ColVI during development and in adult tissues. The data thus far available for the *Col6a1* gene indicate a complex pattern of tissue-specific transcription that is dependent on different enhancer and silencer elements spread over a large genomic region and controlling expression in skin, tendons, joints, peripheral nerves, and skeletal muscles, respectively (Braghetta et al. 1996). Muscle interstitial fibroblasts are the main source of ColVI in the ECM of skeletal muscle (Zou et al. 2008). A study revealed that transcription of the *Col6a1* gene in skeletal muscle is under the control of a muscle-specific enhancer, which is strictly required for activating the expression and synthesis of ColVI in muscle fibroblasts (Braghetta et al. 2008). Remarkably, activation of this enhancer requires signals relayed by myogenic cells, whose presence is a prerequisite for inducing the deposition of ColVI by interstitial fibroblasts (Braghetta et al. 2008). In vitro and in vivo studies have shown that ColVI is synthesized by interstitial fibroblasts; once released in the ECM, the protein contacts myogenic cells and myofibers (Zou et al. 2008; Palma et al. 2009).

COLLAGEN VI-RELATED MUSCLE DISEASES

Deficiency of ColVI in humans gives rise to three main syndromes, Bethlem myopathy

Cite this article as *Cold Spring Harb Perspect Biol* doi: 10.1101/cshperspect.a011387

(BM), Ullrich congenital muscular dystrophy (UCMD) (Lampe and Bushby 2005; Bönnemann 2011), and the recently identified congenital myosclerosis (Merlini et al. 2008b).

BM (MIM #158810; http://www.metalife.com/OMIM) is characterized by axial and proximal muscle weakness with finger joint contractures, proximal muscles being more involved than distal, and extensors more than flexors (Bethlem and Wijngaarden 1976). The hallmark of the disease is the presence of contractures of the interphalangeal joints of the last four fingers, and early flexion contractures of the elbow and ankles are common (Merlini et al. 1994). BM is a very heterogeneous disorder, and patients show a wide range of clinical features, from mild myopathy to more severe cases with early onset and features of slowly progressive muscular dystrophy. BM inheritance is usually autosomal dominant (Jöbsis et al. 1996; Sasaki et al. 2000; Scacheri et al. 2002), but recessive cases were also described (Foley et al. 2009; Gualandi et al. 2009). Immunohistochemistry shows apparently normal or mildly reduced levels of ColVI in the endomysium of most BM patients, and quantitative or qualitative ColVI defects can be detected in cultured fibroblasts derived from skin biopsies.

UCMD (MIM #254090) is a severe congenital muscular dystrophy characterized by early onset, generalized and rapidly progressive muscle wasting and weakness, proximal joint contractures, and distal joint hyperflexibility. The rapid progression of the clinical symptoms usually leads to early death, as a result of respiratory failure (Camacho Vanegas et al. 2001; Demir et al. 2002). Usually, UCMD shows an autosomal recessive inheritance with homozygous or compound heterozygous mutations of the *COL6* genes, but several cases of UCMD with dominant mutation in only one allele were also reported (Pan et al. 2003; Baker et al. 2005; Angelin et al. 2007). Patients with a UCMD phenotype but without mutations in *COL6* genes were described, suggesting a possible genetic heterogeneity for the disease. ColVI appears to be strongly reduced or absent in muscle biopsies from UCMD patients, suggesting that UCMD mutations severely affect the synthesis and secretion of ColVI in muscle. Cultured skin fibroblasts of UCMD patients usually show either a markedly decreased secretion of ColVI, or lack of the characteristic filamentous network in the ECM, suggesting that UCMD mutations severely affect the synthesis and secretion of ColVI (Camacho Vanegas et al. 2001; Zhang et al. 2002; Squarzoni et al. 2006).

Congenital myosclerosis (MIM #255600) is a rare inherited disorder characterized by slender muscles with firm "woody" consistency, and restricted movement of many joints because of diffuse muscle contractures. Reduced expression of ColVI was detected in muscle biopsies and cultured fibroblasts from affected sibs of an Italian family, and the disease is linked to a homozygous mutation of the *COL6A2* gene (Merlini et al. 2008b).

Mutations in any of the *COL6A1-COL6A3* genes have been identified in BM and in UCMD (Lampe and Bushby 2005; Bönnemann 2011). Different *COL6* gene mutations may have different impacts on the mRNA and protein level by predominantly impairing ColVI synthesis, assembly, secretion, or function. Based on the mutation, different mechanisms can be predicted such as: (1) loss-of-function for mutations perturbing triple helix formation, intracellular assembly or ColVI secretion; (2) dominant negative effect for mutations giving rise to abnormal ColVI polypeptides secreted in the ECM; and (3) haploinsufficiency for mutations affecting mRNA stability, with a decreased synthesis of normal ColVI. The effect of specific *COL6* mutations has been investigated by in vitro studies (Lamande et al. 1999, 2002; Sasaki et al. 2000; Zhang et al. 2002; Jimenez-Mallebrera et al. 2006; Baker et al. 2007; Tooley et al. 2010), which provided relevant clues on their effect at the protein level. However, there is no conclusive evidence on the genotype–phenotype correlation for BM and UCMD, which may represent a clinical continuum rather than strictly separate entities (Pepe et al. 2002; Lampe and Bushby 2005; Bönnemann 2011).

As for many other genetic diseases, creation of animal models of ColVI deficiency proved to be the key to understanding disease pathophysiology, and to devise and test potential therapies. Several years ago, one of us produced a mutant mouse with targeted inactivation of the *Col6a1* gene, which codes for the α1(VI) chain (Bonaldo et al. 1998). In the absence of the α1 chain, ColVI does not assemble and is not secreted in the ECM, so that *Col6a1*$^{-/-}$ mice completely lack ColVI and are affected by a myopathic phenotype with weakness and histological changes of skeletal muscle, characterized by early onset and slow or no progression (Bonaldo et al. 1998). These defects are found in various skeletal muscles, in particular the diaphragm, as revealed by in vivo Evans blue dye staining. Similar but milder defects are present in *Col6a1*$^{+/-}$ mice, indicating functional haploinsufficiency for ColVI. Muscle weakness was confirmed by strength measurements ex vivo, showing significantly reduced tension in *Col6a1*$^{-/-}$ muscles (Irwin et al. 2003). Although *Col6a1*-null mice are a model of human ColVI diseases, their phenotype is much closer to the benign myopathy of BM than to the severe dystrophic pathology of UCMD. A recent advance has been the generation of zebrafish models of UCMD and BM, which reproduce the key features of the human diseases (Telfer et al. 2010).

PATHOGENESIS OF COLLAGEN VI MUSCULAR DYSTROPHIES

Ultrastructural analysis of skeletal muscle from *Col6a1*$^{-/-}$ mice revealed the presence of marked dilations of the sarcoplasmic reticulum (SR) and of mitochondrial alterations in 20%–30% of the fibers, while T-tubules were normal (Irwin et al. 2003). Mitochondrial abnormalities ranged from tubular cristae, to electron-dense matrix inclusions, to swelling; myofibers with mitochondrial-SR alterations also displayed nuclear features of apoptosis, suggesting a link between organelle changes and increased incidence of cell death (Irwin et al. 2003). Mitochondrial ultrastructural alterations are observed in many forms of muscle pathology; whether these are the cause or consequence of the disease (or even a side effect of no pathogenic relevance) was the key question that needed to be answered.

Mitochondria in the Pathogenesis of Collagen VI Muscular Dystrophies

We monitored the mitochondrial membrane potential in skeletal muscle fibers and myoblast/fibroblast cultures of *Col6a1*$^{-/-}$ and wild-type mice, and in muscle cell cultures from patients affected by ColVI myopathies, or from normal donors. The resting mitochondrial membrane potential was not affected by lack of ColVI, yet mitochondria depolarized after addition of the F1FO ATP synthase inhibitor oligomycin or of the respiratory chain complex I inhibitor rotenone only in cells from ColVI-deficient individuals (Irwin et al. 2003; Angelin et al. 2007, 2008; Palma et al. 2009; Tiepolo et al. 2009; Sabatelli et al. 2012). Neither response is normal because, in the short term at least, (1) inhibition of ATP synthesis by oligomycin should cause hyperpolarization, and (2) inhibition of respiration should still allow the membrane potential to be maintained by proton pumping by the ATP synthase running "in reverse" and utilizing glycolytic ATP as fuel (see, e.g., Porcelli et al. 2009). Of note, cells from ColVI-defective genotypes also displayed increased rates of apoptosis that could be prevented by plating on ColVI or by addition of cyclosporin A (CsA), treatments that also normalized the mitochondrial response to oligomycin and/or rotenone (Irwin et al. 2003; Angelin et al. 2007, 2008; Palma et al. 2009; Tiepolo et al. 2009; Sabatelli et al. 2012). These findings, which we shall discuss further below, (1) suggest the existence of a latent (compensated) mitochondrial dysfunction that may worsen (and decompensate) with time, which is consistent with the chronic and progressive features of ColVI muscular dystrophies; and (2) indicate that the underlying mitochondrial dysfunction, and possibly the disease, can be cured with CsA.

The pharmacology of CsA is complex. This drug binds to and inhibits cyclophilins (CyPs), a family of cellular peptidyl–prolyl *cis–trans* isomerases (Galat 2003). The complex of CsA with the abundant cytosolic CyPA acquires the ability to bind and inhibit the cytosolic phosphatase calcineurin, preventing the dephosphorylation of its substrates (Liu et al. 1991),

 Cite this article as *Cold Spring Harb Perspect Biol* doi: 10.1101/cshperspect.a011387

which include NFAT and Drp1. Dephosphorylation is essential for the nuclear translocation of NFAT, which is key to activation of the T lymphocytes involved in the immune response to organ transplants; by preventing NFAT nuclear translocation, CsA prevents transplant rejection (Liu et al. 1991). Dephosphorylation of Drp1 is essential for its translocation to mitochondria, where it causes fission and fragmentation, events that are prevented by CsA (Cereghetti et al. 2008). Neither effect is involved in the protective mitochondrial effects of CsA in ColVI diseases because the rescue from mitochondrial dysfunction cannot be mimicked by the calcineurin inhibitor FK506 (Irwin et al. 2003), which acts through its specific receptor FKBP but does not bind CyPs (Liu et al. 1991). As we originally suggested (Irwin et al. 2003), the protective effects of CsA are rather a result of inhibition of CyPD, the unique mitochondrial isoform of CyPs (encoded in the mouse by the *Ppif* gene). Indeed, the protective effects can also be seen with CsA analogs that inhibit CyPs but not calcineurin, like $MeAla^3EtVal^4$-cyclosporin (Debio 025) (Angelin et al. 2007; Tiepolo et al. 2009). *Col6a1*$^{-/-}$ *Ppif*$^{-/-}$ mice lacking both ColVI and CyPD display a striking recovery of both the mitochondrial lesions and the muscle defects (Palma et al. 2009).

The most convincing explanation for the protective effects of CsA and Debio 025 (and of CyPD ablation) in ColVI myopathies is desensitization of the permeability transition pore (PTP), a high-conductance inner membrane mitochondrial channel of unknown molecular identity involved in several forms of cell death (Bernardi et al. 2006; Rasola and Bernardi 2011). CyPD favors PTP opening, and matrix Ca^{2+} is an important permissive factor sensitizing the PTP to a variety of inducing factors. The pore is modulated by the proton electrochemical gradient, both through the voltage (depolarization favors opening) and pH (matrix acidification favors closure), and is finely tuned by the redox poise (a more oxidized state favors pore opening). A detailed coverage of PTP regulation is beyond the scope of this review, and we refer the reader to previous publications on the topic (Bernardi et al. 2006; Rasola and Bernardi 2011). The important points to be considered here are that: (1) short-lived PTP openings cause reversible mitochondrial depolarization (Hüser and Blatter 1999; Petronilli et al. 1999) and may be involved in physiological Ca^{2+} homeostasis (Bernardi and von Stockum 2012); under these conditions, PTP-dependent cytochrome *c* release can be an indirect consequence of cristae remodeling (Scorrano et al. 2002) and matrix swelling (Petronilli et al. 1994) that cooperate with tBID in the absence of outer membrane rupture (Bernardi et al. 2001); and (2) long-lasting PTP openings cause energy dissipation, release of pyridine nucleotides (Vinogradov et al. 1972; Di Lisa et al. 2001), outer membrane rupture, and massive release of cytochrome *c* (Petronilli et al. 2001).

Our current interpretation of the mitochondrial pathogenesis of ColVI muscular dystrophies is that, initially at least, mitochondrial dysfunction is latent rather than overt, and consists in an alteration of the threshold voltage for PTP opening, which (through a still undefined mechanism that may involve matrix Ca^{2+} overload) is poised dangerously close to the resting potential. This shift of the voltage threshold can be unmasked in vitro by the addition of rotenone or oligomycin, which act via different mechanisms. (1) Rotenone inhibits complex I and start a process of depolarization. In normal cells, this initial depolarization can be readily compensated by the mitochondrial ATP synthase (which hydrolyzes glycolytic ATP) because the threshold voltage for PTP opening is not reached, and therefore inner membrane permeability remains low. In ColVI defective cells, the voltage threshold is so close to the resting potential that addition of rotenone causes PTP opening; ATP hydrolysis by the ATP synthase in this case cannot restore the membrane potential as long as the pore is open. (2) Oligomycin inhibits the ATP synthase, and this causes an initial hyperpolarization with parallel decrease of the ATP levels; this would cause a further shift of the threshold voltage to reach the value of the resting potential through two mechanisms, that is, increased ROS production because of the state 3 → state 4 transition, and decreased ability of the SR to take up

Ca^{2+}, causing a secondary increase of matrix mitochondrial Ca^{2+}. The ensuing PTP opening would then cause the observed depolarization, which is very often preceded by the expected hyperpolarization (Irwin et al. 2003; Angelin et al. 2007).

We believe that the voltage threshold shift may explain disease progression in vivo as well. Because the mitochondrial membrane potential decreases during ATP synthesis, in the ColVI defective genotypes increased PTP flickering could occur during increased muscle activity with loss of pyridine nucleotides (Vinogradov et al. 1972; Di Lisa et al. 2001) and progressive decrease of the respiratory reserve. It should be noted that ADP generated by ATP hydrolysis inhibits the PTP; thus, ATP hydrolysis may lead to pore closure and to at least partial restoration of respiration. The maximal attainable rate would be lower than normal, however, and determined by the residual matrix levels of pyridine nucleotides attained after each "cycle" of PTP opening. As long as the energy demand can be matched the fiber would behave normally, but dysfunction would be precipitated by increased workload and/or by a further increase of PTP flickering, events that will eventually lead to individual muscle fiber death as observed both in vivo and in cultured muscle fibers. The recently identified defect of autophagy would then worsen the situation by preventing removal of dysfunctional mitochondria (Grumati et al. 2010).

Defective Autophagy in the Pathogenesis of Collagen VI Muscular Dystrophies

Autophagy is a highly conserved homeostatic mechanism used for the degradation and recycling of bulk cytoplasm, long-lived proteins, and organelles through the lysosomal machinery (Mizushima et al. 2008). We recently demonstrated that autophagy plays a protective role against myofiber defects and muscle wasting in *Col6a1*$^{-/-}$ mice. A failure of the autophagic machinery is responsible for the inefficient removal and persistence of altered organelles in myofibers of *Col6a1*$^{-/-}$ mice and of BM/UCMD patients (Grumati et al. 2010). The ensuing accumulation of dysfunctional mitochondria triggers myofiber apoptosis, which in turn leads to the development of the myopathic pathology. In *Col6a1*-null mice, the dramatic consequences of the autophagy inactivation are even more evident in conditions of muscle stress, such as physical exercise. When *Col6a1*$^{-/-}$ mice were subjected to an intense work, a condition in which autophagy is required for the continuous energy need and the rapid elimination of exhausted organelles, the inefficient autophagy flux determined a massive degeneration of myofibers with a marked exacerbation of the myopathic phenotype (Grumati et al. 2011b). Notably, forced activation of autophagy by dietary, genetic, and pharmacological tools was able to eliminate altered organelles and restored myofiber homeostasis in *Col6a1*$^{-/-}$ mice, with recovery from the myopathic phenotype, thus showing a promising therapeutic potential for counteracting muscle atrophy and weakness in these diseases. The autophagic failure of *Col6a1*$^{-/-}$ mice strictly depends on impairment of the autophagy flux, which determines a decrease in autophagosome formation and inefficient removal of dysfunctional organelles (Grumati et al. 2010). Lack of ColVI has a remarkable impact on molecules involved in the regulation of autophagy, with decreased Beclin 1 protein levels and persistent activation of the Akt/mTOR pathway, even during starvation (Grumati et al. 2010). Although the molecular pathways transducing ColVI signals from the ECM to the autophagy machinery remain to be elucidated, the Beclin 1 complex and the AKT/mTOR pathway are markedly affected by lack of ColVI, and these alterations appear to be the main cause for the autophagy inhibition. Notably, the autophagic failure of *Col6a1*$^{-/-}$ muscles is also associated with a decreased activity of Bnip3, a molecule playing a key role in the selective removal of damaged or dysfunctional mitochondria with a process know as "mitophagy" (Grumati et al. 2011a). Most interestingly, Beclin 1 and Bnip3 levels are also lower in muscle biopsies of patients affected by ColVI diseases, and the amount of Beclin 1 seems to correlate with the severity of the phenotype, being much lower in the severe

UCMD than in the milder BM (Grumati et al. 2010). Our studies represented the first evidence that an impairment of the autophagic machinery plays a crucial role in the pathogenesis of muscular dystrophies, thus opening new venues for therapeutic approaches and paving the way for investigating autophagy defects in other muscular dystrophies (Tolkovsky 2010). Indeed, a recent study revealed that the dy^{3K}/dy^{3K} mouse, which is affected by laminin α2 deficiency and represents a model of human MDC1A, also has an alteration of autophagy. Interestingly, at variance from *Col6a1*$^{-/-}$ animals, dy^{3}/dy^{3K} mice displayed a general upregulation of the autophagic machinery and inhibition of autophagy significantly improved their dystrophic phenotype (Carmignac et al. 2011). This latter observation is not surprising and suggests that, as recently proposed in other pathological conditions, an appropriate balance of the autophagic flux is essential for the maintenance of skeletal muscle health.

OF MICE, FISH, AND MEN

Establishing whether the mitochondrial pathogenesis of the myopathy of *Col6a1*$^{-/-}$ mice also holds for the genetically and clinically heterogeneous UCMD and BM represented a major challenge. To fill this gap, we have systematically studied muscle biopsies and cell cultures from patients affected by ColVI muscular dystrophies, and established that patients affected by UCMD and BM display an increased rate of apoptosis in skeletal muscle in vivo and in cultures derived from biopsies (Angelin et al. 2007, 2008; Merlini et al. 2008a; Sabatelli et al. 2012). The muscle cell cultures also displayed a measurable fraction of altered mitochondria, with morphological abnormalities ranging from shape changes to overt swelling, and the presence of a latent mitochondrial dysfunction that could be unmasked by the depolarizing effects of oligomycin and/or rotenone (Angelin et al. 2007, 2008; Sabatelli et al. 2012). The mitochondrial defect was present in primary cultures from UCMD and BM patients irrespective of whether the primary genetic defect was in the *COL6A1* or *COL6A3* gene, and both in homozygous and heterozygous mutations (Angelin et al. 2007), and could be rescued by both CsA and Debio 025 (Angelin et al. 2007).

Whether patient-derived cells can be safely used as a diagnostic material in ColVI diseases has recently been questioned (Hicks et al. 2009). While confirming our finding that myoblasts from patients affected by UCMD display a latent mitochondrial dysfunction that can be unmasked by oligomycin, Hicks et al. concluded that PTP dysregulation may be a particular characteristic of these cells in culture rather than being specific to the ColVI defect, because PTP dysregulation was also found in muscle-derived cells from patients with Limb Girdle Muscular Dystrophy (LGMD) 2B, but not BM, Merosin Deficient Congenital Muscular Dystrophy (MDC) 1A, LGMD2A, Duchenne Muscular Dystrophy (DMD), and Leigh syndrome (Hicks et al. 2009). As we already noted (Bernardi et al. 2009), mitochondria are causally involved in a variety of degenerative diseases, and it is therefore not surprising that a "mitochondrial phenotype" caused by a PTP opening can be found in other muscular dystrophies, including DMD (Millay et al. 2008; Reutenauer et al. 2008) and MDC1A (Millay et al. 2008), and in Leber hereditary optic neuropathy, a finding that extends the potential use of CsA to mitochondrial DNA diseases (Porcelli et al. 2008). The apparent discrepancy about BM was recently resolved with the demonstration that the mitochondrial phenotype of muscle cell cultures from both UCMD and BM patients was lost after nine passages in culture, which matched a spontaneous decrease of the incidence of apoptosis (Sabatelli et al. 2012). Thus, it is essential to study materials derived from patient muscle biopsies at early passages.

The mitochondrial pathogenesis of ColVI myopathies has been recently confirmed in zebrafish, an organism where the PTP has regulatory features indistinguishable from those of mammals (Azzolin et al. 2010). Injection of morpholino to exon 13 of *Col6a1* caused an in-frame deletion in the carboxy terminal part of the ColVI triple helical domain, a typical dominant BM mutation (Lucioli et al. 2005; Lampe et al. 2008), and resulted in a mild myopathy with late-onset motor deficits and obvious histopathologic abnormalities (Telfer et al. 2010). Injection of morpholino to exon 9 of *Col6a1* caused an in-frame deletion in the amino terminal part of the triple helical domain, a dominant mutation in UCMD (Pepe et al. 2006; Lampe et al. 2008), and caused the predicted severe myopathy with early-onset motor deficits and severe ultrastructural changes (Telfer et al. 2010). In both cases, mitochondrial abnormalities and increased apoptosis were observed, and treatment with CsA improved these pathological signs together with rescue of the motor deficit (Telfer et al. 2010).

Based on the positive results obtained with CsA on their cells in culture (Angelin et al. 2007), a pilot trial on five patients affected by UCMD or BM was carried out. Prior to treatment, all patients displayed mitochondrial dysfunction and increased frequency of apoptosis, as determined in muscle biopsies. Both these pathological signs were largely normalized after 1 month of oral CsA administration, which also increased muscle regeneration (Merlini et al. 2008a). The study was then extended to six individuals with UCMD receiving 3–5 mg/kg CsA daily for 1–3.2 years (Merlini et al. 2011). The primary outcome measure was muscle strength evaluated with a myometer. The score showed significant improvement in five of the six patients, while motor function did not change and respiratory function deteriorated in all. As in the short-term treatment, CsA corrected mitochondrial dysfunction, increased muscle regeneration, and decreased the number of apoptotic nuclei (Merlini et al. 2011). These results indicate that long-term treatment with CsA is well tolerated and ameliorates performance in the limbs, but not in the respiratory muscles; they also suggest that treatment with CsA or one of its nonimmunosuppressive derivatives should be started as early as possible, when diaphragm function is not yet compromised. In summary, although much remains to be done to understand the biochemical and pathophysiological bases of ColVI diseases, we believe that the discoveries of the mitochondrial pathogenesis and autophagic defect in UCMD and BM are opening new perspectives for their therapeutic treatment.

ACKNOWLEDGMENTS

The financial support of Telethon Italy (grant GGP11082) is gratefully acknowledged.

REFERENCES

Angelin A, Tiepolo T, Sabatelli P, Grumati P, Bergamin N, Golfieri C, Mattioli E, Gualandi F, Ferlini A, Merlini L, et al. 2007. Mitochondrial dysfunction in the pathogenesis of Ullrich congenital muscular dystrophy and prospective therapy with cyclosporins. *Proc Natl Acad Sci* **104:** 991–996.

Angelin A, Bonaldo P, Bernardi P. 2008. Altered threshold of the mitochondrial permeability transition pore in Ullrich congenital muscular dystrophy. *Biochim Biophys Acta Bioenergetics* **1777:** 893–896.

Azzolin L, Basso E, Argenton F, Bernardi P. 2010. Mitochondrial Ca^{2+} transport and permeability transition in zebrafish (*Danio rerio*). *Biochim Biophys Acta* **1797:** 1775–1779.

Baker NL, Morgelin M, Peat R, Goemans N, North KN, Bateman JF, Lamande SR. 2005. Dominant collagen VI mutations are a common cause of Ullrich congenital muscular dystrophy. *Hum Mol Genet* **14:** 279–293.

Baker NL, Morgelin M, Pace RA, Peat RA, Adams NE, Gardner RJ, Rowland LP, Miller G, De JP, Ceulemans B, et al. 2007. Molecular consequences of dominant Bethlem myopathy collagen VI mutations. *Ann Neurol* **62:** 390–405.

Bernardi P, von Stockum S. 2012. The permeability transition pore as a Ca^{2+} release channel: New answers to an old question. *Cell Calcium* **52:** 22–27.

Bernardi P, Petronilli V, Di Lisa F, Forte M. 2001. A mitochondrial perspective on cell death. *Trends Biochem Sci* **26:** 112–117.

Bernardi P, Krauskopf A, Basso E, Petronilli V, Blachly-Dyson E, Di Lisa F, Forte MA. 2006. The mitochondrial permeability transition from in vitro artifact to disease target. *FEBS J* **273:** 2077–2099.

Bernardi P, Bonaldo P, Maraldi NM, Merlini L, Sabatelli P. 2009. On the pathogenesis of collagen VI muscular dystrophies—comment on article of Hicks et al. *BRAIN* **132:** e121.

Bethlem J, Wijngaarden GK. 1976. Benign myopathy, with autosomal dominant inheritance. A report on three pedigrees. *BRAIN* **99:** 91–100.

Bidanset DJ, Guidry C, Rosenberg LC, Choi HU, Timpl R, Hook M. 1992. Binding of the proteoglycan decorin to collagen type VI. *J Biol Chem* **267:** 5250–5256.

Bonaldo P, Russo V, Bucciotti F, Bressan GM, Colombatti A. 1989. α1 Chain of chick type VI collagen. The complete cDNA sequence reveals a hybrid molecule made of one short collagen and three von Willebrand factor type A-like domains. *J Biol Chem* **264:** 5575–5580.

Bonaldo P, Russo V, Bucciotti F, Doliana R, Colombatti A. 1990. Structural and functional features of the α3 chain indicate a bridging role for chicken collagen VI in connective tissues. *Biochemistry* **29:** 1245–1254.

Bonaldo P, Braghetta P, Zanetti M, Piccolo S, Volpin D, Bressan GM. 1998. Collagen VI deficiency induces early onset myopathy in the mouse: An animal model for Bethlem myopathy. *Hum Mol Genet* **7:** 2135–2140.

Bönnemann CG. 2011. The collagen VI-related myopathies: Muscle meets its matrix. *Nat Rev Neurol* **7:** 379–390.

Braghetta P, Fabbro C, Piccolo S, Marvulli D, Bonaldo P, Volpin D, Bressan GM. 1996. Distinct regions control transcriptional activation of the α1(VI) collagen promoter in different tissues of transgenic mice. *J Cell Biol* **135:** 1163–1177.

Braghetta P, Ferrari A, Fabbro C, Bizzotto D, Volpin D, Bonaldo P, Bressan GM. 2008. An enhancer required for transcription of the Col6a1 gene in muscle connective tissue is induced by signals released from muscle cells. *Exp Cell Res* **314:** 3508–3518.

Burg MA, Tillet E, Timpl R, Stallcup WB. 1996. Binding of the NG2 proteoglycan to type VI collagen and other extracellular matrix molecules. *J Biol Chem* **271:** 26110–26116.

Camacho Vanegas O, Bertini E, Zhang RZ, Petrini S, Minosse C, Sabatelli P, Giusti B, Chu ML, Pepe G. 2001. Ullrich scleroatonic muscular dystrophy is caused by recessive mutations in collagen type VI. *Proc Natl Acad Sci* **98:** 7516–7521.

Carmignac V, Svensson M, Korner Z, Elowsson L, Matsumura C, Gawlik KI, Allamand V, Durbeej M. 2011. Autophagy is increased in laminin α2 chain-deficient muscle and its inhibition improves muscle morphology in a mouse model of MDC1A. *Hum Mol Genet* **20:** 4891–4902.

Cereghetti GM, Stangherlin A, Martins de Brito O, Chang CR, Blackstone C, Bernardi P, Scorrano L. 2008. Dephosphorylation by calcineurin regulates translocation of Drp1 to mitochondria. *Proc Natl Acad Sci* **105:** 15803–15808.

Chu ML, Zhang RZ, Pan TC, Stokes D, Conway D, Kuo HJ, Glanville R, Mayer U, Mann K, Deutzmann R, et al. 1990. Mosaic structure of globular domains in the human type VI collagen α3 chain: Similarity to von Willebrand factor, fibronectin, actin, salivary proteins and aprotinin type protease inhibitors. *EMBO J* **9:** 385–393.

Colombatti A, Bonaldo P. 1987. Biosynthesis of chick type VI collagen. II. Processing and secretion in fibroblasts and smooth muscle cells. *J Biol Chem* **262:** 14461–14466.

Colombatti A, Bonaldo P. 1991. The superfamily of proteins with von Willebrand factor type A-like domains: One theme common to components of extracellular matrix, hemostasis, cellular adhesion, and defense mechanisms. *Blood* **77:** 2305–2315.

Colombatti A, Bonaldo P, Ainger K, Bressan GM, Volpin D. 1987. Biosynthesis of chick type VI collagen. I. Intracellular assembly and molecular structure. *J Biol Chem* **262:** 14454–14460.

Colombatti A, Mucignat MT, Bonaldo P. 1995. Secretion and matrix assembly of recombinant type VI collagen. *J Biol Chem* **270:** 13105–13111.

Demir E, Sabatelli P, Allamand V, Ferreiro A, Moghadaszadeh B, Makrelouf M, Topaloglu H, Echenne B, Merlini L, Guicheney P. 2002. Mutations in COL6A3 cause severe and mild phenotypes of Ullrich congenital muscular dystrophy. *Am J Hum Genet* **70:** 1446–1458.

Di Lisa F, Menabò R, Canton M, Barile M, Bernardi P. 2001. Opening of the mitochondrial permeability transition pore causes depletion of mitochondrial and cytosolic NAD^+ and is a causative event in the death of myocytes in postischemic reperfusion of the heart. *J Biol Chem* **276:** 2571–2575.

Doliana R, Bonaldo P, Colombatti A. 1990. Multiple forms of chicken α3(VI) collagen chain generated by alternative splicing in type A repeated domains. *J Cell Biol* **111:** 2197–2205.

Foley AR, Hu Y, Zou Y, Columbus A, Shoffner J, Dunn DM, Weiss RB, Bonnemann CG. 2009. Autosomal recessive inheritance of classic Bethlem myopathy. *Neuromuscul Disord* **19:** 813–817.

Furthmayr H, Wiedemann H, Timpl R, Odermatt E, Engel J. 1983. Electron-microscopical approach to a structural model of intima collagen. *Biochem J* **211:** 303–311.

Galat A. 2003. Peptidylprolyl *cis/trans* isomerases (immunophilins): Biological diversity–targets–functions. *Curr Top Med Chem* **3:** 1315–1347.

Gara SK, Grumati P, Urciuolo A, Bonaldo P, Kobbe B, Koch M, Paulsson M, Wagener R. 2008. Three novel collagen VI chains with high homology to the α3 chain. *J Biol Chem* **283:** 10658–10670.

Gara SK, Grumati P, Squarzoni S, Sabatelli P, Urciuolo A, Bonaldo P, Paulsson M, Wagener R. 2011. Differential and restricted expression of novel collagen VI chains in mouse. *Matrix Biol* **30:** 248–257.

Grumati P, Coletto L, Sabatelli P, Cescon M, Angelin A, Bertaggia E, Blaauw B, Urciuolo A, Tiepolo T, Merlini L, et al. 2010. Autophagy is defective in collagen VI muscular dystrophies, and its reactivation rescues myofiber degeneration. *Nat Med* **16:** 1313–1320.

Grumati P, Coletto L, Sandri M, Bonaldo P. 2011a. Autophagy induction rescues muscular dystrophy. *Autophagy* **7:** 426–428.

Grumati P, Coletto L, Schiavinato A, Castagnaro S, Bertaggia E, Sandri M, Bonaldo P. 2011b. Physical exercise stimulates autophagy in normal skeletal muscles but is detrimental for collagen VI-deficient muscles. *Autophagy* **7:** 1415–1423.

Gualandi F, Urciuolo A, Martoni E, Sabatelli P, Squarzoni S, Bovolenta M, Messina S, Mercuri E, Franchella A, Ferlini A, et al. 2009. Autosomal recessive Bethlem myopathy. *Neurology* **73:** 1883–1891.

Hicks D, Lampe AK, Laval SH, Allamand V, Jimenez-Mallebrera C, Walter MC, Muntoni F, Quijano-Roy S, Richard P, Straub V, et al. 2009. Cyclosporine A treatment for Ullrich congenital muscular dystrophy: A cellular study of mitochondrial dysfunction and its rescue. *BRAIN* **132:** 147–155.

Hüser J, Blatter LA. 1999. Fluctuations in mitochondrial membrane potential caused by repetitive gating of the permeability transition pore. *Biochem J* **343:** 311–317.

Irwin WA, Bergamin N, Sabatelli P, Reggiani C, Megighian A, Merlini L, Braghetta P, Columbaro M, Volpin D, Bressan GM, et al. 2003. Mitochondrial dysfunction and apoptosis in myopathic mice with collagen VI deficiency. *Nat Genet* **35:** 267–271.

Jimenez-Mallebrera C, Maioli MA, Kim J, Brown SC, Feng L, Lampe AK, Bushby K, Hicks D, Flanigan KM, Bonnemann C, et al. 2006. A comparative analysis of collagen

VI production in muscle, skin and fibroblasts from 14 Ullrich congenital muscular dystrophy patients with dominant and recessive COL6A mutations. *Neuromuscul Disord* **16:** 571–582.

Jöbsis GJ, Keizers H, Vreijling JP, de Visser M, Speer MC, Wolterman RA, Baas F, Bolhuis PA. 1996. Type VI collagen mutations in Bethlem myopathy, an autosomal dominant myopathy with contractures. *Nat Genet* **14:** 113–115.

Keene DR, Engvall E, Glanville RW. 1988. Ultrastructure of type VI collagen in human skin and cartilage suggests an anchoring function for this filamentous network. *J Cell Biol* **107:** 1995–2006.

Kuo HJ, Maslen CL, Keene DR, Glanville RW. 1997. Type VI collagen anchors endothelial basement membranes by interacting with type IV collagen. *J Biol Chem* **272:** 26522–26529.

Lamande SR, Shields KA, Kornberg AJ, Shield LK, Bateman JF. 1999. Bethlem myopathy and engineered collagen VI triple helical deletions prevent intracellular multimer assembly and protein secretion. *J Biol Chem* **274:** 21817–21822.

Lamande SR, Morgelin M, Selan C, Jobsis GJ, Baas F, Bateman JF. 2002. Kinked collagen VI tetramers and reduced microfibril formation as a result of Bethlem myopathy and introduced triple helical glycine mutations. *J Biol Chem* **277:** 1949–1956.

Lampe AK, Bushby KM. 2005. Collagen VI related muscle disorders. *J Med Genet* **42:** 673–685.

Lampe AK, Zou Y, Sudano D, O'Brien KK, Hicks D, Laval SH, Charlton R, Jimenez-Mallebrera C, Zhang RZ, Finkel RS, et al. 2008. Exon skipping mutations in collagen VI are common and are predictive for severity and inheritance. *Hum Mutat* **29:** 809–822.

Liu J, Farmer JDJ, Lane WS, Friedman J, Weissman I, Schreiber SL. 1991. Calcineurin is a common target of cyclophilin–cyclosporin A and FKBP–FK506 complexes. *Cell* **66:** 807–815.

Lucioli S, Giusti B, Mercuri E, Vanegas OC, Lucarini L, Pietroni V, Urtizberea A, Ben Yaou R, de Visser M, van der Kooi AJ, et al. 2005. Detection of common and private mutations in the COL6A1 gene of patients with Bethlem myopathy. *Neurology* **64:** 1931–1937.

Merlini L, Morandi L, Granata C, Ballestrazzi A. 1994. Bethlem myopathy: Early-onset benign autosomal dominant myopathy with contractures. Description of two new families. *Neuromuscul Disord* **4:** 503–511.

Merlini L, Angelin A, Tiepolo T, Braghetta P, Sabatelli P, Zamparelli A, Ferlini A, Maraldi NM, Bonaldo P, Bernardi P. 2008a. Cyclosporin A corrects mitochondrial dysfunction and muscle apoptosis in patients with collagen VI myopathies. *Proc Natl Acad Sci* **105:** 5225–5229.

Merlini L, Martoni E, Grumati P, Sabatelli P, Squarzoni S, Urciuolo A, Ferlini A, Gualandi F, Bonaldo P. 2008b. Autosomal recessive myosclerosis myopathy is a collagen VI disorder. *Neurology* **71:** 1245–1253.

Merlini L, Sabatelli P, Armaroli A, Gnudi S, Angelin A, Grumati P, Michelini ME, Franchella A, Gualandi F, Bertini E, et al. 2011. Cyclosporine A in Ullrich congenital muscular dystrophy: Long-term results. *Oxid Med Cell Longev* **2011:** 139194.

Millay DP, Sargent MA, Osinska H, Baines CP, Barton ER, Vuagniaux G, Sweeney HL, Robbins J, Molkentin JD. 2008. Genetic and pharmacologic inhibition of mitochondrial-dependent necrosis attenuates muscular dystrophy. *Nat Med* **14:** 442–447.

Mizushima N, Levine B, Cuervo AM, Klionsky DJ. 2008. Autophagy fights disease through cellular self-digestion. *Nature* **451:** 1069–1075.

Palma E, Tiepolo T, Angelin A, Sabatelli P, Maraldi NM, Basso E, Forte MA, Bernardi P, Bonaldo P. 2009. Genetic ablation of cyclophilin D rescues mitochondrial defects and prevents muscle apoptosis in collagen VI myopathic mice. *Hum Mol Genet* **18:** 2024–2031.

Pan TC, Zhang RZ, Sudano DG, Marie SK, Bonnemann CG, Chu ML. 2003. New molecular mechanism for Ullrich congenital muscular dystrophy: A heterozygous in-frame deletion in the COL6A1 gene causes a severe phenotype. *Am J Hum Genet* **73:** 355–369.

Pepe G, Bertini E, Bonaldo P, Bushby K, Giusti B, de Visser M, Guicheney P, Lattanzi G, Merlini L, Muntoni F, et al. 2002. Bethlem myopathy (BETHLEM) and Ullrich scleroatonic muscular dystrophy: 100th ENMC international workshop, 23–24 November 2001, Naarden, The Netherlands. *Neuromuscul Disord* **12:** 984–993.

Pepe G, Lucarini L, Zhang RZ, Pan TC, Giusti B, Quijano-Roy S, Gartioux C, Bushby KM, Guicheney P, Chu ML. 2006. COL6A1 genomic deletions in Bethlem myopathy and Ullrich muscular dystrophy. *Ann Neurol* **59:** 190–195.

Petronilli V, Nicolli A, Costantini P, Colonna R, Bernardi P. 1994. Regulation of the permeability transition pore, a voltage-dependent mitochondrial channel inhibited by cyclosporin A. *Biochim Biophys Acta* **1187:** 255–259.

Petronilli V, Miotto G, Canton M, Brini M, Colonna R, Bernardi P, Di Lisa F. 1999. Transient and long-lasting openings of the mitochondrial permeability transition pore can be monitored directly in intact cells by changes in mitochondrial calcein fluorescence. *Biophys J* **76:** 725–734.

Petronilli V, Penzo D, Scorrano L, Bernardi P, Di Lisa F. 2001. The mitochondrial permeability transition, release of cytochrome *c* and cell death. Correlation with the duration of pore openings in situ. *J Biol Chem* **276:** 12030–12034.

Pfaff M, Aumailley M, Specks U, Knolle J, Zerwes HG, Timpl R. 1993. Integrin and Arg-Gly-Asp dependence of cell adhesion to the native and unfolded triple helix of collagen type VI. *Exp Cell Res* **206:** 167–176.

Porcelli AM, Angelin A, Ghelli A, Mariani E, Martinuzzi A, Carelli V, Petronilli V, Bernardi P, Rugolo M. 2008. Respiratory complex I dysfunction due to mitochondrial DNA mutations shifts the voltage threshold for opening of the permeability transition pore toward resting levels. *J Biol Chem* **284:** 2045–2052.

Porcelli AM, Angelin A, Ghelli A, Mariani E, Martinuzzi A, Carelli V, Petronilli V, Bernardi P, Rugolo M. 2009. Respiratory complex I dysfunction due to mitochondrial DNA mutations shifts the voltage threshold for opening of the permeability transition pore toward resting levels. *J Biol Chem* **284:** 2045–2052.

Rasola A, Bernardi P. 2011. Mitochondrial permeability transition in Ca^{2+}-dependent apoptosis and necrosis. *Cell Calcium* **50:** 222–233.

Reutenauer J, Dorchies OM, Patthey-Vuadens O, Vuagniaux G, Ruegg UT. 2008. Investigation of Debio 025, a cyclophilin inhibitor, in the dystrophic mdx mouse, a model for Duchenne muscular dystrophy. *Br J Pharmacol* **155:** 574–584.

Sabatelli P, Bonaldo P, Lattanzi G, Braghetta P, Bergamin N, Capanni C, Mattioli E, Columbaro M, Ognibene A, Pepe G, et al. 2001. Collagen VI deficiency affects the organization of fibronectin in the extracellular matrix of cultured fibroblasts. *Matrix Biol* **20:** 475–486.

Sabatelli P, Gara SK, Grumati P, Urciuolo A, Gualandi F, Curci R, Squarzoni S, Zamparelli A, Martoni E, Merlini L, et al. 2011. Expression of the collagen VI α5 and α6 chains in normal human skin and in skin of patients with collagen VI-related myopathies. *J Invest Dermatol* **131:** 99–107.

Sabatelli P, Palma E, Angelin A, Squarzoni S, Urciuolo A, Pellegrini C, Tiepolo T, Bonaldo P, Gualandi F, Merlini L, et al. 2012. Critical evaluation of the use of cell cultures for inclusion in clinical trials of patients affected by collagen VI myopathies. *J Cell Physiol* **227:** 2927–2935.

Sasaki T, Hohenester E, Zhang RZ, Gotta S, Speer MC, Tandan R, Timpl R, Chu ML. 2000. A Bethlem myopathy Gly to Glu mutation in the von Willebrand factor A domain N2 of the collagen α3(VI) chain interferes with protein folding. *FASEB J* **14:** 761–768.

Scacheri PC, Gillanders EM, Subramony SH, Vedanarayanan V, Crowe CA, Thakore N, Bingler M, Hoffman EP. 2002. Novel mutations in collagen VI genes: Expansion of the Bethlem myopathy phenotype. *Neurology* **58:** 593–602.

Schessl J, Zou Y, Bonnemann CG. 2006. Congenital muscular dystrophies and the extracellular matrix. *Semin Pediatr Neurol* **13:** 80–89.

Scorrano L, Ashiya M, Buttle K, Weiler S, Oakes SA, Mannella CA, Korsmeyer SJ. 2002. A distinct pathway remodels mitochondrial cristae and mobilizes cytochrome *c* during apoptosis. *Dev Cell* **2:** 55–67.

Squarzoni S, Sabatelli P, Bergamin N, Guicheney P, Demir E, Merlini L, Lattanzi G, Ognibene A, Capanni C, Mattioli E, et al. 2006. Ultrastructural defects of collagen VI filaments in an Ullrich syndrome patient with loss of the α3(VI) N10-N7 domains. *J Cell Physiol* **206:** 160–166.

Telfer WR, Busta AS, Bonnemann CG, Feldman EL, Dowling JJ. 2010. Zebrafish models of collagen VI-related myopathies. *Hum Mol Genet* **19:** 2433–2444.

Tiepolo T, Angelin A, Palma E, Sabatelli P, Merlini L, Nicolosi L, Finetti F, Braghetta P, Vuagniaux G, Dumont JM, et al. 2009. The cyclophilin inhibitor Debio 025 normalizes mitochondrial function, muscle apoptosis and ultrastructural defects in *Col6a1*$^{-/-}$ myopathic mice. *Br J Pharmacol* **157:** 1045–1052.

Tolkovsky AM. 2010. Autophagy thwarts muscle disease. *Nat Med* **16:** 1188–1190.

Tooley LD, Zamurs LK, Beecher N, Baker NL, Peat RA, Adams NE, Bateman JF, North KN, Baldock C, Lamande SR. 2010. Collagen VI microfibril formation is abolished by an α2(VI) von Willebrand factor type A domain mutation in a patient with Ullrich congenital muscular dystrophy. *J Biol Chem* **285:** 33567–33576.

Vinogradov A, Scarpa A, Chance B. 1972. Calcium and pyridine nucleotide interaction in mitochondrial membranes. *Arch Biochem Biophys* **152:** 646–654.

Wiberg C, Hedbom E, Khairullina A, Lamande SR, Oldberg A, Timpl R, Morgelin M, Heinegard D. 2001. Biglycan and decorin bind close to the N-terminal region of the collagen VI triple helix. *J Biol Chem* **276:** 18947–18952.

Zhang RZ, Sabatelli P, Pan TC, Squarzoni S, Mattioli E, Bertini E, Pepe G, Chu ML. 2002. Effects on collagen VI mRNA stability and microfibrillar assembly of three COL6A2 mutations in two families with Ullrich congenital muscular dystrophy. *J Biol Chem* **277:** 43557–43564.

Zou Y, Zhang RZ, Sabatelli P, Chu ML, Bonnemann CG. 2008. Muscle interstitial fibroblasts are the main source of collagen VI synthesis in skeletal muscle: Implications for congenital muscular dystrophy types Ullrich and Bethlem. *J Neuropathol Exp Neurol* **67:** 144–154.

Relevance of Mitochondrial Genetics and Metabolism in Cancer Development

Giuseppe Gasparre[1], Anna Maria Porcelli[2], Giorgio Lenaz[3], and Giovanni Romeo[1]

[1]Department of Medical and Surgical Sciences, Unit of Medical Genetics, University of Bologna Medical School, 40138 Bologna, Italy

[2]Department of Pharmacy and Biotechnologies, University of Bologna, 40126 Bologna, Italy

[3]Department of Biochemistry "Giovanni Moruzzi," University of Bologna Medical School, 40126 Bologna, Italy

Correspondence: romeo@eurogene.org

Cancer cells are characterized in general by a decrease of mitochondrial respiration and oxidative phosphorylation, together with a strong enhancement of glycolysis, the so-called Warburg effect. The decrease of mitochondrial activity in cancer cells may have multiple reasons, related either to the input of reducing equivalents to the electron transfer chain or to direct alterations of the mitochondrial respiratory complexes. In some cases, the depression of respiratory activity is clearly the consequence of disruptive mitochondrial DNA (mtDNA) mutations and leads as a consequence to enhanced generation of reactive oxygen species (ROS). By acting both as mutagens and cellular mitogens, ROS may contribute directly to cancer progression. On the basis of our experimental evidence, we suggest a deep implication of the supercomplex organization of the respiratory chain as a missing link between oxidative stress, energy failure, and tumorigenesis. We speculate that under conditions of oxidative stress, a dissociation of mitochondrial supercomplexes occurs, with destabilization of complex I and secondary enhanced generation of ROS, thus leading to a vicious circle amplifying mitochondrial dysfunction. An excellent model to dissect the role of pathogenic, disassembling mtDNA mutations in tumor progression and their contribution to the metabolic reprogramming of cancer cells (glycolysis vs. respiration) is provided by an often underdiagnosed subset of tumors, namely, the oncocytomas, characterized by disruptive mutations of mtDNA, especially of complex I subunits. Such mutations almost completely abolish complex I activity, which slows down the Krebs cycle, favoring a high ratio of α-ketoglutarate/succinate and consequent destabilization of hypoxia inducible factor 1α (HIF1α). On the other hand, if complex I is partially defective, the levels of NAD^+ may be sufficient to implement the Krebs cycle with higher levels of intermediates that stabilize HIF1α, thus favoring tumor malignancy. The threshold model we propose, based on the population-like dynamics of mitochondrial genetics (heteroplasmy vs. homoplasmy), implies that below threshold complex I is present and functioning correctly, thus favoring tumor growth, whereas above threshold, when complex I is not assembled, tumor growth is arrested. We have therefore termed "oncojanus" the mtDNA genes whose disruptive mutations have such a double-edged effect.

Cite this article as *Cold Spring Harb Perspect Biol* doi: 10.1101/cshperspect.a011411

Cancer cells are characterized by uncontrolled proliferation due to gain-of-function of oncogenes and loss of function of tumor-suppressor genes. Among the hallmarks of the cancer phenotype, there is a metabolic reprogramming (DeBerardinis et al. 2008). Cancer cells are characterized in general by a decrease of mitochondrial respiration and oxidative phosphorylation, together with a strong enhancement of glycolysis, the so-called Warburg effect, first reported by Otto Warburg (Warburg 1956). Owing to space limitations, for this particular topic we refer you to recent excellent reviews (Moreno-Sanchez et al. 2007; Mayevsky 2009; Mathupala et al. 2010; Koppenol et al. 2011). Nevertheless, it is not clear which of the two processes (i.e., decreased respiration and enhanced glycolysis) is the primary event, or even if they are concomitant consequences of a common causal event.

Although the decrease of mitochondrial oxidative phosphorylation is considered by many investigators as a universal feature of neoplastic cells, there are numerous publications in which cancer cells have been reported to show normal or even high respiratory activity (Zu and Guppy 2004; Weinberg and Chandel 2009). This apparent paradox can be explained by the hypothesis advanced by Smolkova et al (2010) that cancer progression is characterized by waves of gene expression, where initial waves suppressing mitochondrial function and stimulating glycolysis are followed by a resumption of mitochondrial oxidative phosphorylation when glutamine becomes the leading substrate for energy production and the major carbon source for anabolic processes. Each malignant tumor is characterized by a distinct metabolic phenotype reflecting the history of the different waves during the sequence of carcinogenic mutations.

The enhanced aerobic glycolysis characterizing cancer cells (and, in general, highly proliferating cells) at least in some periods of their progression toward malignancy may be vital to provide, besides energy, also the starting materials for active biosynthetic processes (Cuezva et al. 2009). Indeed, it was recognized early that incomplete activity of the Krebs cycle allows the use of citrate as a starting intermediate for the synthesis of fatty acids and cholesterol (Parlo and Coleman 1986; Kuhajda et al. 1994), whereas glycolysis may provide the majority of energy required for the anabolic processes. The supply of citrate for lipid biosynthetic processes may be provided by reductive carboxylation of α-ketoglutarate by reversal of the isocitrate dehydrogenase reaction when the major carbon source is glutamine (Holleran et al. 1995; Yoo et al. 2008) and pyruvate dehydrogenase activity is depressed by the activation of pyruvate dehydrogenase kinase (Kim et al. 2006).

During the cell cycle, the genes for glycolytic enzymes are overexpressed at the time of active DNA synthesis (Klevecz et al. 2004). Likewise, several oncogenes (such as c-myc, HIF1α, Akt) or tumor suppressors (p53) are involved in the transactivation of glycolytic enzymes (Cuezva et al. 2009). The activation of these genes may involve mitochondrial dysregulation, for example, through unbalance of Krebs cycle intermediates (see below).

DECREASED MITOCHONDRIAL RESPIRATION IN CANCER CELLS

The decrease of mitochondrial activity in cancer cells may have multiple reasons:

1. Early in the 1920s, it was observed that enhanced glycolysis, as in cancer cells, inhibits respiration (the so-called Crabtree effect); it was suggested that respiratory inhibition is caused by competition of glycolysis and oxidative phosphorylation for Pi and ADP (Ibsen 1961).
2. Decrease of oxidative phosphorylation (OXPHOS) may also derive from hypoxia because O_2 may fall below the K_m for cytochrome oxidase (Brahimi-Horn and Pouyssegur 2007) and thus induce the Pasteur effect (stimulation of phosphofructokinase by AMP when mitochondrial ATP synthesis is deficient).
3. On the other hand, decreased OXPHOS becomes operative during hypoxia because of the stabilization of hypoxia inducible factor 1α (HIF1α) (Semenza 2003). This factor

at normal oxygen tension undergoes hydroxylation of two key proline residues by HIF-prolyl hydroxylases (PHDs), which allow the recognition of the factor by pVHL, the product of the von Hippel–Lindau gene, that addresses the protein to ubiquitin-mediated degradation. The factor becomes stabilized during hypoxia by inactivation of PHD and binds to hypoxia-responsive elements in the DNA, stimulating a large array of genes (Semenza 2007; Kaluz et al. 2008). Besides glycolytic enzymes, HIF1α enhances the expression of pyruvate dehydrogenase kinase (PDK), thus inhibiting pyruvate dehydrogenase (PDH) complex and decreasing the input of reducing equivalents to the respiratory chain (Kim et al. 2006). The decreased input of carbon into the Krebs cycle must necessarily be limited, otherwise the lack of citrate would impair the synthesis of lipids and cholesterol that are required for the formation of new cellular membranes during tumor growth. This is the reason that glutamine becomes a preferred substrate for replenishing the Krebs cycle for both energetic and biosynthetic purposes.

It is of interest that HIF1α is also stabilized by several nonhypoxic stimuli, for example, by succinate (that inhibits prolyl hydroxylase) (King et al. 2006), but also by pyruvate and oxaloacetate (Lu et al. 2005; McFate et al. 2008) and, interestingly, by reactive oxygen species (ROS) (Patten et al. 2010), although the activating effect of ROS on HIF is still a controversial issue (Bell et al. 2008). ROS could act directly but also by stimulating PDK, thus leading to accumulation of pyruvate (Sun et al. 2009; Patten et al. 2010). Thus, derangement of the Krebs cycle may lead to changes in steady-state concentrations of important metabolites that are able to control the activity of HIF1α.

4. Direct alterations of respiratory activity in cancer cells have not been frequently investigated. In general, a depression of respiratory activity is observed in cancer cells (Putignani et al. 2008; Park et al. 2009; Jahnke et al. 2010; Ma et al. 2010). In the case of Ras-transformed fibroblasts, we have observed a direct alteration of the respiratory chain, in particular, a 50% decrease of complex I content and activity, and consequently of NAD-linked respiration and ATP synthesis (Baracca et al. 2010). This seems to be due in part to decreased expression as seen by transcriptome analysis. However, we do not know yet if there are disruptive mitochondrial DNA (mtDNA) mutations as seen in oncocytic tumors (see below). The effect of the adenylate cyclase activator forskolin in reversing the changes accompanying transformation suggests that the mitochondrial changes may be caused by impairment of cAMP-dependent pathways (L Alberghina, pers. comm.).
5. In some cases, the depression of respiratory activity is clearly the consequence of disruptive mtDNA mutations (see below).

Tumorigenic mtDNA mutations affect the respiratory chain complexes and enhance the production of ROS (Brandon et al. 2006); however, the pathological relevance of mtDNA mutations in cancer cells is controversial (Frezza and Gottlieb 2009). Nonetheless, a clear-cut correlation between occurrence of pathogenic mtDNA mutations and mitochondrial energetic impairment is a well-demonstrated feature of oncocytic tumors (Bonora et al. 2006a; Porcelli et al. 2010), which is dealt with in detail below.

Accumulation of NADH due to either hypoxia or alterations of the electron transfer chain inhibits isocitrate dehydrogenase and the Krebs cycle when implemented by acetyl CoA may help the exit of citrate to support lipid synthesis (fatty acids and cholesterol). On the other hand, the entry of reducing equivalents via glutamine may not be inhibited, thus sustaining a truncated Krebs cycle via glutamate, α-ketoglutarate, and the intermediates of the Krebs cycle, thus reaching malate that is converted to pyruvate by the malic enzyme after release in the cytosol. Pyruvate would contribute to stabilization of HIF1α and may also be converted to lactate by lactate dehydrogenase. The extent of this pathway of glutamine oxidation is determined by the residual activity of the respiratory chain, because both glutamate dehydrogenase and α-ketoglutarate dehydrogenase require oxidized NAD^+, and both these enzymes together with succinate dehydrogenase require an intact respiratory chain for their turnover.

THE ROLE OF ROS

An overwhelming body of evidence accumulated in the last decades has shown that mitochondria have a central role in the etiology and pathogenesis of most major chronic diseases and in aging (Wallace 2005; Lenaz et al. 2006; Lenaz and Genova 2010). The involvement of mitochondria in disease, which has generated the term "mitochondrial medicine" (DiMauro et al. 2006), has been largely ascribed to their central role in production of ROS and to the damaging effect of the latter on these organelles. In particular, mtDNA mutations would induce alterations of the encoded polypeptides in the respiratory complexes, with consequent decrease of electron transfer activity, leading to further production of ROS, and thus establishing a vicious circle of oxidative stress and energetic decline (Ozawa 1997). It must be underlined that mtDNA mutations leading to increased ROS production are those allowing at least a partial assembly of the respiratory complexes. On the contrary, for instance, oncocytic tumors display homoplasmic disassembling complex I mutations, which may not be ROS-generating mutations, if one considers that the main ROS production site might be lacking as a whole (Koopman et al. 2007; Moran et al. 2010; Porcelli et al. 2010; Tuppen et al. 2010). Moreover, the mutant load ought to be considered when analyzing functional effects of mtDNA mutations. The threshold model we previously proposed implies that, below threshold, complex I is present and functioning correctly, in contrast to the case of a missense mutation leading to the coexistence of nonmutant along with mutant ROS producing complex I (Park et al. 2009).

Decreased mitochondrial activity is considered to be tumorigenic, mainly because of the enhanced ROS production. H_2O_2 exported to the nucleus enhances the transcription of selected genes that favor tumor progression (Wallace 2005). ROS act both as mutagens and cellular mitogens (Klaunig and Kamendulis 2004). It has been suggested that oncogenesis and neurodegeneration may share common pathogenic features as differential responses to different insults during different phases of the cell cycle, leading either to cell death or to cell proliferation (Morris et al. 2010).

Any lowering of the electron transfer chain activity, either caused by mtDNA mutations and hence structural lesions of the respiratory complexes or depending directly or indirectly on hypoxia, would enhance generation of ROS; it is known that ROS generation is enhanced in State 4 (controlled state in absence of ADP), when complex I is inhibited, and paradoxically under hypoxic conditions.

ROS may contribute to further mitochondrial dysfunction by several mechanisms: (1) direct alterations of respiratory complexes, for example, complex I, particularly on FeS clusters; (2) peroxidation of mitochondrial phospholipids, in particular, cardiolipin (Paradies et al. 2010), that are required for proper assembly and activity of mitochondrial complexes (I, III, IV); (3) (further) mtDNA mutations affecting respiratory complexes; and (4) stabilization of HIF1α (Patten et al. 2010).

In this scenario, it is easy to foresee a deep implication of supercomplex organization as the missing link between oxidative stress and energy failure (Lenaz and Genova 2007). It is tempting to speculate that under conditions of oxidative stress a dissociation of complex I–III aggregates occurs, with loss of facilitated electron channeling and resumption of a less efficient random diffusional behavior with electron transfer depending on the collisional encounters of the free ubiquinone molecules with the partner complexes.

As predicted by Lenaz and Genova (2007), dissociation of supercomplexes might have further deleterious consequences, such as disassembly of complex I and III subunits and loss of electron transfer and/or proton translocation; the consequent alteration of electron transfer may elicit further induction of ROS generation. Following this line of thought, the different susceptibility of different types of cells and tissues to ROS damage may be a consequence of the extent and tightness of supercomplex organization in their respiratory chains, which depend on phospholipid content and composition of their mitochondrial membranes. These changes may have deep metabolic consequences, as depicted in the scheme in Figure 1. An initial enhanced ROS generation due to different possible reasons and originating in different districts of the cell besides mitochondria (Lenaz and Strocchi 2009) would induce supercomplex disorganization eventually leading to possible decrease of complex I assembly; both the lack of efficient electron channeling and the loss of complex I would decrease NAD-linked respiration and ATP synthesis. Here we briefly summarize the experimental evidence pertaining to this hypothesis.

Peroxidized Phospholipids Prevent Supercomplex Formation

Two roles of phospholipid have been distinguished: (1) a dispersive solubilization effect that can be duplicated by appropriate detergents, and (2) a catalytic effect that can be specifically fulfilled only by cardiolipin (Vik and Capaldi 1977; Fry and Green 1980, 1981; Robinson et al. 1980; Lee 2004). The phospholipids in closest vicinity to the protein surface, as well as those in the free bilayer, are actually highly mobile and free to exchange, but cardiolipin was indicated as tightly bound, being more likely buried within the protein complexes (Kang et al. 1979; Sedlak and Robinson 1999; Lange et al. 2001). Recent results seem also to indicate that cardiolipin stabilizes respiratory chain supercomplexes as well as the individual complexes. The availability of a cardiolipin-lacking yeast mutant provided the opportunity to show that mitochondrial membranes still contained the III_2–IV_2 supercomplex, but that it was significantly less stable than supercomplexes in the parental strain (Zhang et al. 2002; Pfeiffer et al. 2003).

Mutations of tafazzin, an acyl transferase involved in the synthesis of mature tetralinoleyl cardiolipin (Neuwald 1997), result in Barth syndrome, a cardioskeletal myopathy with neutropenia, characterized by respiratory chain dysfunction. The cardiolipin defect in Barth

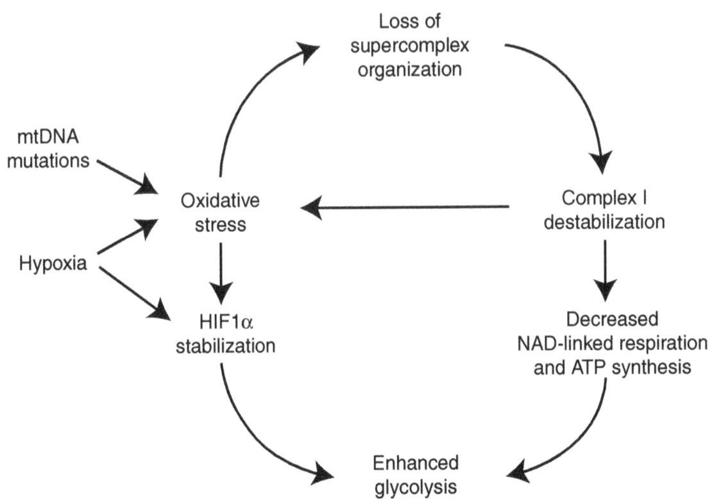

Figure 1. The possible effects of loss of supercomplex organization on mitochondrial function during oncogenesis. Membrane phospholipid peroxidation and consequent loss of supercomplex organization may occur because of oxidative stress induced by genetic changes (i.e., mitochondrial DNA mutations) or by exogenous factors (e.g., hypoxia); the ensuing destabilization of complex I results in OXPHOS deficiency and further oxidative stress. As a consequence of these changes, cells are forced to rely on glycolysis for energy production. See text for further details.

syndrome results in destabilization of the supercomplexes by weakening the interactions between respiratory complexes (McKenzie et al. 2006).

It is well documented that exposure of mitochondria to ROS can affect the respiratory activity via oxidative damage of cardiolipin, which is required for the optimal functioning of the enzyme complexes (Paradies et al. 2000, 2002, 2010; Petrosillo et al. 2003, 2009). We have shown that the maintenance of a I–III supercomplex after reconstitution into phospholipid vesicles is abolished if lipid peroxidation is induced before reconstitution (Bianchi et al. 2003; Genova et al. 2008).

Evidently, the distortion of the lipid bilayer induced by peroxidation and the alteration of the tightly bound phospholipids determine dissociation of the supercomplex originally present in the lipid-poor preparation.

Supercomplex Disassembly Abolishes CoQ Channeling and Resumes Less Efficient Pool Behavior

Studies of respiration in pathological conditions (Rosca et al. 2008; van Raam et al. 2008) showed that electron transfer in absence of supercomplex organization is lost even if activity of the individual complexes is normal. In reconstitution studies of complexes I and III in phospholipid vesicles (Genova et al. 2008), electron transfer between complex I and complex III (NADH cytochrome *c* reductase) shows rates higher than those predicted by random behavior in proteoliposomes enriched in complexes I and III at a protein-to-phospholipid ratio of 1:1 (when kinetic testing according to flux control analysis indicates the presence of a supercomplex I–III). However, at a ratio of 1:30 (when the supercomplex is dissociated), the rate of NADH cytochrome *c* reductase is exactly that predicted by the pool equation on the basis of the individual activities of complex I and complex III. This is a demonstration that supercomplex formation, indeed, enhances the rate of electron transfer above that occurring via a ubiquinone pool in the membrane (Bianchi et al. 2003; Genova et al. 2008), and loss of supercomplex organization resumes less efficient pool activity of the quinone.

Loss of Supercomplexes Prevents Stability and Assembly of Individual Complexes

Analysis of the state of supercomplexes in patients with an isolated deficiency of single complexes (Schagger et al. 2004) and in cultured cell models harboring cytochrome *b* mutations (Acin-Perez et al. 2004; Blakely et al. 2005; D'Aurelio et al. 2006; Diaz et al. 2006; McKenzie et al. 2006) provided evidence that the formation of respirasomes is essential for the assembly/stability of complex I. Genetic alterations leading to a loss of complex III prevented respirasome formation and led to secondary loss of complex I, and therefore primary complex III assembly deficiencies presented as complex III/I defects. Conversely, complex III stability was not influenced by the absence of complex I.

From these findings, supercomplex assembly emerged as a necessary step for respiration, its defect setting the threshold for respiratory impairment in mtDNA mutant cells.

Disassembly of Supercomplex Organization Enhances Generation of ROS

Indirect circumstantial evidence suggests that supercomplex assembly may limit the extent of superoxide generation by the respiratory chain (Lenaz and Genova 2010). A direct study on a mitochondrial fraction containing complexes I and III and reconstituted with different amounts of phospholipids shows that at protein:phospholipids 1:30, when the supercomplex is not formed, the production of superoxide is 20-fold higher than at a 1:1 ratio, when most units are assembled in a supercomplex (ML Genova, G Barbero, E Maranzana, et al., unpubl.).

Although the molecular structure of the individual complexes does not allow us to envision a close apposition of the matrix arm of complex I, where FMN is localized, with either complex III or IV, the actual shape of the $I_1III_2IV_1$ supercomplexes from bovine heart (Schafer et al. 2007) suggests a slightly different conformation of complex I in the supercomplexes with a smaller angle of the matrix arm with the membrane arm showing a higher bending toward the membrane (and presumably complex III), in line with the notion that complex I may undergo important conformational changes (Radermacher et al. 2006).

Do Cancer Cells Lose Supercomplex Association?

The validation of the above working hypothesis requires direct studies on cancer tissues and neoplastic cells. A general decrease of respiratory chain activity is a common feature of several cancers. Nevertheless, very few studies have addressed the respiratory chain supramolecular organization.

Baracca et al. (2010) reported that down-regulation of the respiratory chain in Ras-transformed fibroblasts is operated through strong decrease of complex I activity and content, probably because of lack of correct assembly of the subunits as a consequence of altered supercomplex organization, as shown by loss of the highest-molecular-weight $I_1III_2IV_{1\text{-}2}$ supercomplex. Significantly, ROS generation was strongly enhanced in the Ras cells. In the XTC-UC1 oncocytic cell line (see below), we have also found (Genova et al. 2008) by blue-native electrophoresis of mitochondrial proteins, a complete absence of high-molecular-weight aggregates containing either complex I or complex IV, that are instead present in a control cell line from a nononcocytic thyroid tumor. If the absence of supercomplexes comprising complex I is a consequence of its disassembly due to lack of ND1 (see below), the absence of aggregated complex IV is in line with the working hypothesis.

The loss of supercomplex assembly seems to accompany a decrease/loss of complex I activity; if this decrease is found to be common to most tumors, the consequences would depend on the extent of such a decrease. Disruptive mtDNA mutations in the ND subunits of complex I (as in oncocytomas; see below) almost completely abolish complex I activity, which presumably inactivates the Krebs cycle, that requires NAD^+, favoring a high-ratio α-ketoglutarate/succinate and consequent destabilization of HIF1α. On the other hand, if complex I is partially defective (as in RAS-mutated fibroblasts), the levels of NAD^+ may be sufficient to implement the Krebs cycle with higher levels of intermediates that stabilize HIF1α, thus favoring tumor malignancy.

mtDNA MUTATIONS IN CANCER: THE PECULIAR CASE OF ONCOCYTIC TUMORS

An excellent model to dissect out the role of pathogenic, disassembling mtDNA mutations in tumor progression and their contribution to the metabolic reprogramming of cancer cells (glycolysis vs. respiration) is provided by an often underdiagnosed subset of tumors, namely, the oncocytomas. These neoplasms contain a high percentage or they are completely made of cells characterized by an aberrant mitochondrial hyperplasia. They originate from epithelial tissues and occur mainly in endocrine and exocrine organs, although the oncocytic phenotype has been described in many anatomical districts (for review, see Gasparre et al. 2011a). The majority of oncocytic tumors are considered as benign, displaying low invasiveness and showing a usually favorable prognosis. In the thyroid, nonetheless, the best indicator for prognosis remains the degree of differentiation of neoplastic cells, which defines adenoma or carcinoma, regardless of the occurrence of the oncocytic phenotype.

The most striking feature of oncocytic tumors is appreciated through histochemical and ultrastructural analyses, which highlight cells packed with enlarged globular or ovate mitochondria with a stack of lamelliform, tubular, or flat cristae and occupying up to 60% of the cytoplasm (Tallini 1998; Ambu et al. 2000; Riva and Tandler 2000). Such heterogeneity in mitochondrial morphology has long been an indicator of a functional along with a structural alteration of the organelles in oncocytic cells.

The cause of the mitochondrial hyperplasia in oncocytic tumors has been suspected to consist in an aberrant mitochondrial biogenesis triggered by a mitochondrial dysfunction, which may attempt to compensate for an impaired energy production. Evidence that a compensatory mechanism is responsible for oncocytic phenotype occurrence has accumulated in recent years. It is now clear that the mitochondrial respiration deficiency originally hypothesized (Muller-Hocker et al. 1998a) is, indeed, an event occurring in oncocytic tumors and mainly concerns respiratory complex I (Zielke et al. 1998; Simonnet et al. 2003). NADH-dehydrogenase activity is, in fact, reduced concomitantly with an increase in other respiratory complex abundance and, overall, in mitochondrial mass as shown by measurements of citrate synthase activity and of mtDNA copy number (Zielke et al. 1998; Simonnet et al. 2003; Gasparre et al. 2009; Porcelli et al. 2010).

In vitro studies with transmitochondrial cell hybrids (cybrids) have shown that the defective mitochondrial respiration of thyroid oncocytic cells is transferred to normally respiring osteosarcoma cells when mitochondria, not nuclei, are imported (Bonora et al. 2006a). From this observation, as well as from studies in which the mitochondrial genome was screened for oncocytic cancer-specific genetic lesions such as small and large deletions, depletion, or point mutations (Tallini et al. 1994; Maximo et al. 1998, 2002; Maximo and Sobrinho-Simoes 2000; Costa-Guda et al. 2007; Gasparre et al. 2007, 2008, 2009; Mayr et al. 2008; Zimmermann et al. 2009; Porcelli et al. 2010; Bartoletti-Stella et al. 2011), it is now generally accepted that the genetic hallmark of the oncocytic phenotype is the occurrence of disruptive mtDNA mutations, particularly in complex I subunits. This holds true for virtually all types of oncocytic tumors analyzed to date (Gasparre et al. 2007, 2008, 2009; Porcelli et al. 2010; Bartoletti-Stella et al. 2011), and, inversely, association between the occurrence of disruptive mutations and that of oncocytic foci has been shown, for instance, in endometrial carcinoma (Guerra et al. 2011). Within the plethora of disruptive mutations, both frameshift and pathogenic missense changes are included, which ultimately lead to complex I impairment in terms of function and/or assembly. Hence, the direct consequence of these mutations impinges on respiration, whose impairment may subsequently trigger a retrograde signaling from the organelles to the nucleus, leading in a vicious circle to mitochondrial hyperplasia. An analysis of the entire mitochondrial genomes sequenced in oncocytic tumor samples has allowed identification of MTND1 as a mutational hotspot (Gasparre et al. 2007). Less importantly than MTND1, at least two of the remaining six mitochondria-coded complex I genes display a high rate of pathogenic mutations accumulation, namely, MTND4 and MTND5, which are also among the longest mitochondrial genes and richest in homopolymeric stretches, where many frameshift mutations occur. Complex IV (cytochrome *c* oxidase; COX) mitochondrial subunits appear to be the least involved in contributing to the mitochondrial damage leading to oncocytic transformation, although some cases have been reported of oncocytomas harboring mutations in COX genes (Maximo et al. 2002; Costa-Guda et al. 2007) or being COX-negative upon immunohistochemical analysis (Porcelli et al. 2010). These findings suggest that complex I disruption may affect, in certain cases, the organization of supercomplexes.

Although for most types of oncocytic tumors such as kidney and pituitary oncocytoma (Gasparre et al. 2008; Porcelli et al. 2010), up to 70%–100% of cases have been shown to harbor pathogenic mtDNA mutations, in heterogeneous neoplasms such as those of the thyroid, a lower percentage display the genetic features described. Although studies on microdissected tumors are currently lacking, several oncocytic cancers do not necessarily harbor mtDNA mutations, which suggests that nuclear genetic lesions may create a phenocopy. It is plausible to think that mutations in nuclear-encoded respiratory complexes subunits (38 belonging to NADH-dehydrogenase), or in chaperones, may well trigger the same bioenergetic damage and, consequently, induce the same phenotype as mtDNA mutations. Nuclear *GRIM-19* (alias *NDUFA13*) has been one of the candidate genes, in which somatic mutations in sporadic forms of thyroid oncocytic cancer were reported. However, no *GRIM-19* mutations were found to segregate in familial thyroid oncocytoma (TCO) (Maximo et al. 2005). The mitochondrial protein import system subunit *TIMM44* has been reported to harbor functional mutations in familial TCO (Bonora et al. 2006b). However, preliminary investigation of about 40 oncocytic thyroid tumors has not shown a high incidence of potentially pathogenic variants in nuclear-encoded complex I subunits (E Bonora, G Gasparre, G Romeo et al., unpubl.). Further investigation of possible nuclear-coded complex I mutations in microdissected homogeneous oncocytic tissues are still warranted before the involvement of nuclear subunits can be ruled out.

ENERGETIC COMPETENCE IN ONCOCYTIC TUMORS

Mitochondrial Hyperplasia and the Compensatory Effect Hypothesis

Retrograde signaling in oncocytic tumors from the mitochondrion to the nucleus has been hypothesized to affect several pathways (Butow and Avadhani 2004). A metabolic stress in tumor cells may induce a nuclear response leading to the activation of mitochondrial biogenesis pathways to restore a defective respiration.

Increased mitochondrial biogenesis has been evaluated in oncocytic tumors as well as in the only existing model of oncocytic cancer, the XTC.UC1 cell line. In both models, a defective mitochondrial ATP synthesis and an overexpression of uncoupling proteins and of the biogenesis regulator PGC-1-related coactivator (PRC) have been reported (Savagner et al. 2001a,b, 2003; Higgins et al. 2003; Baris et al. 2004; Finley et al. 2004; Jacques et al. 2005). From these data, it was initially concluded that the defective ATP synthesis may explain the mitochondrial hyperplasia and hence justify the compensatory effect, caused by an oxidative phosphorylation coupling defect of unknown cause (Savagner et al. 2001b). Moreover, the up-regulation of mitochondrial proteins has been described both in thyroid and kidney oncocytomas (Higgins et al. 2003; Baris et al. 2004; Finley et al. 2004). In kidney oncocytomas, the increase in mitochondrial proteins and in the activity of respiratory complexes II, III, IV, and V as well as of citrate synthase seems to be a common feature, in correlation with their benign behavior (Simonnet et al. 2003). In apparent contrast, in oncocytes of both normal and hyperfunctional parathyroids, a deficiency of cytochrome *c* oxidase has been observed (Muller-Hocker 1992; Muller-Hocker et al. 1998b). In oncocytic cells of the same tissues, a higher mtDNA content was reported without difference in expression of mitochondrial transcription factor TFAM or mtDNA γ polymerase (POLG), suggesting a compensatory effect due to a complex IV deficiency (Muller-Hocker et al. 1998a). Although the compensatory effect has been described as a predominant feature in oncocytic tumors, the trigger for such a mechanism has not been identified. Godinot's group reported that the presence of complex I deficiency might increase mitochondrial biogenesis to compensate a respiratory dysfunction (Simonnet et al. 2003), and our group described a mitochondrial energetic impairment in the XTC.UC1 cell line originally characterized by Zielke et al. (1998). In this cell line we reported a decrease in both complex I and III activity and a strong reduction of ATP synthesis, driven by complex I substrates, and mitochondrial membrane potential (Bonora et al. 2006a; Porcelli et al. 2009). We also showed that the energetic dysfunction can be transferred along with the mtDNA in a transmitochondrial cybrid model and hence concluded that the complex I defect must be due to an mtDNA mutation. These functional results were followed by an increased level of most respiratory chain complexes subunits, whereas complex I NDUFA9 and ND6 were strongly reduced. Moreover, we observed that complex I mitochondria-coded ND1 subunit was absent (Bonora et al. 2006a).

The feature of a mitochondrial energetic impairment and decrease in complex I subunits was also reported in a rare case of nasopharynx oncocytoma (Gasparre et al. 2009) and in a peculiar case of Warthin tumor (Porcelli et al. 2010). Altogether these data point to a dysfunction of at least respiratory complex I as the main trigger for oncocytic transformation with a compensatory increase in other mitochondrial proteins leading to the high mitochondrial proliferation in oncocytic cells.

Structural and Functional Consequences of mtDNA Mutations

The common occurrence of mtDNA mutations in complex I genes should result in incomplete or partial assembly and/or function of this enzyme depending on the mtDNA mutation type. Our group showed that the presence of homoplasmic truncating ND mutations mainly cause a loss of complex I subunits in oncocytic neoplasia (Gasparre et al. 2009). Moreover, we took advantage of XTC.UC1 cells bearing the m.3571insC ND1 mutation to clarify the impact of the homoplasmic truncating mtDNA mutations on the respiratory complex content/assembly and energetic function. In these cells, the ND1 subunit was ablated, and the levels of other complex I subunits (i.e., NDUFA9 and ND6) were significantly reduced (Bonora et al. 2006a). Similar results were extended to a survey of a large panel of pituitary, head-and-neck, and thyroid oncocytic tumors for the presence of homoplasmic mtDNA truncating mutations (Porcelli et al. 2010).

The presence of an incomplete or partially assembled complex I in oncocytic tumors raises the question of whether this may contribute to ROS generation. Lack of data on the role of ROS in oncocytic tumors is mainly because ROS measurements in tumor biopsies are not feasible, and cell models for in vitro studies are very scarce. In XTC.UC1 cells, different degrees of heteroplasmy of the 3571insC *ND1* mutation did not influence ROS amounts, probably because of a differential expression of ROS detoxifying enzymes (i.e., manganese superoxide dismutase and catalase) in the presence/absence of complex (Porcelli et al. 2010). Kofler's group speculated that a lack of complex I in oncocytic thyroid may increase ROS production inhibiting pro-apoptotic pathways (Zimmermann et al. 2009). Further studies are warranted to understand whether ROS may influence the proliferative potential and the accumulation of mutations in oncocytic tumors, where most of the mutations reported are indeed homoplasmic and complex I appears to be disassembled.

Because elevated ROS have been proposed to induce apoptosis, additional studies are required to determine the role of apoptosis in regulating the survival and proliferation of oncocytic cells. It has been reported that mtDNA mutations impairing mitochondrial energetic function may protect cells from apoptosis, providing a growth advantage (Shidara et al. 2005; Park et al. 2009), but whether mtDNA mutations prevent apoptosis in oncocytic tumors is still poorly defined.

Alteration of the oxidative metabolism resulting from mtDNA mutations may trigger the Warburg effect permitting a shift toward glycolysis, hence favoring tumor progression (Chen et al. 2007; Smolkova et al. 2010). Several observations have linked respiratory complex dysfunctions to alteration of upstream mitochondrial metabolism, that is, Krebs cycle, in cancer. Mutations in enzymes such as succinate dehydrogenase (SDH) and fumarate hydratase (FH) are actively involved in tumorigenesis through imbalance of the Krebs cycle and stabilization of HIF1α (Pollard et al. 2005; King et al. 2006). Mutations in the B, C, and D subunits of SDH have been shown to lead to the accumulation of the Krebs metabolite succinate at the expense of α-ketoglutarate. Mutations in FH lead to increase of fumarate, slowing down the Krebs cycle (Pollard et al. 2005; Pasini and Stratakis 2009). The accumulation of succinate and fumarate are thought to inhibit PHD, key activators of HIF1α degradation (Boulahbel et al. 2009).

The complex I dysfunction described in oncocytic tumors raises the question of whether a parallel may exist between these neoplasms and SDH/FH mutated cancers, and what is the role played by complex I in regulating cancer metabolism. We recently reported that in the XTC.UC1 cells bearing the homoplasmic truncating *ND1* mutation, the ratio between succinate and α-ketoglutarate is opposite to that reported in SDH and FH mutated tumors, with increase of α-ketoglutarate at the expense of succinate (Porcelli et al. 2010). It is likely that the absence of a functional complex I in these cells leads to accumulation of NADH, which may, in turn, inhibit α-ketoglutarate dehydrogenase and prevent succinate production with a consequent α-ketoglutarate increase.

Because α-ketoglutarate is the main reagent feeding PHD, HIF1α should undergo a chronic destabilization, even in the hypoxic tumor environment. We have, indeed, shown this to be the case in oncocytic tumors harboring homoplasmic disruptive mtDNA mutations causing disassembly of complex I (Porcelli et al. 2010).

CONCLUDING REMARKS: GENERAL IMPLICATIONS OF THE ONCOCYTIC TUMOR MODEL FOR CANCER RESEARCH

During the last few years, the oncocytic tumor model has been used to ask some basic questions regarding the functional effects of mtDNA mutations. First of all, it has been possible to establish a univocal correlation between mtDNA mutations and the corresponding biochemical phenotype thanks to the only available oncocytic cell line, the XTC.UC1, and the cybrids derived from it (Bonora et al. 2006a). Using these tools, it was realized that the mtDNA physiological polyploidy has a strong impact on the disease phenotype according to the threshold effect model. This feature of mitochondrial genetics has not always been taken into consideration when evaluating some basic questions regarding the pro-tumorigenic or metastatic potential of mtDNA mutations (Petros et al. 2005; Ishikawa et al. 2008). A further question that is closely correlated with this typical feature is the possible selective disadvantage or advantage that mtDNA mutations have in vitro as well as in vivo (Gasparre et al. 2007) as a result of the "population dynamics" that might characterize the evolution of mtDNA mutations during tumorigenesis. The possibility of a selective advantage is strongly supported by recent findings (Gasparre et al. 2009), which describe disruptive frameshift mtDNA mutations affecting the ND5 and ND1 subunits of complex I present in homoplasmy in a nasopharyngeal oncocytic tumor. In both cases, the mutations were inherited at a low degree of heteroplasmy through the germline in the patient's family (Gasparre et al. 2009) or were germline de novo events (Pradella et al. 2011) but became homoplasmic in the patient's oncocytic tumor. Immunohistochemistry has shown a perfect correlation between presence of a homoplasmic mutation and lack of the mitochondria-coded ND6 subunit, which suggests impaired complex I assembly. In the described nasopharynx oncocytoma, a few oncocytic areas from the same tumor, expressing ND6 and heteroplasmic for the ND5 mutation, harbored instead a de novo homoplasmic ND1 mutation. Because the shift to homoplasmy of both ND1 (somatic) and ND5 (germinal) mutations occurred exclusively in the tumor cells of the patient, we can conclude that these complex I mutations had a selective advantage that accompanied the oncocytic change. This last conclusion warrants, in turn, a revisitation of the hypotheses regarding the origin and fate of oncocytic cells. As shown in Figure 2, a general mechanism can be hypothesized in which a mitochondrial hit (either in the mtDNA or in the nuclear-coded subunits of complex I) may occur at any stage during the multistep process of tumorigenesis. This particular mitochondrial hit, which may also be inherited, will determine whether the tumor becomes oncocytic or nononcocytic. The necessity of a double hit, one mitochondrial and the main hit likely being a nuclear one conferring the transformed phenotype to a normal cell, seems to date the most plausible hypothesis to explain oncocytic transformation, as also discussed in recent reviews (Maximo et al. 2009). The implications of this model on the selective advantages or disadvantages of the oncocytic cells during tumorigenesis demand also a critical reevaluation of the functional significance of the impact of mtDNA-coded genes in cancer, in analogy to nuclear-encoded mitochondrial genes such as *SDH* and *FH*. The role of metabolic genes in tumorigenesis is still very controversial (Thompson 2009). *SDH* and *FH* are nowadays considered tumor-suppressor genes, as it has been proposed for complex I by Kofler's group (Zimmermann et al. 2011), although exclusively on the basis, primarily, of the observation of mutations detected in oncocytic tumors. Nonetheless, in our opinion, correlation studies are not sufficient to assign complex I genes the status of tumor-suppressor genes,

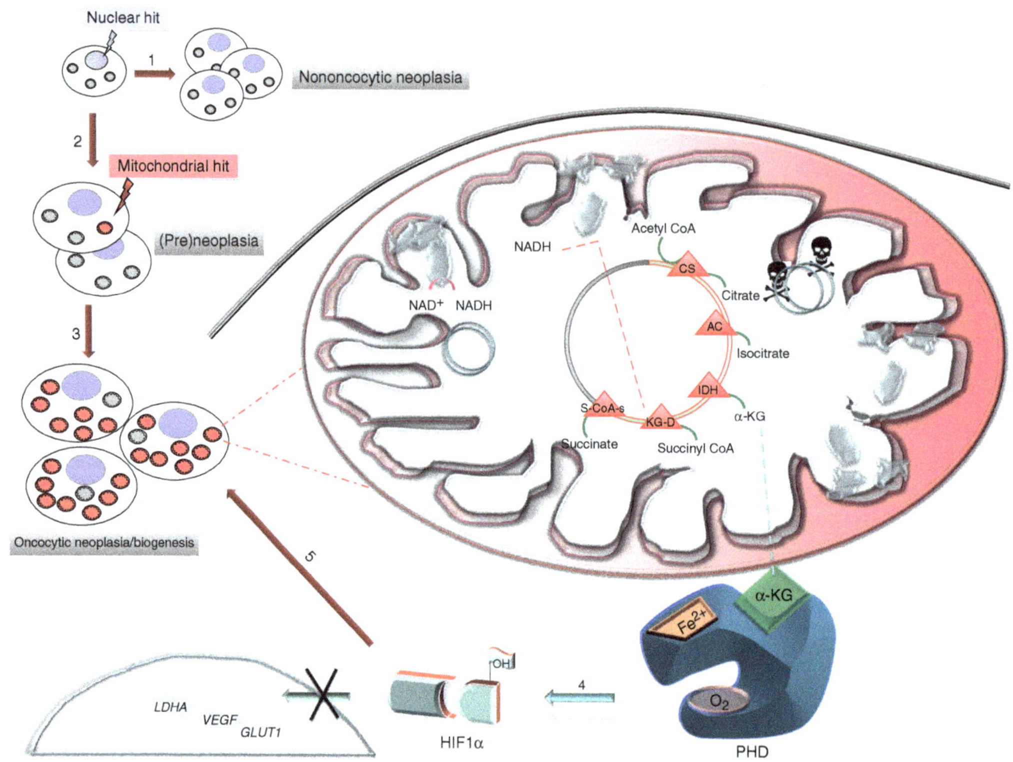

Figure 2. The origin and fate of the oncocytic cell. Nuclear hits (light blue lightning arrow) presiding to cellular transformation contribute to the initiation stage of neoplastic or preneoplastic development. 1. No mtDNA mutations occur, or such mutations are detrimental to cell replication and mutated mitochondria are selected against, preventing the development of an oncocytic phenotype. 2. During cell replication a mutation (red lightning arrow) may occur in mtDNA (mitochondrial hit) either conferring a metabolic advantage by contributing to cell adaptation through the Warburg effect or remaining neutral until a mutation load threshold is reached. 3. Positive selection or random drift may set in through mitochondrial biogenesis, generating the oncocytic phenotype.The pink dotted lines lead to an enlarged dynamic view of an "oncocytic" mitochondrion schematically divided into two parts: a white area showing wild-type mtDNA molecules (gray circles) and correctly assembled complex I, and a shaded pink area with mutated mtDNA and partially or fully disassembled complex I. An mtDNA mutation disassembling complex I, such as those typical of oncocytic tumors, may not allow NADH consumption, shifting the NADH/NAD^+ ratio toward NADH and leading to an increase of the α-KG/SA ratio. Such Krebs cycle alteration may foster activity of PHD with subsequent HIF1α destabilization/degradation even in low oxygen tension conditions (pseudonormoxia). 4. Inactivation of this transcription factor may functionally lead, on one hand, to the inhibition of expression of pro-tumorigenic and glycolytic genes (i.e., *LDHA*, *GLUT1*, and *VEGF*) and, on the other hand, to the down-regulation of glycolysis needed to compensate for the defective mitochondrial respiration (light blue arrow). This scenario does not allow the metabolic adaptation of tumor cells (Warburg effect), hampering the tumorigenic potential of cells, explaining the low-proliferative and low-aggressive features of oncocytic tumors. 5. Furthermore, HIF1α destabilization may not prevent the inhibition of mitochondrial biogenesis that should occur upon hypoxia sensing, hence feeding a vicious loop ultimately leading to the pathological mitochondrial hyperplasia of oncocytic cells (red arrow). CS, citrate synthase; AC, aconitase; IDH: isocitrate dehydrogenase; α-KG, α-ketoglutarate; KG-D, α-ketoglutarate dehydrogenase; S-CoA-s, succinyl-CoA synthase; PHD, prolyl hydroxylase; HIF1α, hypoxia-inducible factor 1α; GLUT1, glucose trasporter 1; VEGF, vascular endothelial growth factor; LDHA, lactate dehydrogenase A.

especially because such definition overlooks the fundamental differences between mitochondrial genetics (whose dynamics are analogous to those of population genetics) and Mendelian genetics. The typical homoplasmic complex I mutations found in oncocytic tumors and shown to generate the phenotype in vivo hamper the tumorigenic potential of cells that harbor them (Park et al. 2009; Gasparre et al. 2011b), accounting for the low-proliferative and low-aggressive behavior of most oncocytic tumors. This effect has been shown by our group to be independent from apoptosis and likely due to the lack of HIF1α stabilization described above and pictured in Figure 2. Such pseudonormoxic status may also explain why only a complete lack of complex I may determine the oncocytic phenotype. HIF1α is a pivotal intracellular oxygen sensor, whose role, when stabilized during hypoxia/anoxia, is envisioned to block mitochondrial biogenesis when O_2 is limiting. The release of a HIF1α block may reinforce the induction of the compensatory mitochondrial biogenesis in complex I-lacking cells, finally leading to mitochondrial hyperplasia.

Last, we have defined the threshold mutation load needed to induce an anti-tumorigenic phenotype, which suggests that these genes are more likely to be lethality associated than canonical tumor suppressors. Mutated mtDNA genes that cause complex I disassembly may hence be anti-tumorigenic, in close dependence on their mutant load, which needs to be carefully measured (Kurelac et al. 2011). Such feature does not generally apply to nuclear genes. We have therefore proposed the denomination of "oncojanus," reminiscent of the double-faceted Roman god Janus, to indicate the threshold-dependent double-edged properties of CI mtDNA-encoded genes (Gasparre et al. 2011b). The understanding of the different properties of mutations affecting complex I genes and the dissection of the mechanisms leading to pseudonormoxia following complex I disassembly should provide fundamental insights linking cancer metabolism to hypoxia adaptation and will likely lead to the discovery of novel approaches in cancer therapy.

ACKNOWLEDGMENTS

This work is supported by the Italian Ministry of University and Research (MIUR) grant FIRB "Futuro in Ricerca" and by Fondazione Umberto Veronesi grant "Disco trip" to G.G., by MIUR grant PRIN 2008 to A.M.P., and by the Associazione Italiana Ricerca sul Cancro (AIRC) to G.R.

REFERENCES

Acin-Perez R, Bayona-Bafaluy MP, Fernandez-Silva P, Moreno-Loshuertos R, Perez-Martos A, Bruno C, Moraes CT, Enriquez JA. 2004. Respiratory complex III is required to maintain complex I in mammalian mitochondria. *Mol Cell* **13:** 805–815.

Ambu R, Riva A, Lai ML, Loffredo F, Riva FT, Tandler B. 2000. Scanning electron microscopy of the interior of cells in Hurthle cell tumors. *Ultrastruct Pathol* **24:** 211–219.

Baracca A, Chiaradonna F, Sgarbi G, Solaini G, Alberghina L, Lenaz G. 2010. Mitochondrial complex I decrease is responsible for bioenergetic dysfunction in K-ras transformed cells. *Biochim Biophys Acta* **1797:** 314–323.

Baris O, Savagner F, Nasser V, Loriod B, Granjeaud S, Guyetant S, Franc B, Rodien P, Rohmer V, Bertucci F, et al. 2004. Transcriptional profiling reveals coordinated up-regulation of oxidative metabolism genes in thyroid oncocytic tumors. *J Clin Endocrinol Metab* **89:** 994–1005.

Bartoletti-Stella A, Salfi NC, Ceccarelli C, Attimonelli M, Romeo G, Gasparre G. 2011. Mitochondrial DNA mutations in oncocytic adnexal lacrimal glands of the conjunctiva. *Arch Ophthalmol* **129:** 664–666.

Bell EL, Klimova T, Chandel NS. 2008. Targeting the mitochondria for cancer therapy: Regulation of hypoxia-inducible factor by mitochondria. *Antioxid Redox Signal* **10:** 635–640.

Bianchi C, Fato R, Genova ML, Parenti Castelli G, Lenaz G. 2003. Structural and functional organization of complex I in the mitochondrial respiratory chain. *Biofactors* **18:** 3–9.

Blakely EL, Mitchell AL, Fisher N, Meunier B, Nijtmans LG, Schaefer AM, Jackson MJ, Turnbull DM, Taylor RW. 2005. A mitochondrial cytochrome *b* mutation causing severe respiratory chain enzyme deficiency in humans and yeast. *FEBS J* **272:** 3583–3592.

Bonora E, Porcelli AM, Gasparre G, Biondi A, Ghelli A, Carelli V, Baracca A, Tallini G, Martinuzzi A, Lenaz G, et al. 2006a. Defective oxidative phosphorylation in thyroid oncocytic carcinoma is associated with pathogenic mitochondrial DNA mutations affecting complexes I and III. *Cancer Res* **66:** 6087–6096.

Bonora E, Evangelisti C, Bonichon F, Tallini G, Romeo G. 2006b. Novel germline variants identified in the inner mitochondrial membrane transporter TIMM44 and their role in predisposition to oncocytic thyroid carcinomas. *Br J Cancer* **4:** 1529–1536.

Boulahbel H, Duran RV, Gottlieb E. 2009. Prolyl hydroxylases as regulators of cell metabolism. *Biochem Soc Trans* **37:** 291–294.

Brahimi-Horn MC, Pouyssegur J. 2007. Oxygen, a source of life and stress. *FEBS Lett* **581:** 3582–3591.

Brandon M, Baldi P, Wallace DC. 2006. Mitochondrial mutations in cancer. *Oncogene* **25:** 4647–4662.

Butow RA, Avadhani NG. 2004. Mitochondrial signaling: The retrograde response. *Mol Cell* **14:** 1–15.

Chen Z, Lu W, Garcia-Prieto C, Huang P. 2007. The Warburg effect and its cancer therapeutic implications. *J Bioenerg Biomembr* **39:** 267–274.

Costa-Guda J, Tokura T, Roth SI, Arnold A. 2007. Mitochondrial DNA mutations in oxyphilic and chief cell parathyroid adenomas. *BMC Endocr Disord* **7:** 8.

Cuezva JM, Ortega AD, Willers I, Sanchez-Cenizo L, Aldea M, Sanchez-Arago M. 2009. The tumor suppressor function of mitochondria: Translation into the clinics. *Biochim Biophys Acta* **1792:** 1145–1158.

D'Aurelio M, Gajewski CD, Lenaz G, Manfredi G. 2006. Respiratory chain supercomplexes set the threshold for respiration defects in human mtDNA mutant cybrids. *Hum Mol Genet* **15:** 2157–2169.

DeBerardinis RJ, Lum JJ, Hatzivassiliou G, Thompson CB. 2008. The biology of cancer: Metabolic reprogramming fuels cell growth and proliferation. *Cell Metab* **7:** 11–20.

Diaz F, Fukui H, Garcia S, Moraes CT. 2006. Cytochrome *c* oxidase is required for the assembly/stability of respiratory complex I in mouse fibroblasts. *Mol Cell Biol* **26:** 4872–4881.

DiMauro S, Hirano M, Schon EA. 2006. *Mitochondrial medicine.* Informa Healthcare, London.

Finley DJ, Zhu B, Fahey TJ 3rd. 2004. Molecular analysis of Hurthle cell neoplasms by gene profiling. *Surgery* **136:** 1160–1168.

Frezza C, Gottlieb E. 2009. Mitochondria in cancer: Not just innocent bystanders. *Semin Cancer Biol* **19:** 4–11.

Fry M, Green DE. 1980. Cardiolipin requirement by cytochrome oxidase and the catalytic role of phospholipid. *Biochem Biophys Res Commun* **93:** 1238–1246.

Fry M, Green DE. 1981. Cardiolipin requirement for electron transfer in complex I and III of the mitochondrial respiratory chain. *J Biol Chem* **256:** 1874–1880.

Gasparre G, Porcelli AM, Bonora E, Pennisi LF, Toller M, Iommarini L, Ghelli A, Moretti M, Betts CM, Martinelli GN, et al. 2007. Disruptive mitochondrial DNA mutations in complex I subunits are markers of oncocytic phenotype in thyroid tumors. *Proc Natl Acad Sci* **104:** 9001–9006.

Gasparre G, Hervouet E, de Laplanche E, Demont J, Pennisi LF, Colombel M, Mege-Lechevallier F, Scoazec JY, Bonora E, Smeets R, et al. 2008. Clonal expansion of mutated mitochondrial DNA is associated with tumor formation and complex I deficiency in the benign renal oncocytoma. *Hum Mol Genet* **17:** 986–995.

Gasparre G, Iommarini L, Porcelli AM, Lang M, Ferri GG, Kurelac I, Zuntini R, Mariani E, Pennisi LF, Pasquini E, et al. 2009. An inherited mitochondrial DNA disruptive mutation shifts to homoplasmy in oncocytic tumor cells. *Hum Mutat* **30:** 391–396.

Gasparre G, Romeo G, Rugolo M, Porcelli AM. 2011a. Learning from oncocytic tumors: Why choose inefficient mitochondria? *Biochim Biophys Acta* **1807:** 633–642.

Gasparre G, Kurelac I, Capristo M, Iommarini L, Ghelli A, Ceccarelli C, Nicoletti G, Nanni P, De Giovanni C, Scotlandi K, et al. 2011b. A mutation threshold distinguishes the anti-tumorigenic effects of the mitochondrial gene MTND1, an oncojanus function. *Cancer Res* **71:** 6220–6229.

Genova ML, Baracca A, Biondi A, Casalena G, Faccioli M, Falasca AI, Formiggini G, Sgarbi G, Solaini G, Lenaz G. 2008. Is supercomplex organization of the respiratory chain required for optimal electron transfer activity? *Biochim Biophys Acta* **1777:** 740–746.

Guerra F, Kurelac I, Cormio A, Zuntini R, Amato LB, Ceccarelli C, Santini D, Cormio G, Fracasso F, Selvaggi L, et al. 2011. Placing mitochondrial DNA mutations within the progression model of type I endometrial carcinoma. *Hum Mol Genet* **20:** 2394–2405.

Higgins JP, Shinghal R, Gill H, Reese JH, Terris M, Cohen RJ, Fero M, Pollack JR, van de Rijn M, Brooks JD. 2003. Gene expression patterns in renal cell carcinoma assessed by complementary DNA microarray. *Am J Pathol* **162:** 925–932.

Holleran AL, Briscoe DA, Fiskum G, Kelleher JK. 1995. Glutamine metabolism in AS-30D hepatoma cells. Evidence for its conversion into lipids via reductive carboxylation. *Mol Cell Biochem* **152:** 95–101.

Ibsen KH. 1961. The Crabtree effect: A review. *Cancer Res* **21:** 829–841.

Ishikawa K, Takenaga K, Akimoto M, Koshikawa N, Yamaguchi A, Imanishi H, Nakada K, Honma Y, Hayashi J. 2008. ROS-generating mitochondrial DNA mutations can regulate tumor cell metastasis. *Science* **320:** 661–664.

Jacques C, Baris O, Prunier-Mirebeau D, Savagner F, Rodien P, Rohmer V, Franc B, Guyetant S, Malthiery Y, Reynier P. 2005. Two-step differential expression analysis reveals a new set of genes involved in thyroid oncocytic tumors. *J Clin Endocrinol Metab* **90:** 2314–2320.

Jahnke VE, Sabido O, Defour A, Castells J, Lefai E, Roussel D, Freyssenet D. 2010. Evidence for mitochondrial respiratory deficiency in rat rhabdomyosarcoma cells. *PLoS ONE* **5:** e8637.

Kaluz S, Kaluzova M, Stanbridge EJ. 2008. Rational design of minimal hypoxia-inducible enhancers. *Biochem Biophys Res Commun* **370:** 613–618.

Kang SY, Gutowsky HS, Hsung JC, Jacobs R, King TE, Rice D, Oldfield E. 1979. Nuclear magnetic resonance investigation of the cytochrome oxidase–phospholipid interaction: A new model for boundary lipid. *Biochemistry* **18:** 3257–3267.

Kim JW, Tchernyshyov I, Semenza GL, Dang CV. 2006. HIF-1-mediated expression of pyruvate dehydrogenase kinase: A metabolic switch required for cellular adaptation to hypoxia. *Cell Metab* **3:** 177–185.

King A, Selak MA, Gottlieb E. 2006. Succinate dehydrogenase and fumarate hydratase: Linking mitochondrial dysfunction and cancer. *Oncogene* **25:** 4675–4682.

Klaunig JE, Kamendulis LM. 2004. The role of oxidative stress in carcinogenesis. *Annu Rev Pharmacol Toxicol* **44:** 239–267.

Klevecz RR, Bolen J, Forrest G, Murray DB. 2004. A genome-wide oscillation in transcription gates DNA replication and cell cycle. *Proc Natl Acad Sci* **101:** 1200–1205.

Koopman WJ, Verkaart S, Visch HJ, van Emst-de Vries S, Nijtmans LG, Smeitink JA, Willems PH. 2007. Human NADH:ubiquinone oxidoreductase deficiency: Radical changes in mitochondrial morphology? *Am J Physiol Cell Physiol* **293:** C22–C29.

Koppenol WH, Bounds PL, Dang CV. 2011. Otto Warburg's contributions to current concepts of cancer metabolism. *Nat Rev Cancer* **11:** 325–337.

Kuhajda FP, Jenner K, Wood FD, Hennigar RA, Jacobs LB, Dick JD, Pasternack GR. 1994. Fatty acid synthesis: A potential selective target for antineoplastic therapy. *Proc Natl Acad Sci* **91:** 6379–6383.

Kurelac I, Lang M, Zuntini R, Calabrese C, Simone D, Vicario S, Santamaria M, Attimonelli M, Romeo G, Gasparre G. 2011. Searching for a needle in the haystack: Comparing six methods to evaluate heteroplasmy in difficult sequencecontext. *Biotechnol Adv* **30:** 363–371.

Lange C, Nett JH, Trumpower BL, Hunte C. 2001. Specific roles of protein–phospholipid interactions in the yeast cytochrome bc1 complex structure. *EMBO J* **20:** 6591–6600.

Lee AG. 2004. How lipids affect the activities of integral membrane proteins. *Biochim Biophys Acta* **1666:** 62–87.

Lenaz G, Genova ML. 2007. Kinetics of integrated electron transfer in the mitochondrial respiratory chain: Random collisions vs. solid state electron channeling. *Am J Physiol Cell Physiol* **292:** C1221–C1239.

Lenaz G, Genova ML. 2010. Structure and organization of mitochondrial respiratory complexes: A new understanding of an old subject. *Antioxid Redox Signal* **12:** 961–1008.

Lenaz G, Strocchi P. 2009. Reactive oxygen species in the induction of toxicity. In *General and applied toxicology review* (ed. Ballantyne B, et al.), Vol. 2, pp. 367–410. Wiley, Chichester, UK.

Lenaz G, Baracca A, Fato R, Genova ML, Solaini G. 2006. New insights into structure and function of mitochondria and their role in aging and disease. *Antioxid Redox Signal* **8:** 417–437.

Lu H, Dalgard CL, Mohyeldin A, McFate T, Tait AS, Verma A. 2005. Reversible inactivation of HIF-1 prolyl hydroxylases allows cell metabolism to control basal HIF-1. *J Biol Chem* **280:** 41928–41939.

Ma Y, Bai RK, Trieu R, Wong LJ. 2010. Mitochondrial dysfunction in human breast cancer cells and their transmitochondrial cybrids. *Biochim Biophys Acta* **1797:** 29–37.

Mathupala SP, Ko YH, Pedersen PL. 2010. The pivotal roles of mitochondria in cancer: Warburg and beyond and encouraging prospects for effective therapies. *Biochim Biophys Acta* **1797:** 1225–1230.

Maximo V, Sobrinho-Simoes M. 2000. Mitochondrial DNA "common" deletion in Hurthle cell lesions of the thyroid. *J Pathol* **192:** 561–562.

Maximo V, Sores P, Rocha AS, Sobrinho-Simoes M. 1998. The common deletion of mitochondrial DNA is found in goiters and thyroid tumors with and without oxyphil cell change. *Ultrastruct Pathol* **22:** 271–273.

Maximo V, Soares P, Lima J, Cameselle-Teijeiro J, Sobrinho-Simoes M. 2002. Mitochondrial DNA somatic mutations (point mutations and large deletions) and mitochondrial DNA variants in human thyroid pathology: A study with emphasis on Hurthle cell tumors. *Am J Pathol* **160:** 1857–1865.

Maximo V, Botelho T, Capela J, Soares P, Lima J, Taveira A, Amaro T, Barbosa AP, Preto A, Harach HR, et al. 2005. Somatic and germline mutation in *GRIM-19*, a dual function gene involved in mitochondrial metabolism and cell death, is linked to mitochondrion-rich (Hurthle cell) tumours of the thyroid. *Br J Cancer* **23:** 1892–1898.

Maximo V, Lima J, Soares P, Sobrinho-Simoes M. 2009. Mitochondria and cancer. *Virchows Arch* **454:** 481–495.

Mayevsky A. 2009. Mitochondrial function and energy metabolism in cancer cells: Past overview and future perspectives. *Mitochondrion* **9:** 165–179.

Mayr JA, Meierhofer D, Zimmermann F, Feichtinger R, Kogler C, Ratschek M, Schmeller N, Sperl W, Kofler B. 2008. Loss of complex I due to mitochondrial DNA mutations in renal oncocytoma. *Clin Cancer Res* **14:** 2270–2275.

McFate T, Mohyeldin A, Lu H, Thakar J, Henriques J, Halim ND, Wu H, Schell MJ, Tsang TM, Teahan O, et al. 2008. Pyruvate dehydrogenase complex activity controls metabolic and malignant phenotype in cancer cells. *J Biol Chem* **283:** 22700–22708.

McKenzie M, Lazarou M, Thorburn DR, Ryan MT. 2006. Mitochondrial respiratory chain supercomplexes are destabilized in Barth syndrome patients. *J Mol Biol* **361:** 462–469.

Moran M, Rivera H, Sanchez-Arago M, Blazquez A, Merinero B, Ugalde C, Arenas J, Cuezva JM, Martin MA. 2010. Mitochondrial bioenergetics and dynamics interplay in complex I–deficient fibroblasts. *Biochim Biophys Acta* **1802:** 443–453.

Moreno-Sanchez R, Rodriguez-Enriquez S, Marin-Hernandez A, Saavedra E. 2007. Energy metabolism in tumor cells. *FEBS J* **274:** 1393–1418.

Morris LG, Veeriah S, Chan TA. 2010. Genetic determinants at the interface of cancer and neurodegenerative disease. *Oncogene* **29:** 3453–3464.

Muller-Hocker J. 1992. Random cytochrome-c-oxidase deficiency of oxyphil cell nodules in the parathyroid gland. A mitochondrial cytopathy related to cell ageing? *Pathol Res Pract* **188:** 701–706.

Muller-Hocker J, Jacob U, Seibel P. 1998a. Hashimoto thyroiditis is associated with defects of cytochrome-*c* oxidase in oxyphil Askanazy cells and with the common deletion (4,977) of mitochondrial DNA. *Ultrastruct Pathol* **22:** 91–100.

Muller-Hocker J, Schafer S, Copeland WC, Wiesner R, Seibel P. 1998b. Immunohistochemical detection of human mtDNA polymerase γ and of human mitochondrial transcription factor A in cytochrome-*c*-oxidase-deficient oxyphil cells of hyperfunctional parathyroids. *Virchows Arch* **433:** 529–536.

Neuwald AF. 1997. Barth syndrome may be due to an acyltransferase deficiency. *Curr Biol* **7:** R465–R466.

Ozawa T. 1997. Genetic and functional changes in mitochondria associated with aging. *Physiol Rev* **77:** 425–464.

Paradies G, Petrosillo G, Pistolese M, Ruggiero FM. 2000. The effect of reactive oxygen species generated from the mitochondrial electron transport chain on the cytochrome *c* oxidase activity and on the cardiolipin content in bovine heart submitochondrial particles. *FEBS Lett* **466:** 323–326.

Paradies G, Petrosillo G, Pistolese M, Ruggiero FM. 2002. Reactive oxygen species affect mitochondrial electron transport complex I activity through oxidative cardiolipin damage. *Gene* **286:** 135–141.

Paradies G, Petrosillo G, Paradies V, Ruggiero FM. 2010. Oxidative stress, mitochondrial bioenergetics, and cardiolipin in aging. *Free Radic Biol Med* **48:** 1286–1295.

Park JS, Sharma LK, Li H, Xiang R, Holstein D, Wu J, Lechleiter J, Naylor SL, Deng JJ, Lu J, et al. 2009. A heteroplasmic, not homoplasmic, mitochondrial DNA mutation promotes tumorigenesis via alteration in reactive oxygen species generation and apoptosis. *Hum Mol Genet* **18:** 1578–1589.

Parlo RA, Coleman PS. 1986. Continuous pyruvate carbon flux to newly synthesized cholesterol and the suppressed evolution of pyruvate-generated CO_2 in tumors: Further evidence for a persistent truncated Krebs cycle in hepatomas. *Biochim Biophys Acta* **886:** 169–176.

Pasini B, Stratakis CA. 2009. SDH mutations in tumorigenesis and inherited endocrine tumours: Lesson from the phaeochromocytoma-paraganglioma syndromes. *J Intern Med* **266:** 19–42.

Patten DA, Lafleur VN, Robitaille GA, Chan DA, Giaccia AJ, Richard DE. 2010. Hypoxia-inducible factor-1 activation in nonhypoxic conditions: The essential role of mitochondrial-derived reactive oxygen species. *Mol Biol Cell* **21:** 3247–3257.

Petros JA, Baumann AK, Ruiz-Pesini E, Amin MB, Sun CQ, Hall J, Lim S, Issa MM, Flanders WD, Hosseini SH, et al. 2005. mtDNA mutations increase tumorigenicity in prostate cancer. *Proc Natl Acad Sci* **102:** 719–724.

Petrosillo G, Ruggiero FM, Di Venosa N, Paradies G. 2003. Decreased complex III activity in mitochondria isolated from rat heart subjected to ischemia and reperfusion: Role of reactive oxygen species and cardiolipin. *FASEB J* **17:** 714–716.

Petrosillo G, Matera M, Moro N, Ruggiero FM, Paradies G. 2009. Mitochondrial complex I dysfunction in rat heart with aging: Critical role of reactive oxygen species and cardiolipin. *Free Radic Biol Med* **46:** 88–94.

Pfeiffer K, Gohil V, Stuart RA, Hunte C, Brandt U, Greenberg ML, Schagger H. 2003. Cardiolipin stabilizes respiratory chain supercomplexes. *J Biol Chem* **278:** 52873–52880.

Pollard PJ, Briere JJ, Alam NA, Barwell J, Barclay E, Wortham NC, Hunt T, Olpin S, Moat SJ, Hargreaves IP, et al. 2005. Accumulation of Krebs cycle intermediates and over-expression of HIF1α in tumours which result from germline FH and SDH mutations. *Hum Mol Genet* **14:** 2231–2239.

Porcelli AM, Angelin A, Ghelli A, Mariani E, Martinuzzi A, Carelli V, Petronilli V, Bernardi P, Rugolo M. 2009. Respiratory complex I dysfunction due to mitochondrial DNA mutations shifts the voltage threshold for opening of the permeability transition pore toward resting levels. *J Biol Chem* **284:** 2045–2052.

Porcelli A, Ghelli A, Ceccarelli C, Lang M, Cenacchi G, Capristo M, Pennisi LF, Ciccarelli E, Melcarne A, Bartoletti-Stella A, et al. 2010. The genetic and metabolic signature of oncocytic transformation implicates HIF1α destabilization. *Hum Mol Genet* **19:** 1019–1032.

Pradella LM, Zuntini R, Magini P, Ceccarelli C, Neri I, Cerasoli S, Graziano C, Gasparre G, Turchetti D. 2011. Two distinct thyroid tumours in a patient with Cowden syndrome carrying both a 10q23 and a mitochondrial DNA germline deletion. *J Med Genet* **48:** 779–782.

Putignani L, Raffa S, Pescosolido R, Aimati L, Signore F, Torrisi MR, Grammatico P. 2008. Alteration of expression levels of the oxidative phosphorylation system (OXPHOS) in breast cancer cell mitochondria. *Breast Cancer Res Treat* **110:** 439–452.

Radermacher M, Ruiz T, Clason T, Benjamin S, Brandt U, Zickermann V. 2006. The three-dimensional structure of complex I from Yarrowia lipolytica: A highly dynamic enzyme. *J Struct Biol* **154:** 269–279.

Riva A, Tandler B. 2000. Three-dimensional structure of oncocyte mitochondria in human salivary glands: A scanning electron microscope study. *Ultrastruct Pathol* **24:** 145–150.

Robinson NC, Strey F, Talbert L. 1980. Investigation of the essential boundary layer phospholipids of cytochrome *c* oxidase using Triton X-100 delipidation. *Biochemistry* **19:** 3656–3661.

Rosca MG, Vazquez EJ, Kerner J, Parland W, Chandler MP, Stanley W, Sabbah HN, Hoppel CL. 2008. Cardiac mitochondria in heart failure: Decrease in respirasomes and oxidative phosphorylation. *Cardiovasc Res* **80:** 30–39.

Savagner F, Chevrollier A, Loiseau D, Morgan C, Reynier P, Clark O, Stepien G, Malthiery Y. 2001a. Mitochondrial activity in XTC.UC1 cells derived from thyroid oncocytoma. *Thyroid* **11:** 327–333.

Savagner F, Franc B, Guyetant S, Rodien P, Reynier P, Malthiery Y. 2001b. Defective mitochondrial ATP synthesis in oxyphilic thyroid tumors. *J Clin Endocrinol Metab* **86:** 4920–4925.

Savagner F, Mirebeau D, Jacques C, Guyetant S, Morgan C, Franc B, Reynier P, Malthiery Y. 2003. PGC-1-related coactivator and targets are upregulated in thyroid oncocytoma. *Biochem Biophys Res Commun* **310:** 779–784.

Schafer E, Dencher NA, Vonck J, Parcej DN. 2007. Three-dimensional structure of the respiratory chain supercomplex I1III2IV1 from bovine heart mitochondria. *Biochemistry* **46:** 12579–12585.

Schagger H, de Coo R, Bauer MF, Hofmann S, Godinot C, Brandt U. 2004. Significance of respirasomes for the assembly/stability of human respiratory chain complex I. *J Biol Chem* **279:** 36349–36353.

Sedlak E, Robinson NC. 1999. Phospholipase A(2) digestion of cardiolipin bound to bovine cytochrome *c* oxidase alters both activity and quaternary structure. *Biochemistry* **38:** 14966–14972.

Semenza GL. 2003. Targeting HIF-1 for cancer therapy. *Nat Rev Cancer* **3:** 721–732.

Semenza GL. 2007. Oxygen-dependent regulation of mitochondrial respiration by hypoxia-inducible factor 1. *Biochem J* **405:** 1–9.

Shidara Y, Yamagata K, Kanamori T, Nakano K, Kwong JQ, Manfredi G, Oda H, Ohta S. 2005. Positive contribution of pathogenic mutations in the mitochondrial genome to the promotion of cancer by prevention from apoptosis. *Cancer Res* **65:** 1655–1663.

Simonnet H, Demont J, Pfeiffer K, Guenaneche L, Bouvier R, Brandt U, Schagger H, Godinot C. 2003. Mitochondrial complex I is deficient in renal oncocytomas. *Carcinogenesis* **24:** 1461–1466.

Smolkova K, Plecita-Hlavata L, Bellance N, Benard G, Rossignol R, Jezek P. 2010. Waves of gene regulation suppress and then restore oxidative phosphorylation in cancer cells. *Int J Biochem Cell Biol* **43:** 950–968.

Sun W, Zhou S, Chang SS, McFate T, Verma A, Califano JA. 2009. Mitochondrial mutations contribute to HIF1α accumulation via increased reactive oxygen species and upregulated pyruvate dehydrogenease kinase 2 in head and neck squamous cell carcinoma. *Clin Cancer Res* **15:** 476–484.

Tallini G. 1998. Oncocytic tumours. *Virchows Arch* **433:** 5–12.

Tallini G, Ladanyi M, Rosai J, Jhanwar SC. 1994. Analysis of nuclear and mitochondrial DNA alterations in thyroid and renal oncocytic tumors. *Cytogenet Cell Genet* **66:** 253–259.

Thompson CB. 2009. Metabolic enzymes as oncogenes or tumor suppressors. *N Engl J Med* **360:** 813–815.

Tuppen HA, Blakely EL, Worgan L, Al-Dosary M, Saretzki G, Alston CL, Morris AA, Clarke M, Jones S, Devlin AM, et al. 2010. The p.M292T NDUFS2 mutation causes complex I–deficient Leigh syndrome in multiple families. *Brain* **133:** 2952–2963.

van Raam BJ, Sluiter W, de Wit E, Roos D, Verhoeven AJ, Kuijpers TW. 2008. Mitochondrial membrane potential in human neutrophils is maintained by complex III activity in the absence of supercomplex organisation. *PLoS ONE* **3:** e2013.

Vik SB, Capaldi RA. 1977. Lipid requirements for cytochrome *c* oxidase activity. *Biochemistry* **16:** 5755–5759.

Wallace DC. 2005. A mitochondrial paradigm of metabolic and degenerative diseases, aging, and cancer: A dawn for evolutionary medicine. *Annu Rev Genet* **39:** 359–407.

Warburg O. 1956. On the origin of cancer cells. *Science* **123:** 309–314.

Weinberg F, Chandel NS. 2009. Mitochondrial metabolism and cancer. *Ann NY Acad Sci* **1177:** 66–73.

Yoo H, Antoniewicz MR, Stephanopoulos G, Kelleher JK. 2008. Quantifying reductive carboxylation flux of glutamine to lipid in a brown adipocyte cell line. *J Biol Chem* **283:** 20621–20627.

Zhang M, Mileykovskaya E, Dowhan W. 2002. Gluing the respiratory chain together. Cardiolipin is required for supercomplex formation in the inner mitochondrial membrane. *J Biol Chem* **277:** 43553–43556.

Zielke A, Tezelman S, Jossart GH, Wong M, Siperstein AE, Duh QY, Clark OH. 1998. Establishment of a highly differentiated thyroid cancer cell line of Hurthle cell origin. *Thyroid* **8:** 475–483.

Zimmermann FA, Mayr JA, Neureiter D, Feichtinger R, Alinger B, Jones ND, Eder W, Sperl W, Kofler B. 2009. Lack of complex I is associated with oncocytic thyroid tumours. *Br J Cancer* **100:** 1434–1437.

Zimmermann FA, Mayr JA, Feichtinger R, Neureiter D, Lechner R, Koegler C, Ratschek M, Rusmir H, Sargsyan K, Sperl W, et al. 2011. Respiratory chain complex I is a mitochondrial tumor suppressor of oncocytic tumors. *Front Biosci* **1:** 315–325.

Zu XL, Guppy M. 2004. Cancer metabolism: Facts, fantasy, and fiction. *Biochem Biophys Res Commun* **313:** 459–465.

Mitochondrial Metabolism, Sirtuins, and Aging

Michael N. Sack and Toren Finkel

Center for Molecular Medicine, National Heart, Lung and Blood Institute, National Institutes of Health, Bethesda, Maryland 20892

Correspondence: finkelt@nih.gov

The sirtuins are a family of proteins that act predominantly as nicotinamide adenine dinucleotide (NAD)-dependent deacetylases. In mammals seven sirtuin family members exist, including three members, Sirt3, Sirt4, and Sirt5, that localize exclusively within the mitochondria. Although originally linked to life-span regulation in simple organisms, this family of proteins appears to have various and diverse functions in higher organisms. One particular property that is reviewed here is the regulation of mitochondrial number, turnover, and activity by various mitochondrial and nonmitochondrial sirtuins. An emerging consensus from these recent studies is that sirtuins may act as metabolic sensors, using intracellular metabolites such as NAD and short-chain carbon fragments such as acetyl coenzyme A to modulate mitochondrial function to match nutrient supply.

THE SIRTUIN FAMILY OF PROTEINS

Originally characterized in yeast as regulators of life span (Kaeberlein et al. 1999), the sirtuins are an evolutionarily conserved family of proteins that appear to exert a wide range of biological functions (Finkel et al. 2009). In mammals seven sirtuin family members exist. Sirt1 appears to be the closest mammalian homolog to yeast Sir2, the first member of the sirtuins linked to aging. Because of this homology, initial studies of mammalian sirtuins focused predominantly on the biology of Sirt1. Insights from these studies have implicated Sirt1 in the regulation of a variety of metabolic phenotypes including insulin secretion (Moynihan et al. 2005; Bordone et al. 2006), lipid mobilization from adipocytes (Picard et al. 2004), and regulation of glucose tolerance (Rodgers et al. 2005). With that said, the role of sirtuins in mammalian aging remains an open question, and even some of the earlier work in lower organisms involving Sir2 and life span has recently been questioned (Burnett et al. 2011).

Although there has been considerable attention directed toward Sirt1 biology, there is also a growing interest in understanding the function of the related family members. It is clear that each of the mammalian sirtuins has a distinct subcellular localization. Sirt1, Sirt6, and Sirt7 are nuclear proteins, although a fraction of Sirt1 can be found in the cytosol. Sirt2, on the other hand, is predominantly cytosolic, although again it can be found in the nucleus in certain situations (North and Verdin 2007). Finally, three sirtuins—Sirt3, Sirt4, and Sirt5—appear to be found exclusively in the mitochondria. From a human genetics point of view, the

Cite this article as *Cold Spring Harb Perspect Biol* doi: 10.1101/cshperspect.a013102

strongest association between aging and sirtuins is the association between polymorphisms in the mitochondrial Sirt3 and longevity (Rose et al. 2003).

The ability of sirtuins to influence metabolism and potentially life span is believed to revolve around the ability of sirtuin family members to function as protein deacetylases. In addition to this enzymatic function, Sirt4 can further act to ADP ribosylate target proteins. Unlike other protein deacetylases, sirtuins require nicotinamide adenine dinucleotide (NAD) as a cofactor in the deacetylation reaction (Imai et al. 2000). The link among NAD, NADH, and sirtuin activity has led many to believe that this family of proteins acts in some fashion as a sensor of energetic status. This may particularly be true in the mitochondria, where levels of NAD and NADH are high and where a disproportionate fraction of proteins appear to be acetylated (Kim et al. 2006).

In this review we will examine the link between sirtuin function and mitochondrial metabolism. While we highlight the role of those sirtuin family members that are uniquely mitochondrial in their localization, we will also discuss other sirtuins that can influence mitochondrial function, biogenesis, and turnover. Finally, although we discuss each sirtuin individually, there is a growing realization that one sirtuin family member can affect the function of other family members. A recent example is Sirt1-dependent regulation of Sirt6 expression in the control of hepatic metabolism (Kim et al. 2010b).

Sirt1 AND MITOCHONDRIAL FUNCTION

Evidence suggests that mitochondrial biogenesis is regulated at least in part by proliferator-activated receptor coactivator-1α (PGC-1α), a transcriptional coactivator of peroxisome proliferator-activated receptor-γ (PPARγ) as well as other transcription factors (Fernandez-Marcos and Auwerx 2011). It was therefore of considerable interest when it was shown that PGC-1α was in fact a deacetylation target of Sirt1 and that acetylation regulated PGC-1α activity (Nemoto et al. 2005; Rodgers et al. 2005). There are at least 13 lysine residues on PGC-1α that appear to be reversibly acetylated (Rodgers et al. 2005). Site-directed mutants that lack all 13 of these sites alter the ability of PGC-1α to regulate gene expression, although it remains unclear if all, or only a subset, of PGC-1α's acetylation sites are truly regulatory in nature. Sirt1 appears to be the predominant in vitro and in vivo regulator of PGC-1α deacetylation. For instance, in vitro knockdown of Sirt1 in hepatic cells leads to increased PGC-1α acetylation with a corresponding reduction in a set of genes that are the rate-limiting enzymes responsible for hepatic gluconeogenesis (Rodgers et al. 2005). Similarly, both overexpression and knockdown studies support a role for Sirt1 in regulating PGC-1α activity through reversible deacetylation, which in turn has dramatic effects on in vivo hepatic glucose and lipid metabolism (Rodgers and Puigserver 2007; Erion et al. 2009). A similar relationship appears to exist in skeletal muscle. In particular, in skeletal muscle, fasting was shown to lead to a Sirt1-dependent deacetylation of PGC-1α, and this deacetylation appeared to be required for PGC-1α-dependent gene expression, including gene products required for effective mitochondrial biogenesis (Gerhart-Hines et al. 2007). Together these studies link Sirt1 and PGC-1α activities in metabolically active tissues such as the liver and skeletal muscle (Fig. 1).

The above studies suggest that mitochondrial biogenesis might be regulated by tissue energetic status and that the sirtuins would represent important energy sensors in this homeostatic loop. Indeed, the notion that PGC-1α acetylation and function, and by extension mitochondrial activity, are regulated in a nutrient-dependent fashion by Sirt1 is appealing. Nonetheless, the concept that Sirt1 is in turn responding to nutrient-sensitive changes in basal NAD levels, although often invoked, has until recently had little experimental support. The difficulty in proving the supposition is that measurement and manipulation of NAD levels in various subcellular compartments is experimentally challenging. One recent approach is to carefully examine mice with a deletion of a major NAD-consuming enzyme, poly(ADP-ribose) polymerase-1 (PARP-1). These mice appear to have

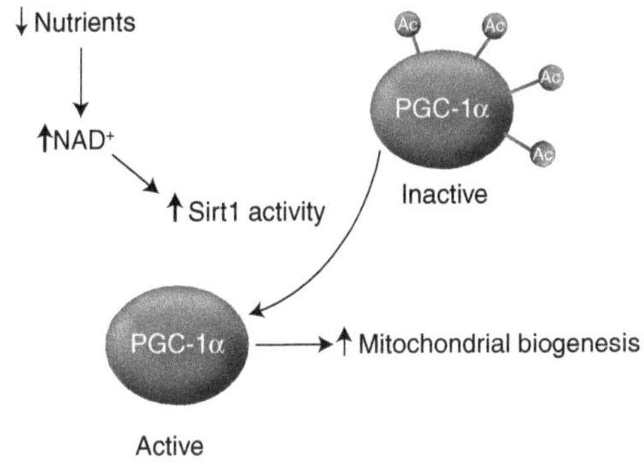

Figure 1. Regulation of PGC-1α acetylation and activity by Sirt1. The transcriptional coactivator and regulator of mitochondrial biogenesis PGC-1α is, at least in part, regulated by lysine acetylation. In the setting of low nutrient availability, the intracellular and particularly the nuclear levels of NAD are believed to increase and lead to activation of Sirt1 enzymatic activity. This leads to PGC-1α deacetylation, resulting in increased PGC-1α activity and hence, ultimately, in an increase in mitochondrial number.

elevated NAD levels along with increased Sirt1 activity (Bai et al. 2011). Furthermore, consistent with increased NAD levels leading to increased PGC-1α activity, PARP-1$^{-/-}$ mice have increased mitochondrial content (Bai et al. 2011). Another recent report suggested that adiponectin, a secreted adipokine, could regulate intracellular NAD levels (Iwabu et al. 2010). Again, in these studies the addition of adiponectin to cells appeared to increase mitochondrial content through a Sirt1- and PGC-1α-dependent pathway. These results might be particularly important because metabolic diseases are often associated with low adiponectin levels (Hotta et al. 2000), as well as with mitochondrial dysfunction (Petersen et al. 2004), although the link between these two observations was previously unknown. Interestingly, these observations would also seem to support the growing link between genetic variants of Sirt1 and a person's risk for developing obesity (Peeters et al. 2008; Zillikens et al. 2009).

The generation of new mitochondria through a Sirt1/PGC-1α-regulated pathway is complemented by another important connection between Sirt1 and the mitochondria. In particular, it would appear that Sirt1 is an important regulator of removing damaged mitochondria through the process of autophagy (Lee et al. 2008). The field of autophagy represents a rapidly expanding area of study, and a full review of the subject is not possible. Suffice it to say that autophagy is an evolutionarily conserved process present in organisms ranging from yeast to mammals. One of the major intracellular roles of autophagy is the removal of damaged organelles such as mitochondria. Evidence suggests that Sirt1 can stimulate autophagy and that Sirt1$^{-/-}$ tissues appear to accumulate abnormal-appearing mitochondria, consistent with what is seen in autophagy-deficient tissues (Lee et al. 2008). A cytoplasm-restricted mutant of Sirt1 can still stimulate autophagy, suggesting that this activity represents an extranuclear function of the protein (Morselli et al. 2011). The molecular basis for how Sirt1 stimulates autophagy is not clear. There is evidence that key molecules required for autophagy including Atg5 and Atg7 are direct targets of sirtuin-dependent deacetylation (Lee et al. 2008). In addition, the FoxO family of transcription factors, known targets of Sirt1 deacetylation as well as regulators of autophagy, has also been implicated (Hariharan et al. 2010; Kume et al. 2010). Nonetheless, although details remain to be elucidated, the notion that Sirt1 can regulate both the creation of new mitochondria as well as the removal of old mitochondria suggests a role for sirtuins in overall mitochondrial flux and in the maintenance of what may be viewed as "youthful" mitochondria in the cell.

A final area that we wish to discuss in which the biology of Sirt1 and the mitochondria intersect is the secretion of insulin by the pancreatic β cell. Previous studies have suggested that mitochondrial uncoupling protein-2 (UCP2) was an important negative regulator of insulin secretion by the β cell (Zhang et al. 2001). Indeed, UCP2 appears to be up-regulated in obese animals and UCP2-deficient mice have increased glucose-stimulated insulin release (Zhang et al. 2001). Subsequent studies identified Sirt1 as an important repressor of UCP2 expression in the pancreas (Moynihan et al. 2005; Bordone et al. 2006). It is believed that starvation may induce

UCP2 expression in the β cell via a reduction in tissue NAD levels (Bordone et al. 2006). There is also evidence that a recently described extracellular, circulating form of the enzyme nicotinamide phosphoribosyltransferase (Nampt) may be an important in vivo mediator of insulin secretion (Revollo et al. 2007). The Nampt enzyme is a key enzyme in NAD biosynthesis and is responsible for the conversion of nicotinamide to nicotinamide mononucleotide, a metabolite that can in turn be converted directly to NAD. Again, these observations highlight the potential function of Sirt1 in linking metabolites such as NAD to the maintenance of overall metabolic homeostasis.

Sirt3 AND MITOCHONDRIAL FUNCTION

The studies we have reviewed indicating that the predominantly nuclear Sirt1 was an important regulator of mitochondrial function spurred interest in the biology of those sirtuin family members that directly localize to this organelle. Comparative analysis of total liver mitochondrial protein acetylation following distinct genetic knockout of each of the three mitochondrial-enriched sirtuins—Sirt3, Sirt4, and Sirt5—showed that Sirt3 is the major mitochondrial deacetylase (Lombard et al. 2007). At the same time, the dynamic flux in mitochondrial protein acetylation in response to changes in caloric load as illustrated by feeding and fasting (Kim et al. 2006), caloric restriction (Schwer et al. 2009), and caloric excess (Hirschey et al. 2011; Kendrick et al. 2011) suggest that, as is the case with Sirt1, Sirt3 may possess a nutrient-sensing regulatory role governing mitochondrial protein function.

Over the last few years, many laboratories have identified and functionally characterized mitochondrial targets of Sirt3 deacetylation and shown that Sirt3 does indeed modulate numerous mitochondrial pathways via protein lysine-residue deacetylation. As mitochondria are most readily identified as the "powerhouse" of the cell, we initially review the role of Sirt3 in mitochondrial bioenergetics. Sirt3 deacetylates and activates multiple steps in substrate catabolism, starting with the oxidation of fatty acids. As an example, Sirt3 has been shown under nutrient-restricted conditions to deacetylate and hence activate long-chain acyl coenzyme A (acyl-CoA) dehydrogenase to increase fatty acid β-oxidation (Hirschey et al. 2010; Hallows et al. 2011). Additional substrate catabolic pathway targets include glutamate dehydrogenase, which facilitates the oxidative deamination of glutamate to α-ketoglutarate, and the citric acid cycle enzyme isocitrate dehydrogenase 2 (IDH2) (Lombard et al. 2007; Schlicker et al. 2008). In keeping with its role as the final common denominator in mitochondrial energy production, numerous proteins in the electron transport chain have been shown to be directly deacetylated by Sirt3. Sirt3 activation has been shown to increase oxidative phosphorylation (Ahn et al. 2008; Bao et al. 2010a; Cimen et al. 2010) and to deacetylate and activate enzymes in complexes I and II of the electron transport chain (Ahn et al. 2008; Cimen et al. 2010; Finley et al. 2011; Kendrick et al. 2011), with additional deacetylation of proteins in complex V (Schlicker et al. 2008; Bao et al. 2010b).

An additional aspect of mitochondrial bioenergetics operational under caloric restriction or fasting conditions is the conversion of acetate to acetyl-CoA, a necessary step required for energy production in extrahepatic tissues and for the generation of ketones in the liver. Again, Sirt3 appears to have an important regulatory role in these pathways, as both acetyl-CoA synthetase 2 (Hallows et al. 2006; Schwer et al. 2006) and 3-hydroxy-3-methylglutaryl-CoA synthase 2 (Shimazu et al. 2010) are targets of Sirt3 deacetylation. A final metabolic pathway that is modulated by Sirt3 is the urea cycle, responsible for the detoxification of ammonia during amino acid metabolism. Here, too, Sirt3 deacetylates and activates a key enzyme, ornithine transcarbamoylase (Hallows et al. 2011).

Sirt3 is also emerging as a regulatory protein in the modulation of additional mitochondrial programs, including the deacetylation and inhibition of the mitochondrial ribosomal protein L10 (MRPL10) (Yang et al. 2009). This results in an NAD-dependent inhibition of mitochondrial protein synthesis, which might function as an energy-sparing response under

 Cite this article as *Cold Spring Harb Perspect Biol* doi: 10.1101/cshperspect.a013102

nutrient-restricted conditions. Another mitochondrial protein that is inhibited by Sirt3 deacetylation is the mitochondrial matrix peptidyl–prolyl isomerase cyclophilin D (Ppif) (Hafner et al. 2010; Shulga and Pastorino 2010; Shulga et al. 2010). Our understanding of the function of cyclophilin D has expanded in recent years to include a role not only in increasing susceptibility to mitochondrial permeability transition (Baines et al. 2005; Nakagawa et al. 2005), but also in regulating mitochondrial calcium efflux with the concordant regulation of Ca^{2+}-dependent mitochondrial enzyme activities (Elrod et al. 2010). An additional, albeit indirect, effect of Sirt3-dependent inactivation of cyclophilin D is the dissociation of hexokinase II from the mitochondria, which plays a role in the promotion of oxidative phosphorylation by Sirt3 (Shulga et al. 2010).

The mitochondrial sirtuins also appear to play an important role in the control of reactive oxygen species. This regulatory role may be particularly relevant to modulating the development of age-associated degenerative conditions. At the direct substrate level, the reactive oxygen species scavenging enzyme MnSOD is activated by Sirt3, and numerous lysine residues have been implicated in mediating this induction of enzyme activity (Qiu et al. 2010; Tao et al. 2010; Chen et al. 2011). Via a more indirect mechanism, the caloric restriction-associated activation of IDH2 has been shown to increase NADH, which is proposed to facilitate the increase in reduced glutathione levels found in association with Sirt3-mediated activation of IDH2 (Someya et al. 2010). The emerging role of Sirt3 in modulating various pathways in metabolism and stress modulatory programs is shown in Figure 2.

Finally, a recently defined target of Sirt3, namely, mitochondrial aldehyde dehydrogenase 2 (ALDH2), has uncovered an additional biological role for the acetylation of lysine residues (Lu et al. 2011). In contrast to the prior targets of Sirt3 discussed above, the change in acetylation status of ALDH2 did not alter the activity of this dehydrogenase. However, because the reactive metabolite of the analgesic agent acetaminophen disrupts proteins by binding to lysine residues, the role of Sirt3 in acetaminophen metabolite binding to ALDH2 was explored (Lu et al. 2011). This analysis confirmed that lysine-residue acetylation can indeed function as an allosteric inhibitor of xenobiotic-reactive metabolite binding to mitochondrial proteins by exposing lysine residues to other posttranslational modifiers, and that in this instance Sirt3 activation has detrimental consequences (Silberman and Mostoslavsky 2011). These results might help explain the clinical observation that acetaminophen liver injury is exacerbated by fasting (Lu et al. 2011)

The investigations of the targets of Sirt3 collectively show that in response to caloric

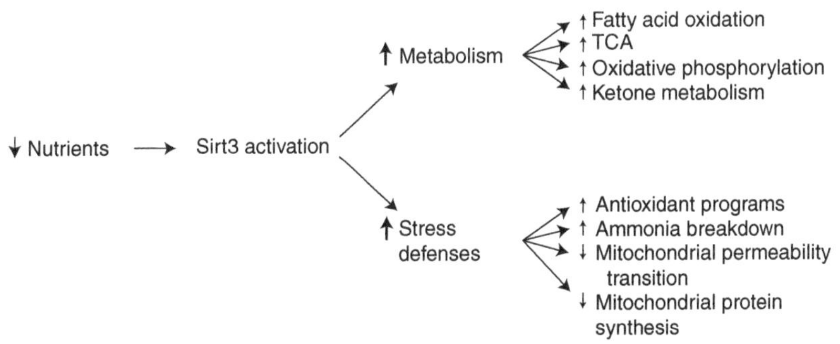

Figure 2. A role for Sirt3 in metabolic adaptation and stress defense in the setting of low nutrients. The sirtuin family member Sirt3 is the predominant mitochondrial deacetylase. Multiple targets of Sirt3 have been described. These targets involve enzymes linked to substrate utilization as well as core components of the electron transport chain. In addition, Sirt3 has been linked to a wide array of other stress-related programs including the maintenance of redox homeostasis. See text for additional details.

restriction or fasting, Sirt3 deacetylates a vast array of mitochondrial proteins, with a resulting panoply of effects including activation, inhibition, and allosteric modification of protein functioning. The physiological impact of these effects is beginning to be explored in a range of age-related or nutrient-sensitive diseases. On balance, recent observations suggest that Sirt3 activation is generally ameliorative with respect to the development of age-associated diseases such as cardiac dysfunction (Sundaresan et al. 2009; Hafner et al. 2010), hearing loss (Someya et al. 2010), metabolic syndromes and diabetes (Hirschey et al. 2011; Jing et al. 2011; Kendrick et al. 2011), and cancer (Kim et al. 2010a).

Sirt5 AND MITOCHONDRIAL FUNCTION

The identification of targets of Sirt5 in the mitochondria has been more limited, although the urea cycle enzyme carbamoyl phosphate synthetase 1 (CPS-1) has been identified as a Sirt5 deacetylation target (Nakagawa et al. 2009). The activation of CPS-1 catalyzes ammonia to urea and would be expected to have ameliorative effects via the elimination of oxidative stress-promoting ammonium.

Recent work shows that in addition to acetyl groups, additional short-chain carbon fragments covalently attached to specific amino acids may be targets of Sirt5. These include the removal of succinyl and malonyl groups from lysine residues (Du et al. 2011; Peng et al. 2011). Succinyl-CoA and malonyl-CoA are metabolic intermediates of the tricarboxylic acid (TCA) cycle (Peng et al. 2011; Zhang et al. 2011). To date, very few specific protein targets of Sirt5-mediated desuccinylation or demalonylation have been identified. However, the levels of hepatic protein succinylation and malonylation are highly enriched in Sirt5 knockout mice (Du et al. 2011; Peng et al. 2011). Although these novel posttranslational modifications attributed to Sirt5 need to be fleshed out, they are compatible with the concept that the sirtuin proteins target lysine-residue modifications to alter metabolic functioning in response to changes in levels of various intracellular nutrients.

CONCLUSIONS

Although originally described in yeast, the mammalian sirtuins represent an intriguing family of proteins that appear to function as sensors and regulators of metabolic status. Included among this family's diverse function is the coordinated control of mitochondrial activity. This regulation includes the creation and targeted destruction of mitochondria by Sirt1, as well as the regulation of substrate utilization and oxidative phosphorylation by Sirt3. Table 1 shows the metabolic substrates of Sirt1, -3, and -5 discussed in this review and summarizes the metabolic effects of these deacetylation reactions.

Table 1. List of described sirtuin substrates, their subcellular localization, and the effect of deacetylation on protein function

Sirtuin	Sirtuin substrate	Localization	Result of sirtuin action	References
Sirt1	PGC-1α	Nucleus	Activates transcriptional coactivation	Nemoto et al. 2005; Rodgers et al. 2005
	FoxO	Nucleus	Activates transcription and autophagy	Hariharan et al. 2010; Kume et al. 2010
	Atg5/7	Cytosol	Deacetylation to promote autophagy	Lee et al. 2008
	UCP2	Nucleus	Repression of transcript	Zhang et al. 2001
Sirt3	LCAD	Mitochondria	Activates enzyme, promotes lipid processing	Hirschey et al. 2010; Hallows et al. 2011
	GDH	Mitochondria	Activates enzyme, oxidative deamination of glutamate	Lombard et al. 2007

Continued

Table 1. *Continued*

Sirtuin	Sirtuin substrate	Localization	Result of sirtuin action	References
	NDUFA9	Mitochondria	Activates enzyme, up-regulates ETC activity	Ahn et al. 2008
	SDHB	Mitochondria	Activates enzyme, up-regulates ETC activity	Cimen et al. 2010
	ATP5A	Mitochondria	Activates enzyme, up-regulates ETC activity	Schlicker et al. 2008; Bao et al. 2010b
	OTC	Mitochondria	Activates enzyme, reduces ammonia toxicity and ROS	Hallows et al. 2011
	ACS2, HMGCoAS2	Mitochondria	Activate ketone metabolism	Hallows et al. 2006; Schwer et al. 2006; Shimazu et al. 2010
	SOD2	Mitochondria	Activates enzyme, reduces ROS	Qui et al. 2010; Tao et al. 2010; Chen et al. 2011
	IDH2	Mitochondria	Activates enzyme, reduces ROS	Schlicker et al. 2008; Someya et al. 2010
	CypD (*PPIF*)	Mitochondria	Deactivates enzyme, prevents interaction with mPTP	Hafner et al. 2010; Shulga and Pastorino 2010; Shulga et al. 2010
	ALDH2	Mitochondria	Deacetylation by Sirt3 allows NAPQI binding to ALDH2, reducing its activity	Lu et al. 2011
	MRPL10	Mitochondria	Inhibition of mitochondrial ribosome	Yang et al. 2009
Sirt5	CPS-1	Mitochondria	Activates enzyme, reduces ammonia toxicity and ROS	Nakagawa et al. 2009

PGC-1α, proliferator-activated receptor coactivator-1α; UCP2, uncoupling protein-2; LCAD, long-chain acyl-CoA dehydrogenase; GDH, glutamate dehydrogenase; NDUFA9, NADH dehydrogenase subcomplex A9; SHDB, succinate dehydrogenase B subunit; ATP5A, ATP synthase subunit 5A; OTC, ornithine transcarbamoylase; ACS2, acyl-CoA synthetase 2; HMGCoAS2, 3-hydroxy-3-methylglutaryl-CoA synthase 2; SOD2, superoxide dismutase 2; IDH2, isocitrate dehydrogenase 2; CypD, cyclophilin D; ALDH2, aldehyde dehydrogenase 2; MRPL10, mitochondrial ribosomal protein L10; CPS-1, carbomoyl phosphate synthetase 1; ETC, electron transport chain; ROS, reactive oxygen species; mPTP, mitochondrial permeability transition pore; NAPQ1, *N*-acetyl-*p*-benzoquinoneimine.

In addition to these biochemical properties, there are tantalizing clues that sirtuins may play an important physiological role in overall metabolic homeostasis and perhaps in modulating age-related metabolic pathologies. Finally, although the predominant function of the sirtuin family revolves around NAD-dependent lysine deacetylation, other less-characterized enzymatic activities including ADP ribosylation, as well as other lysine modifications (e.g., demalonylation and desuccinylation), are just beginning to be explored. Although considerable gaps exist in our understanding, further dissection of sirtuin biology promises to provide important insight into how metabolic supply is coupled to mitochondrial activity.

REFERENCES

Ahn BH, Kim HS, Song S, Lee IH, Liu J, Vassilopoulos A, Deng CX, Finkel T. 2008. A role for the mitochondrial deacetylase Sirt3 in regulating energy homeostasis. *Proc Natl Acad Sci* **105:** 14447–14452.

Bai P, Canto C, Oudart H, Brunyanszki A, Cen Y, Thomas C, Yamamoto H, Huber A, Kiss B, Houtkooper RH, et al. 2011. PARP-1 inhibition increases mitochondrial metabolism through SIRT1 activation. *Cell Metab* **13:** 461–468.

Baines CP, Kaiser RA, Purcell NH, Blair NS, Osinska H, Hambleton MA, Brunskill EW, Sayen MR, Gottlieb RA, Dorn GW, et al. 2005. Loss of cyclophilin D reveals a critical role for mitochondrial permeability transition in cell death. *Nature* **434:** 658–662.

Bao J, Lu Z, Joseph JJ, Carabenciov D, Dimond CC, Pang L, Samsel L, McCoy JP Jr, Leclerc J, Nguyen P, et al. 2010a. Characterization of the murine SIRT3 mitochondrial localization sequence and comparison of mitochondrial enrichment and deacetylase activity of long and short SIRT3 isoforms. *J Cell Biochem* **110:** 238–247.

Bao J, Scott I, Lu Z, Pang L, Dimond CC, Gius D, Sack MN. 2010b. SIRT3 is regulated by nutrient excess and modulates hepatic susceptibility to lipotoxicity. *Free Radic Biol Med* **49:** 1230–1237.

Bordone L, Motta MC, Picard F, Robinson A, Jhala US, Apfeld J, McDonagh T, Lemieux M, McBurney M, Szilvasi A, et al. 2006. Sirt1 regulates insulin secretion by repressing UCP2 in pancreatic β cells. *PLoS Biol* **4:** e31.

Burnett C, Valentini S, Cabreiro F, Goss M, Somogyvari M, Piper MD, Hoddinott M, Sutphin GL, Leko V, McElwee JJ, et al. 2011. Absence of effects of Sir2 overexpression on lifespan in *C. elegans* and *Drosophila*. *Nature* **477:** 482–485.

Chen Y, Zhang J, Lin Y, Lei Q, Guan KL, Zhao S, Xiong Y. 2011. Tumour suppressor SIRT3 deacetylates and activates manganese superoxide dismutase to scavenge ROS. *EMBO Rep* **12:** 534–541.

Cimen H, Han MJ, Yang Y, Tong Q, Koc H, Koc EC. 2010. Regulation of succinate dehydrogenase activity by SIRT3 in mammalian mitochondria. *Biochemistry* **49:** 304–311.

Du J, Zhou Y, Su X, Yu JJ, Khan S, Jiang H, Kim J, Woo J, Kim JH, Choi BH, et al. 2011. Sirt5 is a NAD-dependent protein lysine demalonylase and desuccinylase. *Science* **334:** 806–809.

Elrod JW, Wong R, Mishra S, Vagnozzi RJ, Sakthievel B, Goonasekera SA, Karch J, Gabel S, Farber J, Force T, et al. 2010. Cyclophilin D controls mitochondrial pore-dependent Ca^{2+} exchange, metabolic flexibility, and propensity for heart failure in mice. *J Clin Invest* **120:** 3680–3687.

Erion DM, Yonemitsu S, Nie Y, Nagai Y, Gillum MP, Hsiao JJ, Iwasaki T, Stark R, Weismann D, Yu XX, et al. 2009. SirT1 knockdown in liver decreases basal hepatic glucose production and increases hepatic insulin responsiveness in diabetic rats. *Proc Natl Acad Sci* **106:** 11288–11293.

Fernandez-Marcos PJ, Auwerx J. 2011. Regulation of PGC-1α, a nodal regulator of mitochondrial biogenesis. *Am J Clin Nutr* **93:** 884S–890S.

Finkel T, Deng CX, Mostoslavsky R. 2009. Recent progress in the biology and physiology of sirtuins. *Nature* **460:** 587–591.

Finley LW, Haas W, Desquiret-Dumas V, Wallace DC, Procaccio V, Gygi SP, Haigis MC. 2011. Succinate dehydrogenase is a direct target of sirtuin 3 deacetylase activity. *PLoS ONE* **6:** e23295.

Gerhart-Hines Z, Rodgers JT, Bare O, Lerin C, Kim SH, Mostoslavsky R, Alt FW, Wu Z, Puigserver P. 2007. Metabolic control of muscle mitochondrial function and fatty acid oxidation through SIRT1/PGC-1α. *EMBO J* **26:** 1913–1923.

Hafner AV, Dai J, Gomes AP, Xiao CY, Palmeira CM, Rosenzweig A, Sinclair DA. 2010. Regulation of the mPTP by SIRT3-mediated deacetylation of CypD at lysine 166 suppresses age-related cardiac hypertrophy. *Aging (Albany NY)* **2:** 914–923.

Hallows WC, Lee S, Denu JM. 2006. Sirtuins deacetylate and activate mammalian acetyl-CoA synthetases. *Proc Natl Acad Sci* **103:** 10230–10235.

Hallows WC, Yu W, Smith BC, Devires MK, Ellinger JJ, Someya S, Shortreed MR, Prolla T, Markley JL, Smith LM, et al. 2011. Sirt3 promotes the urea cycle and fatty acid oxidation during dietary restriction. *Mol Cell* **41:** 139–149.

Hariharan N, Maejima Y, Nakae J, Paik J, Depinho RA, Sadoshima J. 2010. Deacetylation of FoxO by Sirt1 plays an essential role in mediating starvation-induced autophagy in cardiac myocytes. *Circ Res* **107:** 1470–1482.

Hirschey MD, Shimazu T, Goetzman E, Jing E, Schwer B, Lombard DB, Grueter CA, Harris C, Biddinger S, Ilkayeva OR, et al. 2010. SIRT3 regulates mitochondrial fatty-acid oxidation by reversible enzyme deacetylation. *Nature* **464:** 121–125.

Hirschey MD, Shimazu T, Jing E, Grueter CA, Collins AM, Aouizerat B, Stancakova A, Goetzman E, Lam MM, Schwer B, et al. 2011. SIRT3 deficiency and mitochondrial protein hyperacetylation accelerate the development of the metabolic syndrome. *Mol Cell* **44:** 177–190.

Hotta K, Funahashi T, Arita Y, Takahashi M, Matsuda M, Okamoto Y, Iwahashi H, Kuriyama H, Ouchi N, Maeda K, et al. 2000. Plasma concentrations of a novel, adipose-specific protein, adiponectin, in type 2 diabetic patients. *Arterioscler Thromb Vasc Biol* **20:** 1595–1599.

Imai S, Armstrong CM, Kaeberlein M, Guarente L. 2000. Transcriptional silencing and longevity protein Sir2 is an NAD-dependent histone deacetylase. *Nature* **403:** 795–800.

Iwabu M, Yamauchi T, Okada-Iwabu M, Sato K, Nakagawa T, Funata M, Yamaguchi M, Namiki S, Nakayama R, Tabata M, et al. 2010. Adiponectin and AdipoR1 regulate PGC-1α and mitochondria by Ca^{2+} and AMPK/SIRT1. *Nature* **464:** 1313–1319.

Jing E, Emanuelli B, Hirschey MD, Boucher J, Lee KY, Lombard D, Verdin EM, Kahn CR. 2011. Sirtuin-3 (Sirt3) regulates skeletal muscle metabolism and insulin signaling via altered mitochondrial oxidation and reactive oxygen species production. *Proc Natl Acad Sci* **108:** 14608–14613.

Kaeberlein M, McVey M, Guarente L. 1999. The SIR2/3/4 complex and SIR2 alone promote longevity in *Saccharomyces cerevisiae* by two different mechanisms. *Genes Dev* **13:** 2570–2580.

Kendrick AA, Choudhury M, Rahman SM, McCurdy CE, Friederich M, Van Hove JL, Watson PA, Birdsey N, Bao J, Gius D, et al. 2011. Fatty liver is associated with reduced SIRT3 activity and mitochondrial protein hyperacetylation. *Biochem J* **433:** 505–514.

Kim SC, Sprung R, Chen Y, Xu Y, Ball H, Pei J, Cheng T, Kho Y, Xiao H, Xiao L, et al. 2006. Substrate and functional diversity of lysine acetylation revealed by a proteomics survey. *Mol Cell* **23:** 607–618.

Kim HS, Patel K, Muldoon-Jacobs K, Bisht KS, Aykin-Burns N, Pennington JD, van der Meer R, Nguyen P, Savage J, Owens KM, et al. 2010a. SIRT3 is a mitochondria-localized tumor suppressor required for maintenance of mitochondrial integrity and metabolism during stress. *Cancer Cell* **17:** 41–52.

Kim HS, Xiao C, Wang RH, Lahusen T, Xu X, Vassilopoulos A, Vazquez-Ortiz G, Jeong WI, Park O, Ki SH, et al. 2010b. Hepatic-specific disruption of SIRT6 in mice results in fatty liver formation due to enhanced glycolysis and triglyceride synthesis. *Cell Metab* **12:** 224–236.

Kume S, Uzu T, Horiike K, Chin-Kanasaki M, Isshiki K, Araki S, Sugimoto T, Haneda M, Kashiwagi A, Koya D. 2010. Calorie restriction enhances cell adaptation to hypoxia through Sirt1-dependent mitochondrial autophagy in mouse aged kidney. *J Clin Invest* **120:** 1043–1055.

Lee IH, Cao L, Mostoslavsky R, Lombard DB, Liu J, Bruns NE, Tsokos M, Alt FW, Finkel T. 2008. A role for the NAD-dependent deacetylase Sirt1 in the regulation of autophagy. *Proc Natl Acad Sci* **105:** 3374–3379.

Lombard DB, Alt FW, Cheng HL, Bunkenborg J, Streeper RS, Mostoslavsky R, Kim J, Yancopoulos G, Valenzuela D, Murphy A, et al. 2007. Mammalian Sir2 homolog SIRT3 regulates global mitochondrial lysine acetylation. *Mol Cell Biol* **27:** 8807–8814.

Lu Z, Bourdi M, Li JH, Aponte AM, Chen Y, Lombard DB, Gucek M, Pohl LR, Sack MN. 2011. SIRT3-dependent deacetylation exacerbates acetaminophen hepatotoxicity. *EMBO Rep* **12:** 840–846.

Morselli E, Marino G, Bennetzen MV, Eisenberg T, Megalou E, Schroeder S, Cabrera S, Benit P, Rustin P, Criollo A, et al. 2011. Spermidine and resveratrol induce autophagy by distinct pathways converging on the acetylproteome. *J Cell Biol* **192:** 615–629.

Moynihan KA, Grimm AA, Plueger MM, Bernal-Mizrachi E, Ford E, Cras-Meneur C, Permutt MA, Imai S. 2005. Increased dosage of mammalian Sir2 in pancreatic β cells enhances glucose-stimulated insulin secretion in mice. *Cell Metab* **2:** 105–117.

Nakagawa T, Shimizu S, Watanabe T, Yamaguchi O, Otsu K, Yamagata H, Inohara H, Kubo T, Tsujimoto Y. 2005. Cyclophilin D-dependent mitochondrial permeability transition regulates some necrotic but not apoptotic cell death. *Nature* **434:** 652–658.

Nakagawa T, Lomb DJ, Haigis MC, Guarente L. 2009. SIRT5 deacetylates carbamoyl phosphate synthetase 1 and regulates the urea cycle. *Cell* **137:** 560–570.

Nemoto S, Fergusson MM, Finkel T. 2005. SIRT1 functionally interacts with the metabolic regulator and transcriptional coactivator PGC-1α. *J Biol Chem* **280:** 16456–16460.

North BJ, Verdin E. 2007. Interphase nucleo-cytoplasmic shuttling and localization of SIRT2 during mitosis. *PLoS ONE* **2:** e784.

Peeters AV, Beckers S, Verrijken A, Mertens I, Roevens P, Peeters PJ, Van Hul W, Van Gaal LF. 2008. Association of *SIRT1* gene variation with visceral obesity. *Hum Genet* **124:** 431–436.

Peng C, Lu Z, Xie Z, Cheng Z, Chen Y, Tan M, Luo H, Zhang Y, He W, Yang K, et al. 2011. The first identification of lysine malonylation substrates and its regulatory enzyme. *Mol Cell Proteomics* **110:** M111.012658.

Petersen KF, Dufour S, Befroy D, Garcia R, Shulman GI. 2004. Impaired mitochondrial activity in the insulin-resistant offspring of patients with type 2 diabetes. *N Engl J Med* **350:** 664–671.

Picard F, Kurtev M, Chung N, Topark-Ngarm A, Senawong T, Machado De Oliveira R, Leid M, McBurney MW, Guarente L. 2004. Sirt1 promotes fat mobilization in white adipocytes by repressing PPAR-γ. *Nature* **429:** 771–776.

Qiu X, Brown K, Hirschey MD, Verdin E, Chen D. 2010. Calorie restriction reduces oxidative stress by SIRT3-mediated SOD2 activation. *Cell Metab* **12:** 662–667.

Revollo JR, Korner A, Mills KF, Satoh A, Wang T, Garten A, Dasgupta B, Sasaki Y, Wolberger C, Townsend RR, et al. 2007. Nampt/PBEF/visfatin regulates insulin secretion in β cells as a systemic NAD biosynthetic enzyme. *Cell Metab* **6:** 363–375.

Rodgers JT, Puigserver P. 2007. Fasting-dependent glucose and lipid metabolic response through hepatic sirtuin 1. *Proc Natl Acad Sci* **104:** 12861–12866.

Rodgers JT, Lerin C, Haas W, Gygi SP, Spiegelman BM, Puigserver P. 2005. Nutrient control of glucose homeostasis through a complex of PGC-1α and SIRT1. *Nature* **434:** 113–118.

Rose G, Dato S, Altomare K, Bellizzi D, Garasto S, Greco V, Passarino G, Feraco E, Mari V, Barbi C, et al. 2003. Variability of the *SIRT3* gene, human silent information regulator *Sir2* homologue, and survivorship in the elderly. *Exp Gerontol* **38:** 1065–1070.

Schlicker C, Gertz M, Papatheodorou P, Kachholz B, Becker CF, Steegborn C. 2008. Substrates and regulation mechanisms for the human mitochondrial sirtuins Sirt3 and Sirt5. *J Mol Biol* **382:** 790–801.

Schwer B, Bunkenborg J, Verdin RO, Andersen JS, Verdin E. 2006. Reversible lysine acetylation controls the activity of the mitochondrial enzyme acetyl-CoA synthetase 2. *Proc Natl Acad Sci* **103:** 10224–10229.

Schwer B, Eckersdorff M, Li Y, Silva JC, Fermin D, Kurtev MV, Giallourakis C, Comb MJ, Alt FW, Lombard DB. 2009. Calorie restriction alters mitochondrial protein acetylation. *Aging Cell* **8:** 604–606.

Shimazu T, Hirschey MD, Hua L, Dittenhafer-Reed KE, Schwer B, Lombard DB, Li Y, Bunkenborg J, Alt FW, Denu JM, et al. 2010. SIRT3 deacetylates mitochondrial 3-hydroxy-3-methylglutaryl CoA synthase 2 and regulates ketone body production. *Cell Metab* **12:** 654–661.

Shulga N, Pastorino JG. 2010. Ethanol sensitizes mitochondria to the permeability transition by inhibiting deacetylation of cyclophilin-D mediated by sirtuin-3. *J Cell Sci* **123:** 4117–4127.

Shulga N, Wilson-Smith R, Pastorino JG. 2010. Sirtuin-3 deacetylation of cyclophilin D induces dissociation of hexokinase II from the mitochondria. *J Cell Sci* **123:** 894–902.

Silberman DM, Mostoslavsky R. 2011. SIRT3 deacetylase: The Jekyll and Hyde sirtuin. *EMBO Rep* **12:** 746–747.

Someya S, Yu W, Hallows WC, Xu J, Vann JM, Leeuwenburgh C, Tanokura M, Denu JM, Prolla TA. 2010. Sirt3 mediates reduction of oxidative damage and prevention of age-related hearing loss under caloric restriction. *Cell* **143:** 802–812.

Sundaresan NR, Gupta M, Kim G, Rajamohan SB, Isbatan A, Gupta MP. 2009. Sirt3 blocks the cardiac hypertrophic response by augmenting Foxo3a-dependent antioxidant defense mechanisms in mice. *J Clin Invest* **119:** 2758–2771.

Tao R, Coleman MC, Pennington JD, Ozden O, Park SH, Jiang H, Kim HS, Flynn CR, Hill S, Hayes McDonald W, et al. 2010. Sirt3-mediated deacetylation of evolutionarily conserved lysine 122 regulates MnSOD activity in response to stress. *Mol Cell* **40:** 893–904.

Yang Y, Cimen H, Han MJ, Shi T, Deng JH, Koc H, Palacios OM, Montier L, Bai Y, Tong Q, et al. 2009. NAD^+-dependent deacetylase SIRT3 regulates mitochondrial protein synthesis by deacetylation of the ribosomal protein MRPL10. *J Biol Chem* **285:** 7417–7429.

Zhang CY, Baffy G, Perret P, Krauss S, Peroni O, Grujic D, Hagen T, Vidal-Puig AJ, Boss O, Kim YB, et al. 2001. Uncoupling protein-2 negatively regulates insulin secretion and is a major link between obesity, β cell dysfunction, and type 2 diabetes. *Cell* **105:** 745–755.

Zhang Z, Tan M, Xie Z, Dai L, Chen Y, Zhao Y. 2011. Identification of lysine succinylation as a new post-translational modification. *Nat Chem Biol* **7:** 58–63.

Zillikens MC, van Meurs JB, Rivadeneira F, Amin N, Hofman A, Oostra BA, Sijbrands EJ, Witteman JC, Pols HA, van Duijn CM, et al. 2009. *SIRT1* genetic variation is related to BMI and risk of obesity. *Diabetes* **58:** 2828–2834.

 Cite this article as *Cold Spring Harb Perspect Biol* doi: 10.1101/cshperspect.a013102

Index

D

N

www.ingramcontent.com/pod-product-compliance
Lightning Source LLC
Chambersburg PA
CBHW042347230926
52025CB00009B/48
9781936113354